Study and Solutions Guide for

# Calculus

## ALTERNATE FOURTH EDITION
## LARSON/HOSTETLER

## DAVID E. HEYD

The Pennsylvania State University

The Behrend College

D. C. Heath and Company
Lexington, Massachusetts     Toronto

# Preface

This guide is designed as a supplement to *Calculus with Analytic Geometry*, Alternate Fourth Edition, by Roland E. Larson and Robert P. Hostetler. All references to chapters, theorems, and exercises relate to the main text. Although this supplement is not a substitute for poor study habits, it can be of great value when incorporated into a well planned course of study. The following suggestions are given to assist you in the use of the text, your lecture notes, and this guide.

*Read the section in the text for general content before class.* You will be surprised how much more you will acquire from the lecture if you are aware of the objectives of the section and the types of problems that will be solved. If you are familiar with the topic, you will understand more of the lecture and you will be able to take fewer (and better) notes.

*As soon after class as possible, work problems from the exercise set.* The exercise sets in the text are divided into groups of similar problems and are presented in approximately the same order as the section topics. Try to get an overall picture of the various types of problems in the set. As you work your way through the exercise set, reread your class notes and the portion of the section that covers each type of problem. Pay particular attention to the solved examples.

*Learning calculus takes much practice.* You cannot learn calculus merely by reading any more than you can learn to play the piano or to bowl merely by reading. Only after you have practiced the techniques of a section and have discovered your weak points can you make good use of the supplementary solutions in this guide. Solutions in the guide are usually given in greater algebraic detail than the solutions in the text. Students who need additional help with algebra will find this especially helpful.

During many years of teaching I have found that good study habits are essential for success in mathematics. My students have found the following additional suggestions to be helpful in making the best use of their time.

1. *Write neatly in pencil.* A notebook filled with unorganized scribbling is of little value.

2. *Work at a deliberate and methodical pace without skipping steps.* When you hurry through a problem you are more apt to make careless arithmetic or algebraic errors that, in the long run, waste time.

3. *Keep up with the work.* This suggestion is crucial because calculus is a very structured topic. If you cannot do the problems in one section, you are not likely to be able to do the problems in the next. The night before a quiz or test is not the time to start working problems. In some instances cramming may help you pass an examination, but it is an inferior way to learn and retain essential concepts.

4. After working some of the assigned exercises with access to the examples and answers, *try at least one of each type of exercise with the book closed.* This will increase your confidence on quizzes and tests.

5. Do not be overly concerned with finding the most efficient way to solve a problem. Your first goal is to find one way that works. Short cuts and clever methods come later.

6. A set of true-false questions for each chapter is found in Chapter 19. These questions are designed to identify common errors, emphasize the hypotheses and conclusions of theorems, and reinforce calculus concepts. The answers to all true-false questions are also included.

7. If you have trouble with the algebra of calculus, refer to the algebra review at the beginning of this guide.

The author wishes to acknowledge several people whose help and encouragement were invaluable in the production of this supplementary guide. First, I am grateful to Roland E. Larson and Robert P. Hostetler for the privilege of working with them on the main text. I also wish to thank Ann R. Kraus for proofreading the manuscript, Randall R. Hammond for the computer graphics, and the staff at D. C. Heath and Company. I am grateful for the encouragement from my wife Jean and for the assistance of our children Ed, Ruth and Andy. I am especially thankful that Ed and Ruth were available to type this supplementary guide.

*David E. Heyd*

# Contents

Chapter 0  Algebra Review  1

Chapter 1  The Cartesian Plane and Functions  11

Chapter 2  Limits and their Properties  33

Chapter 3  Differentiation  47

Chapter 4  Applications of Differentiation  77

Chapter 5  Integration  115

Chapter 6  Applications of Integration  141

Chapter 7  Exponential and Logarithmic Functions  175

Chapter 8  Trigonometric Functions and Inverse Trigonometric Functions  207

Chapter 9  Integration Techniques and Improper Integrals  239

Chapter 10  Infinite Series  281

Chapter 11  Conic Sections  321

Chapter 12  Plane Curves, Parametric Equations, and Polar Coordinates  347

Chapter 13  Vectors and Curves in the Plane  377

Chapter 14  Solid Analytic Geometry and Vectors in Space  405

Chapter 15  Functions of Several Variables  435

Chapter 16  Multiple Integration  479

Chapter 17  Vector Analysis  515

Chapter 18  Differential Equations  545

Chapter 19  True or False Questions for Review  575

# 0 ALGEBRA REVIEW

## 0.1 Monomial Factors

Factor as indicated:

(a) $3x^4 + 4x^3 - x^2 = x^2($ $\quad )$

(b) $2\sqrt{x} + 6x^{3/2} = 2\sqrt{x}($ $\quad )$

(c) $e^{-x} - xe^{-x} + 2x^2 e^{-x} = e^{-x}($ $\quad )$

(d) $x^{-1} - 2 + x = x^{-1} = x^{-1}($ $\quad )$

(e) $\dfrac{x}{2} - 6x^2 = \dfrac{x}{2}($ $\quad )$

(f) $\sin x + \tan x = \sin x($ $\quad )$

(g) $\dfrac{1}{2x^2 + 4x} = \dfrac{1}{2x}\left( \quad \right)$

**Solution:**

(a) $3x^4 + 4x^3 - x^2 = x^2(3x^2 + 4x - 1)$

(b) $2\sqrt{x} + 6x^{3/2} = 2\sqrt{x}(1 + 3x)$

(c) $e^{-x} - xe^{-x} + 2x^2 e^{-x} = e^{-x}(1 - x + 2x^2)$

(d) $x^{-1} - 2 + x = x^{-1} = x^{-1}(1 - 2x + x^2)$

(e) $\dfrac{x}{2} - 6x^2 = \dfrac{x}{2}(1 - 12x)$

(f) $\sin x + \tan x = \sin x + \dfrac{\sin x}{\cos x} = \sin x\left(1 + \dfrac{1}{\cos x}\right)$

$\qquad\qquad\qquad\qquad\qquad = \sin x(1 + \sec x)$

$$\text{(g)} \quad \frac{1}{2x^2 + 4x} = \frac{1}{2x}\left(\frac{1}{x+2}\right)$$

## 0.2   Binomial Factors

Factor as indicated:

(a)   $(x-1)^2(x) - (x-1) = (x-1)(\qquad)$

(b)   $3(x^2+4)(x^2+1) + 6(x^2+4)^2 = 3(x^2+4)(\qquad)$

(c)   $\sqrt{x^2+1} - \dfrac{x^2}{\sqrt{x^2+1}} = \dfrac{1}{\sqrt{x^2+1}}(\qquad)$

(d)   $(x-3)^3(x+2) - 2(x-3)^2(x+2)^2$
$$= (x-3)^2(x+2)(\qquad)$$

(e)   $(2x+1)^{3/2}(x^{1/2}) + (2x+1)^{5/2}(x^{-1/2})$
$$= (2x+1)^{3/2}(x^{-1/2})(\qquad)$$

**Solution:**

(a)   $(x-1)^2(x) - (x-1) = (x-1)[(x-1)x - 1]$
$$= (x-1)(x^2 - x - 1)$$

(b)   $3(x^2+4)(x^2+1) + 6(x^2+4)^2$
$$= 3(x^2+4)[(x^2+1) + 2(x^2+4)]$$
$$= 3(x^2+4)(3x^2+9)$$

(c)   $\sqrt{x^2+1} - \dfrac{x^2}{\sqrt{x^2+1}} = (x^2+1)^{1/2} - x^2(x^2+1)^{-1/2}$
$$= (x^2+1)^{-1/2}[(x^2+1) - x^2]$$
$$= \frac{1}{\sqrt{x^2+1}}$$

(d)   $(x-3)^3(x+2) - 2(x-3)^2(x+2)^2$
$$= (x-3)^2(x+2)[(x-3) - 2(x+2)]$$
$$= (x-3)^2(x+2)(-x-7)$$

(e)   $(2x+1)^{3/2}(x^{1/2}) + (2x+1)^{5/2}(x^{-1/2})$
$$= (2x+1)^{3/2}(x^{-1/2})[x + (2x+1)]$$
$$= (2x+1)^{3/2}(x^{-1/2})(3x+1)$$

## 0.3   Factoring Quadratic Expressions

Factor as indicated:

(a)   $x^2 - 3x + 2 = ($   $)($   $)$

(b)   $x^2 - 9 = ($   $)($   $)$

(c)   $x^2 + 5x - 6 = ($   $)($   $)$

(d)   $x^2 + 5x + 6 = ($   $)($   $)$

(e)   $2x^2 + 5x - 3 = ($   $)($   $)$

(f)   $e^{2x} + 2 + e^{-2x} = ($   $)^2$

(g)   $x^4 - 7x^2 + 12 = ($   $)($   $)($   $)$

(h)   $1 - \sin^2 x = ($   $)($   $)$

**Solution:**

(a)   $x^2 - 3x + 2 = (x - 2)(x - 1)$

(b)   $x^2 - 9 = (x + 3)(x - 3)$

(c)   $x^2 + 5x - 6 = (x + 6)(x - 1)$

(d)   $x^2 + 5x + 6 = (x + 2)(x + 3)$

(e)   $2x^2 + 5x - 3 = (2x - 1)(x + 3)$

(f)   $e^{2x} + 2 + e^{-2x} = (e^x + e^{-x})^2$

(g)   $x^4 - 7x^2 + 12 = (x^2 - 3)(x^2 - 4) = (x^2 - 3)(x + 2)(x - 2)$

(h)   $1 - \sin^2 x = (1 + \sin x)(1 - \sin x)$

## 0.4   Cancellation

Reduce each expression to lowest terms:

(a)   $\dfrac{3x + 9}{6x}$     (b)   $\dfrac{x^2}{x^{1/2}}$

(c) $\dfrac{(x+1)^3(x-2)+3(x+1)^2}{(x+1)^4}$

(d) $\dfrac{x^{1/2}-x^{1/3}}{x^{1/6}}$

(e) $\dfrac{\sqrt{x-1}+(x-1)^{3/2}}{\sqrt{x-1}}$

(f) $\dfrac{1-(\sin x+\cos x)^2}{2\sin x}$

**Solution:**

(a) $\dfrac{3x+9}{6x}=\dfrac{3(x+3)}{3(2x)}=\dfrac{x+3}{2x}$

(b) $\dfrac{x^2}{x^{1/2}}=\dfrac{(x^{1/2})(x^{3/2})}{x^{1/2}}=x^{3/2}$

(c) $\dfrac{(x+1)^3(x-2)+3(x+1)^2}{(x+1)^4}=\dfrac{(x+1)^2[(x+1)(x-2)+3]}{x+1)^4}$

$$=\dfrac{x^2-x+1}{(x+1)^2}$$

(d) $\dfrac{x^{1/2}-x^{1/3}}{x^{1/6}}=\dfrac{x^{1/6}(x^{2/6}-x^{1/6})}{x^{1/6}}=x^{1/3}-x^{1/6}$

(e) $\dfrac{\sqrt{x-1}+(x-1)^{3/2}}{\sqrt{x-1}}=\dfrac{\sqrt{x-1}[1+(x-1)]}{\sqrt{x-1}}=x$

(f) $\dfrac{1-(\sin x+\cos x)^2}{2\sin x}=\dfrac{1-(\sin^2 x+2\sin x\cos x+\cos^2 x)}{2\sin x}$

$$=\dfrac{1-(\sin^2 x+\cos^2 x)-2\sin x\cos x}{2\sin x}$$

$$=\dfrac{1-1-2\sin x\cos x}{2\sin x}=-\cos x$$

## 0.5   Quadratic Formula

|  | *Equation* | *Solve for* |
|---|---|---|
| (a) | $x^2-4x-1=0$ | $x$ |
| (b) | $2x^2+x-3=0$ | $x$ |
| (c) | $\cos^2 x+3\cos x+2=0$ | $\cos x$ |
| (d) | $x^2-xy-(1+y^2)=0$ | $x$ |
| (e) | $x^4-4x^2+2=0$ | $x^2$ |

**Solution:**

(a)   $x = \dfrac{4 \pm \sqrt{16 + 4}}{2} = \dfrac{4 \pm \sqrt{20}}{2} = \dfrac{4 \pm 2\sqrt{5}}{2} = 2 \pm \sqrt{5}$

(b)   $x = \dfrac{-1 \pm \sqrt{1 + 24}}{4} = \dfrac{-1 \pm 5}{4}$

   $x = \dfrac{4}{4} = 1$   or   $x = -\dfrac{6}{4} = -\dfrac{3}{2}$

(c)   $\cos x = \dfrac{-3 \pm \sqrt{9 - 8}}{2} = \dfrac{-3 \pm 1}{2}$

   $\cos x = -\dfrac{2}{2} = -1$   or   $\cos x = -\dfrac{4}{2} = -2$

(d)   $x = \dfrac{y \pm \sqrt{y^2 + 4(1 + y^2)}}{2} = \dfrac{y \pm \sqrt{y^2 + 4 + 4y^2}}{2}$

   $= \dfrac{y \pm \sqrt{5y^2 + 4}}{2}$

(e)   $x^2 = \dfrac{4 \pm \sqrt{16 - 8}}{2} = \dfrac{4 \pm \sqrt{8}}{2} = \dfrac{4 \pm 2\sqrt{2}}{2} = 2 \pm \sqrt{2}$

## 0.6   Synthetic Division

Use synthetic division to factor as indicated:

(a)   $x^3 - 4x^2 + 2x + 1 = (x - 1)(\qquad)$

(b)   $2x^3 + 5x + 7 = (x + 1)(\qquad)$

(c)   $x^4 - 3x^3 + x^2 + x + 2 = (x - 2)(\qquad)$

(d)   $4x^4 + 3x^2 - 1 = (2x - 1)(\qquad)$

**Solution:**

(a)   $x^3 - 4x^2 + 2x + 1$

$$
\begin{array}{r|rrrr}
1 & 1 & -4 & 2 & 1 \\
  &   & 1 & -3 & -1 \\
\hline
  & 1 & -3 & -1 & 0
\end{array}
$$

   $x^3 - 4x^2 + 2x + 1 = (x - 1)(x^2 - 3x - 1)$

(b) $\qquad\qquad 2x^3 + 5x + 7$

$$
\begin{array}{r|rrrr}
-1 & 2 & 0 & 5 & 7 \\
 & & -2 & 2 & -7 \\
\hline
 & 2 & -2 & 7 & 0
\end{array}
$$

$$2x^3 + 5x + 7 = (x+1)(2x^2 - 2x + 7)$$

(c) $\qquad\qquad x^4 - 3x^3 + x^2 + x + 2$

$$
\begin{array}{r|rrrrr}
2 & 1 & -3 & 1 & 1 & 2 \\
 & & 2 & -2 & -2 & -2 \\
\hline
 & 1 & -1 & -1 & -1 & 0
\end{array}
$$

$$x^4 - 3x^3 + x^2 + x + 2 = (x-2)(x^3 - x^2 - x - 1)$$

(d) $\qquad\qquad 4x^4 + 3x^2 - 1$

$$
\begin{array}{r|rrrrr}
\frac{1}{2} & 4 & 0 & 3 & 0 & -1 \\
 & & 2 & 1 & 2 & 1 \\
\hline
 & 4 & 2 & 4 & 2 & 0
\end{array}
$$

$$
\begin{aligned}
4x^4 + 3x^2 - 1 &= \left(x - \frac{1}{2}\right)(4x^3 + 2x^2 + 4x + 2) \\
&= (2x - 1)(2x^3 + x^2 + 2x + 1)
\end{aligned}
$$

## 0.7   Special Products

Factor completely (into linear or irreducible quadratic factors):

(a)   $x^3 - 27$                   (b)   $x^3 - 3x^2 + 3x - 1$

(c)   $x^3 + 6x^2 + 12x + 8$      (d)   $x^4 - 25$

(e)   $x^4 - 8x^3 + 24x^2 - 32x + 16$

**Solution:**

(a)   $x^3 - 27 = (x - 3)(x^2 + 3x + 9)$

(b)   $x^3 - 3x^2 + 3x - 1 = (x - 1)^3$

(c) $x^3 + 6x^2 + 12x + 8 = x^3 + 3(2)x^2 + 3(2^2)x + 2^3 = (x + 2)^3$

(d) $x^4 - 25 = (x^2 + 5)(x^2 - 5) = (x^2 + 5)(x + \sqrt{5})(x - \sqrt{5})$

(e) $x^4 - 8x^3 + 24x^2 - 32x + 16$

$= x^4 - 4(2)x^3 + 6(2^2)x^2 - 4(2^3)x + 2^4 = (x - 2)^4$

## 0.8   Factoring by Grouping

Factor completely (into linear or irreducible quadratic factors):

(a) $x^3 + 4x^2 - 2x - 8$

(b) $x^3 + 2x^2 + 3x + 6$

(c) $5\cos^2 x - 5\sin^2 x + \sin x + \cos x$

(d) $\cos^2 x + 4\cos x + 4 - \tan^2 x$

**Solution:**

(a) $x^3 + 4x^2 - 2x - 8 = x^2(x + 4) - 2(x + 4)$

$= (x^2 - 2)(x + 4)$

$= (x + \sqrt{2})(x - \sqrt{2})(x + 4)$

(b) $x^3 + 2x^2 + 3x + 6 = x^2(x + 2) + 3(x + 2)$

$= (x^2 + 3)(x + 2)$

(c) $5\cos^2 x - 5\sin^2 x + \sin x + \cos x$

$= 5(\cos^2 x - \sin^2 x) + (\sin x + \cos x)$

$= 5(\cos x + \sin x)(\cos x - \sin x) + (\cos x + \sin x)$

$= (\cos x + \sin x)[5(\cos x - \sin x) + 1]$

(d) $\cos^2 x + 4\cos x + 4 - \tan^2 x$

$= (\cos x + 2)^2 - \tan^2 x$

$= (\cos x + 2 + \tan x)(\cos x + 2 - \tan x)$

## 0.9   Simplifying

Rewrite each of the following in simplest form:

(a) $\dfrac{(x-1)(x+3)-(x+1)^2}{x+1}$

(b) $\dfrac{\sqrt{x^2+1}-\dfrac{1}{\sqrt{x^2+1}}}{x^2+1}$

(c) $\dfrac{x^2-5x+6}{x^2-4x+4}$

(d) $\dfrac{1}{x+1}-\dfrac{1}{x-1}-\dfrac{1}{x^2-1}$

(e) $\dfrac{x(-2x)}{2\sqrt{1-x^2}}+\sqrt{1-x^2}+\dfrac{1}{\sqrt{1-x^2}}$

**Solution:**

(a) $\dfrac{(x-1)(x+3)-(x+1)^2}{x+1}=\dfrac{(x^2+2x-3)-(x^2+2x+1)}{x+1}$

$$=\dfrac{-4}{x+1}$$

(b) $\dfrac{\sqrt{x^2+1}-\dfrac{1}{\sqrt{x^2+1}}}{x^2+1}=\dfrac{\dfrac{1}{\sqrt{x^2+1}}(x^2+1-1)}{x^2+1}$

$$=\dfrac{x^2+1-1}{\sqrt{x^2+1}(x^2+1)}=\dfrac{x^2}{(x^2+1)^{3/2}}$$

(c) $\dfrac{x^2-5x+6}{x^2-4x+4}=\dfrac{(x-2)(x-3)}{(x-2)^2}=\dfrac{x-3}{x-2}$

(d) $\dfrac{1}{x+1}-\dfrac{1}{x-1}-\dfrac{2}{x^2-1}=\dfrac{(x-1)-(x+1)-2}{x^2-1}=\dfrac{-4}{x^2-1}$

(e) $\dfrac{x(-2x)}{2\sqrt{1-x^2}}+\sqrt{1-x^2}+\dfrac{1}{\sqrt{1-x^2}}$

$$=\dfrac{-x^2}{\sqrt{1-x^2}}+\dfrac{1-x^2}{\sqrt{1-x^2}}+\dfrac{1}{\sqrt{1-x^2}}$$

$$=\dfrac{2-2x^2}{\sqrt{1-x^2}}$$

$$=\dfrac{2(1-x^2)}{\sqrt{1-x^2}}=2\sqrt{1-x^2}$$

## 0.10   Rationalizing

Remove the sum or difference from the denominator by multiplying the numerator and denominator by the conjugate of the denominator.

(a) $\dfrac{1}{1 - \cos x}$     (b) $\dfrac{x}{1 - \sqrt{x^2 + 1}}$     (c) $\dfrac{2}{x + \sqrt{x^2 + 1}}$

**Solution:**

(a)
$$\frac{1}{1 - \cos x} = \left(\frac{1}{1 - \cos x}\right)\left(\frac{1 + \cos x}{1 + \cos x}\right)$$
$$= \frac{1 + \cos x}{1 - \cos^2 x} = \frac{1 + \cos x}{\sin^2 x}$$

(b)
$$\left(\frac{x}{1 - \sqrt{x^2 + 1}}\right)\left(\frac{1 + \sqrt{x^2 + 1}}{1 + \sqrt{x^2 + 1}}\right)$$
$$= \frac{x(1 + \sqrt{x^2 + 1})}{1 - (x^2 + 1)}$$
$$= \frac{x(1 + \sqrt{x^2 + 1})}{-x^2} = \frac{1 + \sqrt{x^2 + 1}}{-x}$$

(c)
$$\left(\frac{2}{x + \sqrt{x^2 + 1}}\right)\left(\frac{x - \sqrt{x^2 + 1}}{x - \sqrt{x^2 + 1}}\right)$$
$$= \frac{2(x - \sqrt{x^2 + 1})}{x^2 - (x^2 + 1)} = -2(x - \sqrt{x^2 + 1})$$

## 0.11  Algebraic Errors to Avoid

| Error | Correct form | Comment |
|---|---|---|
| $a - (x - b) \neq a - x - b$ | $a - (x - b) = a - x + b$ | Change all signs when distributing negative through parentheses. |
| $(a + b)^2 \neq a^2 + b^2$ | $(a + b)^2 = a^2 + 2ab + b^2$ | Don't forget middle term when squaring binomials. |
| $\left(\frac{1}{2}a\right)\left(\frac{1}{2}b\right) \neq \frac{1}{2}ab$ | $\left(\frac{1}{2}a\right)\left(\frac{1}{2}b\right) = \frac{1}{4}(ab)$ | 1/2 occurs twice as a factor. |
| $\frac{a}{x + b} \neq \frac{a}{x} + \frac{a}{b}$ | Leave as $\frac{a}{x + b}$ | Don't add denominators when adding fractions. |
| $\frac{1}{a} + \frac{1}{b} \neq \frac{1}{a + b}$ | $\frac{1}{a} + \frac{1}{b} = \frac{a + b}{ab}$ | Use definition for adding fractions. |
| $\frac{\frac{x}{a}}{b} \neq \frac{bx}{a}$ | $\frac{\frac{x}{a}}{b} = \left(\frac{x}{a}\right)\left(\frac{1}{b}\right) = \frac{x}{ab}$ | Multiply by reciprocal of the denominator. |
| $\frac{1}{3x} \neq \frac{1}{3}x$ | $\frac{1}{3x} = \frac{1}{3} \cdot \frac{1}{x}$ | Use definition for multiplying fractions. |
| $1/x + 2 \neq \frac{1}{x + 2}$ | $1/x + 2 = \frac{1}{x} + 2$ | Be careful when using a slash to denote division. |
| $(x^2)^3 \neq x^5$ | $(x^2)^3 = x^{2 \cdot 3} = x^6$ | Multiply exponents when an exponential form is raised to a power. |
| $2x^3 \neq (2x)^3$ | $2x^3 = 2(x^3)$ | Exponents have priority over coeffiecients. |
| $\frac{1}{x^2 + x^3} \neq x^{-2} + x^{-3}$ | Leave as $\frac{1}{x^2 + x^3}$ | Don't shift term-by-term from denominator to numerator. |
| $\sqrt{5x} \neq 5\sqrt{x}$ | $\sqrt{5x} = \sqrt{5}\sqrt{x}$ | Radicals apply to every factor inside radical. |
| $\sqrt{x^2 + a^2} \neq x + a$ | Leave as $\sqrt{x^2 + a^2}$ | Don't apply radicals term by term. |
| $\frac{a + bx}{a} \neq 1 + bx$ | $\frac{a + bx}{a} = 1 + \frac{b}{a}x$ | Cancel common factors, *not* common terms. |
| $\frac{a + ax}{a} \neq a + x$ | $\frac{a + ax}{a} = 1 + x$ | Factor *before* canceling. |

# 1 THE CARTESIAN PLANE AND FUNCTIONS

## 1.1 Real Numbers and the Real Line

---

**7.** Determine whether the real number $\sqrt[3]{64}$ is rational or irrational.

**Solution:**

Since $4^3 = 64$, it follows that $\sqrt[3]{64} = 4$ and is therefore rational.

**13.** Express $0.297\overline{297}$ as a ratio of two integers.

**Solution:**

Let $x = 0.297297\ldots$. Since the repeating pattern occurs every three decimal places, we multiply both sides of the equation by 1000 and obtain

$$
\begin{aligned}
1000x &= 297.297297\ldots \qquad \\
x &= \phantom{297.}0.297297\ldots \qquad \text{Subtract} \\
\hline
999x &= 297 \\
x &= \frac{297}{999} = \frac{11}{37}
\end{aligned}
$$

**19.** Solve the inequality $x - 5 \geq 7$ and graph the solution on the real line.

**Solution:**

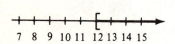

$$x - 5 \geq 7$$
$$x - 5 + 5 \geq 7 + 5$$
$$x \geq 12$$

**25.** Solve the inequality $-4 < 2x - 3 < 4$ and graph the solution on the real line.

**Solution:**

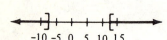

$$-4 < \quad 2x - 3 \quad < 4$$
$$-4 + 3 < 2x - 3 + 3 < 4 + 3$$
$$-1 < \quad 2x \quad < 7$$
$$-\frac{1}{2} < \quad x \quad < \frac{7}{2}$$

**33.** Solve the inequality $\left| \dfrac{x - 3}{2} \right| \geq 5$ and graph the solution on the real number line.

**Solution:**

Since $\left| \dfrac{x - 3}{2} \right| \geq 5$, we have

$$-\frac{x - 3}{2} \geq 5 \qquad \text{or} \qquad \frac{x - 3}{2} \geq 5$$
$$-\frac{x - 3}{2}(-2) \leq 5(-2) \qquad \text{or} \qquad \frac{x - 3}{2}(2) \geq 5(2)$$
$$x - 3 \leq -10 \qquad \text{or} \qquad x - 3 \geq 10$$
$$x \leq -7 \qquad \text{or} \qquad x \geq 13$$

**55.** Use absolute values to define the following intervals:

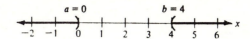

**Solution:**

From the accompanying figure we have

$$x < 0 \qquad \text{or} \qquad x > 4$$
$$x - 2 < -2 \qquad \text{or} \qquad x - 2 > 4 - 2 \qquad \text{Centered at 2}$$
$$x - 2 < -2 \qquad \text{or} \qquad x - 2 > 2$$

Therefore, the magnitude of $x - 2$ must be greater than 2 or

$$|x - 2| > 2.$$

**60.** In the manufacture and sale of a certain product, the revenue for selling $x$ units is $R = 115.95x$, and the cost of producing $x$ units is $C = 95x + 750$. In order for a profit to be realized, $R$ must be greater than $C$. For what values of $x$ will this product return a profit?

**Solution:**

Since the revenue $R$ must be greater than the cost $C$ for the product to return a profit we have

$$R > C$$
$$115.95x > 95x + 750$$
$$115.95x - 95x > 750$$
$$20.95x > 750$$
$$x > 35.7995 \qquad \text{or} \qquad x \geq 36 \text{ units}$$

**62.** The heights, $h$, of two-thirds of the members of a certain population satisfy the inequality

$$\left|\frac{h - 68.5}{2.7}\right| \leq 1$$

Solve for $h$.

**Solution:**

$$\left|\frac{h - 68.5}{2.7}\right| \leq 1$$

$$-1 \leq \frac{h - 68.5}{2.7} \leq 1$$

$$-2.7 \leq h - 68.5 \leq 2.7$$

$$65.8 \leq h \leq 71.2$$

**67.** Prove that $|ab| = |a||b|$.

**Solution:**

Case 1:    $a > 0,\ b > 0,$    and    $ab > 0$.

$$|ab| = ab = |a||b|$$

Case 2:    $a < 0,\ b < 0,$    and    $ab > 0$.

$$|ab| = ab = (-a)(-b) = |a||b|$$

Case 3:    $a > 0,\ b < 0,$    and    $ab < 0$.

$$|ab| = -(ab) = a(-b) = |a||b|$$

Case 4:    $a < 0,\ b > 0,$    and    $ab < 0$.

$$|ab| = -(ab) = (-a)(b) = |a||b|$$

## 1.2   The Cartesian plane

1. Plot the points $(2,1)$ and $(4,5)$. Find the distance between the points and find the midpoint of the line segment joining the points.

**Solution:**

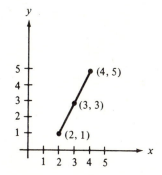

Let $(2,1) = (x_1, y_1)$   and   $(4,5) = (x_2, y_2)$. Then

$$d = \sqrt{(x_2 - x_1)^2 + (y_2 - y_1)^2} = \sqrt{(4-2)^2 + (5-1)^2}$$
$$= \sqrt{2^2 + 4^2} = \sqrt{20} = 2\sqrt{5}$$

$$\text{midpoint} = \left(\frac{x_1 + x_2}{2}, \frac{y_1 + y_2}{2}\right) = \left(\frac{2+4}{2}, \frac{1+5}{2}\right) = (3,3)$$

7. Show that the points $(4,0), (2,1)$, and $(-1,-5)$ are vertices of a right triangle.

**Solution:**

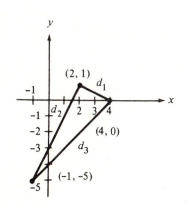

Let $d_1 =$ distance between $(4,0)$ and $(2,1)$.

$$d_1{}^2 = (2-4)^2 + (1-0)^2 = (-2)^2 + 1^2 = 5$$

Let $d_2 =$ distance between $(2,1)$ and $(-1,-5)$.

$$d_2{}^2 = (-1-2)^2 + (-5-1)^2 = (-3)^2 + (-6)^2 = 45$$

Let $d_3 =$ distance between $(4,0)$ and $(-1,-5)$.

$$d_3{}^2 = (-1-4)^2 + (-5-0)^2 = (-5)^2 + (-5)^2 = 50$$

Since $d_1{}^2 + d_2{}^2 = 5 + 45 = 50 = d_3{}^2$, the triangle must be a right triangle.

**13.** Use the distance formula to determine if the points $(-2, 1), (-1, 0)$, and $(2, -2)$ lie on a straight line.

**Solution:**

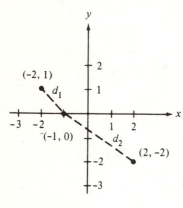

Let $d_1$ = distance between $(-2, 1)$ and $(-1, 0)$. Then

$$d_1 = \sqrt{[-1 - (-2)]^2 + (0 - 1)^2} = \sqrt{1^2 + (-1)^2} = \sqrt{2}$$

Let $d_2$ = distance between $(-1, 0)$ and $(2, -2)$. Then

$$d_2 = \sqrt{[2 - (-1)]^2 + (-2 - 0)^2} = \sqrt{3^2 + (-2)^2} = \sqrt{13}$$

Let $d_3$ = distance between $(-2, 1)$ and $(2, -2)$. Then

$$d_3 = \sqrt{[2 - (-2)]^2 + (-2 - 1)^2} = \sqrt{4^2 + (-3)^2} = \sqrt{25} = 5$$

The points $(-2, 1), (-1, 0)$, and $(2, -2)$ lie on a line only if $d_1 + d_2 = d_3$. Since $\sqrt{2} + \sqrt{13} \approx 5.02 \neq 5$, the points are *not* collinear.

**19.** Find the relationship betwwn $x$ and $y$ so that $(x, y)$ is equidistant from $(4, -1)$ and $(-2, 3)$.

**Solution:**

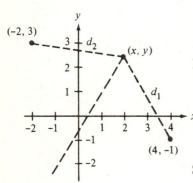

Let $d_1$ = distance between $(4, -1)$ and $(x, y)$. Then

$$d_1 = \sqrt{(x - 4)^2 + [y - (-1)]^2}$$
$$= \sqrt{x^2 - 8x + 16 + y^2 + 2y + 1}$$
$$= \sqrt{x^2 - 8x + y^2 + 2y + 17}$$

Let $d_2$ = distance between $(-2, 3)$ and $(x, y)$. Then

$$d_2 = \sqrt{[x - (-2)]^2 + (y - 3)^2}$$
$$= \sqrt{x^2 + 4x + 4 + y^2 - 6y + 9}$$
$$= \sqrt{x^2 + 4x + y^2 - 6y + 13}$$

Setting $d_1$ equal to $d_2$, we have

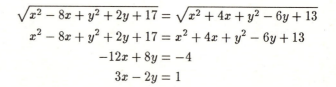

$$\sqrt{x^2 - 8x + y^2 + 2y + 17} = \sqrt{x^2 + 4x + y^2 - 6y + 13}$$
$$x^2 - 8x + y^2 + 2y + 17 = x^2 + 4x + y^2 - 6y + 13$$
$$-12x + 8y = -4$$
$$3x - 2y = 1$$

**21.** Use the Midpoint Rule successively to find the three points that divide the line segment joining $(x_1, y_1)$ and $(x_2, y_2)$ into four equal parts.

**Solution:**

The midpoint of the given line segment is $\left( \dfrac{x_1 + x_2}{2}, \dfrac{y_1 + y_2}{2} \right)$.

The midpoint between $(x_1, y_1)$ and $\left( \dfrac{x_1 + x_2}{2}, \dfrac{y_1 + y_2}{2} \right)$ is

$$\left( \frac{x_1 + \dfrac{x_1 + x_2}{2}}{2}, \frac{y_1 + \dfrac{y_1 + y_2}{2}}{2} \right)$$

$$= \left( \frac{1}{2} \left( \frac{2x_1 + x_1 + x_2}{2} \right), \frac{1}{2} \left( \frac{2y_1 + y_1 + y_2}{2} \right) \right)$$

$$= \left( \frac{3x_1 + x_2}{4}, \frac{3y_1 + y_2}{4} \right)$$

The midpoint between $\left( \dfrac{x_1 + x_2}{2}, \dfrac{y_1 + y_2}{2} \right)$ and $(x_2, y_2)$ is

$$\left( \frac{\dfrac{x_1 + x_2}{2} + x_2}{2}, \frac{\dfrac{y_1 + y_2}{2} + y_2}{2} \right)$$

$$= \left( \frac{1}{2} \left( \frac{x_1 + x_2 + 2x_2}{2} \right), \frac{1}{2} \left( \frac{y_1 + y_2 + 2y_2}{2} \right) \right)$$

$$= \left( \frac{x_1 + 3x_2}{4}, \frac{y_1 + 3y_2}{4} \right)$$

Thus the three points are

$$\left( \frac{3x_1 + x_2}{4}, \frac{3y_1 + y_2}{4} \right), \left( \frac{x_1 + x_2}{2}, \frac{y_1 + y_2}{2} \right)$$

$$\left( \frac{x_1 + 3x_2}{4}, \frac{y_1 + 3y_2}{4} \right)$$

**33.** Write the general equation of the circle with center at $(2, -1)$ and radius 4.

**Solution:**

Let $(2, -1) = (h, k)$ and $r = 4$. Then using the standard form of the equation of a circle, we have

$$(x - h)^2 + (y - k)^2 = r^2$$
$$(x - 2)^2 + [y - (-1)]^2 = 4^2$$
$$x^2 - 4x + 4 + y^2 + 2y + 1 = 16$$
$$x^2 + y^2 - 4x + 2y - 11 = 0$$

**39.** Write the general equation of the circle passing through the points $(0, 0), (0, 8)$, and $(6, 0)$.

**Solution:**

Since the general form of the equation of the circle is

$$x^2 + y^2 + Dx + Ey + F = 0$$

we must find the coefficients $D, E,$ and $F$ such that the given points are solution points.

| Solution Point | Resulting equation | Coefficient |
|---|---|---|
| $(0, 0)$ | $(0)^2 + (0)^2 + D(0) + E(0) + F = 0$ | $F = 0$ |
| $(0, 8)$ | $(0)^2 + (8)^2 + D(0) + E(8) + F = 0$ | $E = -8$ |
| $(6, 0)$ | $(6)^2 + (0)^2 + D(6) + E(0) + F = 0$ | $D = -6$ |

Therefore, the general equation is $x^2 + y^2 - 6x - 8y = 0$.

**45.** Write the equation $2x^2 + 2y^2 - 2x - 2y - 3 = 0$ in standard form and sketch its graph.

**Solution:**

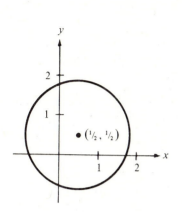

$$2x^2 + 2y^2 - 2x - 2y - 3 = 0$$

$$x^2 + y^2 - x - y = \frac{3}{2}$$

$$(x^2 - x + \quad) + (y^2 - y + \quad) = \frac{3}{2}$$

$$\left(x^2 - x + \frac{1}{4}\right) + \left(y^2 - y + \frac{1}{4}\right) = \frac{3}{2} + \frac{1}{4} + \frac{1}{4}$$

$$\left(x - \frac{1}{2}\right)^2 + \left(y - \frac{1}{2}\right)^2 = 2$$

Thus the circle is centered at $(\frac{1}{2}, \frac{1}{2})$ with a radius of $\sqrt{2}$.

**51.** Find an equation of the circle passing through the points $(4, 3), (-2, -5)$, and $(5, 2)$.

**Solution:**

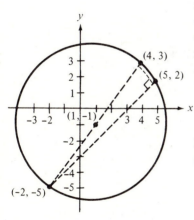

First, we observe that the points $(4, 3), (-2, -5$, and $(5, 2)$ form the vertices of a right triangle. Therefore, we can use the result from geometry that if a right triangle is inscribed in a circle, the hypotenuse of the triangle must lie on a diameter of the circle. Thus, the center of the circle must be the midpoint of the line segment joining $(4, 3)$ and $(-2, -5)$ (see the accompanying figure).

$$(h, k) = \left(\frac{4 - 2}{2}, \frac{3 - 5}{2}\right) = (1, -1)$$

Furthermore, the radius is the distance between $(4, 3)$ and $(1, -1)$.

$$r = \sqrt{(4 - 1)^2 + [3 - (-1)]^2} = \sqrt{9 + 16} = 5$$

Therefore, an equation of the circle is

$$(x - 1)^2 + [y - (-1)]^2 = 5^2$$

$$(x - 1)^2 + (y + 1)^2 = 25$$

**53.** Sketch the set of points satisfying the inequality

$$x^2 + y^2 - 4x + 2y + 1 \le 0.$$

**Solution:**

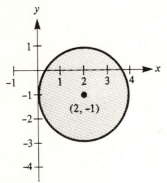

$$x^2 + y^2 - 4x + 2y + 1 \le 0$$
$$(x^2 - 4x + 4) + (y^2 + 2y + 1) \le -1 + 4 + 1$$
$$(x - 2)^2 + (y + 1)^2 \le 4$$

Therefore, the inequality is satisfied by the set of all points lying on the boundary and in the interior of the circle with center $(2, -1)$ and radius 2.

**61.** Prove that an angle inscribed in a semicircle is a right angle.

**Solution:**

For simplicity, we assume the semicircle is centered at the origin with a radius $r$ (see the accompanying figure). Since $(x, y)$ lies on the semicircle, it must satisfy the equation

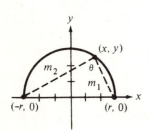

$$x^2 + y^2 = r^2$$
$$y^2 = r^2 - x^2$$
$$y = \sqrt{r^2 - x^2}$$

The slope of the line containing $(x, y)$ and $(r, 0)$ is

$$m_1 = \frac{0 - y}{r - x} = \frac{-\sqrt{r^2 - x^2}}{r - x}$$
$$= \frac{-\sqrt{(r + x)(r - x)}}{r - x} = \frac{-\sqrt{r + x}}{\sqrt{r - x}}$$

The slope of the line containing $(x, y)$ and $(-r, 0)$ is

$$m_2 = \frac{0 - y}{-r - x} = \frac{-\sqrt{r^2 - x^2}}{-r - x}$$
$$= \frac{\sqrt{(r + x)(r - x)}}{r + x} = \frac{\sqrt{r - x}}{\sqrt{r + x}}$$

Since $m_1 = -1/m_2$, the lines joining $(r, 0)$ and $(-r, 0)$ to $(x, y)$ must be perpendicular. Hence, the angle $\theta$ is a right angle.

## 1.3   Graphs of Equations

**9.** Find the intercepts of the graph of $y = x^2 + x - 2$.

**Solution:**

To find the $x$-intercepts, let $y = 0$. Then
$$x^2 + x - 2 = 0$$
and by factoring (or by the quadratic formula),
$$(x + 2)(x - 1) = 0.$$
Therefore, $y = 0$ when $x = -2$ or $x = 1$ and the $x$-intercepts are $(-2, 0)$ and $(1, 0)$. To find the $y$-intercepts, let $x = 0$. Then $y = 0^2 + 0 - 2 = -2$, and the $y$-intercept is $(0, -2)$.

**15.** Find the intercepts of the graph of $x^2y - x^2 + 4y = 0$.

**Solution:**

To find the $x$-intercepts we let $y = 0$. Then $x^2(0) - x^2 + 4(0) = -x^2 = 0$, which implies that $x = 0$. Letting $y = 0$ yields the same result and the only intercept is $(0, 0)$.

**21.** Check $y^2 = x^3 - 4x$ for symmetry about both axes and the origin.

**Solution:**

*No symmetry* about the $y$-axis since replacing $x$ with $-x$ in the equation yields
$$y^2 = (-x)^3 - 4(-x) = -x^3 + 4x$$
which is *not* equivalent to the original equation.

*Symmetry* about the $x$-axis since replacing $y$ with $-y$ in the equation yields
$$(-y)^2 = x^3 - 4x \qquad \text{or} \qquad y^2 = x^3 - 4x$$
which *is* equivalent to the original equation.

*No symmetry* about the origin since replacing $x$ with $-x$ and $y$ with $-y$ in the equation yields
$$(-y)^2 = (-x)^3 - 4(-x) \qquad \text{or} \qquad y^2 = -x^3 + 4x$$
which is *not* equivalent to the original equation.

**25.** Check $y = x/(x^2 + 1)$ for symmetry about both axes and the origin.

**Solution:**

*No symmetry* about the $y$-axis since replacing $x$ with $-x$ in the equation yields

$$y = \frac{-x}{(-x)^2 + 1} = \frac{-x}{x^2 + 1}$$

which is *not* equivalent to the original equation.

*No symmetry* about the $x$-axis since replacing $y$ with $-y$ in the equation yields

$$-y = \frac{x}{x^2 + 1}$$

which is *not* equivalent to the original equation.

*Symmetry* with respect to the origin since replacing $x$ with $-x$ and $y$ with $-y$ in the equation yields

$$-y = \frac{-x}{(-x)^2 + 1} \qquad \text{or} \qquad y = \frac{x}{x^2 + 1}$$

which *is* equivalent to the original equation.

**41.** Use the methods of this section to sketch the graph of the equation $x^2 + 4y^2 = 4$. Identify the intercepts and test for symmetry.

**Solution:**

To find the $x$-intercepts, let $y = 0$. Then $x^2 = 4$ which implies tht the $x$-intercepts are $(-2, 0)$ and $(2, 0)$. To find the $y$-intercepts, let $x = 0$. Then $4y^2 = 4$, which implies that the $y$-intercepts are $(0, -1)$ and $(0, 1)$.

There is symmetry with respect to the $x$-axis, the $y$-axis, and the origin since each of the variables is raised only to an even power. Some first-quadrant solution points are:

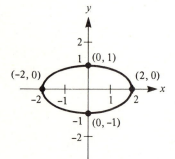

| $x$ | $\dfrac{1}{2}$ | $1$ | $\dfrac{3}{2}$ |
|---|---|---|---|
| $y$ | $\dfrac{\sqrt{15}}{4}$ | $\dfrac{\sqrt{3}}{2}$ | $\dfrac{\sqrt{7}}{4}$ |

**49.** Find the points of intersection of the graphs of $x + y = 7$ and $3x - 2y = 11$.

**Solution:**

To solve the two equations

$$x + y = 7$$
$$3x - 2y = 11$$

simultaneously, we multiply the first equation by 2 and add. Thus

$$2x + 2y = 14$$
$$(+) \quad 3x - 2y = 11$$
$$\overline{\qquad 5x \qquad\quad = 25}$$
$$x = 5$$

Substituting $x = 5$ into the first equation, we have

$$5 + y = 7 \qquad \text{or} \qquad y = 2$$

Thus the point of intersection is $(5, 2)$.

**51.** Find the points of intersection of the graphs of $x^2 + y^2 = 5$ and $x - y = 1$.

**Solution:**

To solve the two equations

$$x^2 + y^2 = 5$$
$$x - y = 1$$

simultaneously, we solve the second equation for $x$ and obtain

$$x = y + 1.$$

Substituting into the first equation, we have

$$(y + 1)^2 + y^2 = 5$$
$$(y^2 + 2y + 1) + y^2 = 5$$
$$2y^2 + 2y - 4 = 0$$
$$2(y + 2)(y - 1) = 0$$

which implies that $y = -2$ or $y = 1$. Therefore,

$$x = (-2) + 1 = -1 \qquad \text{or} \qquad x = 1 + 1 = 2$$

and the points of intersection are $(-1, -2)$ and $(2, 1)$.

**55.** Find the points of intersection of the graphs of
$y = x^3 - 2x^2 + x - 1$ and $y = -x^2 + 3x - 1$.

**Solution:**

By equating the two expressions for $y$, we have

$$x^3 - 2x^2 + x - 1 = -x^2 + 3x - 1$$
$$x^3 - 2x^2 + x^2 + x - 3x - 1 + 1 = 0$$
$$x^3 - x^2 - 2x = 0$$
$$x(x - 2)(x + 1) = 0$$

Thus $x = 0, 2$, or $-1$, and substituting these values into the second (or first) equation, we have

$$y = -(0)^2 + 3(0) - 1 = -1$$
$$y = -(2)^2 + 3(2) - 1 = \;\;1$$
$$y = -(-1)^2 + 3(-1) - 1 = -5$$

Therefore, the points of intersection are $(0, -1), (2, 1)$, and $(-1, -5)$.

**63.** For what values of $k$ does the graph of $y = kx^3$ pass through the points (a) $(1, 4)$, (b) $(-2, 1)$, (c) $(0, 0)$, and (d) $(-1, -1)$?

**Solution:**

(a) If $x = 1$ and $y = 4$, then $4 = k(1)^3$ and we obtain $k = 4$.

(b) If $x = -2$ and $y = 1$, then $1 = k(-2)^3$ and we obtain $k = -\frac{1}{8}$.

(c) If $x = y = 0$, then $0 = k(0)^3$ and $k$ may have *any value*.

(d) If $x = y = -1$, then $-1 = k(-1)^3$ and we obtain $k = 1$.

## 1.4   Lines in the Plane

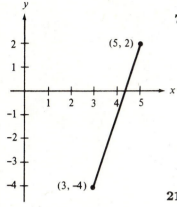

**7.** Plot the points $(3, -4)$ and $(5, 2)$ and find the slope of the line passing through them.

**Solution:**

Let $(3, -4) = (x_1, y_1)$ and $(5, 2) = (x_2, y_2)$. The slope of the line passing through $(x_1, y_1)$ and $(x_2, y_2)$ is

$$m = \frac{y_2 - y_1}{x_2 - x_1} = \frac{2 - (-4)}{5 - 3} = \frac{6}{2} = 3$$

**21.** Find an equation of the line passing through $(2, 1)$ and $(0, -3)$ and sketch its graph

**Solution:**

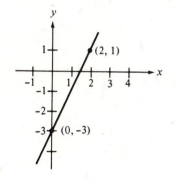

Let $(2, 1) = (x_1, y_1)$ and $(0, -3) = (x_2, y_2)$. The slope of the line passing through $(x_1, y_1)$ and $(x_2, y_2)$ is

$$m = \frac{y_2 - y_1}{x_2 - x_1} = \frac{-3 - 1}{0 - 2} = 2.$$

Using the point-slope form of the equation of a line, we have

$$y - y_1 = m(x - x_1)$$
$$y - 1 = 2(x - 2)$$
$$y - 1 = 2x - 4$$
$$0 = 2x - y - 3 \qquad \text{General form}$$

**27.** Find an equation of the line passing through $(0, 3)$ with a slope of $m = \frac{3}{4}$ and sketch its graph.

**Solution:**

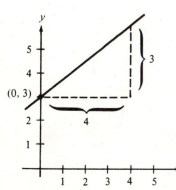

Using the slope-intercept form of the equation of a line, we have

$$y = mx + b$$
$$y = \frac{3}{4}x + 3$$
$$4y = 3x + 12$$
$$0 = 3x - 4y + 12$$

**35.** Write an equation of the line whose $x$-intercept is $(2, 0)$ and whose $y$-intercept is $(0, 3)$.

**Solution:**

From Exercise 34, the two-intercept form of the equation of a line is given by

$$\frac{x}{a} + \frac{y}{b} = 1$$

where $a$ is the $x$-intercept and $b$ is the $y$-intercept. Since $a = 2$ and $b = 3$, we have

$$\frac{x}{2} + \frac{y}{3} = 1$$
$$3x + 2y = 6$$
$$3x + 2y - 6 = 0$$

**41.** Write an equation of the line passing through $(2, 1)$ (a) parallel and (b) perpendicular to the line $4x - 2y = 3$.

**Solution:**

The line given by $4x - 2y = 3$ has a slope of 2 since

$$4x - 2y = 3$$
$$-2y = -4x + 3$$
$$y = 2x - \frac{3}{2} = mx + b$$

(a) The line through $(2, 1)$ parallel to $4x - 2y = 3$ must also have a slope of 2. Thus its equation must be

$$y - y_1 = m(x - x_1)$$
$$y - 1 = 2(x - 2)$$
$$-2x + y + 3 = 0$$
$$2x - y - 3 = 0$$

(b) The line through $(2, 1)$ perpendicular to $4x - 2y = 3$ must have a slope of $m = -\frac{1}{2}$. Thus its equation must be

$$y - y_1 = m(x - x_1)$$
$$y - 1 = -\frac{1}{2}(x - 2)$$
$$2y - 2 = -x + 2$$
$$x + 2y - 4 = 0$$

**55.** Use slope to determine if the points $(-2, 1), (-1, 0)$, and $(2, -2)$ are collinear.

**Solution:**

Let $(-2, 1) = (x_1, y_1), (-1, 0) = (x_2, y_2)$, and $(2, -2) = (x, y)$. The point $(x, y)$ lies on the line passing through $(x_1, y_1)$ and $(x_2, y_2)$ if and only if

$$\frac{y - y_1}{x - x_1} = m = \frac{y - y_2}{x - x_2}$$

Since

$$\frac{-2 - 1}{2 - (-2)} = -\frac{3}{4} \neq -\frac{2}{3} = \frac{-2 - 0}{2 - (-1)}$$

the three points are *not* collinear.

**65.** A small business purchases a piece of equipment for \$875. After 5 years the equipment will be obsolete and have no value. Write a linear equation giving the value $y$ of the equipment during the five years it will be used. (Let $t$ represent the time in years.)

**Solution:**

Two solution points to the linear equation are $(t_1, y_1) = (0, 875)$ and $(t_2, y_2) = (5, 0)$. Therefore, the slope (rate of depreciation per year) is

$$m = \frac{0 - 875}{5 - 0} = -\$175 \text{ per year}$$

and the equation is

$$y = mt + b$$
$$y = -175t + 875 \qquad \text{where} \qquad 0 \leq t \leq 5$$

**71.** Find the distance between the point $(-2, 1)$ and the line $x - y - 2 = 0$.

**Solution:**

Letting $(-2, 1) = (x_1, y_1)$ and $x - y - 2 = Ax + By + C = 0$, we have

$$d = \frac{|Ax_1 + By_1 + C|}{\sqrt{A^2 + B^2}}$$
$$= \frac{|1(-2) + (-1)(1) - 2|}{\sqrt{1^2 + 1^2}}$$
$$= \frac{|-5|}{\sqrt{2}} = \frac{5\sqrt{2}}{2}$$

**73.** Find the distance between the parallel lines $x + y = 1$ and $x + y = 5$.

**Solution:**

A point on the line $x + y = 1$ is $(2, -1)$. The distance between the given parallel lines is equal to the distance from $(2, -1)$ to the line $x + y = 5$. Letting $(2, -1) = (x_1, y_1)$ and $x + y - 5 = Ax + By + C = 0$, we have

$$d = \frac{|Ax_1 + By_1 + C|}{\sqrt{A^2 + B^2}}$$
$$= \frac{|1(2) + 1(-1) - 5|}{\sqrt{1^2 + 1^2}}$$
$$= \frac{|-4|}{\sqrt{2}} = \frac{4}{\sqrt{2}} = 2\sqrt{2}$$

## 1.5   Functions

**3.** Given $f(x) = \sqrt{x + 3}$, find:

(a) $f(-2)$      (b) $f(6)$      (c) $f(c)$      (d) $f(x + \Delta x)$.

**Solution:**

(a) $f(-2) = \sqrt{-2 + 3} = \sqrt{1} = 1$ (b) $f(6) = \sqrt{6 + 3} = \sqrt{9} = 3$

(c) $f(c) = \sqrt{c + 3}$            (d) $f(x + \Delta x) = \sqrt{x + \Delta x + 3}$

**9.** Given $f(x) = x^3$, find $\dfrac{f(x + \Delta x) - f(x)}{\Delta x}$.

**Solution:**

$$
\begin{aligned}
\frac{f(x + \Delta x) - f(x)}{\Delta x} &= \frac{(x + \Delta x)^3 - x^3}{\Delta x} \\
&= \frac{x^3 + 3x^2\,\Delta x + 3x(\Delta x)^2 + (\Delta x)^3 - x^3}{\Delta x} \\
&= \frac{\Delta x[3x^2 + 3x\,\Delta x + (\Delta x)^2]}{\Delta x} \\
&= 3x^2 + 3x\,\Delta x + (\Delta x)^2
\end{aligned}
$$

**19.** Find the domain and range of the function $f(x) = \sqrt{9 - x^2}$ and sketch its graph.

**Solution:**

Since $9 - x^2$ must be nonnegative ($9 - x^2 \geq 0$), the domain is $[-3, 3]$. The range is $[0, 3]$. There is symmetry with respect to the $y$-axis since

$$
y = \sqrt{9 - (-x)^2} = \sqrt{9 - x^2}
$$

is equivalent to the original equation. Squaring both members of the equation, we have

$$
y^2 = 9 - x^2 \qquad \text{or} \qquad x^2 + y^2 = 3^2.
$$

This the standard form of an equation of a circle with center $(0, 0)$ and radius 3. Therefore, the graph of $f(x) = \sqrt{9 - x^2}$ is a semicircle in the first and second quadrants with center $(0, 0)$ and radius 3.

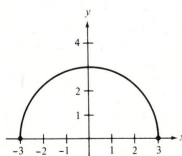

**36.** Determine if $y$ is a function of $x$ for the equation $x^2 y - x^2 + 4y = 0$.

**Solution:**

Solving the equation for $y$, we obtain

$$x^2 y - x^2 + 4y = 0$$
$$(x^2 + 4)y = x^2$$
$$y = \frac{x^2}{x^2 + 4}$$

For each value of the independent variable $x$, there corresponds exactly one value of the dependent variable $y$. Therefore, $y$ is a function of $x$.

**42.** Given $f(x) = 1/x$ and $g(x) = x^2 - 1$, find:

(a) $f[g(2)]$      (b) $g[f(2)]$      (c) $f[g(1/\sqrt{2})]$

(d) $g[f(1/\sqrt{2})]$      (e) $g[f(x)]$      (f) $f[g(x)]$

**Solution:**

(a) $f[g(2)] = f(2^2 - 1) = f(3) = \dfrac{1}{3}$

(b) $g[f(2)] = g\left(\dfrac{1}{2}\right) = \left(\dfrac{1}{2}\right)^2 - 1 = \dfrac{1}{4} - 1 = -\dfrac{3}{4}$

(c) $f\left[g\left(\dfrac{1}{\sqrt{2}}\right)\right] = f\left[\left(\dfrac{1}{\sqrt{2}}\right)^2 - 1\right] = f\left(-\dfrac{1}{2}\right) = \dfrac{1}{-\frac{1}{2}} = -2$

(d) $g\left[f\left(\dfrac{1}{\sqrt{2}}\right)\right] = g\left(\dfrac{1}{1/\sqrt{2}}\right) = g(\sqrt{2}) = (\sqrt{2})^2 - 1 = 1$

(e) $g[f(x)] = g\left(\dfrac{1}{x}\right) = \left(\dfrac{1}{x}\right)^2 - 1 = \dfrac{1}{x^2} - 1 = \dfrac{1 - x^2}{x^2}$

(f) $f[g(x)] = f(x^2 - 1) = \dfrac{1}{x^2 - 1}$

**53.** Determine if the function $f(x) = x(4 - x^2)$ is even, odd, or neither.

**Solution:**

The function is odd since

$$f(-x) = (-x)[4 - (-x)^2] = -x(4 - x^2) = -f(x)$$

**69.** A man is in a boat 2 miles from the nearest point on the coast. He is to go to a point $Q$, 3 miles down the coast and 1 mile inland (see figure). He can row at 2 miles per hour and walk at 4 miles per hour. Express the total time $T$ of the trip as a function of $x$.

**Solution:**

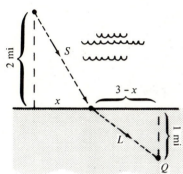

$$S = \text{distance on water} = \sqrt{x^2 + 4}$$
$$L = \text{distance on land} = \sqrt{1 + (3-x)^2} = \sqrt{x^2 - 6x + 10}$$

$$\text{Total time} = (\text{time on water}) + (\text{time on land})$$

$$T = \frac{S}{(\text{rate on water})} + \frac{L}{(\text{rate on land})}$$

$$= \frac{\sqrt{x^2 + 4}}{2} + \frac{\sqrt{x^2 - 6x + 10}}{4}$$

## Review Exercises for Chapter 1

**3.** Sketch the intervals defined by $4 < (x+3)^2$.

**Solution:**

By finding the square root of each member of the inequality $4 < (x+3)^2$ we obtain

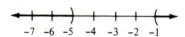

$$2 < x + 3 \qquad \text{or} \qquad -2 > x + 3$$
$$-1 < x \qquad \text{or} \qquad -5 > x$$

**13.** Determine the value of $c$ so that the circle given by the equation $x^2 + y^2 - 6x + 8y = c$ has a radius of 2.

**Solution:**

$$x^2 - 6x + y^2 + 8y = c$$
$$(x^2 - 6x + 9) + (y^2 + 8y + 16) = c + 9 + 16$$
$$(x - 3)^2 + (y + 4)^2 = c + 25$$

If the radius of the circle is 2, then $c + 25 = 2^2$ or $c = -21$.

**33.** Express the value $v$ of a farm at \$850 per acre, with buildings, livestock, and equipment worth \$300,000, as a function of the number of acres $a$, and give its domain.

**Solution:**

The value of the farm is \$300,000 plus the value of the land. If the farm has one acre of land, its value is

$$v = 300,000 + 1(850).$$

If the farm has two acres of land, its value is

$$v = 300,000 + 2(850).$$

Thus, in general, for $a$ acres its value is

$$v = 300,000 + a(850) = 300,000 + 850a.$$

The domain of the function is $a \geq 0$.

**44.** Sketch the graph of $x^3 - y^2 + 1 = 0$ and determine if $y$ is a function of $x$.

**Solution:**

There is symmetry with respect to the $x$-axis since

$$x^3 - (-y)^2 + 1 = 0$$
$$x^3 - y^2 + 1 = 0$$

is equivalent to the original equation.

To find the $y$-intercepts, let $x = 0$. Then

$$-y^2 + 1 = 0$$
$$y^2 = 1$$
$$y = \pm 1$$

and the $y$-intercepts are $(0, 1)$ and $(0, -1)$. Therefore, $y$ is *not* a function of $x$ since for $x = 0$ there corresponds two values of $y$. Solving the equation for $y$, we obtain

$$y^2 = x^3 + 1$$
$$y = \pm\sqrt{x^3 + 1}$$

The domain of the function is $x \geq -1$. The following table gives a a few solution points to the equation:

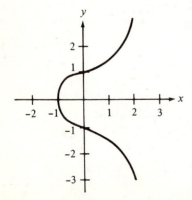

| $x$ | $-1$ | $-\dfrac{1}{2}$ | $0$ | $2$ |
|---|---|---|---|---|
| $y$ | $0$ | $\pm\sqrt{\dfrac{7}{8}} \approx \pm 0.935$ | $\pm 1$ | $\pm 3$ |

# 2 LIMITS AND THEIR PROPERTIES

## 2.1 An Introduction to Limits

**5.** Complete the following table and use the result to estimate the limit

$$\lim_{x \to 3} \frac{\dfrac{1}{x+1} - \dfrac{1}{4}}{x - 3}$$

| $x$ | 2.9 | 2.99 | 2.999 | 3.001 | 3.01 | 3.1 |
|-----|-----|------|-------|-------|------|-----|
| $f(x)$ | | | | | | |

**Solution:**

The table lists values of $f(x)$ at several $x$-values near 3.

| $x$ | 2.9 | 2.99 | 2.999 | 3.001 | 3.01 | 3.1 |
|-----|-----|------|-------|-------|------|-----|
| $f(x)$ | $-0.0641$ | $-0.0627$ | $-0.0625$ | $-0.0625$ | $-0.0623$ | $-0.0610$ |

As $x$ approaches 3 from the left and from the right $f(x)$ approaches $-0.0625$. Therefore, we estimate the limit to be $-\frac{1}{16}$.

**19.** Find $\lim_{x \to 3} \sqrt{x + 1}$.

**Solution:**

$$\lim_{x \to 3} \sqrt{x + 1} = \sqrt{3 + 1} = \sqrt{4} = 2$$

**29.** If $\lim\limits_{x \to c} f(x) = 4$, find:

(a)    $\lim\limits_{x \to c} [f(x)]^3$

(b)    $\lim\limits_{x \to c} \sqrt{f(x)}$

(c)    $\lim\limits_{x \to c} [3f(x)]$

(d)    $\lim\limits_{x \to c} [f(x)]^{3/2}$

**Solution:**

(a)    $\lim\limits_{x \to c} [f(x)]^3 = \left[\lim\limits_{x \to c} f(x)\right]^3 = 4^3 = 64$

(b)    $\lim\limits_{x \to c} \sqrt{f(x)} = \sqrt{\lim\limits_{x \to c} f(x)} = \sqrt{4} = 2$

(c)    $\lim\limits_{x \to c} [3f(x)] = 3\lim\limits_{x \to c} f(x) = 3(4) = 12$

(d)    $\lim\limits_{x \to c} [f(x)]^{3/2} = \left[\lim\limits_{x \to c} f(x)\right]^{3/2} = 4^{3/2} = 8$

## 2.2    Techniques for Evaluating Limits

**5.** Find (if it exists)  $\lim\limits_{x \to -1} \dfrac{x^2 - 1}{x + 1}$.

**Solution:**

$$\lim\limits_{x \to -1} \frac{x^2 - 1}{x + 1} = \lim\limits_{x \to -1} \frac{(x + 1)(x - 1)}{x + 1} = \lim\limits_{x \to -1} (x - 1) = -2$$

**13.** Find (if it exists)

$$\lim\limits_{\Delta x \to 0} \frac{(x + \Delta x)^2 - 2(x + \Delta x) + 1 - (x^2 - 2x + 1)}{\Delta x}.$$

**Solution:**

$$\lim\limits_{\Delta x \to 0} \frac{(x + \Delta x)^2 - 2(x + \Delta x) + 1 - (x^2 - 2x + 1)}{\Delta x}$$
$$= \lim\limits_{\Delta x \to 0} \frac{x^2 + 2x\Delta x + (\Delta x)^2 - 2x - 2\Delta x + 1 - x^2 + 2x - 1}{\Delta x}$$

$$= \lim_{\Delta x \to 0} \frac{2x\Delta x + (\Delta x)^2 - 2\Delta x}{\Delta x}$$

$$= \lim_{\Delta x \to 0} \frac{\Delta x(2x + \Delta x - 2)}{\Delta x}$$

$$= \lim_{\Delta x \to 0} (2x + \Delta x - 2) = 2x - 2$$

**19.** Find (if it exists)  $\displaystyle \lim_{x \to 0} \frac{\sqrt{3 + x} - \sqrt{3}}{x}$

**Solution:**

$$\lim_{x \to 0} \frac{\sqrt{3 + x} - \sqrt{3}}{x} = \lim_{x \to 0} \left( \frac{\sqrt{3 + x} - \sqrt{3}}{x} \right) \left( \frac{\sqrt{3 + x} + \sqrt{3}}{\sqrt{3 + x} + \sqrt{3}} \right)$$

$$= \lim_{x \to 0} \frac{3 + x - 3}{x(\sqrt{3 + x} + \sqrt{3})}$$

$$= \lim_{x \to 0} \frac{1}{\sqrt{3 + x} + \sqrt{3}} = \frac{1}{2\sqrt{3}} = \frac{\sqrt{3}}{6}$$

**21.** Find (if it exists)

$$\lim_{x \to 0} \frac{\dfrac{1}{2 + x} - \dfrac{1}{2}}{x}$$

**Solution:**

$$\lim_{x \to 0} \frac{\dfrac{1}{2 + x} - \dfrac{1}{2}}{x} = \lim_{x \to 0} \frac{\dfrac{2 - (2 + x)}{2(2 + x)}}{x}$$

$$= \lim_{x \to 0} \frac{-x}{x(2)(2 + x)}$$

$$= \lim_{x \to 0} \frac{-1}{2(2 + x)} = -\frac{1}{4}$$

**29.** Use the graph to visually determine:

(a)  $\lim\limits_{x \to c^+} f(x)$ (b)  $\lim\limits_{x \to c^-} f(x)$ (c)  $\lim\limits_{x \to c} f(x)$

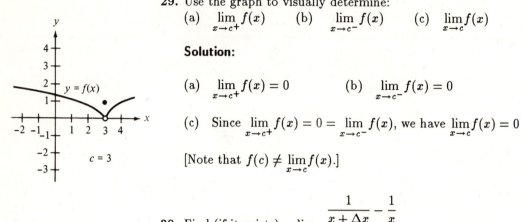

**Solution:**

(a)  $\lim\limits_{x \to c^+} f(x) = 0$ (b)  $\lim\limits_{x \to c^-} f(x) = 0$

(c)  Since $\lim\limits_{x \to c^+} f(x) = 0 = \lim\limits_{x \to c^-} f(x)$, we have $\lim\limits_{x \to c} f(x) = 0$

[Note that $f(c) \neq \lim\limits_{x \to c} f(x)$.]

**39.** Find (if it exists)  $\lim\limits_{\Delta x \to 0^+} \dfrac{\dfrac{1}{x + \Delta x} - \dfrac{1}{x}}{\Delta x}$.

**Solution:**

$$\lim_{\Delta x \to 0^+} \frac{\dfrac{1}{x + \Delta x} - \dfrac{1}{x}}{\Delta x} = \lim_{\Delta x \to 0^+} \frac{\dfrac{x - (x + \Delta x)}{x(x + \Delta x)}}{\Delta x}$$

$$= \lim_{\Delta x \to 0^+} \frac{-\Delta x}{(\Delta x)(x)(x + \Delta x)}$$

$$= \lim_{\Delta x \to 0^+} \frac{-1}{x(x + \Delta x)} = -\frac{1}{x^2}$$

**43.** Find (if it exists)  $\lim\limits_{x \to 3} f(x)$, where

$$f(x) = \begin{cases} \dfrac{x + 2}{2}, & x \leq 3 \\ \dfrac{12 - 2x}{2}, & x > 3 \end{cases}$$

**Solution:**

$$\lim_{x \to 3^-} f(x) = \lim_{x \to 3^-} \frac{x + 2}{5} = \frac{5}{2}$$

$$\lim_{x \to 3^+} f(x) = \lim_{x \to 3^+} \frac{12 - 2x}{3} = 2$$

Since the limit from the left is *not equal* to the limit from the right, the limit does *not* exist.

**45.** Find (if it exists) $\lim\limits_{x \to 1} f(x)$, where

$$f(x) = \begin{cases} x^3 + 1, & x < 1 \\ x + 1, & x \geq 1 \end{cases}$$

**Solution:**

$$\lim_{x \to 1^-} f(x) = \lim_{x \to 1^-} (x^3 + 1) = 2$$

$$\lim_{x \to 1^+} f(x) = \lim_{x \to 1^+} (x + 1) = 2$$

Since the limit from the left is *equal* to the limit from the right, we have $\lim_{x \to 1} f(x) = 2$.

## 2.3  Continuity

**9.** Find the discontinuities (if any) for $f(x) = 1/(x - 1)$. Which of the discontinuities are removable?

**Solution:**

From Theorem 2.9 we know that $f$ is continuous for all x other than $x = 1$. At $x = 1$ the function is discontinuous and the discontinuity is nonremovable since

$$\lim_{x \to 1^-} \frac{1}{x - 1} = -\infty \quad \text{and} \quad \lim_{x \to 1^+} \frac{1}{x - 1} = \infty.$$

**13.** Find the discontinuities (if any) for $f(x) = (x+2)/(x^2 - 3x - 10)$ Which of the discontinuities are removable?

**Solution:**

Since $x^2 - 3x - 10 = (x - 5)(x + 2)$, $x = 5$ and $x = -2$ are not in the domain of $f$. By Theorem 2.9, $f$ is continuous for all $x$ other than $x = 5$ or $x = -2$. At $x = 5$ the function is discontinuous and the discontinuity is nonremovable since

$$\lim_{x \to 5^-} \frac{x + 2}{(x - 5)(x + 2)} = \lim_{x \to 5^-} \frac{1}{x - 5} = -\infty$$

and

$$\lim_{x \to 5^+} \frac{x + 2}{(x - 5)(x + 2)} = \lim_{x \to 5^+} \frac{1}{x - 5} = \infty.$$

At $x = -2$ the function is discontinuous but it is removable since

$$\lim_{x \to -2} \frac{x + 2}{x^2 - 3x - 10} = \lim_{x \to -2} \frac{1}{x - 5} = -\frac{1}{7}.$$

17. Find the discontinuities (if any) for

$$f(x) = \begin{cases} \dfrac{x}{2} + 1, & x \le 2 \\ 3 - x, & x > 2 \end{cases}$$

Which of the discontinuities are removable?

**Solution:**

Since $f$ is linear to the right and left of $x = 2$, it is continuous for all x other than possibly at $x = 2$. At $x = 2$,

$$\lim_{x \to 2^-} f(x) = \lim_{x \to 2^-} \left( \frac{x}{2} + 1 \right) = 2$$

$$\lim_{x \to 2^+} f(x) = \lim_{x \to 2^+} (3 - x) = 1$$

Thus $f$ is discontinuous at $x = 2$ and this discontinuity is nonremovable since the limit from the left is *not equal* to the limit from the right.

41. Use the Intermediate Value Theorem to approximate the zero of the function $f(x) = x^3 + x - 1$ in the interval $[0, 1]$.

(a) Begin by locating the zero in a subinterval of length 0.1.

(b) Refine your approximation by locating the zero in a subinterval of length 0.01.

**Solution:**

Since $f$ is a polynomial, it is continuous. Therefore, by the Intermediate Value Theorem, if $f$ changes signs on the interval $[a, b]$, then a zero of $f$ is in that interval. Since $f(0) = -1$ and $f(1) = 1$, a zero of $f$ is in the interval $[0, 1]$.

(a) The following table gives the sign of the function at each value of x. The table reveals that the zero of $f$ is in the subinterval $[0.6, 0.7]$.

| $x$ | 0.1 | 0.2 | 0.3 | 0.4 | 0.5 | 0.6 | 0.7 | 0.8 | 0.9 |
|------|-----|-----|-----|-----|-----|-----|-----|-----|-----|
| $f(x)$ | − | − | − | − | − | − | + | + | + |

(b) The following table gives the sign of the function at each value of x. The table reveals that the zero of $f$ is in the subinterval $[0.68, 0.69]$.

| $x$ | 0.61 | 0.62 | 0.63 | 0.64 | 0.65 | 0.66 | 0.67 | 0.68 | 0.69 |
|-----|------|------|------|------|------|------|------|------|------|
| $f(x)$ | − | − | − | − | − | − | − | − | + |

**46.** Verify the applicability of the Intermediate Value Theorem and find the value of $c$ in $[\frac{5}{2}, 4]$ guaranteed by the theorem if $f(x) = (x^2 + x)/(x - 1)$ and $f(c) = 6$.

**Solution:**

By Theorem 2.9, $f(x) = (x^2 + x)/(x - 1)$ is continuous for every real number except $x = 1$. Therefore, it is continuous on $[\frac{5}{2}, 4]$. Also,

$$f\left(\frac{5}{2}\right) = \frac{35}{6} < 6 < \frac{20}{3} = f(4)$$

Hence, the Intermediate Value Theorem applies. To find $c$, solve the equation

$$\frac{x^2 + x}{x - 1} = 6$$

$$\left(\frac{x^2 + x}{x - 1}\right)(x - 1) = 6(x - 1)$$

$$x^2 - 5x + 6 = 0$$

$$(x - 3)(x - 2) = 0$$

The solution of this equation is $x = 2$ or $x = 3$. Since 2 is not in $[\frac{5}{2}, 4]$, we have $c = 3$.

**52.** The number of units in inventory in a small company is given by

$$N(t) = 25\left(2\left[\!\left[\frac{t + 2}{2}\right]\!\right] - t\right)$$

where t is the time in months. Sketch the graph of this function and discuss its continuity. How often must this company replenish its inventory?

**Solution:**

Since the number of units in inventory is given by

$$N(t) = 25\left(2\left[\!\left[\frac{t+2}{2}\right]\!\right] - t\right) = 25\left(2\left[\!\left[\frac{t}{2} + 1\right]\!\right] - t\right),$$

we observe that the greatest integer function is discontinuous at every positive even integer. Therefore, $N(t)$ is discontinuous at every positive even integer. We demonstrate this by evaluating the function for several values of $t$ and plotting the points to generate the accompanying figure.

$$N(0) = 25(2[\![1]\!] - 0) = 25[2(1) - 0] = 50$$
$$N(1) = 25(2[\![1.5]\!] - 1) = 25[2(1) - 1] = 25$$
$$N(1.8) = 25(2[\![1.9]\!] - 1.8) = 25[2(1) - 1.8] = 5$$
$$N(2) = 25(2[\![2]\!] - 2) = 25[2(2) - 2] = 50$$
$$N(3) = 25(2[\![2.5]\!] - 3) = 25[2(2) - 3] = 25$$
$$N(3.8) = 25(2[\![2.9]\!] - 3.8) = 25[2(2) - 3.8] = 5$$
$$N(4) = 25(2[\![3]\!] - 4) = 25[2(3) - 4] = 50$$

The company must replenish its inventory every two months.

## 2.4    Infinite Limits

**3.** Determine whether $f(x) = 1/(x^2 - 9)$ approaches $\infty$ or $-\infty$ as $x$ approaches $-3$ from the left and from the right.

**Solution:**

Since the denominator of the function

$$f(x) = \frac{1}{x^2 - 9}$$

is zero when $x = -3$, a vertical asymptote of $f$ is $x = -3$. The behavior of $f$ as $x$ approaches $-3$ from the left and from the right can be seen from the following table:

| $x$ | $-3.5$ | $-3.1$ | $-3.01$ | $-3.001$ | $-2.999$ | $-2.99$ | $-2.9$ | $-2.5$ |
|---|---|---|---|---|---|---|---|---|
| $f(x)$ | 0.31 | 1.64 | 16.64 | 166.64 | $-166.69$ | $-16.69$ | $-1.69$ | $-0.36$ |

Therefore,

$$\lim_{x \to -3^-} \frac{1}{x^2 - 9} = \infty \quad \text{and} \quad \lim_{x \to -3^+} \frac{1}{x^2 - 9} = -\infty$$

**15.** Find the vertical asymptotes (if any) for $f(x) = 1 - 4/x^2$.

**Solution:**

By rewriting the equation for this function as the ratio of two polynomials, we have

$$f(x) = 1 - \frac{1}{x^2} = \frac{x^2 - 4}{x^2}$$

Since the denominator of the function is zero only when $x$ is zero, the only vertical asymptote is $x = 0$.

**17.** Find the vertical asymptotes (if any) for $f(x) = x/(x^2 + x - 2)$.

**Solution:**

By factoring the denominator we have

$$f(x) = \frac{x}{x^2 + x - 2} = \frac{x}{(x-1)(x+2)}.$$

Therefore, we observe that the function has two vertical asymptotes, and they are $x = 1$ and $x = -2$.

**31.** Find $\displaystyle\lim_{x \to 1+} \frac{x^2 + x + 1}{x^3 - 1}$.

**Solution:**

$$\lim_{x \to 1+} \frac{x^2 + x + 1}{x^3 - 1} = \lim_{x \to 1+} \frac{x^2 + x + 1}{(x-1)(x^2 + x + 1)}$$

$$= \lim_{x \to 1+} \frac{1}{x - 1} = \infty$$

**41.** A 25-foot ladder is leaning against a house. If the base of the ladder is pulled away from the house at a rate of 2 feet per second, the top will move down the wall at a rate of

$$r = \frac{2x}{\sqrt{625 - x^2}} \text{ feet per second}$$

(a)   Find the rate when $x$ is 7 feet. (b)   Find the rate when $x$ is 15 feet. (c)   Find the limit of $r$ as $x \to 25^-$.

**Solution:**

(a) When $x = 7$, we have

$$r = \frac{2(7)}{\sqrt{625 - 7^2}} = \frac{14}{\sqrt{576}} = \frac{7}{12} \text{ feet per second.}$$

(b) When $x = 15$, we have

$$r = \frac{2(15)}{\sqrt{625 - 15^2}} = \frac{30}{\sqrt{400}} = \frac{3}{2} \text{ feet per second.}$$

(c) As $x$ approaches 25 from the left, we have

$$\lim_{x \to 25^-} \frac{2x}{\sqrt{625 - x^2}} = \infty.$$

## 2.5   $\epsilon$-$\delta$ Definition of Limits

**1.** Find $L$ such that $\lim\limits_{x \to 2}(3x + 2) = L$ and then find $\delta > 0$ such that $|f(x) - L| < 0.01$ whenever $0 < |x - c| < \delta$.

**Solution:**

We use the definition of limit to verify that

$$\lim_{x \to 2}(3x + 2) = 8.$$

We are required to show that there exists a $\delta$ such that $|f(x) - L| < 0.01$ whenever $0 < |x - 2| < \delta$.

$$|f(x) - L| < 0.01$$
$$|(3x + 2) - 8| < 0.01$$
$$|3x - 6| < 0.01$$
$$3|x - 2| < 0.01$$
$$0 < |x - 2| < \frac{0.01}{3}$$

Therefore, $\delta = \dfrac{0.01}{3}$.

**7.** Find $L$ such that $\lim\limits_{x \to 2}(x+3) = L$. Then for $\epsilon > 0$, find $\delta > 0$ such that $|f(x) - L| < \epsilon$ whenever $0 < |x - c| < \delta$.

**Solution:**

We use the definition of limit to verify that

$$\lim_{x \to 2}(x+3) = 5.$$

Given $\epsilon > 0$,

$$|(x+3) - 5| < \epsilon$$
$$|x - 2| < \epsilon = \delta$$

**11.** Find $L$ such that $\lim\limits_{x \to 0} \sqrt[3]{x} = L$. Then for $\epsilon > 0$, find $\delta > 0$ such that $|f(x) - L| < \epsilon$ whenever $0 < |x - c| < \delta$.

**Solution:**

We use the definition of limit to verify that

$$\lim_{x \to 0} \sqrt[3]{x} = 0.$$

Given $\epsilon > 0$,

$$|\sqrt[3]{x} - 0| < \epsilon$$
$$(|\sqrt[3]{x}|)^3 < \epsilon^3$$
$$|x| < \epsilon^3 = \delta$$

**19.** Find $L$ such that $\lim\limits_{x \to 2}(x^2 - 2) = L$. Then for $\epsilon > 0$, find $\delta > 0$, such that $|f(x) - L| < \epsilon$ whenever $0 < |x - c| < \delta$.

**Solution:**

We use the definition of limit to verify that

$$\lim_{x \to 2}(x^2 - 2) = 2.$$

Given $\epsilon > 0$, we have

$$|(x^2 - 2) - 2| < \epsilon$$
$$|x^2 - 4| < \epsilon$$
$$|(x+2)(x-2)| < \epsilon$$
$$|x+2|\,|x-2| < \epsilon$$

Without loss of generality, we assume that $x$ is in the interval $(1, 3)$. Therefore, $|x + 2| < 5$ and we have

$$|x + 2|\,|x - 2| < 5|x - 2| < \epsilon$$
$$|x - 2| < \frac{\epsilon}{5} = \delta$$

**23.** Use the definition of an infinite limit to verify that

$$\lim_{x \to -1+} \frac{1}{x + 1}.$$

**Solution:**

Given $M > 0$, find $\delta > 0$ such that $1/(x + 1) > M$ whenever $0 < x - (-1) = x + 1 < \delta$. Since $x + 1 > 0$ for $x > -1$, we have

$$\frac{1}{x + 1} > M \quad \Longleftrightarrow \quad x + 1 < \frac{1}{M}$$

Therefore, let $\delta = 1/M$.

**27.** Prove that $f(x) = x^2$ is continuous at $x = 3$.

**Solution:**

We must show that $\lim_{x \to 3} x^2 = f(3) = 9$. Given $\epsilon > 0$,

$$|x^2 - 9| < \epsilon$$
$$|(x - 3)(x + 3)| < \epsilon$$
$$|x - 3|\,|x + 3| < \epsilon$$
$$|x - 3| < \frac{\epsilon}{|x + 3|}$$

If $\delta < 1$, then $2 < x < 4$ and $|x + 3| < 7$. Thus

$$\frac{\epsilon}{7} < \frac{\epsilon}{|x + 3|}$$

whenever $2 < x < 4$. Therefore, we let $\delta = \epsilon/7$ and obtain

$$|x - 3| < \delta = \frac{\epsilon}{7} < \frac{\epsilon}{|x + 3|}.$$

## Review Exercises for Chapter 2

**9.** Evaluate $\lim\limits_{x\to 0} \dfrac{\dfrac{1}{x+1} - 1}{x}$.

**Solution:**

$$\lim_{x\to 0} \frac{\dfrac{1}{x+1} - 1}{x} = \lim_{x\to 0} \frac{\dfrac{1 - (x+1)}{x+1}}{x}$$

$$= \lim_{x\to 0} \frac{1 - x - 1}{x(x+1)}$$

$$= \lim_{x\to 0} \frac{-1}{x+1} = -1$$

**11.** Evaluate $\lim\limits_{x\to -1} \dfrac{x^3 + 1}{x+1}$.

**Solution:**

$$\lim_{x\to -1} \frac{x^3 + 1}{x+1} = \lim_{x\to -1} \frac{(x+1)(x^2 - x + 1)}{x+1}$$

$$= \lim_{x\to -1} (x^2 - x + 1) = 3$$

**25.** Determine whether $\lim\limits_{x\to 0} \dfrac{|x|}{x} = 1$ is true or false.

**Solution:**

$$\lim_{x\to 0^+} \frac{|x|}{x} = \lim_{x\to 0^+} \frac{x}{x} = \lim_{x\to 0^+} 1 = 1$$

$$\lim_{x\to 0^-} \frac{|x|}{x} = \lim_{x\to 0^-} \frac{-x}{x} = \lim_{x\to 0^-} (-1) = -1$$

Since the limit from the right and the limit from the left are *not* equal, the limit does *not* exist and the given statement is false.

**39.** Determine the value of $c$ so that the function

$$f(x) = \begin{cases} x+3, & x \leq 2 \\ cx+6, & x > 2 \end{cases}$$

is continuous on the entire real line.

**Solution:**

Since $f$ is linear to the right and left of 2, it is continuous for all values of $x$ other than possibly at $x = 2$. Furthermore, since $f(2) = 2+3 = 5$, we can make $f$ continuous at $x = 2$ by finding $c$ such that

$$c(2) + 6 = 5$$
$$2c = -1$$
$$c = -\frac{1}{2}$$

Now we have

$$\lim_{x \to 2^-} f(x) = \lim_{x \to 2^-} (x+3) = 2+3 = 5$$

and

$$\lim_{x \to 2^+} f(x) = \lim_{x \to 2^+} \left(-\frac{x}{2} + 6\right) = -1 + 6 = 5.$$

Thus for $c = -1/2$, $f$ is continuous at $x = 2$ and consequently $f$ is continuous for all $x$.

# 3 DIFFERENTIATION

## 3.1 The Derivative and the Tangent Line Problem

**9.** Find the derivative of $f(x) = 2x^2 + x - 1$ by the four-step process.

**Solution:**

$$f(x) = 2x^2 + x - 1$$

(a)
$$f(x + \Delta x) = 2(x + \Delta x)^2 + (x + \Delta x) - 1$$
$$= 2x^2 + 4x\Delta x + 2(\Delta x)^2$$
$$+ x + \Delta x - 1$$

(b)
$$f(x + \Delta x) - f(x) = 2x^2 + 4x\Delta x + 2(\Delta x)^2 + x$$
$$+ \Delta x - 1 - (2x^2 + x - 1)$$
$$= 4x\Delta x + 2(\Delta x)^2 + \Delta x$$

(c)
$$\frac{f(x + \Delta x) - f(x)}{\Delta x} = \frac{\Delta x(4x + 2\,\Delta x + 1)}{\Delta x}$$
$$= 4x + 2\,\Delta x + 1$$

(d)
$$\lim_{\Delta x \to 0} \frac{f(x + \Delta x) - f(x)}{\Delta x} = 4x + 1$$

**11.** Find the derivative of $f(x) = 1/(x - 1)$ by the four-step process.

**Solution:**

$$f(x) = \frac{1}{x-1}$$

(a)  $$f(x + \Delta x) = \frac{1}{(x + \Delta x) - 1}$$

(b)  $$f(x + \Delta x) - f(x) = \frac{1}{(x + \Delta x - 1)} - \frac{1}{x-1}$$

$$= \frac{x - 1 - (x + \Delta x - 1)}{(x + \Delta x - 1)(x - 1)}$$

$$= \frac{-\Delta x}{(x + \Delta x - 1)(x - 1)}$$

(c)  $$\frac{f(x + \Delta x) - f(x)}{\Delta x} = \frac{-\Delta x}{\Delta x(x + \Delta x - 1)(x - 1)}$$

$$= \frac{-1}{(x + \Delta x - 1)(x - 1)}$$

(d)  $$\lim_{\Delta x \to 0} \frac{f(x + \Delta x) - f(x)}{\Delta x} = \frac{-1}{(x - 1)(x - 1)}$$

$$= \frac{-1}{(x - 1)^2}$$

**17.** Use the four-step process to find the derivative of $f(x) = x^3$.
Sketch the graph of $f$ and find the equation of the tangent line
at $(2, 8)$.

**Solution:**

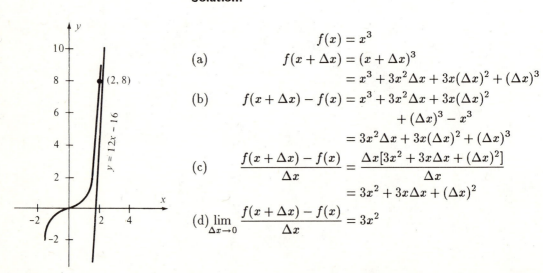

$$f(x) = x^3$$

(a)  $$f(x + \Delta x) = (x + \Delta x)^3$$

$$= x^3 + 3x^2\Delta x + 3x(\Delta x)^2 + (\Delta x)^3$$

(b)  $$f(x + \Delta x) - f(x) = x^3 + 3x^2\Delta x + 3x(\Delta x)^2$$

$$+ (\Delta x)^3 - x^3$$

$$= 3x^2\Delta x + 3x(\Delta x)^2 + (\Delta x)^3$$

(c)  $$\frac{f(x + \Delta x) - f(x)}{\Delta x} = \frac{\Delta x[3x^2 + 3x\Delta x + (\Delta x)^2]}{\Delta x}$$

$$= 3x^2 + 3x\Delta x + (\Delta x)^2$$

(d)  $$\lim_{\Delta x \to 0} \frac{f(x + \Delta x) - f(x)}{\Delta x} = 3x^2$$

Thus, $f'(x) = 3x^2$ and the slope of the tangent line at $(2, 8)$ is $f'(2) = 3(2)^2 = 12$. Finally, by the point-slope equation of a line, we have

$$y - y_1 = m(x - x_1)$$
$$y - 8 = 12(x - 2)$$
$$y = 12x - 16$$

**19.** Use the four-step process to find the derivative of $f(x) = \sqrt{x + 1}$. Sketch the graph of $f$ and find the equation of the tangent line at $(3, 2)$.

**Solution:**

$$f(x) = \sqrt{x + 1}$$

(a)
$$f(x + \Delta x) = \sqrt{x + \Delta x + 1}$$

(b)
$$f(x + \Delta x) - f(x) = \sqrt{x + \Delta x + 1} - \sqrt{x + 1}$$
$$= (\sqrt{x + \Delta x + 1} - \sqrt{x + 1})$$
$$\left( \frac{\sqrt{x + \Delta x + 1} + \sqrt{x + 1}}{\sqrt{x + \Delta x + 1} + \sqrt{x + 1}} \right)$$
$$= \frac{(x + \Delta x + 1) - (x + 1)}{\sqrt{x + \Delta x + 1} + \sqrt{x + 1}}$$
$$= \frac{\Delta x}{\sqrt{x + \Delta x + 1} + \sqrt{x + 1}}$$

(c)
$$\frac{f(x + \Delta x) - f(x)}{\Delta x} = \frac{\Delta x}{\Delta x(\sqrt{x + \Delta x + 1} + \sqrt{x + 1})}$$
$$= \frac{1}{\sqrt{x + \Delta x + 1} + \sqrt{x + 1}}$$

(d) $\lim\limits_{\Delta x \to 0} \dfrac{f(x + \Delta x) - f(x)}{\Delta x} = \dfrac{1}{\sqrt{x + 0 + 1} + \sqrt{x + 1}}$
$$= \frac{1}{2\sqrt{x + 1}}$$

Thus, $f'(x) = 1/2\sqrt{x + 1}$ and the slope of the tangent line at $(3, 2)$ is $f'(3) = 1/2\sqrt{3 + 1} = \frac{1}{4}$. Finally, by the point-slope equation of a line, we have

$$y - y_1 = m(x - x_1)$$
$$y - 2 = \frac{1}{4}(x - 3)$$
$$4y - 8 = x - 3$$
$$4y = x + 5 \quad \text{or} \quad y = \frac{x}{4} + \frac{5}{4}.$$

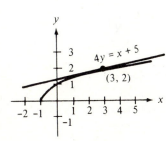

**23.** Use the alternative limit form to find the derivative (if it exists) of $f(x) = x^3 + 2x^2 + 1$ at $x = -2$.

**Solution:**

$$
\begin{aligned}
f'(-2) &= \lim_{x \to -2} \frac{f(x) - f(-2)}{x - (-2)} \\
&= \lim_{x \to -2} \frac{(x^3 + 2x^2 + 1) - 1}{x + 2} \\
&= \lim_{x \to -2} \frac{x^2(x + 2)}{x + 2} \\
&= \lim_{x \to -2} x^2 = 4
\end{aligned}
$$

**31.** Find every point at which the function $f(x) = (x - 3)^{2/3}$ is differentiable.

**Solution:**

The graph of $f(x) = (x - 3)^{2/3}$ is continuous at $x = 3$. However, the one-sided limits

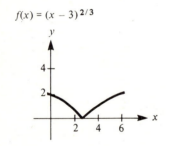

$f(x) = (x - 3)^{2/3}$

$$
\lim_{x \to 3^+} \frac{f(x) - f(3)}{x - 3} = \lim_{x \to 3^+} \frac{(x - 3)^{2/3}}{x - 3} = \lim_{x \to 3^+} \frac{1}{(x - 3)^{1/3}} = \infty
$$

$$
\lim_{x \to 3^-} \frac{f(x) - f(3)}{x - 3} = \lim_{x \to 3^-} \frac{(x - 3)^{2/3}}{x - 3} = \lim_{x \to 3^-} \frac{1}{(x - 3)^{1/3}} = -\infty
$$

are not equal. Therefore, $f$ is not differentiable at $x = 3$ and the point $(3, 0)$ is called a **cusp**. Hence, $f$ is differentiable in the intervals $(-\infty, 3)$ and $(3, \infty)$.

**41.** Find the equation of a line that is tangent to the curve $y = x^3$ and is parallel to the line $3x - y + 1 = 0$.

**Solution:**

The slope of the line given by $3x - y + 1 = 0$ is 3 since

$$
3x - y + 1 = 0
$$
$$
y = 3x + 1 = mx + b
$$

Thus we wish to find a point (or points) on the graph of $y = x^3$ such that the tangent line at that point has a slope of 3. From

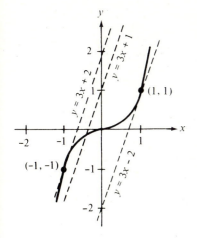

Exercise 17, we know that $dy/dx = 3x^2$ is the slope at any point on the graph of $y = x^3$. Therefore, we have

$$\frac{dy}{dx} = 3x^2 = 3$$

$$x^2 = 1 \qquad \text{or} \qquad x = \pm 1$$

Finally, we conclude that the slope of the graph of $y = x^3$ is 3 at the points $(1,1)$ and $(-1,-1)$. The tangent lines at these two points are

$$y - 1 = 3(x - 1) \qquad y - (-1) = 3[x - (-1)]$$

$$y = 3x - 2 \qquad \qquad y = 3x + 2$$

**43.** Find the equations of the two tangent lines to the graph of $f(x) = 4x - x^2$ that pass through the point $(2, 5)$.

**Solution:**

To begin, we find $dy/dx$ as follows:

$$y = f(x) = 4x - x^2$$

(a)
$$f(x + \Delta x) = 4(x + \Delta x) - (x + \Delta x)^2$$
$$= 4x + 4\Delta x - x^2$$
$$\quad - 2x\Delta x - (\Delta x)^2$$

(b)
$$f(x + \Delta x) - f(x) = 4x + 4\Delta x - x^2 - 2x\Delta x$$
$$\quad - (\Delta x)^2 - 4x + x^2$$
$$= 4\Delta x - 2x\Delta x - (\Delta x)^2$$

(c)
$$\frac{f(x + \Delta x) - f(x)}{\Delta x} = \frac{\Delta x(4 - 2x - \Delta x)}{\Delta x}$$
$$= 4 - 2x - \Delta x$$

(d)
$$\lim_{\Delta x \to 0} \frac{f(x + \Delta x) - f(x)}{\Delta x} = 4 - 2x$$

Now let $(x, y)$ be a point on the graph of $y = 4x - x^2$. Since $dy/dx = 4 - 2x$, the slope of the tangent line at $(x, y)$ is $m = 4 - 2x$. On the other hand, if the tangent line at $(x, y)$ passes through the point $(2, 5)$, then its slope must be

$$m = \frac{y - 5}{x - 2} = \frac{4x - x^2 - 5}{x - 2}$$

Equating these two values of $m$, we have

$$\frac{4x - x^2 - 5}{x - 2} = 4 - 2x$$

$$4x - x^2 - 5 = (4 - 2x)(x - 2)$$

$$4x - x^2 - 5 = 4x - 2x^2 - 8 + 4x$$

$$x^2 - 4x + 3 = 0$$

$$(x - 3)(x - 1) = 0$$

$$x = 3 \quad \text{or} \quad x = 1$$

If $x = 3$, then $y = 4(3) - 3^2 = 3$ and $m = 4 - 2(3) = -2$. Therefore,

$$y - 3 = -2(x - 3) \quad \text{or} \quad y = -2x + 9.$$

If $x = 1$, then $y = 4(1) - 1^2 = 3$ and $m = 4 - 2(1) = 2$. Therefore,

$$y - 3 = 2(x - 1) \quad \text{or} \quad y = 2x + 1.$$

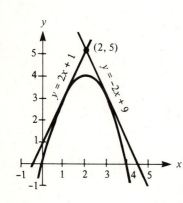

## 3.2 Velocity, Acceleration, and Other Rates of Change

**3.** Find the average rate of change of $f(x) = 1/(x + 1)$ between $(0, 1)$ and $(3, \frac{1}{4})$. Compare this average rate of change to the instantaneous rate of change at each point.

**Solution:**

The average rate of change is given by

$$\frac{\Delta y}{\Delta x} = \frac{\frac{1}{4} - 1}{3 - 0} = \frac{-\frac{3}{4}}{3} = -\frac{1}{4}$$

To find the instantaneous rate of change, $f'(x)$, we use the four-step process as follows:

$$f(x) = \frac{1}{x+1}$$

(a)
$$f(x + \Delta x) = \frac{1}{x + \Delta x + 1}$$

(b)
$$f(x + \Delta x) - f(x) = \frac{1}{x + \Delta x + 1} - \frac{1}{x+1}$$
$$= \frac{x + 1 - (x + \Delta x + 1)}{(x + \Delta x + 1)(x+1)}$$
$$= \frac{-\Delta x}{(x + \Delta x + 1)(x+1)}$$

(c)
$$\frac{f(x + \Delta x) - f(x)}{\Delta x} = \frac{-\Delta x}{\Delta x(x + \Delta x + 1)(x+1)}$$
$$= \frac{-1}{(x + \Delta x + 1)(x+1)}$$

(d) $\lim\limits_{\Delta x \to 0}$
$$\frac{f(x + \Delta x) - f(x)}{\Delta x} = \frac{-1}{(x + 0 + 1)(x+1)} = \frac{-1}{(x+1)^2}$$

Thus the instantaneous rate of change at $(0, 1)$, is given by

$$f'(0) = \frac{-1}{(0+1)^2} = -1$$

and the instantaneous rate of chage at $(3, \frac{1}{4})$ is given by

$$f'(3) = \frac{-1}{(3+1)^2} = -\frac{1}{16}$$

7. The height $s$ at time $t$ of a silver dollar dropped from the World Trade Center is given by $s(t) = -16t^2 + 1350$, where $s$ is measured in feet and $t$ is measured in seconds. $[s'(t) = -32t]$
   (a)   Find the average velocity on the interval $[1, 2]$.
   (b)   Find the instantaneous velocity when $t = 1$ and $t = 2$.
   (c)   How long will it take the dollar to hit the ground?
   (d)   Find the velocity of the dollar when it hits the ground.

**Solution:**

(a) The average velocity is given by

$$\frac{s(2) - s(1)}{2 - 1} = 1286 - 1334 = -48 \text{ ft/sec}$$

(b) Since $v(t) = s'(t) = -32t$, the instantaneous velocity at time $t = 1$ is

$$s'(1) = -32(1) = -32 \text{ ft/sec}$$

and at time $t = 2$ is

$$s'(2) = -32(2) = -64 \text{ ft/sec}$$

(c) The dollar will be at ground level when

$$s(t) = -16t^2 + 1350 = 0$$
$$t^2 = \frac{1350}{16}$$
$$t = \frac{1}{4}\sqrt{1350} = \frac{15\sqrt{6}}{4} \approx 9.2 \text{ sec}$$

(d) The velocity when it hits the ground is

$$v\left(\frac{1}{4}\sqrt{1350}\right) = -32\left(\frac{1}{4}\sqrt{1350}\right) = -8\sqrt{1350}$$
$$\approx -293.9 \text{ ft/sec.}$$

**21.** Find $f'''(x)$ if $f''(x) = (2x - 2)/x$.

**Solution:**

We rewrite the second derivative as

$$f''(x) = \frac{2x - 2}{x} = 2 - \frac{2}{x}$$

Then

$$f'''(x) = \lim_{\Delta x \to 0} \frac{f''(x + \Delta x) - f''(x)}{\Delta x}$$
$$= \lim_{\Delta x \to 0} \frac{2 - \dfrac{2}{x + \Delta x} - \left(2 - \dfrac{2}{x}\right)}{\Delta x}$$
$$\doteq \lim_{\Delta x \to 0} \frac{2(x + \Delta x) - 2x}{x(\Delta x)(x + \Delta x)}$$
$$= \lim_{\Delta x \to 0} \frac{2}{x(x + \Delta x)} = \frac{2}{x^2}$$

## 3.3   Differentiation Rules for Sums, Constants Multiples, and Powers

**11.** Differentiate $s(t) = t^3 - 2t + 4$.

**Solution:**

$$s(t) = t^3 - 2t + 4$$
$$s'(t) = 3t^2 - 2$$

**17.** Differentiate $y = (2x + 1)^2$ and evaluate the derivative at $(0, 1)$.

**Solution:**

$$y = (2x + 1)^2 = 4x^2 + 4x + 1$$
$$y' = 8x + 4$$

At the point $(0, 1)$ the derivative is

$$y' = 8(0) + 4 = 4$$

**19.** Find $f'(x)$ if $f(x) = x^2 - (4/x)$.

**Solution:**

$$f(x) = x^2 - \frac{4}{x} = x^2 - 4x^{-1}$$
$$f'(x) = 2x - (-1)4x^{-2} = 2x + \frac{4}{x^2}$$

**23.** Find $f'(x)$ if $f(x) = (x^3 - 3x^2 + 4)/x^2$.

**Solution:**

$$f(x) = \frac{x^3 - 3x^2 + 4}{x^2} = \frac{x^3}{x^2} - \frac{3x^2}{x^2} + \frac{4}{x^2} = x - 3 + 4x^{-2}$$
$$f'(x) = 1 + 4(-2)x^{-3} = 1 - \frac{8}{x^3} = \frac{x^3 - 8}{x^3}$$

**29.** Find $f'(x)$ if $f(x) = \sqrt[3]{x} + \sqrt[5]{x}$.

**Solution:**

$$f(x) = \sqrt[3]{x} + \sqrt[5]{x} = x^{1/3} + x^{1/5}$$

$$f'(x) = \left(\frac{1}{3}\right)x^{-2/3} + \left(\frac{1}{5}\right)x^{-4/5} = \frac{1}{3x^{2/3}} + \frac{1}{5x^{4/5}}$$

**34.** Find $f'(x)$ if $f(x) = \pi/(3x)^2$.

**Solution:**

Function:    $f(x) = \dfrac{\pi}{(3x)^2}$

Rewrite:    $f(x) = \left(\dfrac{\pi}{9}\right)x^{-2}$

Derivative:    $f'(x) = \left(\dfrac{\pi}{9}\right)(-2)x^{-3}$

Simplify:    $f'(x) = \dfrac{-2\pi}{9x^3}$

**37.** Find the equation of the line tangent to $y = x^4 - 3x^2 + 2$ at the point $(1, 0)$.

**Solution:**

$$y = x^4 - 3x^2 + 2$$
$$y' = 4x^3 - 6x$$

Thus the slope of the tangent line at $(1, 0)$ is

$$y' = 4(1)^3 - 6(1) = 4 - 6 = -2$$

The equation of the tangent line at $(1, 0)$ is

$$y - y_1 = m(x - x_1)$$
$$y - 0 = -2(x - 1)$$
$$2x + y - 20$$

**39.** At what points, if any, does $y = x^4 - 3x^2 + 2$ have horizontal tangents?

**Solution:**

A tangent line is horizontal if the derivative (the slope) at the point of tangency is zero. Since $y' = 4x^3 - 6x = 2x(2x^2 - 3)$, we must find all values of $x$ that satisfy the equation $y' = 2x(2x^2 - 3) = 0$. The solutions to this equation are $x = 0$ and $x = \pm\sqrt{3/2}$. At $x = 0$, we have

$$y = (0)^4 - 3(0)^2 + 2 = 2.$$

At $x = \pm\sqrt{3/2}$, we have

$$y = \left(\pm\sqrt{\frac{3}{2}}\right)^4 - 3\left(\pm\sqrt{\frac{3}{2}}\right)^2 + 2$$

$$= \frac{9}{4} - \frac{9}{2} + 2 = \frac{9 - 18 + 8}{4} = -\frac{1}{4}$$

Thus the points of horizontal tangency are:

$$(0, 2), \ \left(\sqrt{\tfrac{3}{2}}, -\tfrac{1}{4}\right), \ \text{and} \ \left(-\sqrt{\tfrac{3}{2}}, -\tfrac{1}{4}\right).$$

**43.** Sketch the graphs of the two equations $y = x^2$ and $y = -x^2 + 6x - 5$ and sketch the two lines that are tangent to both graphs. Find equations for these lines.

**Solution:**

Let $(x_1, y_1)$ and $(x_2, y_2)$ be the points of tangency on the graphs of $y = x^2$ and $y = -x^2 + 6x - 5$, respectively. We know that $y_1 = x_1^2$ and $y_2 = -x_2^2 + 6x_2 - 5$. Let $m$ be the slope of the tangent line. Since the line passes through $(x_1, y_1)$ and $(x_2, y_2)$, we have

(1) $$m = \frac{y_2 - y_1}{x_2 - x_1} = \frac{-x_2^2 + 6x_2 - 5 - x_1^2}{x_2 - x_1}$$

Since the line is tangent to $y = x^2$ at $(x_1, y_1)$ and the derivative of this curve is $y' = 2x$, we have

$$m = 2x_1$$

Since the line is tangent to $y = -x^2 + 6x - 5$ at $(x_2, y_2)$ and the derivative of this curve is $y' = -2x_2 + 6$, we have

$$m = -2x_2 + 6$$

Thus, from the preceding two equations, we have

$$m = 2x_1 = -2x_2 + 6$$

(2)
$$x_1 = -x_2 + 3$$

From the equations (1) and (2) we have

$$m = -2x_2 + 6 = \frac{-x_2^2 + 6x_2 - 5 - x_1^2}{x_2 - x_1}$$

$$-2x_2 + 6 = \frac{-x_2^2 + 6x_2 - 5 - (-x_2 + 3)^2}{x_2 - (-x_2 + 3)}$$

$$(2x_2 - 3)(-2x_2 + 6) = -x_2^2 + 6x_2 - 5 - (x_2^2 - 6x_2 + 9)$$

$$-4x_2^2 + 18x_2 - 18 = -2x_2^2 + 12x_2 - 14$$

$$-2x_2^2 + 6x_2 - 4 = 0$$

$$x_2^2 - 3x_2 + 2 = 0$$

$$(x_2 - 2)(x_2 - 1) = 0$$

$$x_2 = 1 \quad \text{or} \quad x_2 = 2$$

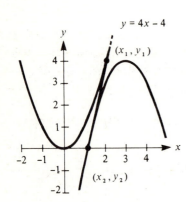

$y = 4x - 4$

$(x_1, y_1)$

$(x_2, y_2)$

If $x_2 = 1$, then $y_2 = -1^2 + 6(1) - 5 = 0$, $x_1 = -1 + 3 = 2$, and $y_1 = 2^2 = 4$. Thus the line containing $(1, 0)$ and $(2, 4)$ is tangent to both curves. The equation of this line is

$$y - 0 = \left(\frac{4 - 0}{2 - 1}\right)(x - 1) \quad \text{or} \quad y = 4x - 4$$

If $x_2 = 2$, then $y_2 = -2^2 + 6(2) - 5 = -4 + 12 - 5 = 3$, $x_1 = -2 + 3 = 1$, and $y_1 = 1^2 = 1$. Thus the line containing $(2, 3)$ and $(1, 1)$ is tangent to both curves. The equation of this line is

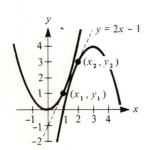

$y = 2x - 1$

$(x_2, y_2)$

$(x_1, y_1)$

$$y - 1 = \left(\frac{3 - 1}{2 - 1}\right)(x - 1)$$

$$= 2x - 2$$

$$y = 2x - 1$$

**47.** Suppose a certain company finds that by charging $p$ dollars per units, its monthly revenue $R$ will be

$$R = 12,000p - 1,000p^2, \qquad 0 \le p \le 12$$

(Note that the revenue is zero when $p = 12$ since no one is willing to pay that much.) find the rate of change of $R$ with respect to $p$ when:

(a)  $p = 1$      (b)  $p = 4$      (c)  $p = 6$      (d)  $p = 10$

**Solution:**

$$R(p) = 12,000p - 1,000p^2$$
$$R'(p) = 12,000 - 2,000p$$

Therefore, the rate of change of $R$ with respect to $p$ is as follows:

(a)          $R'(1) = 12,000 - 2,000(1) = 10,000$

(b)          $R'(4) = 12,000 - 2,000(4) = 4,000$

(c)          $R'(6) = 12,000 - 2,000(6) = 0$

(d)          $R'(10) = 12,000 - 2,000(10) = -8,000$

## 3.4    Differentiation Rules for Products, and Quotients

**5.** Differentiate $f(x) = (x^3 - 3x)(2x^2 + 3x + 5)$ and find $f'(0)$.

**Solution:**

$$f(x) = (x^3 - 3x)(2x^2 + 3x + 5)$$
$$f'(x) = (x^3 - 3x)(4x + 3) + (2x^2 + 3x + 5)(3x^2 - 3)$$
$$= 4x^4 + 3x^3 - 12x^2 - 9x + 6x^4 - 6x^2 + 9x^3 - 9x + 15x^2 - 15$$
$$= 10x^4 + 12x^3 - 3x^2 - 18x - 15$$
$$f'(0) = -15$$

**11.** Differentiate $f(x) = \dfrac{3 - 2x - x^2}{x^2 - 1}$.

**Solution:**

$$f(x) = \frac{3 - 2x - x^2}{x^2 - 1}$$

$$= -\frac{x^2 + 2x - 3}{x^2 - 1} = -\frac{(x+3)(x-1)}{(x+1)(x-1)} = -\frac{x+3}{x+1}$$

$$f'(x) = -\frac{(x+1)(1) - (x+3)(1)}{(x+1)^2} = \frac{2}{(x+1)^2}$$

**19.** Differentiate $g(x) = \left(\dfrac{x+1}{x+2}\right)(2x - 5)$.

**Solution:**

$$g(x) = \frac{(x+1)(2x-5)}{x+2} = \frac{2x^2 - 3x - 5}{x+2}$$

$$g'(x) = \frac{(x+2)(4x-3) - (2x^2 - 3x - 5)(1)}{(x+2)^2}$$

$$= \frac{4x^2 + 5x - 6 - 2x^2 + 3x + 5}{(x+2)^2} = \frac{2x^2 + 8x - 1}{(x+2)^2}$$

**21.** Differentiate $f(x) = (3x^3 + 4x)(x - 5)(x + 1)$.

**Solution:**

$$f(x) = [(3x^3 + 4x)(x - 5)](x + 1)$$
$$f'(x) = [(3x^3 + 4x)(x - 5)](1)$$
$$+ (x + 1)[(3x^3 + 4x)(1) + (x - 5)(9x^2 + 4)]$$
$$= (3x^3 + 4x)(x - 5) + (x + 1)(3x^3 + 4x)$$
$$+ (x + 1)(x - 5)(9x^2 + 4)$$
$$= 15x^4 - 48x^3 - 33x^2 - 32x - 20$$

**35.** Find an equation of the line tangent to the graph of $f(x) = x/(x-1)$ at the point $(2,2)$.

**Solution:**

$$f(x) = \frac{x}{x-1}$$

$$f'(x) = \frac{(x-1)(1) - (x)(1)}{(x-1)^2} = \frac{x-1-x}{(x-1)^2} = \frac{-1}{(x-1)^2}$$

Therefore, the slope of the tangent line at $(2,2)$ is $f'(2) = -1/(2-1)^2 = -1$ and an equation of the tangent line at $(2,2)$ is

$$y - 2 = -1(x-2)$$
$$= -x + 2$$
$$y = -x + 4$$

**43.** A population of 500 bacteria is introduced into a culture and grows in number according to the equation

$$P(t) = 500\left(1 + \frac{4t}{50 + t^2}\right)$$

where $t$ is measured in hours. Find the rate at which the population is growing when $t = 2$.

**Solution:**

$$P'(t) = 500\left[\frac{(50 + t^2)(4) - (4t)(2t)}{(50 + t^2)^2}\right]$$
$$= \frac{500(200 + 4t^2 - 8t^2)}{(50 + t^2)^2} = \frac{2000(50 - t^2)}{(50 + t^2)^2}$$

Therefore, the rate of population growth when $t = 2$ is

$$P'(2) = \frac{2000(50 - 4)}{(50 + 4)^2} \approx 31.55$$

## 3.5    The Chain Rule

**13.** Find the first derivative of $f(t) = \left(\dfrac{1}{t-3}\right)^2$.

**Solution:**

$$f(t) = \frac{1}{(t-3)^2} = (t-3)^{-2}$$

$$f'(t) = (-2)(t-3)^{-3}(1) = \frac{-2}{(t-3)^3}$$

**17.** Find the first derivative of $f(x) = x^2(x-2)^4$.

**Solution:**

$$f(x) = x^2(x-2)^4$$
$$f'(x) = x^2(4)(x-2)^3(1) + (x-2)^4(2x)$$
$$= 2x(x-2)^3(2x + x - 2) = 2x(x-2)^3(3x - 2)$$

**23.** Find the first derivative of $y = \sqrt[3]{9x^2 + 4}$.

**Solution:**

$$y = \sqrt[3]{9x^2 + 4} = (9x^2 + 4)^{1/3}$$
$$\frac{dy}{dx} = \left(\frac{1}{3}\right)(9x^2 + 4)^{-2/3}(18x) = \frac{6x}{(9x^2 + 4)^{2/3}}$$

**29.** Find the first derivative of $y = \dfrac{1}{\sqrt{x+2}}$.

**Solution:**

$$y = \frac{1}{\sqrt{x+2}} = (x+2)^{-1/2}$$
$$y' = \left(-\frac{1}{2}\right)(x+2)^{-3/2}(1) = \frac{-1}{2(x+2)^{3/2}}$$

**39.** Find the first derivative of $g(t) = \dfrac{3t^2}{\sqrt{t^2 + 2t - 1}}$.

**Solution:**

$$g(t) = \frac{3t^2}{\sqrt{t^2 + 2t - 1}} = \frac{3t^2}{(t^2 + 2t - 1)^{1/2}}$$

$$g'(t) = \frac{(t^2 + 2t - 1)^{1/2}(6t) - (3t^2)(\frac{1}{2})(t^2 + 2t - 1)^{-1/2}(2t + 2)}{t^2 + 2t - 1}$$

$$= \frac{(3t)(t^2 + 2t - 1)^{-1/2}[2(t^2 + 2t - 1) - (t)(t + 1)]}{t^2 + 2t - 1}$$

$$= \frac{3t(2t^2 + 4t - 2 - t^2 - t)}{(t^2 + 2t - 1)^{3/2}}$$

$$= \frac{3t(t^2 + 3t - 2)}{(t^2 + 2t - 1)^{3/2}}$$

**49.** Find the first derivative of $f(x) = \sqrt{x^2 + x + 1}$.

**Solution:**

$$f(x) = \sqrt{x^2 + x + 1} = (x^2 + x + 1)^{1/2}$$

$$f'(x) = \frac{1}{2}(x^2 + x + 1)^{-1/2}(2x + 1)$$

$$= \frac{2x + 1}{2\sqrt{x^2 + x + 1}}$$

**56.** The speed $S$ of blood that is $r$ centimeters from the center of an artery is given by $S = C(R^2 - r^2)$, where $C$ is a constant, $R$ is the radius of the artery, and $S$ is measured in centimeters per second. Suppose a drug is administered and the artery begins dilating at the rate $dR/dt$. At a constant distance $r$, find the rate at which $S$ changes with respect to t for $C = 1.76 \times 10^5$, $R = 1.2 \times 10^{-2}$, and $dR/dt = 10^{-5}$.

**Solution:**

Using the Chain Rule, we have

$$S = C(R^2 - r^2)$$

$$\frac{dS}{dt} = C\left(2R\frac{dR}{dt} - 2r\frac{dr}{dt}\right)$$

Since $r$ is constant, we have $\dfrac{dr}{dt} = 0$ and

$$\frac{dS}{dt} = (1.76 \times 10^5)[2(1.2 \times 10^{-2})(10^{-5}) - 0] = 4.224 \times 10^{-2}$$

## 3.6   Implicit Differentiation

**3.** Given $xy = 4$, find $dy/dx$ by implicit differentiation and evaluate the derivative at $(-4, -1)$.

**Solution:**

$$xy = 4$$
$$xy' + y(1) = 0$$
$$xy' = -y$$
$$y' = -\frac{y}{x}$$

At $(-4, -1)$, $\dfrac{dy}{dx} = -\dfrac{-1}{-4} = -\dfrac{1}{4}$.

**7.** Given $x^3 - xy + y^2 = 4$, find $dy/dx$ by implicit differentiation and evaluate the derivative at $(0, -2)$.

**Solution:**

$$x^3 - xy + y^2 = 4$$
$$3x^2 - [xy' + y(1)] + 2yy' = 0$$
$$3x^2 - xy' - y + 2yy' = 0$$
$$y'(2y - x) = y - 3x^2$$
$$y' = \frac{y - 3x^2}{2y - x}$$

At $(0, -2)$, $\dfrac{dy}{dx} = \dfrac{-2 - 3(0^2)}{2(-2) - 0} = \dfrac{-2}{-4} = \dfrac{1}{2}$.

**23.** Sketch the graph of the equation $9x^2 + 16y^2 = 144$. Find $dy/dx$ implicitly and explicitly and show that the two results are equivalent.

**Solution:**

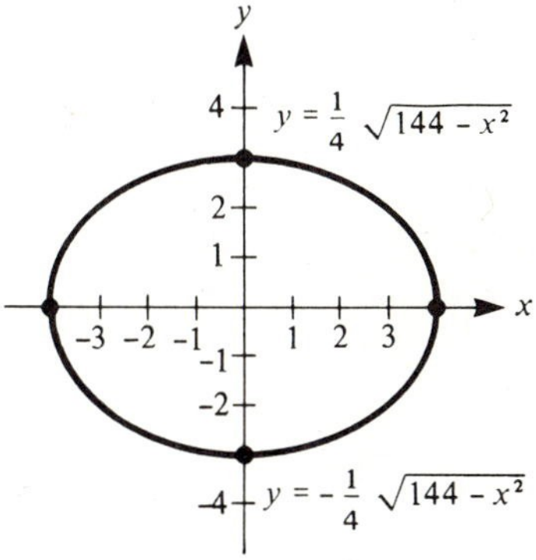

Differentiating implicitly we have

$$9x^2 + 16y^2 = 144$$
$$18x + 32yy' = 0$$
$$y' = \frac{-9x}{16y}$$

Solving the equation for $y$, we obtain

$$9x^2 + 16y^2 = 144$$
$$16y^2 = 144 - 9x^2$$
$$y^2 = \frac{1}{16}(144 - 9x^2)$$
$$y = \pm\frac{1}{4}\sqrt{144 - 9x^2}, \qquad -4 \le x \le 4$$

We now differentiate explicitly to obtain

$$\frac{dy}{dx} = \pm\left(\frac{1}{4}\right)\left(\frac{1}{2}\right)(144 - 9x^2)^{-1/2}(-18x)$$
$$= \frac{\mp 9x}{4\sqrt{144 - 9x^2}} = \frac{\mp 9x}{16(\frac{1}{4})\sqrt{144 - 9x^2}} = \frac{-9x}{16y}$$

**27.** Given $x^2 - y^2 = 16$, find $d^2y/dx^2$ in terms of $x$ and $y$.

**Solution:**

$$x^2 - y^2 = 16$$
$$2x - 2yy' = 0$$
$$-2yy' = -2x$$
$$y' = \frac{x}{y}$$

By differentiating again, we have

$$y'' = \frac{y(1) - (x)y'}{y^2} = \frac{y - x(x/y)}{y^2} = \frac{y^2 - x^2}{y^3} = -\frac{16}{y^3}$$

The graph (left margin):

$y = \frac{1}{4}\sqrt{144 - x^2}$

$y = -\frac{1}{4}\sqrt{144 - x^2}$

**35.** Show that the graphs of the equations $2x^2 + y^2 = 6$ and $y^2 = 4x$ are orthogonal (the curves intersect at right angles). Sketch the graph of each equation.

**Solution:**

To find the points of intersection, we set $y^2 = 6 - 2x^2$ and $y^2 = 4x$ equal to each other.

$$4x = 6 - 2x^2$$
$$2x^2 + 4x - 6 = 0$$
$$x^2 + 2x - 3 = 0$$
$$(x + 3)(x - 1) = 0$$
$$x = -3 \quad \text{and} \quad x = 1$$

When $x = 1, y = \pm 2$, and when $x = -3, y$ is undefined. Thus the two points of intersection are $(1, 2)$ and $(1, -2)$. For $2x^2 + y^2 = 6$

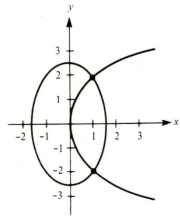

$$4x + 2yy' = 0$$
$$y' = -\frac{2x}{y}$$

and for $y^2 = 4x$

$$2yy' = 4$$
$$y' = \frac{2}{y}$$

Thus at $(1, 2)$ the slopes of the two curves are

$$\frac{-2(1)}{2} = -1 \quad \text{and} \quad \frac{2}{2} = 1$$

which implies that the tangent lines at this point are perpendicular. Finally, at $(1, -2)$ the slopes of the two curves are

$$\frac{-2(1)}{-2} = 1 \quad \text{and} \quad \frac{2}{-2} = -1$$

and the tangent lines at this point are also perpendicular.

**40.** Show that the normal line (the line perpendicular to the tangent line to a curve) at any point on the circle $x^2 + y^2 = r^2$ passes through the origin.

**Solution:**

By implicit differentiation,

$$x^2 + y^2 = r^2$$
$$2x + 2yy' = 0$$
$$y' = -\frac{x}{y}$$

Thus if $(x, y)$ is a point on the circle $x^2 + y^2 = r^2$, the slope of the tangent line at $(x, y)$ is $-x/y$. On the other hand, the slope of the line passing through $(x, y)$ and $(0, 0)$ is

$$m = \frac{y - 0}{x - 0} = \frac{y}{x}$$

Since this slope is the negative reciprocal of $y'$, the line passing through $(x, y)$ and $(0, 0)$ must be perpendicular to the tangent line at $(x, y)$.

## 3.7   Related Rates

**3.** Assuming that $x$ and $y$ are differentiable functions of $t$ and $xy = 4$ find:

(a)   $\dfrac{dy}{dx}$ when   $x = 8$ if   $\dfrac{dx}{dt} = 10$

(b)   $\dfrac{dx}{dt}$ when   $x = 1$ if   $\dfrac{dy}{dt} = -6$

**Solution:**

$$xy = 4$$

(1)   $$x\frac{dy}{dt} + y\frac{dx}{dt} = 0$$

(a) Solving (1) for $\dfrac{dy}{dt}$, and substituting $y = \dfrac{1}{2}$ and $\dfrac{dx}{dt} = 10$ when $x = 8$, we have

$$\frac{dy}{dt} = -\frac{y}{x}\frac{dx}{dt} = -\frac{\frac{1}{2}}{8}(10) = -\frac{5}{8}.$$

(b) Solving (1) for $\dfrac{dx}{dt}$, and substituting $y = 4$ and $\dfrac{dy}{dt} = -6$ when $x = 1$, we have

$$\frac{dx}{dt} = -\frac{x}{y}\frac{dy}{dt} = -\frac{1}{4}(-6) = \frac{3}{2}.$$

**11.** At a sand and gravel plant, sand is falling off a conveyer and onto a conical pile at the rate of 10 ft$^3$/min. The diameter of the base of the cone is approximately three times the altitude. At what rate is the height of the pile changing when is is 15 ft high?

**Solution:**

Let

$$V = \text{volume of cone} = \frac{1}{3}\pi r^2 h.$$

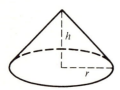

Since the diameter of the base is approximately three times the altitude, we have

$$2r = 3h \qquad \text{or} \qquad r = \frac{3}{2}h$$

Therefore,

$$V = \frac{1}{3}\pi\left(\frac{3}{2}h\right)^2 h = \frac{3\pi}{4}h^3$$

Differentiating with respect to $t$, we have

$$\frac{dV}{dt} = \frac{9\pi}{4}h^2\frac{dh}{dt} \qquad \text{or} \qquad \frac{4(dV/dt)}{9\pi h^2} = \frac{dh}{dt}.$$

Now, letting $h = 15$ and $\dfrac{dV}{dt} = 10$, we obtain the following rate of change of the height of the conical pile.

$$\frac{dh}{dt} = \frac{4(10)}{9\pi(15)^2} = \frac{8}{405\pi} \text{ ft/min.}$$

**19.** A swimming pool is 40 ft long, 20 ft wide, 4 ft deep at the shallow end, and 9 ft deep at the deep end, the bottom being an inclined plane. Assume that the water is being pumped into the pool at 10 ft$^3$/min and there is 4 ft of water at the deep end.

(a)   What percentage of the pool is filled?

(b)   At what rate is the water level rising?

**Solution:**

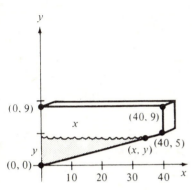

From the figure we see that $x$ and $y$ are related by the equation

$$m = \frac{y - 0}{x - 0} = \frac{5 - 0}{40 - 0}$$

$$\frac{y}{x} = \frac{1}{8}$$

The volume of the inclined portion of the pool is given by

$$V_L = \frac{20xy}{2} = 10xy = 10(8y)y = 80y^2$$

When $y = 5$, the inclined portion of the pool has a volume of

$$V_L = 80(5^2) = 2000 \text{ ft}^3$$

Since the upper rectangular portion of the pool has a volume of

$$V_U = 4(40)(20) = 3200 \text{ ft}^3$$

the total volume of the pool is

$$V = V_L + V_U = 2000 + 3200 = 5200 \text{ ft}^3$$

(a) When $y = 4$, the ratio of the filled portion of the pool to the total volume is

$$\frac{80y^2}{V} = \frac{80(4^2)}{5200} = 24.6\%.$$

(b) When $y = 4$, we can find $dy/dt$ by differentiating $V_L = 80y^2$ and substituting $dV_L/dt = 10$.

$$V_L = 80y^2$$

$$\frac{dV_L}{dt} = 160y\frac{dy}{dt}$$

$$10 = 160(4)\frac{dy}{dt}$$

$$\frac{dy}{dt} = \frac{10}{640} = \frac{1}{64} \text{ ft/min}$$

**21.** A ladder 25 ft long is leaning against a house. If the base of the ladder is pulled away from the house wall at a rate of 2 feet per second, how fast is the top moving down the wall when the base of the ladder is:

(a)   7 feet from the wall?
(b)   15 feet from the wall?
(c)   24 feet from the wall?

**Solution:**

From the figure we see that $x$ and $y$ are related by the equation

$$x^2 + y^2 = (25)^2$$

Differentiating this equation with respect to $t$, we have

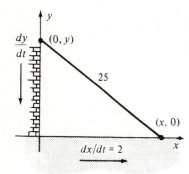

$$2x\frac{dx}{dt} + 2y\frac{dy}{dt} = 0$$

$$y\frac{dy}{dt} = -x\frac{dx}{dt}$$

$$\frac{dy}{dt} = -\frac{x}{y}\frac{dx}{dt}$$

Since $\dfrac{dx}{dt} = 2$, we have $\dfrac{dy}{dt} = -2\dfrac{x}{y}$.

(a)   When $x = 7, y = \sqrt{(25)^2 - 7^2} = \sqrt{576} = 24.$

$$\frac{dy}{dt} = -2\left(\frac{7}{24}\right) = -\frac{7}{12} \approx -0.583 \text{ ft/sec}$$

(b)   When $x = 15, y = \sqrt{(25)^2 - (15)^2} = \sqrt{400} = 20.$

$$\frac{dy}{dt} = -2\left(\frac{15}{20}\right) = -\frac{3}{2} = -1.5 \text{ ft/sec}$$

(c)   When $x = 24, y = \sqrt{(25)^2 - (24)^2} = \sqrt{49} = 7.$

$$\frac{dy}{dt} = -2\left(\frac{24}{7}\right) = -\frac{48}{7} \approx -6.857 \text{ ft/sec}$$

**29.** A man 6 ft tall walks at a rate of 5 ft/s away from a light that is 15 ft above the ground. When the man is 10 ft from the base of the light:

(a)   at what rate is the tip of his shadow moving?

(b)   at what rate is the length of his shadow changing?

**Solution:**

From the figure we see that $x$ and $s$ are related by similar triangles in such a way that

$$\frac{s - x}{6} = \frac{s}{15}$$
$$15s - 15x = 6s$$
$$9s = 15x$$
$$s = \frac{5}{3}x$$

(a)   To find $ds/dt$, given that $dx/dt = 5$, we differentiate with respect to $t$ as follows:

$$\frac{ds}{dt} = \frac{5}{3} \cdot \frac{dx}{dt} = \frac{5}{3}(5) = \frac{25}{3} \approx 8.3 \text{ ft/sec}$$

(b)   The rate at which the shadow is increasing is

$$\frac{ds}{dt} - \frac{dx}{dt} = \frac{25}{3} - 5 = \frac{10}{3} \approx 3.3 \text{ ft/sec}$$

(Note: The measurement 10 feet given in this problem is a "red herring" since the distance from the base of the light does not affect $ds/dt$.)

## Review Excercises for Chapter 3

**9.** Find the derivative of $f(x) = (3x^2 + 7)(x^2 - 2x + 3)$.

**Solution:**

$$f(x) = (3x^2 + 7)(x^2 - 2x + 3)$$
$$f'(x) = (3x^2 + 7)(2x - 2) + (x^2 - 2x + 3)(6x)$$
$$= 6x^3 - 6x^2 + 14x - 14 + 6x^3 - 12x^2 + 18x$$
$$= 2(6x^3 - 9x^2 + 16x - 7)$$

**15.** Find the derivative of $f(x) = \dfrac{x^2 + x - 1}{x^2 - 1}$.

**Solution:**

$$f(x) = \frac{x^2 + x - 1}{x^2 - 1}$$

$$f'x = \frac{(x^2 - 1)(2x + 1) - (x^2 + x - 1)(2x)}{(x^2 - 1)^2}$$

$$= \frac{2x^3 + x^2 - 2x - 1 - 2x^3 - 2x^2 + 2x}{(x^2 - 1)^2}$$

$$= -\frac{x^2 + 1}{(x^2 - 1)^2}$$

**21.** Find the derivative of $f(x) = \dfrac{2x}{\sqrt{x + 1}}$.

**Solution:**

$$f(x) = \frac{2x}{\sqrt{x + 1}} = 2\left[\frac{x}{(x + 1)^{1/2}}\right]$$

$$f'(x) = 2\left[\frac{(x + 1)^{1/2}(1) - x(\frac{1}{2})(x + 1)^{-1/2}(1)}{x + 1}\right]$$

$$= 2\left[\frac{\sqrt{x + 1} - \dfrac{x}{2\sqrt{x + 1}}}{x + 1}\right]$$

$$= \frac{\dfrac{2(x + 1) - x}{\sqrt{x + 1}}}{x + 1}$$

$$= \frac{x + 2}{(x + 1)^{3/2}}$$

**25.** Find the second derivative of $f(x) = \sqrt{x^2 + 9}$.

**Solution:**

$$f(x) = \sqrt{x^2 + 9} = (x^2 + 9)^{1/2}$$

$$f'x = \left(\frac{1}{2}\right)(x^2 + 9)^{-1/2}(2x) = \frac{x}{(x^2 + 9)^{1/2}}$$

$$f''(x) = \frac{(x^2 + 9)^{1/2}(1) - (x)(\frac{1}{2})(x^2 + 9)^{-1/2}(2x)}{x^2 + 9}$$

$$= \left[\frac{(x^2 + 9)^{1/2} - x^2(x^2 + 9)^{-1/2}}{x^2 + 9}\right]\left[\frac{(x^2 + 9)^{1/2}}{(x^2 + 9)^{1/2}}\right]$$

$$= \frac{(x^2 + 9 - x^2)}{(x^2 + 9)^{3/2}} = \frac{9}{(x^2 + 9)^{3/2}}$$

**31.** Use implicit differentiation to find $dy/dx$ for $x^2 + 3xy + y^3 = 10$.

**Solution:**

$$x^2 + 3xy + y^3 = 10$$

$$2x + 3xy' + 3y + 3y^2y' = 0$$

$$(3x + 3y^2)y' = -2x - 3y$$

$$y' = \frac{-(2x + 3y)}{3(x + y^2)}$$

**41.** Find the equations of the tangent line and the normal line to the graph of $y = \sqrt[3]{(x-2)^2}$ at $(3, 1)$.

**Solution:**

$$y = \sqrt[3]{(x-2)^2} = (x-2)^{2/3}$$

$$y' = \left(\frac{2}{3}\right)(x-2)^{-1/3} = \frac{2}{3(x-2)^{1/3}}$$

Thus the slope of the tangent line at $(3, 1)$ is

$$y' = \frac{2}{3(3-2)^{1/3}} = \frac{2}{3}$$

and the equation of the tangent line at $(3, 1)$ is

$$y - 1 = \frac{2}{3}(x - 3)$$
$$3y - 3 = 2x - 6$$
$$-2x + 3y + 3 = 0$$

Since the slope of the tangent line is $\frac{2}{3}$, the slope of the normal line is $-\frac{3}{2}$ and the equation of the normal line is

$$y - 1 = -\frac{3}{2}(x - 3)$$
$$2y - 2 = -3x + 9$$
$$3x + 2y - 11 = 0$$

**53.** What is the smallest initial velocity that is required to throw a stone to the top of a 49-foot silo?

**Solution:**

We assume the stone is thrown from an initial height of $s_0 = 0$. Thus the position equation is

$$s = -16t^2 + v_0 t$$

The maximum value of $s$ occurs when $ds/dt = 0$ and thus we have

$$\frac{ds}{dt} = -32t + v_0 = 0$$
$$-32t = -v_0$$
$$t = \frac{v_0}{32}$$

This means that the maximum height is

$$s = -16\left(\frac{v_0}{32}\right)^2 + v_0\left(\frac{v_0}{32}\right) = \frac{v_0^2}{64}$$

If $s$ is to attain a value of 49, we must have

$$\frac{v_0^2}{64} = 49$$
$$v_0^2 = 3136$$
$$v_0 = 56 \text{ ft/sec}$$

**57.** The path of a projectile, thrown at an angle of 45° with the ground, is given by $y = x - (32/v_0^2)(x^2)$, where the initial velocity is $v_0$ feet per second. Show that doubling the initial velocity of the projectile multiplies both the maximum height and the range by a factor of 4.

**Solution:**

The maximum height occurs when the rate of change of $y$ with respect to $x$ is zero.

$$\frac{dy}{dx} = 1 - \left(\frac{64}{v_0^2}\right)x = 0$$

when $x = v_0^2/64$. Therefore, the maximum height of the projectile is

$$y = \frac{v_0^2}{64} - \left(\frac{32}{v_0^2}\right)\left(\frac{v_0^2}{64}\right)^2 = \frac{v_0^2}{64} - \frac{v_0^2}{128} = \frac{v_0^2}{128}$$

If the initial velocity is doubled, the maximum height is

$$\frac{(2v_0)^2}{128} = 4\left(\frac{v_0^2}{128}\right)$$

of four times the original maximum height.
The projectile is at ground level when $y = 0$ or

$$x - \left(\frac{32}{v_0^2}\right)x^2 = x\left[1 - \left(\frac{32}{v_0^2}\right)x\right] = 0$$

This occurs when $x = 0$ or $x = v_0^2/32$. Therefore, the range is $v_0^2/32$. When the initial velocity is doubled, the range is

$$\frac{(2v_0)^2}{32} = 4\left(\frac{v_0^2}{32}\right)$$

or four times the original range.

**61.** The cross section of a 5-foot trough is an isosceles trapezoid with lower base 2 feet, upper base 3 feet, and altitude 2 feet. Water is running into the trough at the rate of 1 cubic foot per minute. How fast is the water level rising when the water is 1 foot deep?

**Solution:**

The figure is a cross section of the trough when the water is at a depth of $h$ feet.

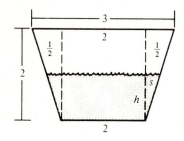

$$\frac{s}{h} = \frac{\frac{1}{2}}{2}$$

$$s = \frac{1}{4}h$$

$A$ = area of cross section of water at depth $h$

$$= 2h + 2\left(\frac{1}{2}sh\right)$$

$$= 2h + \left(\frac{1}{4}h\right)h = 2h + \frac{1}{4}h^2$$

$V$ = volume of water in trough at depth $h$

$$= 5A = 5\left(2h + \frac{1}{4}h^2\right)$$

Differentiating with respect to $t$, we have

$$\frac{dV}{dt} = 5\left(2 + \frac{1}{2}h\right)\frac{dh}{dt} = \frac{5}{2}(4 + h)\frac{dh}{dt}$$

$$\frac{2(dV/dt)}{5(4 + h)} = \frac{dh}{dt}$$

Therefore, when $dV/dt = 1$ and $h = 1$, we have

$$\frac{dh}{dt} = \frac{2(1)}{5(4 + 1)} = \frac{2}{25} \text{ ft/min.}$$

# 4 APPLICATIONS OF DIFFERENTIATION

## 4.1 Extrema on an interval

**9.** Locate the extrema of $f(x) = -x^2 + 3x$ on the interval $[0, 3]$.

**Solution:**

$$f(x) = -x^2 + 3x$$
$$f'(x) = -2x + 3 = 0$$

Therefore, $x = 3/2$ is a critical number in $[0, 3]$.

We determine the extrema of $f$ by evaluating $f$ at the critical number and at the endpoints of $[0, 3]$.

$$f(0) = -0^2 + 3(0) = 0 \qquad \text{Minimum}$$
$$f\left(\frac{3}{2}\right) = -\left(\frac{3}{2}\right)^2 + 3\left(\frac{3}{2}\right) = \frac{9}{4} \qquad \text{Maximum}$$
$$f(3) = -3^2 + 3(3) = 0 \qquad \text{Minimum}$$

**13.** Locate the extrema of $f(x) = 3x^{2/3} - 2x$ on the interval $[-1, 1]$.

**Solution:**

$$f(x) = 3x^{2/3} - 2x$$
$$f'(x) = 2x^{-1/3} - 2$$
$$= 2\left(\frac{1 - \sqrt[3]{x}}{\sqrt[3]{x}}\right)$$

Therefore, $x = 1$ and $x = 0$ are critical numbers in the interval $[-1, 1]$. ($f'(0)$ is undefined.)

We determine the extrema of $f$ by evaluating $f$ at the critical numbers and at the endpoints of the interval $[-1, 1]$.

$$f(-1) = 3(-1)^{2/3} - 2(-1) = 5 \qquad \text{Maximum}$$
$$f(0) = 3(0)^{2/3} - 2(0) = 0 \qquad \text{Minimum}$$
$$f(1) = 3(1)^{2/3} - 2(1) = 1$$

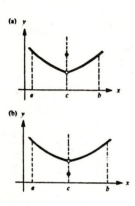

**23.** Determine from the graph of $f$ if $f$ possesses a relative minimum in the interval $(a, b)$.

**Solution:**

(a) Since $f$ is decreasing in the interval $(a, c)$ and increasing in the interval $(c, b)$, the only possible minimum in the interval $(a, b)$ would occur at $x = c$. However, $f(c)$ is greater than $f(x)$ for $x$ near $c$. Thus, there is no minimum.

(b) Since $f(c) \leq f(x)$ for all $x$ in $(a, b)$, $f(c)$ is a minimum.

**29.** The error estimate for the Trapezoid Rule (see Section 5.6) involves the maximum of the absolute value of the second derivative in an interval. Find the maximum value of $|f''(x)|$ for $f(x) = \sqrt{1 + x^3}$ in $[0, 2]$.

**Solution:**

To find the maximum value of $|f''(x)|$ in $[0, 2]$, we select the maximum of $|f''(0)|$, $|f''(2)|$, and $|f''(c)|$, where $c$ is any critical number of $f''(x)$ [i.e., $f'''(c) = 0$ or $f'''(c)$ does not exist].

$$f(x) = \sqrt{1 + x^3} = (1 + x^3)^{1/2}$$
$$f'(x) = \frac{1}{2}(1 + x^3)^{-1/2}(3x^2) = \frac{3}{2}x^2(1 + x^3)^{-1/2}$$
$$f''(x) = \frac{3}{2}\left[x^2\left(-\frac{1}{2}\right)(1 + x^3)^{-3/2}(3x^2) + 2x(1 + x^3)^{-1/2}\right]$$
$$= \frac{3}{2}\left[\frac{3x^4}{2(1 + x^3)^{3/2}} + \frac{2x}{(1 + x^3)^{1/2}}\right]$$
$$= \frac{3x(x^3 + 4)}{4(1 + x^3)^{3/2}}$$

$$f'''(x) = \frac{3}{4}\left[\frac{(1+x^3)^{3/2}(4x^3+4) - (x^4+4x(\frac{3}{2})(1+x^3)^{1/2}(3x^2)}{(1+x^3)^3}\right]$$

$$= \frac{3(1+x^3)^{1/2}[(1+x^3)(4x^3+4) - \frac{3}{2}(x^4+4x)(3x^2)]}{4(1+x^3)^3}$$

$$= \frac{-3(x^6+20x^3-8)}{8(1+x^3)^{5/2}}$$

Therefore, $f'''(x) = 0$ when

$$x^6 + 20x^3 - 8 = 0$$
$$(x^3)^2 + 20(x^3)^1 - 8 = 0$$
$$x^3 = \frac{-20 \pm \sqrt{(20)^2 - 4(1)(-8)}}{2(1)}$$
$$x = \sqrt[3]{\frac{-20 \pm \sqrt{432}}{2}}$$
$$= \sqrt[3]{-10 + \sqrt{108}}$$

which is the critical number in $[0, 2]$.

$$f'''(\sqrt[3]{-10 + \sqrt{108}}) \approx 1.47$$
$$|f''(0)| = 0$$
$$|f''(2)| = \frac{2}{3}$$

The maximum value of $|f''(x)|$ in $[0, 2]$ is

$$|f''(\sqrt[3]{-10 + \sqrt{108}})| \approx |f''(0.732)| \approx 1.47$$

**35.** The formula for the power output $P$ of a battery is given by

$$P = VI - RI^2$$

where $V$ is the electromotive force in volts, $R$ is the resistance, and $I$ is the current. Find the current (measured in amperes) that corresponds to a maximum value of $P$ in a battery for which $V = 12$ volts and $R = 0.5$ ohms. (Assume that a 15 amp fuse bounds the output in the interval $0 \le I \le 15$.)

**Solution:**

Since $V = 12$ and $R = 0.5$, we have

$$P = 12I - \frac{1}{2}I^2$$

$$\frac{dP}{dI} = 12 - I = 0$$

Therefore, $I = 12$ is a critical number on the interval $[0, 15]$.

We determine the maximum of $P$ by evaluating $P$ at the critical number and at the endpoints of the interval $[0, 15]$. Since

$$P(0) = 12(0) - \frac{1}{2}(0)^2 = 0$$

$$P(12) = 12(12) - \frac{1}{2}(12)^2 = 72$$

$$P(15) = 12(15) - \frac{1}{2}(15)^2 = 67.5$$

we conclude the power $P$ is maximum when $I = 12$.

## 4.2   Rolle's Theorem and the Mean Value Theorem

5. Determine whether Rolle's Theorem can be applied to the function $f(x) = (x - 1)(x - 2)(x - 3)$ on the interval $[1, 3]$. If it can be applied, find all values of $c$ in the interval such that $f'(c) = 0$.

**Solution:**

Since $f$ is a polynomial, it is continuous and differentiable for all $x$. Also, the zeros of $f$ are $x = 1, x = 2$, and $x = 3$. Thus the intervals on which Rolle's Theorem can be applied are $[1, 2]$ and $[2, 3]$. Setting $f'(x) = 0$ yields

$$f'(x) = 3x^2 - 12x + 11 = 0$$

$$x = \frac{12 \pm \sqrt{144 - 132}}{6} = \frac{12 \pm 2\sqrt{3}}{6} = \frac{6 \pm \sqrt{3}}{3}$$

Therefore, in the interval $[1, 2]$,

$$f'\left(\frac{6 - \sqrt{3}}{3}\right) = 0 \qquad \text{where} \qquad c = \frac{6 - \sqrt{3}}{3} \approx 1.423$$

and in the interval $[2,3]$,

$$f'\left(\frac{6+\sqrt{3}}{3}\right) = 0 \qquad \text{where} \qquad c = \frac{6+\sqrt{3}}{3} \approx 2.577$$

9. Determine whether Rolle's Theorem can be applied to the function $f(x) = x^{2/3} - 1$ on the interval $[-8, 8]$. If it can be applied, find all values of $c$ such that $f'(c) = 0$.

**Solution:**

We first observe that $f$ is continuous on the specified interval. The zeros of $f$ are found by solving the following equation.

$$f(x) = x^{2/3} - 1 = 0$$
$$x^{2/3} = 1$$
$$x^2 = 1$$
$$x = \pm 1$$

Since
$$f'(x) = \frac{2}{3x^{1/3}}$$

we observe that $f'(0)$ is undefined and therefore, $f$ is not differentiable at $x = 0$. Thus, we cannot apply Rolle's Theorem to this function.

17. Apply the Mean Value Theorem to $f(x) = x/(x+1)$ on the interval $[-\frac{1}{2}, 2]$. Find all values of $c$ in $[-\frac{1}{2}, 2]$ such that

$$f'(c) = \frac{f(2) - f(-\frac{1}{2})}{2 - (-\frac{1}{2})}.$$

**Solution:**

Since $f(x) = x/(x+1)$ is continuous and differentiable for all $x$ other than $x = -1$, we can apply the Mean Value Theorem on

$[-\frac{1}{2}, 2]$.

$$f(x) = \frac{x}{x+1}$$

$$f'(x) = \frac{(x+1)(1) - x(1)}{(x+1)^2} = \frac{f(2) - f(-\frac{1}{2})}{2 - (-\frac{1}{2})}$$

$$\frac{1}{(x+1)^2} = \frac{(\frac{2}{3}) - (-1)}{\frac{5}{2}} = \frac{2}{3}$$

$$(x+1)^2 = \frac{3}{2}$$

$$x + 1 = \pm\frac{\sqrt{3}}{\sqrt{2}}$$

$$x = -1 \pm \frac{\sqrt{6}}{2}$$

Therefore, on the interval $[-\frac{1}{2}, 2]$, $c = -1 + \frac{\sqrt{6}}{2}$.

**27.** Show that $x^{2n+1} + ax + b$ cannot have two real roots, where $a > 0$ and $n$ is any positive integer.

**Solution:**

The polynomial $f(x) = x^{2n+1} + ax + b$ is continuous and differentiable for all $x$. Therefore, by Rolle's Theorem, if $f(x) = 0$ for two distinct values of $x$, there would have to be at least one value of $x$ such that $f'(x) = 0$. However,

$$f(x) = x^{2n+1} + ax + b$$
$$f'(x) = (2n + 1)x^{2n} + a = 0$$
$$(x^n)^2 = -\frac{a}{2n+1}$$

(positive number) = (negative number)

Therefore, $f'(x) = 0$ has no solution and consequently $f(x) = 0$ cannot have two real zeros.

## 4.3   Increasing and Decreasing Functions, and The First Derivative Test

**11.** Find the critical numbers (if any) of $f(x) = 2x^3 + 3x^2 - 12x$, the open intervals on which $f$ is increasing or decreasing, and locate all relative extrema.

**Solution:**

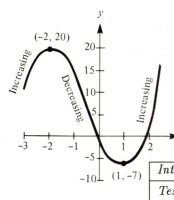

$$f(x) = 2x^3 + 3x^2 - 12x$$
$$f'(x) = 6x^2 + 6x - 12 = 6(x+2)(x-1)$$

Therefore, $f'(x) = 0$ when $x = -2$ or $x = 1$. Since $f$ is a polynomial, it is differentiable for all $x$ and the only critical numbers are $x = -2$ and $x = 1$.

| Interval | $(-\infty, -2)$ | $(-2, 1)$ | $(1, \infty)$ |
|---|---|---|---|
| Test value | $x = -3$ | $x = 0$ | $x = 2$ |
| Sign of $f'(x)$ | $f'(-3) = 24 > 0$ | $f'(0) = -12 < 0$ | $f'(2) = 24 > 0$ |
| Conclusion | $f$ is increasing | $f$ is decreasing | $f$ is increasing |

When $x = -2$, we have $f(-2) = 2(-2)^3 + 3(-2)^2 - 12(-2) = 20$. When $x = 1$, we have $f(1) = 2 + 3 - 12 = -7$. Therefore, we conclude that $(-2, 20)$ is a relative maximum and $(1, -7)$ is a relative minimum.

**17.** Find the critical numbers (if any) of $f(x) = x^{1/3} + 1$, the open intervals on which $f$ is increasing or decreasing, and locate all relative extrema.

**Solution:**

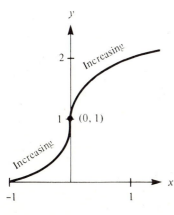

$$f(x) = x^{1/3} + 1$$
$$f'(x) = \left(\frac{1}{3}\right)x^{-2/3} = \frac{1}{3x^{2/3}}$$

Since $f$ is continuous for all $x$ and differentiable for all $x$ other than $x = 0$, the only critical number is $x = 0$. ($f'(0)$ is undefined.) We also observe that $f'(x) > 0$ for all $x$ not equal to zero. Therefore, we conclude that $f$ is increasing for all $x$ and there are **no** relative extrema.

**25.** Find the critical numbers (if any) of $f(x) = (x^2 - 2x + 1)/(x+1)$, the open intervals on which $f$ is increasing or decreasing, and locate all relative extrema.

**Solution:**

The function $f$ is continuous for all $x$ other than $x = -1$. The graph of $f$ has a vertical asymptote at $x = -1$.

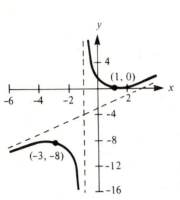

$$f(x) = \frac{x^2 - 2x + 1}{x + 1}$$

$$f'(x) = \frac{x + 1)(2x - 2) - (x^2 - 2x + 1)(1)}{(x+1)^2}$$

$$= \frac{(x+1)(2)(x-1) - (x-1)^2}{(x+1)^2}$$

$$= \frac{(x-1)(2x + 2 - x + 1)}{(x+1)^2} = \frac{(x-1)(x+3)}{(x+1)^2}$$

Since $f$ is differentiable for all $x$ other than $x = -1$, the only critical numbers are $x = 1$ and $x = 3$.

| Interval | $(-\infty, -3)$ | $(-3, -1)$ | $(-1, 1)$ | $(1, \infty)$ |
|---|---|---|---|---|
| Test value | $x = -4$ | $x = -2$ | $x = 0$ | $x = 2$ |
| Sign of $f'(x)$ | $f'(-4) = \dfrac{5}{9} > 0$ | $f'(-2) = -3 < 0$ | $f'(0) = -3 < 0$ | $f'(2) = \dfrac{5}{9} > 0$ |
| Conclusion | $f$ increasing | $f$ decreasing | $f$ decreasing | $f$ increasing |

When $x = -3$, we have $f(-3) = (9 + 6 + 1)/ - 2 = -8$, and when $x = 1$, we have $f(1) = (1 - 2 + 1)/2 = 0$. Therefore, we conclude that $(-3, -8)$ is a relative maximum and $(1, 0)$ is a relative minimum.

**31.** Coughing forces the trachea (windpipe) to contract, which affects the velocity $v$ of the air passing through the trachea. Suppose the velocity of the air during coughing is given by

$$v = k(R - r)r^2$$

where $k$ is a constant, $R$ is the normal radius of the trachea, and $r$ is the radius during coughing. What radius will produce the maximum air velocity?

**Solution:**

$$v = k(R - r)r^2 = k(Rr^2 - r^3)$$

$$\frac{dv}{dr} = k(2Rr - 3r^2) = kr(2R - 3r)$$

Therefore, $\dfrac{dv}{dr} = 0$ when $r = 0$ or $r = \frac{2}{3}R$. Since $v$ is continuous and differentiable for all $r$, the only critical numbers are $r = 0$ and $r = 2R/3$.

| Interval | $(-\infty, 0)$ | $\left(0, \dfrac{2}{3}R\right)$ | $\left(\dfrac{2}{3}R, \infty\right)$ |
|---|---|---|---|
| Test value | $r = -R$ | $r = \dfrac{R}{3}$ | $r = \dfrac{4R}{3}$ |
| Sign of $v'(r)$ | $v'(-R) < 0$ | $v'\left(\dfrac{R}{3}\right) > 0$ | $v'\left(\dfrac{4R}{3}\right) < 0$ |
| Conclusion | $f$ decreasing | $f$ increasing | $f$ decreasing |

We conclude that the velocity $v$ is maximum when $r = \dfrac{2R}{3}$.

**39.** Find $a$, $b$, $c$, and $d$ so that the function $f(x) = ax^3 + bx^2 + cx + d$ has a relative minimum at $(0, 0)$ and a relative maximum at $(2, 2)$.

**Solution:**

In order for $(0, 0)$ and $(2, 2)$ to be solution points to $f$, we must have

$$f(0) = 0 = a(0^3) + b(0^2) + c(0) + d$$

$$0 = d$$

$$f(2) = 2 = a(2^2) + b(2^2) + c(2) + d$$

$$2 = 8a + 4b + 2c + 0$$

(1) $$1 = 4a + 2b + c$$

Also, for $f$ to have relative extrema at $(0, 0)$ and $(2, 2)$, $f'(x)$ must be zero at these points. Since $f'(x) = 3ax^2 + 2bx + c$, we have

$$f'(0) = 0 = 3a(0^2) + 2b(0) + c$$

$$0 = c$$

$$f'(2) = 0 = 3a(2^2) + 2b(2) + c$$

$$0 = 12a + 4b + 0$$

(2) $$0 = 3a + b$$

Letting $c = 0$ in equation 1 and multiplying equation 2 by 2, we have

$$4a + 2b = 1$$
$$\underline{6a + 2b = 0}$$
$$-2a \quad\quad = 1$$
$$a = -\frac{1}{2}$$
$$4\left(-\frac{1}{2}\right) + 2b = 1$$
$$2b = 3$$
$$b = \frac{3}{2}$$

Finally, we conclude that $a = -\frac{1}{2}, b = \frac{3}{2}, c = 0$, and $d = 0$. Thus, $f(x) = -\frac{1}{2}x^3 + \frac{3}{2}x^2$.

## 4.4   Concavity and the Second Derivative Test

**11.** Identify all relative extrema for $f(x) = x^3 - 3x^2 + 3$. Use the Second-Derivative Test when applicable.

**Solution:**

$$f(x) = x^3 - 3x^2 + 3$$
$$f'(x) = 3x^2 - 6x = 3x(x - 2)$$
$$f'(x) = 0 \quad \text{when} \quad x = 0, 2$$
$$f''(x) = 6x - 6$$

At $x = 0$, we have $f(0) = 3$, $f'(0) = 0$, and $f''(0) = -6$. Therefore, by the Second-Derivative Test, $(0, 3)$ is a relative maximum.

At $x = 2$, we have $f(2) = -1$, $f'(2) = 0$, and $f''(2) = 6$. Therefore, by the Second-Derivative Test, $(2, -1)$ is a relative minimum.

**17.** Identify all relative extrema for $f(x) = x + (4/x)$. Use the Second-Derivative Test when applicable.

**Solution:**

$$f(x) = x + \frac{4}{x}$$

$$f'(x) = 1 - \frac{4}{x^2} = \frac{x^2 - 4}{x^2}$$

$$f'(x) = 0 \text{ when } +x = \pm 2$$

$$f''(x) = (-4)(-2)x^{-3} = \frac{8}{x^3}$$

At $x = -2$, we have $f(-2) = -4$, $f'(-2) = 0$, and $f''(-2) = -1$. Therefore, by the Second-Derivative Test, $(-2, -4)$ is a relative maximum.

Since $f$ is symmetrical with respect to the origin, $(2, 4)$ is a relative minimum.

**25.** Sketch the graph of $f(x) = x(x - 4)^3$, and identify all relative extrema and points of inflection.

**Solution:**

$$f(x) = x(x - 4)^3$$

$$f'(x) = x[3(x - 4)^2(1)] + (x - 4)^3(1)$$

$$= (x - 4)^2[3x + (x - 4)] = 4(x - 4)^2(x - 1)$$

$$f'(x) = 0 \quad \text{when} \quad x = 1, 4$$

$$f''(x) = 4[(x - 4)^2(1) + (x - 1)(2)(x - 4)(1)]$$

$$= 4(x - 4)[(x - 4) + 2(x - 1)] = 12(x - 4)(x - 2)$$

$$f''(x) = 0 \quad \text{when} \quad x = 2, 4$$

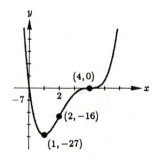

| $x$ | 0 | 1 | 2 | 3 | 4 | 5 |
|---|---|---|---|---|---|---|
| $f(x)$ | 0 | $-27$ | $-16$ | $-3$ | 0 | 5 |
| $f'(x)$ | $-$ | 0 | $+$ | $+$ | 0 | $+$ |
| $f''(x)$ | $+$ | $+$ | 0 | $-$ | 0 | $+$ |

Therefore, $(1, -27)$ is a relative minimum, and $(2, -16)$ and $(4, 0)$ are points of inflection.

**29.** Sketch the graph of the function $f(x) = x\sqrt{x+3}$, and identify all relative extrema and points of inflection.

**Solution:**

$$f(x) = x\sqrt{x+3} \qquad \text{domain:}\ [-3, \infty)$$

$$f'(x) = x\left(\frac{1}{2}\right)(x+3)^{-1/2}(1) + (x+3)^{1/2}(1)$$

$$= \frac{x}{2\sqrt{x+3}} + \sqrt{x+3}$$

$$= \frac{x + 2(x+3)}{2\sqrt{x+3}} = \frac{3(x+2)}{2\sqrt{x+3}}$$

$$f'(x) = 0 \quad \text{when} \quad x = -2$$

$$f''(x) = \frac{3}{2}\left[\frac{\sqrt{x+3}(1) - (x+2)(\frac{1}{2})(x+3)^{-1/2}(1)}{x+3}\right]$$

$$= \frac{3}{2}\left[\frac{\sqrt{x+3} - (x+2)/2\sqrt{x+3}}{x+3}\right]$$

$$= \frac{3(x+4)}{4(x+3)^{3/2}} > 0 \qquad \text{for all } x \text{ in}(-3, \infty)$$

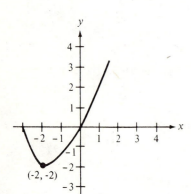

We conclude that the graph is concave upward for each $x$ in the domain of $f$ and therefore that $(-2, -2)$ is a relative minimum.

**43.** Find $a, b, c$ and $d$ so that the function $f(x) = ax^3 + bx^2 + cx + d$ has a relative maximum at $(3, 3)$, a relative minimum at $(5, 1)$, and a point of inflection at $(4, 2)$.

**Solution:**

$$f(x) = ax^3 + bx^2 + cx + d$$
$$f'(x) = 3ax^2 + 2bx + c$$
$$f''(x) = 6ax + 2b = 2(3ax + b)$$

| | | |
|---|---|---|
| $f(3) = 3$ | $\implies$ | $27a + 9b + 3c + d = 3$     (1) |
| $f(4) = 2$ | $\implies$ | $64a + 16b + 4c + d = 2$     (2) |
| $f(5) = 1$ | $\implies$ | $125a + 25b + 5c + d = 1$     (3) |
| $f'(3) = 0$ | $\implies$ | $27a + 6b + c = 0$     (4) |
| $f'(5) = 0$ | $\implies$ | $75a + 10b + c = 0$     (5) |
| $f''(4) = 0$ | $\implies$ | $2(12a + b) = 0$     (6) |

From (6) we get $b = -12a$. Substituting this into (5), we obtain

$$75a + 10(-12a) + c = 0$$
$$-45a + c = 0$$
$$c = 45a$$

Substitution into (1) and (2) yields

$$27a + 9(-12a) + 3(45a) + d = 3$$
$$54a + d = 3 \qquad (7)$$
$$64a + 16(-12a) + 4(45a) + d = 2$$
$$52a + d = 2 \qquad (8)$$

To solve simultaneously, we subtract (8) from (7) and obtain

$$2a = 1 \qquad a = \frac{1}{2}$$

From (7)

$$d = 3 - 54\left(\frac{1}{2}\right) = -24$$
$$c = 45\left(\frac{1}{2}\right) = \frac{45}{2}$$
$$b = -12\left(\frac{1}{2}\right) = -6$$

Therefore,

$$f(x) = \frac{1}{2}x^3 - 6x^2 + \frac{45}{2}x - 24 = \frac{1}{2}(x^3 - 12x^2 + 45x - 48)$$

**46.** Prove that a cubic function with three real zeros has a point of inflection whose $x$-coordinate is the average of the three zeros.

**Solution:**

We assume the three zeros of the cubic are $r_1, r_2,$ and $r_3$. Then

$$f(x) = a(x - r_1)(x - r_2)(x - r_3)$$
$$f'(x) = a[(x - r_1)(x - r_2) + (x - r_1)(x - r_3) + (x - r_2)(x - r_3)]$$
$$f''(x) = a[(x - r_1) + (x - r_2) + (x - r_1) + (x - r_3) + (x - r_2)$$
$$+ (x - r_3)]$$
$$= a[6x - 2(r_1 + r_2 + r_3)]$$

Consequently, $f''(x) = 0$ if

$$x = \frac{2(r_1 + r_2 + r_3)}{6} = \frac{r_1 + r_2 + r_3}{3} = \text{average of } r_1, r_2 \text{ and } r_3$$

## 4.5 Limits at Infinity

**11.** Evaluate $\lim\limits_{x \to \infty} \dfrac{x}{x^2 - 1}$.

**Solution:**

$$\lim_{x \to \infty} \frac{x}{x^2 - 1} = \lim_{x \to \infty} \frac{x/x^2}{(x^2/x^2) - (1/x^2)} = \lim_{x \to \infty} \frac{1/x}{1 - (1/x^2)}$$

$$= \frac{0}{1 - 0} = 0$$

**13.** Evaluate $\lim\limits_{x \to -\infty} \dfrac{5x^2}{x + 3}$.

**Solution:**

Since

$$\lim_{x \to -\infty} \frac{5x^2}{x + 3} = \lim_{x \to -\infty} \frac{5x^2/x^2}{(x/x^2) + (3/x^2)}$$

$$= \lim_{x \to -\infty} \frac{5}{(1/x) + (3/x^2)} = -\infty,$$

the limit does not exist. To determine that the limit is $-\infty$ rather than $+\infty$, we note that $5x^2/(x + 3) < 0$ whenever $x < -3$. (Note that the degree of the numerator is greater than the degree of the denominator. See Exercise 50 in this section.)

**17.** Evaluate $\lim\limits_{x \to -\infty} \left( \dfrac{2x}{x - 1} + \dfrac{3x}{x + 1} \right)$.

**Solution:**

$$\lim_{x \to -\infty} \left( \frac{2x}{x - 1} + \frac{3x}{x + 1} \right) = \lim_{x \to -\infty} \left[ \frac{2x/x}{(x/x) - (1/x)} + \frac{3x/x}{(x/x) + (1/x)} \right]$$

$$= \lim_{x \to -\infty} \left[ \frac{2}{1 - (1/x)} + \frac{3}{1 + (1/x)} \right]$$

$$= \frac{2}{1 - 0} + \frac{3}{1 + 0} = 5$$

(Note that for each of the rational functions in this exercise, the denominator and numberator are of equal degree. See Exercise 50 in this section.)

**19.** Evaluate $\displaystyle\lim_{x \to -\infty} \frac{x}{\sqrt{x^2 - x}}$.

**Solution:**

$$\lim_{x \to -\infty} \frac{x}{\sqrt{x^2 - x}} = \lim_{x \to -\infty} \frac{x/x}{\sqrt{x^2 - x}/(-\sqrt{x^2})} = \lim_{x \to -\infty} \frac{1}{-\sqrt{1 - (1/x)}}$$

$$= -\frac{1}{\sqrt{1 + 0}} = -1$$

(Note: For $x < 0, x = -\sqrt{x^2}$.)

**27.** Evaluate $\displaystyle\lim_{x \to \infty} (x - \sqrt{x^2 + x})$.

**Solution:**

$$\lim_{x \to \infty} (x - \sqrt{x^2 + x}) = \lim_{x \to \infty} (x - \sqrt{x^2 + x}) \frac{x + \sqrt{x^2 + x}}{x + \sqrt{x^2 + x}}$$

$$= \lim_{x \to \infty} \frac{x^2 - (x^2 + x)}{x + \sqrt{x^2 + x}}$$

$$= \lim_{x \to \infty} \frac{-x}{x + \sqrt{x^2 + x}}$$

$$= \lim_{x \to \infty} \frac{-x/x}{(x/x) + \sqrt{x^2 + x}/\sqrt{x^2}}$$

$$= \lim_{x \to \infty} \frac{-1}{1 + \sqrt{1 + (1/x)}}$$

$$= \frac{-1}{1 + \sqrt{1 + 0}} = -\frac{1}{2}$$

(Note: For $x > 0, x = \sqrt{x^2}$.)

**29.** Sketch the graph of $y = (2 + x)/(1 - x)$. As a sketching aid examine the equation for intercepts, symmetry, and asymptotes.

**Solution:**

If $x = 0$, then $y = 2$ and the $y$-intercept occurs at $(0, 2)$. If $y = 0$, then $2 + x = 0$, $x = -2$, and the $x$-intercept is $(-2, 0)$. There is no symmetry with respect to either axis or to the origin. Since the denominator of $(2 + x)/(1 - x)$ is zero when $x = 1$, there is a vertical asymptote at $x = 1$. Furthermore,

$$\lim_{x \to 1^-} \frac{2 + x}{1 - x} = \infty \qquad \text{and} \qquad \lim_{x \to 1^+} \frac{2 + x}{1 - x} = -\infty$$

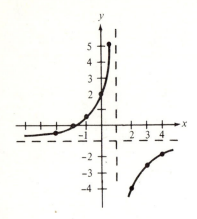

$$\lim_{x \to \pm\infty} \frac{2+x}{1-x} = \lim_{x \to \pm\infty} \frac{2/x + x/x}{1/x - x/x}$$

$$= \lim_{x \to \pm\infty} \frac{\frac{2}{x} + 1}{\frac{1}{x} - 1}$$

$$= \frac{0+1}{0-1} = -1$$

Therefore, there is a horizontal asymptote (to the right and left) at $y = -1$. To complete the graph, we add a few points as shown in the accompanying table.

| $x$ | $-3$ | $-1$ | 0.5 | 2 | 3 | 4 |
|---|---|---|---|---|---|---|
| $y$ | $-0.25$ | 0.5 | 5 | $-4$ | $-2.5$ | $-2$ |

39. Sketch the graph of $y = x^3/\sqrt{x^2 - 4}$. As a sketching aid examine the equation for intercepts, symmetry, and asymptotes.

**Solution:**

Since $x^2 - 4$ must be positive, the domain is $(-\infty, -2)$ and $(2, \infty)$. There is no symmetry with respect to either axis. However, there is symmetry with respect to the origin since

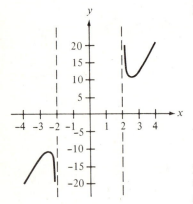

$$(-y) = \frac{(-x)^3}{\sqrt{(-x)^2 - 4}}$$

$$-y = \frac{-x^3}{\sqrt{x^2 - 4}}$$

$$y = \frac{x^3}{\sqrt{x^2 - 4}}$$

which is equivalent to the original equation. Since the denominator of $x^3/\sqrt{x^2 - 4}$ is zero when $x = 2$ or $x = -2$, there are vertical asymptotes at $x = 2$ and $x = -2$.

| $x$ | 2.25 | 2.50 | 2.75 | 3.00 | 4.00 |
|---|---|---|---|---|---|
| $y$ | 11.05 | 10.45 | 11.02 | 12.07 | 18.48 |

## 4.6  A Summary of Curve Sketching

9. Sketch the graph of $y = 3x^4 + 4x^3$, choosing a scale that allows all relative extrema and points of inflection to be identified on the sketch.

**Solution:**

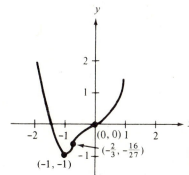

$$y = 3x^4 + 4x^3 = x^3(3x + 4) \qquad \text{Intercepts: } (0,0), (-\tfrac{4}{3}, 0)$$
$$y' = 12x^3 + 12x^2 = 12x^2(x + 1) \qquad \text{Critical numbers:}$$
$$x = 0, \ x = -1$$
$$y'' = 36x^2 + 24x = 12x(3x + 2) \qquad \text{Possible inflection points:}$$
$$(-\tfrac{2}{3}, -\tfrac{16}{27}), \ (0, 0)$$

| $x$ | $y$ | $y'$ | $y''$ | *Shape of graph* |
|---|---|---|---|---|
| $x$ in $(-\infty, -1)$ | | $-$ | $+$ | decreasing, concave up |
| $x = -1$ | $-1$ | $0$ | $+$ | relative minimum |
| $x$ in $(-1, -\tfrac{2}{3}$ | | $+$ | $+$ | increasing, concave up |
| $x = -\tfrac{2}{3}$ | $-\tfrac{16}{27}$ | $+$ | $0$ | point of inflection |
| $x$ in $(-\tfrac{2}{3}, 0)$ | | $+$ | $-$ | increasing, concave down |
| $x = 0$ | $0$ | $0$ | $0$ | point of inflection |
| $x$ in $(0, \infty)$ | | $+$ | $+$ | increasing, concave up |

21. Sketch the graph of $y = x\sqrt{4 - x}$. Label the intercepts, relative extrema, points of inflection, and the domain.

**Solution:**

$$y = x\sqrt{4-x} \qquad \text{Domain: } x \le 4, \quad \text{Intercepts: } (0,0)(4,0)$$

$$y' = x\left(\frac{1}{2}\right)(-1)(4-x)^{-1/2} + (4-x)^{1/2}$$

$$= \frac{-x}{2\sqrt{4-x}} + \sqrt{4-x}$$

$$= \frac{-x+8-2x}{2\sqrt{4-x}}$$

$$= \frac{-3x+8}{2\sqrt{4-x}} \qquad \text{Critical numbers: } \frac{8}{3}, 4$$

$$y'' = \frac{2(4-x)^{1/2}(-3) - (-3x+8)(2)(1/2)(-1)(4-x)^{-1/2}}{4(4-x)}$$

$$= \frac{-6(4-x) + (-3x+8)}{4(4-x)^{3/2}} = \frac{-24+6x-3x+8}{4(4-x)^{3/2}}$$

$$= \frac{3x-16}{4(4-x)^{3/2}}$$

Note: $x = \frac{16}{3}$ is not in the domain of the function.

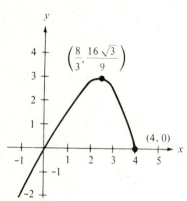

| $x$ | $y$ | $y'$ | $y''$ | *Shape of graph* |
|---|---|---|---|---|
| $x$ in $\left(-\infty, \dfrac{8}{3}\right)$ | | $+$ | $-$ | increasing, concave down |
| $x = \dfrac{8}{3}$ | $\dfrac{16\sqrt{3}}{9}$ | $0$ | $-$ | relative maximum |
| $x$ in $\left(\dfrac{8}{3}, 4\right)$ | | $-$ | $-$ | decreasing, concave down |
| $x = 4$ | $0$ | undefined | undefined | |

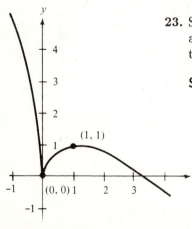

**23.** Sketch the graph of $y = 3x^{2/3} - 2x$, choosing a scale that allows all relative extrema and points of inflection to be identified on the sketch.

**Solution:**

$$y = 3x^{2/3} - 2x = x^{2/3}(3 - 2x^{1/3}) \qquad \text{Intercepts: } (0,0), \left(\frac{27}{8}, 0\right)$$

$$y' = 2x^{-1/3} - 2 = \frac{2(1 - x^{1/3})}{x^{1/3}} \qquad \text{Critical numbers:}$$

$$x = 0, \ x = 1$$

$$y'' = \left(\frac{-1}{3}\right)(2)x^{-4/3} = \frac{-2}{3x^{4/3}}$$

| $x$ | $y$ | $y'$ | $y''$ | Shape of graph |
|---|---|---|---|---|
| $x$ in $(-\infty, 0)$ | | $-$ | $-$ | decreasing, concave down |
| $x = 0$ | $0$ | undefined | undefined | relative minimum |
| $x$ in $(0, 1)$ | | $+$ | $-$ | increasing, concave down |
| $x = 1$ | $1$ | $0$ | $-$ | relative maximum |
| $x$ in $(1, \infty)$ | | $-$ | $-$ | decreasing, concave down |

**27.** Sketch the graph of $y = [1/(x - 2)] - 3$. Label the intercepts, relative extrema, points of inflection, and the domain.

**Solution:**

$$y = \frac{1}{x - 2} - 3$$

$$= \frac{1 - 3(x - 2)}{x - 2}$$

$$= \frac{7 - 3x}{x - 2} \qquad \text{Domain: all } x \neq 2, \quad \text{Intercepts: } \left(\tfrac{7}{3}, 0\right), \left(0, -\tfrac{7}{2}\right)$$

$$y' = \frac{-1}{(x - 2)^2}$$

$$y'' = \frac{2}{(x - 2)^3}$$

The graph has a vertical asymptote at $x = 2$. Since

$$\lim_{x \to \pm\infty} \left(\frac{1}{x - 2} - 3\right) = -3,$$

there is a horizontal asymptote at $y = -3$.

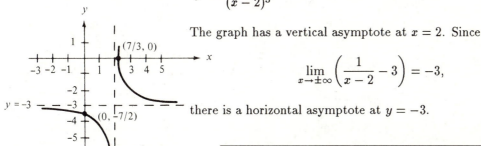

| $x$ | $y'$ | $y''$ | Shape of graph |
|---|---|---|---|
| $x$ in $(-\infty, 2)$ | $-$ | $-$ | decreasing, concave down |
| $x$ in $(2, \infty)$ | $-$ | $+$ | decreasing, concave up |

**35.** Sketch the graph of $y = x^3/(2x^2 - 8)$. Use any of the sketching aids that we have developed, including slant asymptotes.

**Solution:**

The graph is symmetric to the origin and has an intercept at $(0,0)$. There are vertical asymptotes at $x = \pm 2$ and

$$\lim_{x \to 2^-} \frac{x^3}{2x^2 - 8} = -\infty \qquad \text{and} \qquad \lim_{x \to 2^+} \frac{x^3}{2x^2 - 8} = \infty$$

Since the numerator of $x^3/(2x^2 - 8)$ is one degree higher than the denominator, we divide $2x^2 - 8$ into $x^3$ to find the slant asymptote.

Thus

$$\frac{x^3}{2x^2 - 8} = \frac{x}{2} + \frac{4x}{2x^2 - 8}$$

and $y = x/2$ is a slant asymptote. To complete the graph, we add a few points as shown in the accompanying table.

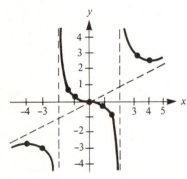

| $x$ | 1 | 1.5 | 3 | 4 |
|---|---|---|---|---|
| $y$ | -0.16 | -0.96 | 2.70 | 2.67 |

By adding these points to the graph and using symmetry, we have the sketch as shown.

**37.** Determine conditions on the coefficients $a, b$, and $c$ such that the graph of $f(x) = ax^3 + bx^2 + cx + d$ will resemble the accompanying graph.

**Solution:**

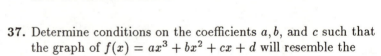

Since $\lim_{x \to \infty} f(x) = -\infty$, $a < 0$. Also, $f(x)$ is a decreasing function, and therefore $f'(x) = 3ax^2 + 2bx + c < 0$ for all $x$. Hence, the discriminant must be negative and we have

$$(2b)^2 - 4(3a)(c) < 0$$
$$4(b^2 - 3ac) < 0$$
$$b^2 < 3ac$$

## 4.7   Optimization Problems

5. Find two positive numbers such that the second number is a reciprocal of the first and their sum in minimum.

**Solution:**

(a) Let $x =$ first number, $y =$ second number, and $S =$ the sum to be minimized.

(b) To minimize $S$, we use the *primary* equation

$$S = x + y$$

(c) Since the second number is the reciprocal of the first we have the *secondary* equation

$$y = \frac{1}{x}$$

and therefore

$$S = x + \frac{1}{x}.$$

(d) Differentiation yields

$$\frac{dS}{dx} = 1 - \frac{1}{x^2}$$

$$\frac{dS}{dx} = 0 \quad \text{when} \quad x = \pm 1$$

$$\frac{d^2 S}{dx^2} = \frac{2}{x^3}$$

Finally, since the second derivative is positive when $x = 1$, we conclude that $S$ is minimum when $x = 1$ and $y = 1$.

**11.** Find the coordinates of the point on the curve $y = \sqrt{x}$ closest to the point $(4, 0)$.

**Solution:**

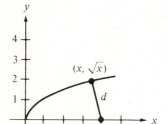

(a) Consider $(x, y)$ to be a point on the graph of $y = \sqrt{x}$.

(b) The distance between $(x, y)$ and the point $(4, 0)$ is given by the *primary* equation

$$d(x) = \sqrt{(x - 4)^2 + (y - 0)^2} = [(x - 4)^2 + y^2]^{1/2}$$

(c) Since $y = \sqrt{x}$, we have

$$d(x) = [(x - 4)^2 + x]^{1/2}$$

(d) To minimize $d(x)$, we solve $d'(x) = 0$ as follows:

$$d'(x) = \frac{1}{2}[(x - 4)^2 + x]^{-1/2}[2(x - 4) + 1]$$

$$= \frac{2(x - 4) + 1}{2\sqrt{(x - 4)^2 + x}}$$

$$= \frac{2x - 7}{2\sqrt{(x - 4)^2 + x}}$$

We observe that $d'(x) = 0$ when

$$2x - 7 = 0$$

$$x = \frac{7}{2} \quad \text{and} \quad y = \sqrt{\frac{7}{2}}$$

Therefore, the required point is $(7/2, \sqrt{7/2})$.

**17.** An open box is to be made from a square piece of material, 12 inches on a side, by cutting equal squares from each corner and turning up the sides.

Find the volume of the largest box that can be made in this manner.

**Solution:**

The volume of the box (see the accompanying figure) is given by

$$V = x(12 - 2x)^2 \qquad (0 < x < 6)$$

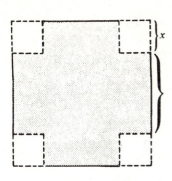

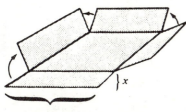

$12 - 2x$

Thus to find the maximum volume, we solve $dV/dx = 0$ as follows:

$$\frac{dV}{dx} = x(2)(12 - 2x)(-2) + (12 - 2x)^2(1)$$

$$= (12 - 2x)(-4x + 12 - 2x)$$

$$= 12(6 - x)(2 - x) = 0$$

Therefore, $dV/dx = 0$ when $x = 6$ or $x = 2$. Since 6 is not in the domain of $V$ ($V = 0$ if $x = 6$), we test $x = 2$ to determine if the volume is a maximum for this value of $x$.

$$\frac{d^2V}{dx^2} = 12[(6 - x)(-1) + (2 - x)(-1)] = 12(2x - 8)$$

When $x = 2$, $\frac{d^2V}{dx^2} = -48 < 0$. Therefore, $V$ is maximum when $x = 2$ and

$$V = 2[12 - 2(2)]^2 = 128 \text{ cubic inches.}$$

**25.** A right triangle is formed in the first quadrant by the $x$- and $y$-axes and a line through the point $(2, 3)$. Find the vertices of the triangle so that its area is minimum.

**Solution:**

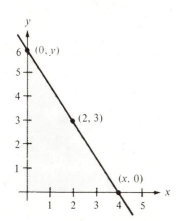

The slope of the line (see the accompanying figure) is given by

$$m = \frac{y - 3}{0 - 2} = \frac{0 - 3}{x - 2}$$

Therefore,

$$y - 3 = \frac{6}{x - 2}$$

$$y = \frac{6}{x - 2} + 3 = \frac{6 + 3x - 6}{x - 2} = \frac{3x}{x - 2}$$

The area of the triangle is

$$A = \frac{1}{2}xy = \frac{1}{2}x\left(\frac{3x}{x - 2}\right) \qquad 2 < x$$

$$= \frac{3x^2}{2(x - 2)}$$

To minimize $A$, we solve $dA/dx = 0$ as follows:

$$\frac{dA}{dx} = \frac{2(x-2)(6x) - (3x^2)(2)}{4(x-2)^2}$$

$$= \frac{6x(2x - 4 - x)}{4(x-2)^2} = \frac{3x(x-4)}{2(x-2)^2} = 0$$

Disregarding $x = 0$, we have

$$x = 4 \quad \text{and} \quad y = \frac{3(4)}{4-2} = 6$$

Thus the vertices are $(0,0), (4,0)$, and $(0,6)$.

**31.** A right circular cylinder is to be designed to hold 12 fluid ounces of a soft drink and to use the minimal amount of material in its construction. Find the required dimensions, assuming that 1 fluid ounce (oz) requires 1.80469 cubic inches.

**Solution:**

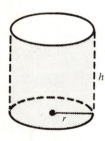

The volume of the cylinder is given by

$$V = \pi r^2 h = 12\text{oz}$$

$$= 12(1.80469)\text{in.}^3 = 21.65628\text{in.}^3$$

The surface area of the cylinder is

$$S = 2(\text{area of base}) + (\text{lateral surface})$$

$$= 2\pi r^2 + 2\pi rh = 2\pi r(r + h)$$

Since $h = 21.65628/\pi r^2$, we have

$$S = 2\pi r\left(r + \frac{21.65628}{\pi r^2}\right) = 2\pi\left(r^2 + \frac{21.65628}{\pi r}\right)$$

$$\frac{dS}{dr} = 2\pi\left(2r - \frac{21.65628}{\pi r^2}\right) = 0$$

$$2r = \frac{21.65628}{\pi r^2}$$

$$r^3 = 3.44670$$

$$r = 1.51 \text{ in.}$$

$$h = \frac{21.65628}{\pi r^2} = 3.02 \text{ in.}$$

(Note that $S$ is minimum when $h = 2r$.)

**43.** A wooden beam has a rectangular cross section of height $h$ and width $w$, as shown in the figure. The strength $S$ of the beam is directly proportional to the width and the square of the height. What are the dimensions of the strongest beam that can be cut from a round log of diameter 24 inches. (Hint: $S = kh^2w$, where $k$ is the proportionality constant.)

**Solution:**

Letting $S$ be the strength and $k$ the constant of proportionality, we have $S = kwh^2$. Since $w^2 + h^2 = 24^2$, we have $h^2 = 24^2 - w^2$.

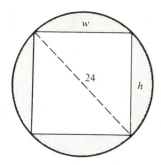

$$S = kw(24^2 - w^2) = k(576w - w^3)$$

Differentiating and solving $dS/dw = 0$, we obtain

$$\frac{dS}{dw} = k(576 - 3w^2) = 0$$

$$3w^2 = 576$$

$$w^2 = 192$$

$$w = \pm 8\sqrt{3}$$

Since $w$ is positive, we conclude that $w = 8\sqrt{3}$ inches and $h = \sqrt{24^2 - (8\sqrt{3})^2} = 8\sqrt{6}$ inches will produce the strongest beam.

**45.** A man is in a boat 2 miles from the nearest point on the coast. He is to go to a point $Q$, 3 miles down the coast and 1 mile inland. If he can row at 2 mi/h and walk at 4 mi/h, toward what point on the coast should he row in order to reach point $q$ in the least time?

**Solution:**

From the accompanying figure, we have

$$S = \text{distance on water} = \sqrt{x^2 + 4}$$

$$L = \text{distance on land} = \sqrt{1 + (3 - x)^2} = \sqrt{x^2 - 6x + 10}$$

$$\text{Total time} = (\text{time on water}) + (\text{time on land})$$

$$T = \frac{S}{(\text{rate on water})} + \frac{L}{(\text{rate on land})}$$

$$= \frac{\sqrt{x^2 + 4}}{2} + \frac{\sqrt{x^2 - 6x + 10}}{4}$$

To minimize $T$, we first find $dT/dx$ as follows:

$$\frac{dT}{dx} = \left(\frac{1}{2}\right)\left(\frac{1}{2}\right)(x^2 + 4)^{-1/2}(2x)$$

$$+ \left(\frac{1}{4}\right)\left(\frac{1}{2}\right)(x^2 - 6x + 10)^{-1/2}(2x - 6)$$

$$= \frac{x}{2\sqrt{x^2 + 4}} + \frac{x - 3}{4\sqrt{x^2 - 6x + 10}}$$

We next solve the equation $dT/dx = 0$ to determine the critical numbers.

$$\frac{x}{2\sqrt{x^2 + 4}} = \frac{3 - x}{4\sqrt{x^2 - 6x + 10}}$$

$$\frac{x}{\sqrt{x^2 + 4}} = \frac{3 - x}{2\sqrt{x^2 - 6x + 10}}$$

$$\frac{x^2}{x^2 + 4} = \frac{9 - 6x + x^2}{4(x^2 - 6x + 10)}$$

$$x^2(4)(x^2 - 6x + 10) = (x^2 + 4)(9 - 6x + x^2)$$

$$4x^4 - 24x^3 + 40x^2 = 9x^2 + 36 - 6x^3 - 24x$$
$$+ x^4 + 4x^2$$

$$3x^4 - 18x^3 + 27x^2 + 24x - 36 = 0$$

$$x^4 - 6x^3 + 9x^2 + 8x - 12 = 0$$

This fourth-degree equation has only one positive real root, $x = 1$. Thus the man should row to a point 1 mile from the nearest point on the coast.

## 4.8   Newton's Method

5. Approximate the zero of $f(x) = x^3 + x - 1$ in the interval $[0, 1]$. Use Newton's Method and continue the process until you are correct to three decimal places.

**Solution:**

$$f(x) = x^3 + x - 1 \qquad f'(x) = 3x^2 + 1$$

| $n$ | $x_n$ | $f(x_n)$ | $f'(x_n)$ | $f(x_n)/f'(x_n)$ | $x_n - [f(x_n)/f'(x_n)]$ |
|---|---|---|---|---|---|
| 1 | 0.5000 | $-0.3750$ | 1.7500 | $-0.2143$ | 0.7143 |
| 2 | 0.7143 | 0.0787 | 2.5306 | 0.0311 | 0.6832 |
| 3 | 0.6832 | 0.0021 | 2.4002 | 0.0009 | 0.6823 |
| 4 | 0.6823 | 0.0000 | 2.3967 | 0.0000 | 0.6823 |

Therefore, we approximate the root to be $x = 0.682$.

7. Approximate the zero of $f(x) = 3\sqrt{x-1} - x$ in the interval $[1, 2]$. Use Newton's Method and continue the process until you are correct to three decimal places.

**Solution:**

$$f(x) = 3\sqrt{x-1} - x \qquad f'(x) = \frac{3}{2\sqrt{x-1}} - 1$$

| $n$ | $x_n$ | $f(x_n)$ | $f'(x_n)$ | $f(x_n)/f'(x_n)$ | $x_n - [f(x_n)/f'(x_n)]$ |
|---|---|---|---|---|---|
| 1 | 1.2000 | 0.1416 | 2.3541 | 0.0602 | 1.1398 |
| 2 | 1.1398 | $-0.0180$ | 3.0113 | $-0.0060$ | 1.1458 |
| 3 | 1.1458 | $-0.0003$ | 2.9283 | $-0.0001$ | 1.1459 |

Therefore, we approximate the zero to be $x = 1.146$.

**19.** Use Newton's Method to obtain a general formula for approximating $\sqrt{a}$. [Hint: Apply Newton's Method to the function $f(x) = x^2 - a$.]

**Solution:**

Let $f(x) = x^2 - a$. Then $f'(x) = 2x$. Since $\sqrt{a}$ is a zero of $f(x) = 0$, we can use Newton's Method to approximate $\sqrt{a}$ as follows:

$$x_{i+1} = x_i - \frac{x_i^2 - a}{2x_i} = \frac{x_i^2 + a}{2x_i}$$

For example, if $a = 2$, and $x_1 = 1$, then we can approximate $\sqrt{2}$ as follows:

$$x_1 = 1$$

$$x_2 = \frac{1^2 + 2}{2(1)} = \frac{1+2}{2} = 1.50000$$

$$x_3 = \frac{(1.5)^2 + 2}{2(1.5)} = \frac{4.25}{3} = 1.41667$$

$$x_4 = \frac{(1.41667)^2 + 2}{2(1.41667)} = \frac{4.00697}{2.8333} = 1.41421$$

(Note: To five decimal places, $\sqrt{2} = 1.41421$.)

## 4.9   Differentials

**9.** Find the differential $dy$ of $y = x\sqrt{1 - x^2}$.

**Solution:**

$$y = x\sqrt{1 - x^2}$$

$$dy = [x(\tfrac{1}{2})(1 - x^2)^{-1/2}(-2x) + (1 - x^2)^{1/2}(1)]dx$$

$$= \left[\frac{-x^2}{\sqrt{1 - x^2}} + \frac{1 - x^2}{\sqrt{1 - x^2}}\right]dx = \frac{1 - 2x^2}{\sqrt{1 - x^2}}dx$$

**23.** The radius of a sphere is claimed to be 6 inches, with a possible error of 0.02 inches. Using differentials, approximate the maximum possible error in calculating the following:

(a) the volume of the sphere

(b) the surface area of the sphere

(c) What is the relative error in parts (a) and (b)?

**Solution:**

The radius of the sphere is given as $r = 6 \pm 0.02$.

(a) The volume of the sphere is given by

$$V = \frac{4}{3}\pi r^3$$

To approximate $\Delta V$ by $dV$, we let $r = 6$ and $dr = \pm 0.02$

$$dV = \frac{4}{3}\pi(3r^2)dr = 4\pi(36)(\pm 0.02) = \pm 2.88\pi \text{ in}^3$$

(b) The surface area of the sphere is given by

$$S = 4\pi r^2$$

To approximate $\Delta S$ by $dS$, we let $r = 6$ and $dr = \pm 0.02$.

$$dS = 8\pi r \, dr = 8\pi(6)(\pm 0.02) = \pm 0.96\pi \text{ in}^2$$

(c) For part (a) the relative error is approximately

$$\frac{dV}{V} = \frac{2.88\pi}{(\frac{4}{3})\pi(6^3)} = \frac{2.88}{288} = 0.01 = 1\%$$

For (b) the relative error is approximately

$$\frac{dS}{S} = \frac{0.96\pi}{4\pi(6^2)} = \frac{0.96}{144} = 0.0067 = \frac{2}{3}\%$$

**25.** The period of a pendulum is given by $T = 2\pi\sqrt{L/g}$, where $L$ is the length of the pendulum in feet, $g$ is the acceleration due to gravity, and $T$ is time in seconds. Suppose that the pendulum has been subjected to an increase in temperature so that the length increases by $\frac{1}{2}\%$.

(a) What is the approximate percentage change in the period?

(b) Using the result of part (a), find the approximate error in this pendulum clock in one day.

**Solution:**

(a)
$$T = 2\pi\sqrt{\frac{L}{g}}$$

$$dT = 2\pi\left(\frac{1}{2}\right)\left(\frac{L}{g}\right)^{-1/2}\left(\frac{1}{g}\right)dL = \frac{\pi}{g\sqrt{L/g}}dL$$

$$\frac{dT}{T}(100) = \text{percentage error}$$

$$= \frac{(\pi/g\sqrt{L/g})dL}{2\pi\sqrt{L/g}}(100) = \frac{1}{2}\left(\frac{dL}{L}100\right)$$

$$= \frac{1}{2}(\text{percentage change in } L) = \frac{1}{2}\left(\frac{1}{2}\right) = \frac{1}{4}\%$$

(b) approximate error $= \left(\frac{1}{4}\%\right)$ (number of seconds per day)

$$= (0.0025)(60)(60)(24)$$

$$= 216 \quad \text{seconds} = 3.6 \quad \text{minutes}$$

## 4.10   Business and Economics Applications

**9.** Find the price per unit $p$ that produces the maximum profit $P$ if the demand function is $p = 90 - x$ and the cost fucntion is $C = 100 + 30x$ where $x$ is the number of units sold.

**Solution:**

The profit is given by

$$P = (\text{price per unit})(\text{number of units}) - (\text{cost})$$

$$= px - C$$

$$= (90 - x)(x) - (100 + 30x) = -x^2 + 60x - 100$$

To maximize $P$, we solve $dP/dx = 0$ and solve as follows:

$$\frac{dP}{dx} = -2x + 60 = 0$$

$$2x = 60 \quad \text{and} \quad x = 30$$

Therefore, the profit is maximum when the price is

$$p = 90 - 30 = 60.$$

**15.** A manufacturer of radios charges $90 per unit when the average production cost per unit is $60. To encourage large order from distributors, the manufacturer will reduce the charge by $0.10 per unit for each unit ordered in excess of 100 (for example, there would be a charge of $88 per radio for an order size of 120). Find the largest order size the manufacturer should allow so as to realize maximum profit.

**Solution:**

Let $x =$ the number of units purchased, $p =$ the price per unit, and $P =$ the total profit. Then

$$p = 90 - (0.10)(x - 100) = 90 - 0.1x + 10 = 100 - 0.1x$$

Since each radio costs the manufacturer $60, the profit per radio is $p - 60$ and the total profit is

$$P = x(p - 60) = x(100 - 0.1x - 60)$$
$$= x(40 - 0.1x) = 40x - 0.1x^2$$

To maximize $P$, we solve $dP/dx = 0$ as follows:

$$\frac{dP}{dx} = 40 - 0.2x = 0$$
$$0.2x = 40$$
$$x = \frac{40}{0.2} = 200$$

Therefore, the manufacturer should not continue the discount for over 200 units.

**19.** Assume that the amount of money deposited in a bank is proportional to the square of the interest rate the bank pays on this money. Furthermore, the bank can reinvest this money at 12%. Find the interest rate the bank should pay to maximize profit. (Use the simple interest formula.)

**Solution:**

Let
$$d = \text{amount in the bank}$$
$$i = \text{interest rate paid by the bank}$$
$$p = \text{profit}$$

The bank can take the deposited money $d$ and reinvest to obtain 12% or $(0.12)d$. Since the bank pays out interest to its depositors, its profit is

$$P = (0.12)d - id$$

Finally, since $d$ is proportional to the square of $i$, we have

$$d = ki^2$$

Thus

$$P = (0.12)(ki^2) - i(ki^2) = k[(0.12)i^2 - i^3]$$

To maximize $P$, we solve $dP/di = 0$ as follows:

$$\frac{dP}{di} = k(0.24i - 3i^2) = 0$$

$$ki(0.24 - 3i) = 0$$

(We disregard the critical number $i = 0$.)

$$i = \frac{0.24}{3} = 0.08$$

Thus the bank can maximize its profit by setting $i = 8\%$.

25. The ordering and transportation cost $C$ of the components used in manufacturing a certain product is given by

$$C = 100\left(\frac{200}{x^2} + \frac{x}{x+30}\right), \quad 1 \le x$$

where $C$ is measured in thousands of dollars and x is the order size in hundreds.  Find the order size that minimizes cost.

**Solution:**

$$C = 100\left(\frac{200}{x^2} + \frac{x}{x+30}\right)$$

$$\frac{dC}{dx} = 100\left(-\frac{400}{x^3} + \frac{30}{(x+30)^2}\right)$$

$$= 1000\left[\frac{-40(x+30)^2 + 3x^3}{x^3(x+30)^2}\right]$$

To find the critical number of $C$ we use Newton's Method to find the zero of the function $f(x) = -40(x + 30)^2 + 3x^3$ using $x_1 = 30$ as our first estimate.

$$f(x) = 3x^3 - 40x^2 - 2400x - 36000 \quad \text{and} \quad f'(x) = 9x^2 - 80x - 2400$$

| $n$ | $x_n$ | $f(x_n)$ | $f'(x_n)$ | $f(x_n)/f'(x_n)$ | $x_n - [f(x_n)/f'(x_n)]$ |
|-----|-------|----------|-----------|------------------|--------------------------|
| 1 | 30 | $-63000$ | 3300 | $-19.091$ | 42.091 |
| 2 | 42.091 | 104,700 | 15,362 | 6.816 | 40.556 |
| 3 | 40.556 | 991.398 | 9158.562 | 0.108 | 40.448 |
| 4 | 40.448 | 3.773 | 9088.305 | 0.0004 | 40.447 |

Therefore, we conclude that the critical number is $x \approx 40.4$ and the minimum cost occurs when 40 units are ordered.

## Review Exercises for Chapter 4

**3.** Make use of domain, range, symmetry, asymptotes, intercepts, relative extrema, or points of inflection to obtain an accurate graph of $f(x) = x\sqrt{16 - x^2}$.

**Solution:**

The domain of $f$ is all real numbers in the interval $[-4, 4]$ and the graph of $f$ is symmetric to the origin since

$$f(-x) = (-x)\sqrt{16 - (-x)^2} = -x\sqrt{16 - x^2} = -f(x)$$

$$f'(x) = x\left(\frac{1}{2}\right)(16 - x^2)^{-1/2}(-2x) + (16 - x^2)^{1/2}$$

$$= \frac{16 - 2x^2}{\sqrt{16 - x^2}}$$

$$f''(x) = \frac{\sqrt{16 - x^2}(-4x) - (16 - 2x^2)(\frac{1}{2})(16 - x^2)^{-1/2}(-2x)}{16 - x^2}$$

$$= \frac{2x(x^2 - 24)}{(16 - x^2)^{3/2}}$$

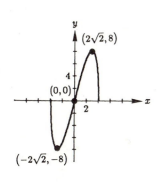

Thus, $f'(x) = 0$ when $x = \pm 2\sqrt{2}$ and undefined when $x = \pm 4$. Since $f''(2\sqrt{2}) < 0$, the graph is concave downward and $(2\sqrt{2}, 8)$ is a maximum. By symmetry, $(-2\sqrt{2}, -8)$ is a minimum. There is a point of inflection at $(0, 0)$.

**13.** Make use of domain, range, symmetry, asymptotes, intercepts, relative extrema, or points of inflection to obtain an accurate graph of $f(x) = x^{1/3}(x+3)^{2/3}$.

**Solution:**

$$f(x) = x^{1/3}(x+3)^{2/3} \qquad \text{Intercepts: } (0,0), (-3,0)$$

$$f'(x) = x^{1/3}(\tfrac{2}{3})(x+3)^{-1/3} + (x+3)^{2/3}(\tfrac{1}{3})x^{-2/3}$$

$$= \frac{2x^{1/3}}{3(x+3)^{1/3}} + \frac{(x+3)^{2/3}}{3x^{2/3}}$$

$$= \frac{2x + x + 3}{3x^{2/3}(x+3)^{1/3}}$$

$$= \frac{x+1}{x^{2/3}(x+3)^{1/3}} = \frac{x+1}{(x^3+3x^2)^{1/3}}$$

Critical numbers: $x = -1$, $x = 0$, $x = -3$

$$f''(x) = \frac{(x^3+3x^2)^{1/3}(1) - (x+1)(\tfrac{1}{3})(x^3+3x^2)^{-2/3}(3x^2+6x)}{(x^3+3x^2)^{2/3}}$$

$$= \frac{(x^3+3x^2) - (x+1)(x^2+2x)}{(x^3+3x^2)^{4/3}}$$

$$= \frac{x^3+3x^2 - x^3 - 2x^2 - x^2 - 2x}{(x+3x^4)^{2/3}}$$

$$= \frac{-2x}{(x^3+3x^2)^{4/3}} = \frac{-2}{x^{5/3}(x+3)^{4/3}}$$

Possible point of inflection: $(0,0)$

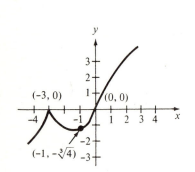

| $x$ | $f(x)$ | $f'(x)$ | $f''(x)$ | *Shape of graph* |
|---|---|---|---|---|
| $x$ in $(-\infty, -3)$ | | $+$ | $+$ | increasing, concave up |
| $x = -3$ | $0$ | undefined | undefined | relative maximum |
| $x$ in $(-3, -1)$ | | $-$ | $+$ | decreasing, concave up |
| $x = -1$ | $-\sqrt[3]{4}$ | $0$ | $+$ | relative minimum |
| $x$ in $(-1, 0)$ | | $+$ | $+$ | increasing, concave up |
| $x = 0$ | $0$ | undefined | undefined | point of inflection |
| $x$ in $(0, \infty)$ | | $+$ | $-$ | increasing, concave down |

**35.** For the function $f(x) = Ax^2 + Bx + C$, determine the value of $c$ guaranteed by the Mean Value Theorem on the interval $[x_1, x_2]$.

**Solution:**

Given the function $f(x) = Ax^2 + Bx + C$, we have

$$f'(x) = 2Ax + B$$
$$f'(c) = 2Ac + B = \frac{f(x_2) - f(x_1)}{x_2 - x_1}$$
$$= \frac{Ax_2^2 + Bx_2 + C - Ax_1^2 - Bx_1 - C}{x_2 - x_1}$$
$$= \frac{A(x_2^2 - x_1^2) + B(x_2 - x_1)}{x_2 - x_1}$$
$$= A(x_2 + x_1) + B$$
$$2c = x_2 + x_1$$
$$c = \frac{x_1 + x_2}{2}$$

**39.** Find the maximum profit if the demand equation is $p = 36 - 4x$ and the total cost is $C = 2x^2 + 6$.

**Solution:**

The profit is given by

$$P = (\text{price per unit})(\text{number of units}) - (\text{cost}) = px - C$$
$$= (36 - 4x)(x) - (2x^2 + 6) = 36x - 4x^2 - 2x^2 - 6$$
$$= -6x^2 + 36x - 6$$

To maximize $P$, we solve $dP/dx = 0$ as follows:

$$\frac{dP}{dx} = -12x + 36 = 0$$
$$12x = 36$$
$$x = 3 \text{ units}$$

Thus the maximum profit is

$$P = -6(3)^2 + 36(3) - 6 = -54 + 108 - 6 = \$48$$

**47.** Find the length of the longest pipe that can be carried level around a right-angle corner if the two intersecting corridors are of widths 4 ft and 6 ft.

**Solution:**

The longest pipe that will go around the corner will have a length equal to the minimum length of the hypotenuse [through the point $(4,6)$] of the triangle whose vertices are $(0,0), (x,0)$, and $(0,y)$. We begin by relating $x$ and $y$ as follows:

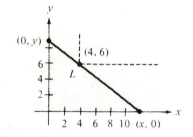

$$m = \frac{y-6}{0-4} = \frac{6-0}{4-x}$$

$$y - 6 = \frac{-24}{4-x}$$

$$y = \frac{24}{x-4} + 6 = \frac{6x}{x-4}$$

[Note that $dy/dx = -24/(x-4)^2$.] Now the length of the hypotenuse is given by

$$L = \sqrt{x^2 + y^2}$$

To minimize $L$, we solve $dL/dx = 0$ as follows.

$$\frac{dL}{dx} = \frac{(\frac{1}{2})[2x + (2y)(dy/dx)]}{\sqrt{x^2 + y^2}} = 0$$

$$x = -y\frac{dy}{dx} = -\left(\frac{6x}{x-4}\right)\left[\frac{-24}{(x-4)^2}\right]$$

$$x(x-4)^3 = 144x$$

$$(x-4)^3 = 144$$

$$x - 4 = \sqrt[3]{144}$$

$$x = \sqrt[3]{144} + 4$$

Therefore, the minimum length of $L$ and the maximum length of pipe are given by

$$L = \sqrt{x^2 + y^2} = \sqrt{x^2 + \frac{36x^2}{(x-4)^2}} = \frac{x}{x-4}\sqrt{(x-4)^2 + 36}$$

$$= \frac{\sqrt[3]{144} + 4}{\sqrt[3]{144}}\sqrt{144^{2/3} + 36} \approx 14.05 \text{ ft}$$

**49.** The cost of inventory depends on ordering cost and storage cost, according to the following inventory model:

$$C = \left(\frac{Q}{x}\right) + \left(\frac{x}{2}\right)r.$$

Determine the order size that will minimize the cost, assuming that sales occur at a constant rate, $Q$ is the number of units sold per year, $r$ is the cost of storing one unit for one year, $s$ is the cost of placing an order, and $x$ is the number of units per order.

**Solution:**

To minimize $C$, we solve $dC/dx = 0$ as follows:

$$C = \left(\frac{Q}{x}\right) + \left(\frac{x}{2}\right)r$$

$$\frac{dC}{dx} = -\frac{Qs}{x^2} + \frac{r}{2} = 0$$

Therefore,

$$\frac{Qs}{x^2} = \frac{r}{2}$$

$$x^2 = \frac{2Qs}{r}$$

$$x = \pm\sqrt{\frac{2Qs}{r}}.$$

Therefore, the inventory cost is minimum when $x = \sqrt{\frac{2Qs}{r}}$.

# 5 INTEGRATION

## 5.1  Antiderivatives and Indefinite Integration

**3.** Complete the following table for the indefinite integral

$$\int \frac{1}{x\sqrt{x}}\,dx.$$

| *Given* | *Rewrite* | *Integrate* | *Simplify* |
|---|---|---|---|

$$\int \frac{1}{x\sqrt{x}}\,dx$$

**Solution:**

| *Given* | *Rewrite* | *Integrate* | *Simplify* |
|---|---|---|---|
| $\int \dfrac{1}{x\sqrt{x}}\,dx$ | $\int x^{-3/2}\,dx$ | $\dfrac{x^{-1/2}}{-\frac{1}{2}} + C$ | $\dfrac{-2}{\sqrt{x}} + C$ |

**9.** Evaluate the indefinite integral $\int (x^{3/2} + 2x + 1)\,dx$ and check your results by differentiation.

**Solution:**

$$\int (x^{3/2} + 2x + 1)\,dx = \frac{x^{5/2}}{\frac{5}{2}} + 2\left(\frac{x^2}{2}\right) + x + C$$

$$= \frac{2x^{5/2}}{5} + x^2 + x + C$$

**Check**

If $y = \dfrac{2x^{5/2}}{5} + x^2 + x + C$, then

$$\frac{dy}{dx} = \left(\frac{2}{5}\right)\left(\frac{5}{2}\right)x^{3/2} + 2x + 1 + 0 = x^{3/2} + 2x + 1$$

**21.** Evaluate the indefinite integral $\displaystyle\int \frac{t^2 + 2}{t^2}\, dt$ and check your results by differentiation.

**Solution:**

$$\int \frac{t^2 + 2}{t^2}\, dt = \int \left(\frac{t^2}{t^2} + \frac{2}{t^2}\right) dt = \int (1 + 2t^{-2})\, dt$$

$$= t + \frac{2(t^{-1})}{-1} + C = t - \frac{2}{t} + C$$

**Check**

If $y = t - \dfrac{2}{t} + C = t - 2t^{-1} + C$, then

$$\frac{dy}{dt} = 1 - 2(-1)t^{-2} + 0 = 1 + \frac{2}{t^2} = \frac{t^2 + 2}{t^2}$$

**27.** Find the equation of the curve such that $dy/dx = 2x - 1$ and the curve passes through the point $(1, 1)$.

**Solution:**

$$\frac{dy}{dx} = 2x - 1$$

$$y = \int (2x - 1)\, dx = x^2 - x + C$$

Since the curve passes through the point $(1, 1)$, we have

$$y = x^2 - x + C$$
$$1 = (1)^2 - (1) + C \qquad \text{or} \qquad C = 1$$

Therefore the required equation is $y = x^2 - x + 1$.

**33.** Find $y = f(x)$ if $f''(x) = x^{-3/2}$, $f'(4) = 2$, and $f(0) = 0$.

**Solution:**

$$f''(x) = x^{-3/2}$$
$$f'(x) = \int x^{-3/2}\,dx$$
$$= \frac{x^{-1/2}}{-\frac{1}{2}} + C_1 = \frac{-2}{\sqrt{x}} + C_1$$
$$f'(4) = \frac{-2}{\sqrt{4}} + C_1 = 2 \quad \Longrightarrow \quad C_1 = 3$$
$$f'(x) = -2x^{-1/2} + 3$$
$$f(x) = \int (-2x^{-1/2} + 3)\,dx = \frac{(-2)x^{1/2}}{\frac{1}{2}} + 3x + C_2$$
$$= -4x^{1/2} + 3x + C_2$$
$$f(0) = -4(0)^{1/2} + 3(0) + C_2 = 0 \quad \Longrightarrow \quad C_2 = 0$$

Therefore
$$f(x) = -4x^{1/2} + 3x = -4\sqrt{x} + 3x$$

**37.** With what initial velocity must an object be thrown upward from ground level to reach a maximum height of 550 ft (approximate height of the Washington Monument)?

**Solution:**

If $s = f(t)$, we know that

$$f''(t) = -32$$

($-32$ ft/sec$^2$ is the acceleration due to gravity.)

$$f'(t) = \int -32\,dt = -32t + C_1 = -32t + v_0$$

where $v_0$ is the initial velocity. Furthermore,

$$f(t) = \int (-32t + v_0)\,dt$$
$$= -16t^2 + v_0 t + C_2 = -16t^2 + v_0 t + s_0$$

where $s_0 = 0$ is the initial height. Thus,

$$s = f(t) = -16t^2 + v_0 t$$

Now since $s$ is a maximum when $f'(t) = 0$, we have

$$-32t + v_0 = 0 \qquad \text{or} \qquad t = \frac{v_0}{32}$$

Finally, in order for $s$ to attain a height of 550 ft, we must have

$$s = -16\left(\frac{v_0}{32}\right)^2 + v_0\left(\frac{v_0}{32}\right) = 550$$

$$\frac{v_0{}^2}{64} = 550$$

$$v_0{}^2 = 35,200$$

$$v_0 = \sqrt{35,200}$$

$$= 40\sqrt{22} \approx 187.617 \text{ ft/sec}$$

**43.** At the instant the traffic light turns green, an automobile that has been waiting at an intersection starts ahead with a constant acceleration of 6 ft/sec$^2$. At the same instant a truck traveling with a constant velocity of 30 ft/sec overtakes and passes the car.

(a) How far beyond its starting point will the automobile overtake the truck?

(b) How fast will it be traveling?

**Solution:**

Let $T(t)$ and $A(t)$ represent the position functions of the truck and auto. Then we know that

$$T'(t) = 30, \quad T(0) = 0$$

$$A''(t) = 6, \quad A'(0) = 0, \quad A(0) = 0$$

For the truck, we have

$$T(t) = \int 30\, dt = 30t + C_1$$

$$T(0) = 30(0) + C_1 = 0 \quad \Longrightarrow \quad C_1 = 0$$

For the auto, we have

$$A'(t) = \int 6\,dt = 6t + C_2$$

$$A'(0) = 6(0) + C_2 = 0 \implies C_2 = 0$$

$$A'(t) = 6t$$

$$A(t) = \int 6t\,dt = 3t^2 + C_3$$

$$A(0) = 3(0)^2 + C_3 = 0 \implies C_3 = 0$$

$$A(t) = 3t^2$$

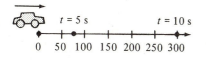

Therefore, when the auto catches up with the truck, we have

$$A(t) = T(t)$$
$$3t^2 = 30t$$
$$3t^2 - 30t = 0$$
$$3t(t - 10) = 0$$
$$t = 10 \text{ sec} \qquad \text{We disregard } t = 0.$$

(a)   When $t = 10$ sec, the auto will have traveled

$$A(10) = 3(10^2) = 300 \text{ ft.}$$

(b)   It will be traveling

$$A'(10) = 60 \text{ ft/sec} = \frac{60(3600)}{5280} \approx 41 \text{ mi/hr.}$$

## 5.2   Area

**3.** Find the sum

$$\sum_{k=0}^{4} \frac{1}{1 + k^2}$$

**Solution:**

$$\sum_{k=0}^{4} \frac{1}{1+k^2} = \frac{1}{1+0^2} + \frac{1}{1+1^2} + \frac{1}{1+2^2} + \frac{1}{1+3^2} + \frac{1}{1+4^2}$$

$$= \frac{1}{1} + \frac{1}{2} + \frac{1}{5} + \frac{1}{10} + \frac{1}{17}$$

$$= \frac{170 + 85 + 34 + 17 + 10}{170}$$

$$= \frac{316}{170} = \frac{158}{85}$$

**15.** Write the sum

$$\left[ \left( \frac{2}{n} \right)^3 - \frac{2}{n} \right] \left( \frac{2}{n} \right) + \left[ \left( \frac{4}{n} \right)^3 - \frac{4}{n} \right] \left( \frac{2}{n} \right) + \cdots + \left[ \left( \frac{2n}{n} \right)^3 - \frac{2n}{n} \right] \left( \frac{2}{n} \right)$$

in sigma notation.

**Solution:**

We begin by noting that the $n$ terms in this sum are each of the form

$$f(i) = \left[ \left( \frac{2i}{n} \right)^3 - \frac{2i}{n} \right] \left( \frac{2}{n} \right)$$

Furthermore, we observe that in the first term $i = 1$, in the second term $i = 2$ and so on until we reach the $n$th term. Thus our index $i$ runs from 1 to $n$, and the sigma notation for the given sum is

$$\sum_{i=1}^{n} f(i) = \sum_{i=1}^{n} \left[ \left( \frac{2i}{n} \right)^3 - \frac{2i}{n} \right] \left( \frac{2}{n} \right)$$

$$= \frac{2}{n} \sum_{i=1}^{n} \left[ \left( \frac{2i}{n} \right)^3 - \frac{2i}{n} \right].$$

**23.** Use the properties of sigms notation and the expressions for the sums of powers of the first $n$ positive integers to evaluate the sum

$$\sum_{i=1}^{15} \frac{1}{n^3}(i-1)^2$$

**Solution:**

$$\sum_{i=1}^{15} \frac{1}{n^3}(i-1)^2 = \frac{1}{n^3}\sum_{i=1}^{15}(i^2 - 2i + 1)$$

$$= \frac{1}{n^3}\left[\sum_{i=1}^{15} i^2 - 2\sum_{i=1}^{15} i + \sum_{i=1}^{15} 1\right]$$

$$= \frac{1}{n^3}\left[\frac{15(16)(31)}{6} - 2\frac{15(16)}{2} + 15\right]$$

$$= \frac{1}{n^3}(1240 - 240 + 15) = \frac{1015}{n^3}$$

**27.** Find the limit of the sequence $s(n)$ as $n \to \infty$ where

$$s(n) = \frac{81}{n^4}\left[\frac{n^2(n+1)^2}{4}\right]$$

**Solution:**

$$\lim_{n\to\infty} s(n) = \lim_{n\to\infty} \frac{81}{n^4}\left[\frac{n^2(n+1)^2}{4}\right]$$

$$= \lim_{n\to\infty} \frac{81}{4}\left(\frac{n^4 + 2n^3 + n^2}{n^4}\right)$$

$$= \lim_{n\to\infty} \frac{81}{4}\left(1 + \frac{2}{n} + \frac{1}{n^2}\right) = \frac{81}{4}$$

**31.** Find

$$\lim_{n \to \infty} \sum_{i=1}^{n} \frac{1}{n^3}(i-1)^2$$

**Solution:**

$$\sum_{i=1}^{n} \frac{1}{n^3}(i-1)^2 = \frac{1}{n^3} \sum_{i=1}^{n}(i^2 - 2i + 1)$$

$$= \frac{1}{n^3}\left[\sum_{i=1}^{n} i^2 - 2\sum_{i=1}^{n} i + \sum_{i=1}^{n} 1\right]$$

$$= \frac{1}{n^3}\left[\frac{n(n+1)(2n+1)}{6} - 2\frac{n(n+1)}{2} + n\right]$$

$$= \frac{1}{n^3}\left(\frac{n^3}{3} - \frac{n^2}{2} + \frac{n}{6}\right)$$

$$= \frac{1}{3} - \frac{1}{2n} + \frac{1}{6n^2}$$

Therefore $\displaystyle\lim_{n \to \infty} \sum_{i=1}^{n} \frac{1}{n^3}(i-1)^2 = \lim_{n \to \infty}\left(\frac{1}{3} - \frac{1}{2n} + \frac{1}{6n^2}\right) = \frac{1}{3}$

**41.** Use the upper and lower sums to approximate the area of the region between the graph of $y = \sqrt{1 - x^2}$ and the $x$-axis over the interval $[0, 1]$. Use five subdivisions.

**Solution:**

Dividing the interval into five parts, we have

$$x_0 = 0, \quad x_1 = 0.2, \quad x_2 = 0.4, \quad x_3 = 0.6,$$
$$x_4 = 0.8, \quad x_5 = 1$$

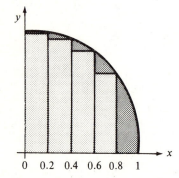

Since $y$ is decreasing from 0 to 1, the lower sum is obtained by using the *right* endpoints of the five subintervals. Thus

$$s = 0.2\sqrt{1 - (0.2)^2} + 0.2\sqrt{1 - (0.4)^2} + 0.2\sqrt{1 - (0.6)^2}$$
$$+ 0.2\sqrt{1 - (0.8)^2} + 0.2\sqrt{1 - (1)^2}$$
$$= 0.2(\sqrt{0.96} + \sqrt{0.84} + \sqrt{0.64} + \sqrt{0.36})$$
$$\approx 0.2(0.9798 + 0.9165 + 0.8 + 0.6) \approx 0.659.$$

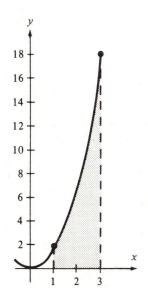

Similarly, the upper sum is obtained by using the *left* endpoints of the five subdivisions. Thus

$$S = 0.2\sqrt{1 - 0^2} + 0.2\sqrt{1 - (0.2)^2} + 0.2\sqrt{1 - (0.4)^2}$$
$$+ 0.2\sqrt{1 - (0.6)^2} + 0.2\sqrt{1 - (0.8)^2}$$
$$= 0.2(1 + \sqrt{0.96} + \sqrt{0.84} + \sqrt{0.64} + \sqrt{0.36})$$
$$\approx 0.2(1 + 0.9798 + 0.9165 + 0.8 + 0.6) \approx 0.859.$$

**49.** Use the limit process to find the area of the region between the graph of $y = 2x^2$ and the $x$-axis over the interval $[1, 3]$. Sketch the region.

**Solution:**

Let $\Delta x = (3 - 1)/n = 2/n$. Choosing right endpoints, we have $c_i = 1 + i(2/n)$. Therefore,

$$S(n) = \sum_{i=1}^{n} f\left(1 + \frac{2i}{n}\right)\left(\frac{2}{n}\right) = \sum_{i=1}^{n} 2\left(1 + \frac{2i}{n}\right)^2\left(\frac{2}{n}\right)$$

$$= \frac{4}{n}\left[\sum_{i=1}^{n}\left(1 + \frac{4i}{n} + \frac{4i^2}{n^2}\right)\right]$$

$$= \frac{4}{n}\left[\sum_{i=1}^{n} 1 + \frac{4}{n}\sum_{i=1}^{n} i + \frac{4}{n^2}\sum_{i=1}^{n} i^2\right]$$

$$= \frac{4}{n}\left[n + \frac{4}{n}\cdot\frac{n(n+1)}{2} + \frac{4}{n^2}\cdot\frac{n(n+1)(2n+1)}{6}\right]$$

$$= \frac{4n}{n} + \frac{8n^2 + 8n}{n^2} + \frac{16n^3 + 24n^2 + 8n}{3n^3}$$

$$= 4 + 8 + \frac{8}{n} + \frac{16}{3} + \frac{8}{n} + \frac{8}{3n^2} = \frac{52}{3} + \frac{16}{n} + \frac{8}{3n^2}$$

Finally, we have

$$\text{area} = \lim_{n\to\infty} S(n) = \lim_{n\to\infty}\left(\frac{52}{3} + \frac{16}{n} + \frac{8}{3n^2}\right) = \frac{52}{3}$$

**53.** Use the limit process to find the area of the region between the graph of $y = x^2 - x^3$ and the $x$-axis over the interval $[-1, 1]$. Sketch the region.

**Solution:**

Let $\Delta x = [1 - (-1)]/n = 2/n$. Choosing right endpoints, we have

$$c_i = -1 + i\left(\frac{2}{n}\right) = -1 + \frac{2i}{n}$$

Therefore,

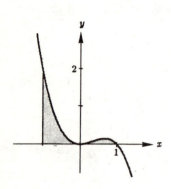

$$\text{area} = \lim_{n \to \infty} \sum_{i=1}^{n} f\left(-1 + \frac{2i}{1}\right)\left(\frac{2}{n}\right)$$

$$= \lim_{n \to \infty} \frac{2}{n} \sum_{i=1}^{n} \left[\left(-1 + \frac{2i}{n}\right)^2 - \left(-1 + \frac{2i}{n}\right)^3\right]$$

$$= \lim_{n \to \infty} \frac{2}{n} \sum_{i=1}^{n} \left[2 - \frac{10i}{n} + \frac{16i^2}{n^2} - \frac{8i^3}{n^3}\right]$$

$$= \lim_{n \to \infty} \left[\frac{2}{n} \sum_{i=1}^{n} 2 - \frac{20}{n^2} \sum_{i=1}^{n} i + \frac{32}{n^3} \sum_{i=1}^{n} i^2 - \frac{16}{n^4} \sum_{i=1}^{n} i^3\right]$$

$$= \lim_{n \to \infty} \left[\frac{2}{n}(2n) - \left(\frac{20}{n^2}\right)\frac{n(n+1)}{2}\right.$$
$$\left. + \left(\frac{32}{n^3}\right)\frac{n(n+1)(2n+1)}{6} - \left(\frac{16}{n^4}\right)\frac{n^2(n+1)^2}{4}\right]$$

$$= \lim_{n \to \infty} \left(4 - 10 - \frac{10}{n} + \frac{32}{3} + \frac{16}{n} + \frac{16}{3n^2} - 4 - \frac{8}{n} - \frac{4}{n^2}\right)$$

$$= 4 - 10 + \frac{32}{3} - 4 = \frac{2}{3}$$

## 5.3    Riemann Sums and the Definite Integral

**15.** Sketch the region whose area is indicated by $\int_0^2 (2x + 5)\, dx$. Then use a geometric formula to evaluate the integral.

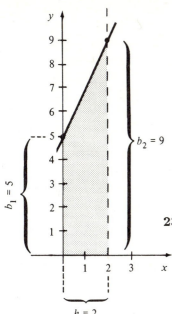

$b_2 = 9$

$b_1 = 5$

$h = 2$

**Solution:**

The region whose area is given by $\int_0^2 (2x+5)\, dx$ is shown by the accompanying figure to be a trapezoid. Since the height of the trapezoid is $h = 2$ and the lengths of the two bases are $b_1 = 5$ and $b_2 = 9$, the area of the trapezoid is

$$A = h\left[\frac{b_1 + b_2}{2}\right] = 2\left[\frac{5+9}{2}\right] = 14$$

**23.** If $\int_2^6 f(x)\, dx = 10$ and $\int_2^6 g(x)\, dx = -2$, find

(a) $\displaystyle\int_2^6 [f(x) + g(x)]\, dx$     (b) $\displaystyle\int_2^6 [g(x) - f(x)]\, dx$

(c) $\displaystyle\int_2^6 2g(x)\, dx$     (d) $\displaystyle\int_2^6 3f(x)\, dx$

**Solution:**

(a) $\displaystyle\int_2^6 [f(x) + g(x)]\, dx = \int_2^6 f(x)\, dx + \int_2^6 g(x)\, dx$

$$= 10 + (-2) = 8$$

(b) $\displaystyle\int_2^6 [g(x) - f(x)]\, dx = \int_2^6 g(x)\, dx - \int_2^6 f(x)\, dx$

$$= -2 - 10 = -12$$

(c) $\displaystyle\int_2^6 2g(x)\, dx = 2\int_2^6 g(x)\, dx = 2(-2) = -4$

(d) $\displaystyle\int_2^6 3f(x)\, dx = 3\int_2^6 f(x)\, dx = 3(10) = 30$

**27.** Evaluate the definite integral $\int_{-1}^{1} x^3 \, dx$.

**Solution:**

Let $\Delta x = [1 - (-1)]/n = 2/n$. Using right-hand endpoints, we have $c_i = -1 + (2i/n)$, and the definite integral is given by the limit

$$\int_{-1}^{1} x^3 \, dx = \lim_{n \to \infty} \sum_{i=1}^{n} \left(-1 + \frac{2i}{n}\right)^3 \left(\frac{2}{n}\right)$$

$$= \lim_{n \to \infty} \sum_{i=1}^{n} \left(\frac{2}{n}\right) \left(-1 + \frac{6i}{n} - \frac{12i^2}{n^2} + \frac{8i^3}{n^3}\right)$$

$$= \lim_{n \to \infty} \left(\frac{2}{n}\right) \left[-\sum_{i=1}^{n} 1 + \frac{6}{n} \sum_{i=1}^{n} i - \frac{12}{n^2} \sum_{i=1}^{n} i^2 + \frac{8}{n^3} \sum_{i=1}^{n} i^3\right]$$

$$= \lim_{n \to \infty} \left(\frac{2}{n}\right) \left[-n + \frac{6n(n+1)}{2n} - \frac{12n(n+1)(2n+1)}{6n^2}\right.$$
$$\left. + \frac{8n^2(n+1)^2}{4n^3}\right]$$

$$= \lim_{n \to \infty} \left[\frac{-2n}{n} + \frac{6n(n+1)}{n^2} - \frac{4n(n+1)(2n+1)}{n^3}\right.$$
$$\left. + \frac{4n^2(n+1)^2}{n^4}\right]$$

$$= -2 + 6 - 8 + 4 = 0$$

**32.** Use Example 1 as a model to evaluate the limit

$$\lim_{n \to \infty} \sum_{i=1}^{n} f(c_i) \Delta x_i$$

over the region bounded by the graphs of $f(x) = \sqrt[3]{x}, y = 0, x = 0$ and $x = 1$.

**Solution:**

If we let $c_i = \dfrac{i^2}{n^3}$ then the width of the $i$th subinterval is given by

$$\Delta x_i = \frac{i^3}{n^3} - \frac{(i-1)^3}{n^3}$$

$$= \frac{i^3 - i^3 + 3i^2 - 3i + 1}{n^3} = \frac{3i^2 - 3i + 1}{n^3}$$

Thus, we have

$$\lim_{n\to\infty} \sum_{i=1}^{n} f(c_i)\Delta x_i = \lim_{n\to\infty} \sum_{i=1}^{n} \sqrt[3]{\frac{i^3}{n^3}} \left(\frac{3i^2 - 3i + 1}{n^3}\right)$$

$$= \lim_{n\to\infty} \sum_{i=1}^{n} \frac{3i^3 - 3i^2 + i}{n^4}$$

$$= \lim_{n\to\infty} \frac{1}{n^4} \left[\frac{3n^2(n+1)^2}{4} - \frac{3n(n+1)(2n+1)}{6}\right.$$
$$\left. + \frac{n(n+1)}{2}\right]$$

$$= \lim_{n\to\infty} \frac{3n^4 + 2n^3 - n^2}{4n^4} = \frac{3}{4}$$

## 5.4   The Fundamental Theorem of Calculus

**7.** Evaluate the definite integral $\int_0^1 (2t-1)^2\, dt$.

**Solution:**

$$\int_0^1 (2t-1)^2\, dt = \int_0^1 (4t^2 - 4t + 1)\, dt = \left[\frac{4t^3}{3} - \frac{4t^2}{2} + t\right]_0^1$$

$$= \left(\frac{4}{3} - \frac{4}{2} + 1\right) - (0 - 0 + 0) = \frac{4}{3} - \frac{6}{3} + \frac{3}{3} = \frac{1}{3}$$

**15.** Evaluate the definite integral $\int_1^4 \frac{u-2}{\sqrt{u}}\, du$.

**Solution:**

$$\int_1^4 \frac{u-2}{\sqrt{u}}\, du = \int_1^4 (u^{1/2} - 2u^{-1/2})\, du = \left[\frac{2}{3}u^{3/2} - 4u^{1/2}\right]_1^4$$

$$= \left[\frac{2}{3}(\sqrt{4})^3 - 4\sqrt{4}\right] - \left(\frac{2}{3} - 4\right) = \frac{2}{3}$$

**23.** Evaluate the definite integral $\int_0^4 |x^2 - 4x + 3|\, dx$.

**Solution:**

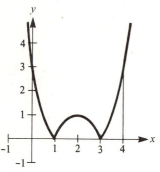

Since $x^2 - 4x + 3 = (x-1)(x-3)$, we know that $x^2 - 4x + 3$ is positive in the interval $(0, 1)$, negative in the interval $(1, 3)$, and positive in the interval $(3, 4)$. Therefore we can omit the absolute value signs by considering the three following intergrals. (Note the change in signs.)

$$\int_0^4 |x^2 - 4x + 3|\, dx$$

$$= \int_0^1 (x^2 - 4x + 3)\, dx + \int_1^3 (-x^2 + 4x - 3)\, dx + \int_3^4 (x^2 - 4x + 3)\, dx$$

$$= \left[\frac{x^3}{3} - \frac{4x^2}{2} + 3x\right]_0^1 + \left[-\frac{x^3}{3} + \frac{4x^2}{2} - 3x\right]_1^3 + \left[\frac{x^3}{3} - \frac{4x^2}{2} + 3x\right]_3^4$$

$$= \left(\frac{1}{3} - 2 + 3\right) - (0) + (-9 + 18 - 9)$$

$$\quad - \left(-\frac{1}{3} + 2 - 3\right) + \left(\frac{64}{3} - 32 + 12\right) - (9 - 18 + 9)$$

$$= \frac{4}{3} - 0 + 0 - \left(-\frac{4}{3}\right) + \frac{4}{3} - 0 = \frac{12}{3} = 4$$

**29.** Determine the area of the region having the boundries shown in the accompanying figure.

**Solution:**

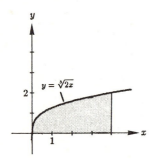

$$\text{area} = \int_0^4 \sqrt[3]{2x}\, dx$$

$$= \sqrt[3]{2} \int_0^4 x^{1/3}\, dx$$

$$= \sqrt[3]{2}\left[\frac{3}{4}x^{4/3}\right]_0^4 = \sqrt[3]{2}\left[\frac{3}{4}(4^{4/3}) - \frac{3}{4}(0)\right] = 6$$

**33.** Find the area of the region bounded by the graphs of
$y = x^3 + x, x = 2$, and $y = 0$.

**Solution:**

From the accompanying figure we see that the area of the
region is given by the definite integral $\int_0^2 (x^3 + x)\, dx$.

$$\text{area} = \int_0^2 (x^3 + x)\, dx$$

$$= \left[\frac{1}{4}x^4 + \frac{1}{2}x^2\right]_0^2$$

$$= (4 + 2) - (0) = 6$$

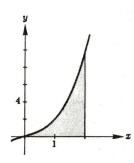

**39.** Sketch the graph of $f(x) = 4 - x^2$ over the interval $[-2, 2]$.
Find the average value of $f(x)$ over this interval and find all
values of $x$ where $f(x)$ equals its average.

**Solution:**

average value of $f(x)$ over $[-2, 2]$

$$= \frac{\int_{-2}^2 (4 - x^2)\, dx}{2 - (-2)}$$

$$= \frac{1}{4} \int_{-2}^2 (4 - x^2)\, dx$$

$$= 2\left(\frac{1}{4}\right) \int_0^2 (4 - x^2)\, dx \qquad \text{(By Symmetry)}$$

$$= \frac{1}{2}\left[4x - \frac{1}{3}x^3\right]_0^2 = \frac{1}{2}\left(8 - \frac{8}{3}\right) = \frac{8}{3}$$

To find the values of $x$ for which $f(x) = 8/3$ in the interval
$[-2, 2]$, we solve the equation

$$f(x) = 4 - x^2 = \frac{8}{3}$$

$$x^2 = \frac{4}{3}$$

$$x = \pm\frac{2}{\sqrt{3}} = \pm\frac{2\sqrt{3}}{3} \approx \pm 1.155$$

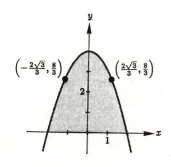

## 5.5    Integration by Substitution

9. Evaluate the indefinite integral $\int x^2(x^3 - 1)^4\,dx$ and check your results by differentiation.

**Solution:**

To evaluate $\int x^2(x^3 - 1)^4\,dx$, we use the method of pattern recognition by letting $g(x) = x^3 - 1$, and $g'(x) = 3x^2$. Thus, by Theorem 5.14, we have

$$\int x^2(x^3 - 1)^4\,dx = \int (x^3 - 1)^4 \left(\frac{1}{3}\right)(3x^2)\,dx$$

$$= \frac{1}{3}\int \underbrace{(x^3 - 1)^4}_{[g(x)]^4}\,\underbrace{(3x^2)}_{g'(x)}\,dx$$

$$= \frac{1}{3}\frac{[g(x)]^5}{5} + C$$

$$= \left(\frac{1}{3}\right)\left[\frac{(x^3 - 1)^5}{5}\right] + C = \frac{1}{15}(x^3 - 1)^5 + C$$

**Check**

If $y = \frac{1}{15}(x^3 - 1)^5 + C$, then

$$\frac{dy}{dx} = \frac{1}{15}(5)(x^3 - 1)^4(3x^2) + 0 = x^2(x^3 - 1)^4$$

11. Evaluate the indefinite integral $\int 5x\sqrt[3]{1 - x^2}\,dx$ and check your results by differentiation.

**Solution:**

To evaluate $\int 5x\sqrt[3]{1 - x^2}\,dx$ , we use the method of pattern recognition by letting $g(x) = 1 - x^2$, and $g'(x) = -2x$.

$$\int 5x\sqrt[3]{1 - x^2}\,dx = -\frac{5}{2}\int \underbrace{(1 - x^2)^{1/3}}_{[g(x)]^{1/3}}\,\underbrace{(-2x)}_{g'(x)}\,dx$$

$$= -\frac{5}{2}\frac{[g(x)]^{4/3}}{4/3} + C$$

$$= -\left(\frac{5}{2}\right)\left[\frac{(1 - x^2)^{4/3}}{4/3}\right] + C = -\frac{15}{8}(1 - x^2)^{4/3} + C$$

**Check**

If $y = \frac{-15}{8}(1 - x^2)^{4/3} + C$, then

$$\frac{dy}{dx} = -\left(\frac{15}{8}\right)\left(\frac{4}{3}\right)(1 - x^2)^{1/3}(-2x) + 0 = 5x(1 - x^2)^{1/3}$$

**19.** Evaluate the indefinite integral

$$\int \left(1 + \frac{1}{t}\right)^3 \left(\frac{1}{t^2}\right) dt$$

and check your results by differentiation.

**Solution:**

We evaluate this integral by changing the variable, and letting $u$ be the function $u = 1 + (1/t)$. Then, we obtain

$$du = -\frac{1}{t^2}\, dt.$$

Therefore,

$$\int \left(1 + \frac{1}{t}\right)^3 \left(\frac{1}{t^2}\right) dt = -\int \left(1 + \frac{1}{t}\right)^3 \left(\frac{-1}{t^2}\right) dt$$

$$= -\int u^3\, du$$

$$= -\frac{u^4}{4} + C$$

$$= -\frac{1}{4}\left(1 + \frac{1}{t}\right)^4 + C$$

**Check**

If $y = -\frac{1}{4}\left(1 + \frac{1}{t}\right)^4 + C$, then

$$\frac{dy}{dt} = \left(-\frac{1}{4}\right)(4)\left(1 + \frac{1}{t}\right)^3 \left(-\frac{1}{t^2}\right) + 0 = \left(1 + \frac{1}{t}\right)^3 \left(\frac{1}{t^2}\right)$$

**21.** Evaluate the indefinite integral $\int (1/\sqrt{2x})\, dx$ and check your results by differentiation.

**Solution:**

$$\int \frac{1}{\sqrt{2x}}\, dx = \int \frac{1}{\sqrt{2}\sqrt{x}}\, dx = \frac{1}{\sqrt{2}} \int x^{-1/2}\, dx$$

$$= \frac{1}{\sqrt{2}} \left( \frac{x^{1/2}}{\frac{1}{2}} \right) + C = \sqrt{2x} + C$$

**Check**

If $y = \sqrt{2x} + C = (2x)^{1/2} + C$, then

$$\frac{dy}{dx} = \frac{1}{2}(2x)^{-1/2}(2) + 0 = \frac{1}{\sqrt{2x}}$$

**31.** Evaluate $\int x^2\sqrt{1-x}\, dx$.

**Solution:**

Let $u = \sqrt{1-x}$. Then $u^2 = 1 - x$, $x = 1 - u^2$, and $dx = -2u\, du$. Thus

$$\int x^2\sqrt{1-x}\, dx = \int (1-u^2)^2 u(-2u)\, du$$

$$= -\int (2u^2 - 4u^4 + 2u^6)\, du$$

$$= -\left( \frac{2u^3}{3} - \frac{4u^5}{5} + \frac{2u^7}{7} \right) + C$$

$$= \frac{-2u^3}{105}(35 - 42u^2 + 15u^4) + C$$

$$= \frac{-2}{105}(1-x)^{3/2}[35 - 42(1-x) + 15(1-x)^2] + C$$

$$= \frac{-2}{105}(1-x)^{3/2}(15x^2 + 12x + 8) + C$$

**33.** Evaluate $\displaystyle\int \frac{x^2 - 1}{\sqrt{2x - 1}}\, dx$.

**Solution:**

Let $u = \sqrt{2x - 1}$. Then $u^2 = 2x - 1$, $x = \dfrac{u^2 + 1}{2}$, and $dx = u\, du$. Thus,

$$
\begin{aligned}
\int \frac{x^2 - 1}{\sqrt{2x - 1}}\, dx &= \int \frac{[(u^2 + 1)/2]^2 - 1}{u}\, (u\, du) \\
&= \frac{1}{4}\int (u^4 + 2u^2 - 3)\, du \\
&= \frac{1}{4}\left(\frac{u^5}{5} + \frac{2u^3}{3} - 3u\right) + C \\
&= \frac{u}{60}(3u^4 + 10u^2 - 45) + C \\
&= \frac{1}{60}\sqrt{2x - 1}[3(2x - 1)^2 \\
&\qquad + 10(2x - 1) - 45] + C \\
&= \frac{1}{60}\sqrt{2x - 1}(12x^2 + 8x - 52) + C \\
&= \frac{1}{15}\sqrt{2x - 1}(3x^2 + 2x - 13) + C
\end{aligned}
$$

**39.** Evaluate the definite integral $\int_{-1}^{1} x(x^2 + 1)^3\, dx$.

**Solution:**

Let $u = x^2 + 1$. Then $du = 2x\, dx$, and $dx = du/2$. Furthermore, if $x = -1$, then $u = 2$ and if $x = 1$, then $u = 2$. Since the upper and lower limits are equal, we have

$$
\int_{-1}^{1} x(x^2 + 1)^3\, dx = \int_{2}^{2} u^3 \left(\frac{1}{2}\right) du = 0.
$$

**43.** Evaluate the definite integral $\int_1^9 \frac{1}{\sqrt{x}(1+\sqrt{x})^2}\,dx$.

**Solution:**

Let $u = 1 + \sqrt{x}$. Then $du = \frac{1}{2\sqrt{x}}\,dx$. Furthermore, if $x = 1$, then $u = 2$, and if $x = 9$, then $u = 4$. Hence,

$$\int_1^9 \frac{1}{\sqrt{x}(1+\sqrt{x})^2}\,dx = 2\int_1^9 \frac{1}{(1+\sqrt{x})^2}\left(\frac{1}{2\sqrt{x}}\right)dx$$

$$= 2\int_2^4 \frac{1}{u^2}\,du$$

$$= 2\int_2^4 u^{-2}\,du$$

$$= \left[\frac{-2}{u}\right]_2^4$$

$$= -\frac{1}{2} - (-1) = \frac{1}{2}$$

**49.** Evaluate the definite integral $\int_0^7 x\sqrt[3]{x+1}\,dx$.

**Solution:**

Let $u = \sqrt[3]{x+1}$. Then $u^3 = x + 1$, $x = u^3 - 1$, and $dx = 3u^2\,du$. Furthermore, if $x = 0$, then $u = 1$, and if $x = 7$, then $u = 2$. Thus

$$\int_0^7 x\sqrt[3]{x+1}\,dx = \int_1^2 (u^3 - 1))(u)(3u^2\,du)$$

$$= 3\int_1^2 (u^6 - u^3)\,du = 3\left[\frac{u^7}{7} - \frac{u^4}{4}\right]_1^2$$

$$= 3\left(\frac{128}{7} - \frac{16}{4} - \frac{1}{7} + \frac{1}{4}\right)$$

$$= 3\left(\frac{127}{7} - \frac{15}{4}\right) = 3\left(\frac{508 - 105}{28}\right) = \frac{1209}{28}$$

## 5.6   Numerical Integration

**5.** Use the Trapezoidal Rule and Simpson's Rule with $n = 8$ to approximate the value of $\int_0^2 x^3\,dx$. Compare these results with the exact value of the definite integral. Round your answers to four decimal places.

**Solution:**

(a)   Trapezoidal Rule $(n = 8)$

$$\int_0^2 x^3\,dx \approx \frac{2}{2(8)}\left[0 + 2\left(\frac{1}{4}\right)^3 + 2\left(\frac{2}{4}\right)^3 + 2\left(\frac{3}{4}\right)^3 + 2\left(\frac{4}{4}\right)^3\right.$$
$$\left. + 2\left(\frac{5}{4}\right)^3 + 2\left(\frac{6}{4}\right)^3 + 2\left(\frac{7}{4}\right)^3 + 2^3\right]$$
$$= \frac{1}{8}\left[\frac{2(1^3 + 2^3 + 3^3 + 4^3 + 5^3 + 6^3 + 7^3)}{4^3} + 8\right]$$
$$= \frac{1}{8}\left[\frac{2(784)}{164} + 8\right] = \frac{65}{16} = 4.0625$$

(b) Simpson's Rule $(n = 8)$

$$\int_0^2 x^3\,dx \approx \frac{2}{3(8)}\left[0 + 4\left(\frac{1}{4}\right)^3 + 2\left(\frac{2}{3}\right)^3 + 4\left(\frac{3}{4}\right)^3 + 2\left(\frac{4}{4}\right)^3\right.$$
$$\left. + 4\left(\frac{5}{4}\right)^3 + 2\left(\frac{6}{4}\right)^3 + 4\left(\frac{7}{4}\right)^3 + 2^3\right]$$
$$= \frac{1}{12}\left[\frac{4(1^3 + 3^3 + 5^3 + 7^3) + 2(2^3 + 4^3 + 6^3)}{4^3} + 8\right]$$
$$= \frac{1}{12}\left[\frac{4(496) + 2(288)}{4^3} + 8\right]$$
$$= \frac{1}{12}\left(\frac{2560}{64} + 8\right) = \frac{1}{12}(48) = 4$$

(c)   In this particular case, we see that Simpson's Rule is exact since

$$\int_0^2 x^3\,dx = \left[\frac{x^4}{4}\right]_0^2 = \frac{16}{4} = 4.$$

15. Approximate $\int_0^1 \sqrt{x}\sqrt{1-x}\,dx$ using (a) the Trapezoidal Rule and (b) Simpson's Rule with $n = 4$.

**Solution:**

(a)  Trapezoidal Rule ($n = 4$)

$$\int_0^1 \sqrt{x}\sqrt{1-x}\,dx$$

$$\approx \frac{1}{2(4)}\left[0 + 2\sqrt{\frac{1}{4}}\sqrt{\frac{3}{4}} + 2\sqrt{\frac{2}{4}}\sqrt{\frac{2}{4}} + 2\sqrt{\frac{3}{4}}\sqrt{\frac{1}{4}} + 0\right]$$

$$= \frac{1}{8}\left[\frac{2\sqrt{3}}{4} + \frac{2(2)}{4} + \frac{2\sqrt{3}}{4}\right] = \frac{1}{8}(1 + \sqrt{3}) \approx 0.342$$

(b)  Simpson's Rule ($n = 4$)

$$\int_0^1 \sqrt{x}\sqrt{1-x}\,dx$$

$$\approx \frac{1}{3(4)}\left[0 + 4\sqrt{\frac{1}{4}}\sqrt{\frac{3}{4}} + 2\sqrt{\frac{2}{4}}\sqrt{\frac{2}{4}} + 4\sqrt{\frac{3}{4}}\sqrt{\frac{1}{4}} + 0\right]$$

$$= \frac{1}{12}\left[\frac{4\sqrt{3}}{4} + \frac{2(2)}{4} + \frac{4\sqrt{3}}{4}\right] = \frac{1}{12}(2\sqrt{3} + 1) \approx 0.372$$

## Review Exercises for Chapter 5

5. Find the indefinite integral $\displaystyle\int \frac{(1+x)^2}{\sqrt{x}}\,dx$.

**Solution:**

$$\int \frac{(1+x)^2}{\sqrt{x}}\,dx = \int \frac{1 + 2x + x^2}{\sqrt{x}}\,dx$$

$$= \int \left(x^{-1/2} + 2x^{1/2} + x^{3/2}\right)dx$$

$$= 2x^{1/2} + \frac{4}{3}x^{3/2} + \frac{2}{5}x^{5/2} + C$$

$$= \frac{2\sqrt{x}}{15}(15 + 10x + 3x^2) + C$$

**7.** Find the indefinite integral $\int \dfrac{x^2}{\sqrt{x^3+3}}\,dx$.

**Solution:**

To find $\int x^2/\sqrt{x^3+3}\,dx$, we let $u = x^3 + 3$, then $du = 3x^2\,dx$.

$$\int \frac{x^2}{\sqrt{x^3+3}}\,dx = \frac{1}{3}\int (x^3+3)^{-1/2}(3x^2)\,dx$$

$$= \frac{1}{3}\int u^{-1/2}\,du$$

$$= \frac{2}{3}u^{1/2} + C$$

$$= \frac{2}{3}(x^3+3)^{1/2} + C$$

$$= \frac{2}{3}\sqrt{x^3+3} + C$$

**25.** Use the Fundamental Theorem of Calculus to evaluate the definite integral

$$\int_0^3 \frac{1}{\sqrt{1+x}}\,dx$$

**Solution:**

We let $u = 1 + x$, then $du = dx$. Also, when $x = 0$, $u = 1$, and when $x = 3$, $u = 4$. Therefore,

$$\int_0^3 \frac{1}{\sqrt{1+x}}\,dx = \int_1^4 u^{-1/2}\,du$$

$$= \left[2u^{1/2}\right]_1^4 = 2(2-1) = 2$$

**29.** Use the Fundamental Theorem of Calculus to evaluate the definite integral

$$2\pi \int_0^1 (y+1)\sqrt{1-y}\,dy$$

**Solution:**

We let $u = \sqrt{1 - y}$. Then $y = 1 - u^2$ and $dy = -2u\,du$.
Furthermore, when $y = 0$, $u = 1$ and when $y = 1$, $u = 0$.

$$2\pi \int_0^1 (y+1)\sqrt{1 - y}\,dy = 2\pi \int_1^0 (2 - u^2)(u)(-2u)\,du$$

$$= -4\pi \int_1^0 (2u^2 - u^4)\,du$$

$$= -4\pi \left[ \frac{2}{3}u^3 - \frac{1}{5}u^5 \right]_1^0$$

$$= \frac{28\pi}{15}$$

**33.** An airplane taking off from a runway travels 3600 feet before lifting off. If it starts from rest, moves with constant acceleration, and makes the run in 30 seconds, with what velocity does it lift off?

**Solution:**

Let the position function of the plane be given by $s(t)$ where $s$ is measured in feet and $t$ is time in seconds. If $t = 0$ is the time the plane starts its take off rate, then we are given $s(0) = 0$, $s(30) = 3600$, $s'(0) = 0$ and $s''(t) = a$ where $a$ is constant.

$$s'(t) = \int a\,dt = at + C_1$$

$$s'(0) = 0 + C_1 = 0 \quad \Longrightarrow \quad C_1 = 0$$

$$s(t) = \int at\,dt = \frac{a}{2}t^2 + C_2$$

$$s(0) = 0 + C_2 = 0 \quad \Longrightarrow \quad C_2 = 0 \quad \Longrightarrow \quad s(t) = \frac{a}{2}t^2$$

$$s(30) = \frac{a}{2}(30)^2 = 3600 \quad \text{or} \quad a = \frac{3600(2)}{30^2} = 8 \text{ ft/sec}^2$$

Therefore

$$s(t) = 4t^2, \;\; s'(t) = v(t) = 8t \quad \text{and} \quad v(30) = 8(30) = 240 \text{ ft/sec.}$$

**45.** Find the average value of $f(x) = 1/\sqrt{x-1}$ over the interval $[5, 10]$. Find the values of $x$ where the function assumes its average value and sketch the graph of the function.

**Solution:**

The average value is given by

$$\frac{1}{10-5} \int_5^{10} \frac{1}{\sqrt{x-1}}\, dx = \frac{1}{5} \int_5^{10} (x-1)^{-1/2}(1)\, dx$$

$$= \left[ \frac{2}{5}(x-1)^{1/2} \right]_5^{10} = \frac{2}{5}$$

To find the value of $x$ where the function assumes its mean value on $[5, 10]$, solve

$$\frac{1}{\sqrt{x-1}} = \frac{2}{5}$$

$$\sqrt{x-1} = \frac{5}{2}$$

$$x - 1 = \frac{25}{4}$$

$$x = \frac{29}{4}$$

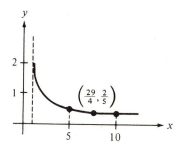

$\left( \dfrac{29}{4}, \dfrac{2}{5} \right)$

# 6 APPLICATIONS OF INTEGRATION

## 6.1 Area of a Region between Two Curves

**9.** Sketch the region bounded by the graphs of $f(x) = x^2 + 2x + 1$ and $g(x) = 3x + 3$ and find the area of this region by means of a definite integral.

**Solution:**

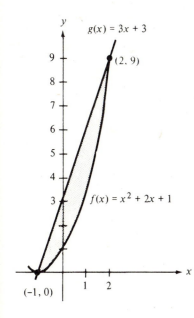

The points of intersection $f$ and $g$ are found by solving

$$f(x) = g(x)$$
$$x^2 + 2x + 1 = 3x + 3$$
$$x^2 - x - 2 = 0$$
$$(x - 2)(x + 1) = 0 \implies x = -1,\ 2$$

Since $x^2 + 2x + 1 \leq 3x + 3$ for $-1 \leq x \leq 2$, we have

$$\text{area} = \int_{-1}^{2} [(3x + 3) - (x^2 + 2x + 1)]dx$$

$$= \int_{-1}^{2} (-x^2 + x + 2)dx$$

$$= \left[ \frac{-x^3}{3} + \frac{x^2}{2} + 2x \right]_{-1}^{2}$$

$$= \left( \frac{-8}{3} + 2 + 4 \right) - \left( \frac{1}{3} + \frac{1}{2} - 2 \right)$$

$$= \frac{-16 + 12 + 24 - 2 - 3 + 12}{6} = \frac{27}{6} = \frac{9}{2}$$

**23.** Sketch the region bounded by the graphs of
$f(y) = y^2 + 1$, $g(y) = 0$, $y = -1$, and $y = 2$ and find the area of
this region by means of a definite integral.

**Solution:**

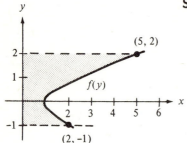

$$\text{area} = \int_{-1}^{2} (y^2 + 1)dy = \left[\frac{y^3}{3} + y\right]_{-1}^{2}$$

$$= \left(\frac{8}{3} + 2\right) - \left(\frac{-1}{3} - 1\right)$$

$$= \frac{8 + 6 + 1 + 3}{3} = 6$$

**27.** Use integration to find the area of the triangle with vertices
$(0,0)$, $(a,0)$, and $(b,c)$.

**Solution:**

From the accompanying figure we observe that the triangular
region is bounded by $y = (c/b)x$, $y = [c/(b-a)](x-a)$, and
$y = 0$. Therefore, the area is given by

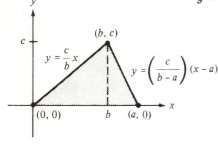

$$\text{area} = \int_{0}^{b} \frac{c}{b}x \, dx + \int_{b}^{a} \left(\frac{c}{b-a}\right)(x-a)dx$$

$$= \frac{c}{b}\left[\frac{x^2}{2}\right]_{0}^{b} + \left(\frac{c}{b-a}\right)\left[\frac{x^2}{2} - ax\right]_{b}^{a}$$

$$= \frac{1}{2}bc + \left(\frac{c}{b-a}\right)\left[\left(\frac{a^2}{2} - a^2\right) - \left(\frac{b^2}{2} - ab\right)\right]$$

$$= \frac{1}{2}bc + \frac{1}{2}(a-b)c = \frac{1}{2}ac$$

**33.** Find the area of the region bounded by the graph of $f(x) = x^3$
and the tangent line to the graph at the point $(1,1)$.

**Solution:**

Since $f(x) = x^3$, $f'(x) = 3x^2$, and $f'(1) = 3$ (slope of the
tangent line), the equation of the tangent line to the graph of $f$
is given by

$$y - 1 = 3(x - 1)$$

$$y = 3x - 2.$$

The $x$-coordinates of the points of intersection of the tangent line and the function of area the solutions to the equation

$$x^3 = 3x - 2$$
$$x^3 - 3x + 2 = 0$$
$$(x - 1)(x^2 + x - 2) = 0$$
$$(x - 1)^2(x + 2) = 0 \implies x = -2, 1$$

Therefore the points of intersection are given by $(1,1)$ and $(-2,-8)$. (See the accompanying figure.)

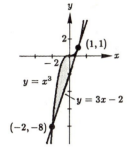

$$\text{area} = \int_{-2}^{1} [x^3 - (3x - 2)]dx$$

$$= \int_{-2}^{1} (x^3 - 3x + 2)dx$$

$$= \left[ \frac{1}{4}x^4 - \frac{3}{2}x^2 + 2x \right]_{-2}^{1}$$

$$= \left( \frac{1}{4} - \frac{3}{2} + 2 \right) - (4 - 6 - 4) = \frac{27}{4}$$

**41.** Find the consumer surplus and producer surplus of the demand function $p_1 = 50 - 0.5x$ and supply function $p_2 = 0.125x$. The consumer surplus and producer surplus are represented by the areas shown in the accompanying figure.

**Solution:**

Solving the equations simultaneously we find the point of equilibrium to be $(80, 10)$. Therefore

$$\text{Consumer Surplus} = \int_{0}^{80} [(50 - 0.5x) - 10]dx$$

$$= \left[ 40x - 0.25x^2 \right]_{0}^{80} = 1600$$

$$\text{Producer Surplus} = \int_{0}^{80} (10 - 0.125x)dx$$

$$= \left[ 10x - 0.0625x^2 \right]_{0}^{80} = 400$$

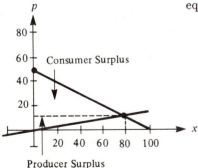

## 6.2   Volume: The Disc Method

**7.** Find the volume of the solid generated by revolving the region bounded by $y = x^2$ and $y = x^3$ about the $x$-axis.

**Solution:**

From the accompanying figure we have

$$R(x) = x^2 \qquad \text{outer radius}$$
$$r(x) = x^3 \qquad \text{inner radius}$$

Now, integrating between 0 and 1, we have

$$V = \pi \int_0^1 \left( [R(x)]^2 - [r(x)]^2 \right) dx$$

$$= \pi \int_0^1 (x^4 - x^6) \, dx$$

$$= \pi \left[ \frac{x^5}{5} - \frac{x^7}{7} \right]_0^1 = \pi \left( \frac{1}{5} - \frac{1}{7} \right) = \frac{2\pi}{35}$$

**13.** Find the volume of the solid generated by revolving the region bounded by $y = \sqrt{x}$, $y = 0$, and $x = 4$ about:
(a)   the $x$-axis
(b)   the $y$-axis
(c)   the line $x = 4$
(d)   the line $x = 6$

**Solution:**

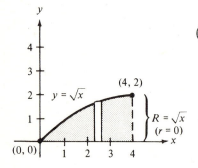

(a)

$$V = \pi \int_0^4 (\sqrt{x})^2 \, dx$$

$$= \pi \int_0^4 x \, dx = \pi \left[ \frac{x^2}{2} \right]_0^4$$

$$= \frac{16\pi}{2} = 8\pi$$

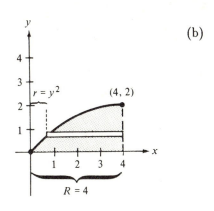

(b)

$$V = \pi \int_0^2 [4^2 - (y^2)^2]\, dy$$

$$= \pi \int_0^2 (16 - y^4)\, dy$$

$$= \left[ 16y - \frac{y^5}{5} \right]_0^2$$

$$= \pi \left[ 32 - \frac{32}{5} \right] = \frac{128\pi}{5}$$

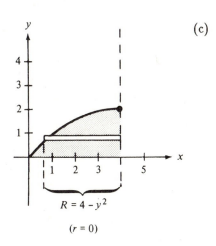

(c)

$$V = \pi \int_0^2 (4 - y^2)^2\, dy$$

$$= \pi \int_0^2 (16 - 8y^2 + y^4)\, dy$$

$$= \pi \left[ 16y - \frac{8y^3}{3} + \frac{y^5}{5} \right]_0^2$$

$$= \pi \left[ 32 - \frac{64}{3} + \frac{32}{5} \right] = \frac{256\pi}{15}$$

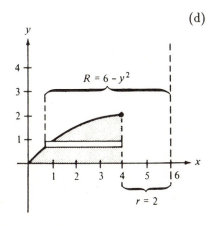

(d)

$$V = \pi \int_0^2 [(6 - y^2)^2 - 2^2]\, dy$$

$$= \pi \int_0^2 (32 - 12y^2 + y^4)\, dy$$

$$= \pi \left[ 32y - 4y^3 + \frac{y^5}{5} \right]_0^2$$

$$= \pi \left[ 64 - 32 + \frac{32}{5} \right] = \frac{192\pi}{5}$$

**15.** Find the volume of the solid formed by revolving the region bounded by $y = x^2$ and $y = 4x - x^2$ about:
(a)   the $x$-axis                                       (b)   the line $y = 6$

**Solution:**

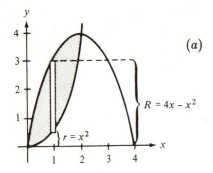

(a) $$V = \pi \int_0^2 \left[(4x - x^2)^2 - (x^2)^2\right] dx$$

$$= \pi \int_0^2 (16x^2 - 8x^3) dx$$

$$= \pi \left[\frac{16}{3}x^3 - 2x^4\right]_0^2$$

$$= 8\pi \left[\frac{16}{3} - 4\right] = \frac{32\pi}{3}$$

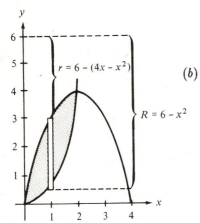

(b) $$V = \pi \int_0^2 \left[(6 - x^2)^2 - (6 - 4x + x^2)^2\right] dx$$

$$= 8\pi \int_0^2 (x^3 - 5x^2 + 6x) dx$$

$$= 8\pi \left[\frac{x^4}{4} - \frac{5}{3}x^3 + 3x^2\right]_0^2$$

$$= 32\pi \left(1 - \frac{10}{3} + 3\right) = \frac{64\pi}{3}$$

**35.** Use the Disc Method to verify that the volume of a sphere of radius $r$ is $\frac{4}{3}\pi r^3$.

**Solution:**

Let $y = \sqrt{r^2 - x^2}$ and let the region bounded by $y = \sqrt{r^2 - x^2}$ and $y = 0$ be revolved about the $x$-axis. The resulting solid of

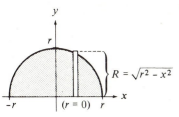

revolution is a sphere with volume

$$V = \pi \int_{-r}^{r} (\sqrt{r^2 - x^2})^2 \, dx$$

$$= \pi \int_{-r}^{r} (r^2 - x^2) \, dx$$

$$= \pi \left[ r^2 x - \frac{x^3}{3} \right]_{-r}^{r}$$

$$= \pi \left[ \left( r^3 - \frac{r^3}{3} \right) - \left( -r^3 + \frac{r^3}{3} \right) \right] = \frac{4\pi r^3}{3}$$

**40.** The tank on a water tower is a sphere of radius 50 ft. Determine the depth of the water when the tank is filled to 21.6% of its total capacity.

**Solution:**

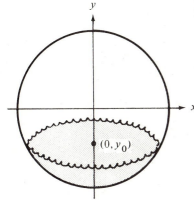

The total volume of the sphere is

$$V = \frac{4\pi(50)^3}{3} = \frac{500,000\pi}{3} \text{ ft}^3$$

The volume of the portion filled with water is

$$(0.216)V = 36,000\pi$$

$$= \pi \int_{-50}^{y_0} (\sqrt{2500 - y^2})^2 \, dy$$

$$= \pi \int_{-50}^{y_0} (2500 - y^2) \, dy = \pi \left[ 2500y - \frac{y^3}{3} \right]_{-50}^{y_0}$$

$$= \pi \left[ \left( 2500y_0 - \frac{y_0^3}{3} \right) - \left( -125,000 + \frac{125,000}{3} \right) \right]$$

$$= \pi \left[ 2500y_0 - \frac{y_0^3}{3} + \frac{250,000}{3} \right]$$

$$108,000 = 7500y_0 - y_0^3 + 250,000$$

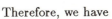

Therefore, we have

$$y_0^3 - 7500y_0 - 142,000 = 0$$

$$(y_0 + 20)(y_0^2 - 20y_0 - 7100) = 0$$

$$y_0 = -20, \ 10 \pm 60\sqrt{2}$$

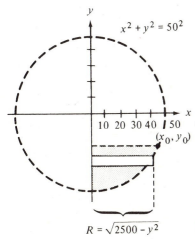

Since we are interested only in values of $y_0$ such that $-50 < y_0 < 50$, we choose $y_0 = -20$ ft and conclude that the depth of the water is $[-20 - (-50)] = 30$ ft.

**47.** The base of a solid is bounded by $y = x^3, y = 0$, and $x = 1$. Find the volume of the solid if the cross sections perpendicular to the $y$- axis are:

(a) squares.

(b) semicircles.

(c) equilateral triangles.

(d) trapezoids for which $h = b_1 = \frac{1}{2}b_2$, where $b_1$ and $b_2$ are the lengths of the upper and lower bases, respectively.

(e) semiellipses whose heights are twice the lenth of their bases.

**Solution:**

The base of the solid is shown in the accompanying figure. Since the cross sections are taken perpendicular to the $y$-axis, the base of each cross section is given by $(1 - x) = (1 - \sqrt[3]{y})$.

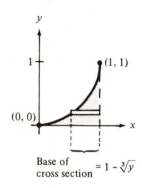

$y$

$1$ — $(1, 1)$

$(0, 0)$

$x$

Base of cross section $= 1 - \sqrt[3]{y}$

(a) The cross sections are squares whose sides are given by

$$s = (1 - \sqrt[3]{y})$$

Thus

$$A(y) = s^2 = (1 - \sqrt[3]{y})^2$$

and

$$V = \int_0^1 (1 - \sqrt[3]{y})^2 \, dy = \int_0^1 (1 - 2y^{1/3} + y^{2/3}) \, dy$$

$$= \left[ y - \frac{2y^{4/3}}{\frac{4}{3}} + \frac{y^{5/3}}{\frac{5}{3}} \right]_0^1 = 1 - \frac{3}{2} + \frac{3}{5} = \frac{1}{10}$$

$s$

$s$

$1 - \sqrt[3]{y}$

(b) The cross sections are semicircles whose radii are given by

$$r = \left( \frac{1}{2} \right)(1 - \sqrt[3]{y})$$

Thus

$$A(y) = \left( \frac{1}{2} \right)\pi \left[ \left( \frac{1}{2} \right)(1 - \sqrt[3]{y}) \right]^2 = \frac{\pi}{8}(1 - \sqrt[3]{y})^2$$

and

$$V = \frac{\pi}{8} \int_0^1 (1 - \sqrt[3]{y})^2 \, dy$$

$$= \frac{\pi}{8}\left( \frac{1}{10} \right) = \frac{\pi}{80} \qquad \text{From part (a)}$$

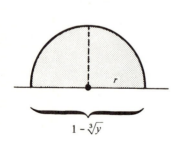

$r$

$1 - \sqrt[3]{y}$

(c) The cross sections are equilateral triangles whose sides are given by

$$s = (1 - \sqrt[3]{y})$$

Thus

$$A(y) = \frac{\sqrt{3}s^2}{4} = \frac{\sqrt{3}}{4}(1 - \sqrt[3]{y})^2$$

and

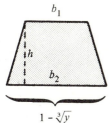

$$V = \frac{\sqrt{3}}{4} \int_0^1 (1 - \sqrt[3]{y})^2 \, dy$$

$$= \frac{\sqrt{3}}{4}\left(\frac{1}{10}\right) = \frac{\sqrt{3}}{40} \qquad \text{From part (a)}$$

(d) The cross sections are trapezoids for which $h = b_1 = b_2/2$, where

$$b_2 = (1 - \sqrt[3]{y})$$

Thus

$$A(y) = \left(\frac{b_1 + b_2}{2}\right)h = \left(\frac{3}{4}\right)(b_2)\left(\frac{1}{2}\right)(b_2) = \frac{3}{8}(1 - \sqrt[3]{y})^2$$

and

$$V = \frac{3}{8} \int_0^1 (1 - \sqrt[3]{y})^2 \, dy$$

$$= \frac{3}{8}\left(\frac{1}{10}\right) = \frac{3}{80} \qquad \text{From part (a)}$$

(e) The cross sections are semiellipses whose heights are twice the lengths of their bases. Thus $h = 2b$, where

$$b = (1 - \sqrt[3]{y})$$

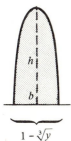

Thus

$$A(y) = \left(\frac{1}{2}\right)\pi(h)\left(\frac{b}{2}\right) = \frac{\pi}{2}b^2 = \frac{\pi}{2}(1 - \sqrt[3]{y})^2$$

and

$$V = \frac{\pi}{2} \int_0^1 (1 - \sqrt[3]{y})^2 \, dy$$

$$= \frac{\pi}{2}\left(\frac{1}{10}\right) = \frac{\pi}{20} \qquad \text{From part (a)}$$

**49.** A wedge is cut from a right circular cylinder of radius $r$ inches by a plane through the diameter of the base, which makes a $45°$ angle with the plane of the base. Find the volume of the wedge cut out.

**Solution:**

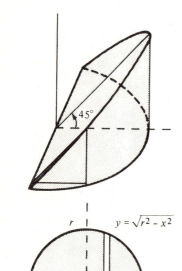

The cross sections are isosceles right triangles for which the base and height are equal. Thus

$$A(x) = \left(\frac{1}{2}\right) bh$$

$$= \left(\frac{1}{2}\right)(\sqrt{r^2 - x^2})(\sqrt{r^2 - x^2}) = \left(\frac{1}{2}\right)(r^2 - x^2)$$

and

$$V = \frac{1}{2} \int_{-r}^{r} (r^2 - x^2)\, dx$$

$$= \frac{1}{2} \left[ r^2 x - \frac{x^3}{3} \right]_{-r}^{r}$$

$$= \frac{1}{2}\left[ \left(r^3 - \frac{r^3}{3}\right) - \left(-r^3 + \frac{r^3}{3}\right) \right] = \frac{2r^3}{3} \text{ in}^3$$

## 6.3   Volume: The Shell Method

**3.** Use the Shell Method to find the volume generated by revolving the region bounded by $y = x$, $y = 0$, and $x = 2$ about the $x$-axis.

**Solution:**

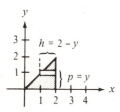

$$V = 2\pi \int_0^2 y(2 - y)\, dy$$

$$= 2\pi \int_0^2 (2y - y^2)\, dy = 2\pi \left[ y^2 - \frac{y^3}{3} \right]_0^2$$

$$= 2\pi\left(4 - \frac{8}{3}\right) = \frac{8\pi}{3}$$

**9.** Use the Shell Method to find the volume generated by revolving the region bounded by $y = x^2$ and $y = 4x - x^2$ about the $y$-axis.

**Solution:**

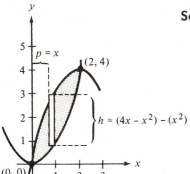

$$V = 2\pi \int_0^2 x[(4x - x^2) - (x^2)]\, dx$$

$$= 2\pi \int_0^2 x(4x - 2x^2)\, dx$$

$$= 4\pi \int_0^2 (2x^2 - x^3)\, dx = 4\pi \left[\frac{2x^3}{3} - \frac{x^4}{4}\right]_0^2$$

$$= 4\pi \left[\frac{16}{3} - 4\right] = \frac{16\pi}{3}$$

**11.** Use the Shell Method to find the volume generated by revolving the region bounded by $y = x^2$ and $y = 4x - x^2$ about the line $x = 4$.

**Solution:**

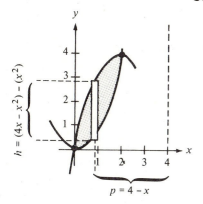

$$V = 2\pi \int_0^2 (4 - x)[(4x - x^2) - (x^2)]\, dx$$

$$= 2\pi \int_0^2 (4 - x)(4x - 2x^2)\, dx$$

$$= 4\pi \int_0^2 (8x - 6x^2 + x^3)\, dx$$

$$= 4\pi \left[4x^2 - 2x^3 + \frac{x^4}{4}\right]_0^2 = 4\pi[16 - 16 + 4] = 16\pi$$

**21.** Use the Disc or Shell Method to find the volume of the solid generated by revolving the region bounded by $y = x^3, y = 0$, and $x = 2$ about:
  (a)   the $x$-axis                              (b)   the $y$-axis
  (c)   the line $x = 4$                          (d)   the line $y = 8$

**Solution:**

(a)  Disc Method

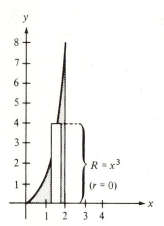

$$V = \pi \int_0^2 [(x^3)^2 - (0)^2]\, dx$$

$$= \pi \int_0^2 x^6\, dx$$

$$= \pi \left[\frac{x^7}{7}\right]_0^2 = \frac{128\pi}{7}$$

(b)  Shell Method

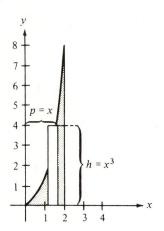

$$V = 2\pi \int_0^2 x(x^3)\, dx$$

$$= 2\pi \int_0^2 x^4\, dx$$

$$= 2\pi \left[\frac{x^5}{5}\right]_0^2 = \frac{64\pi}{5}$$

(c)  Shell Method

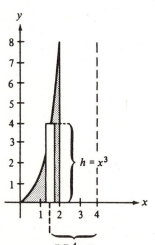

$$V = 2\pi \int_0^2 (4 - x)x^3\, dx$$

$$= 2\pi \int_0^2 (4x^3 - x^4)\, dx$$

$$= 2\pi \left[x^4 - \frac{x^5}{5}\right]_0^2 = \frac{96\pi}{5}$$

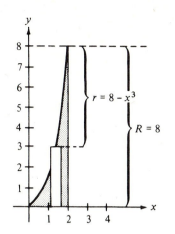

(d)   Disc Method

$$V = \pi \int_0^2 [8^2 - (8 - x^3)^2] \, dx$$

$$= \pi \int_0^2 (16x^3 - x^6) \, dx$$

$$= \pi \left[ 4x^4 - \frac{x^7}{7} \right]_0^2 = \frac{320\pi}{7}$$

**25.** A solid is generated by revolving the region bounded by $y = \frac{1}{2}x^2$ and $y = 2$ about the $x$-axis. A hole, centered along the axis of revolution, is drilled through this solid so that $\frac{1}{4}$ of the volume is removed. Find the diameter of the hole.

**Solution:**

The total volume of the solid is given by

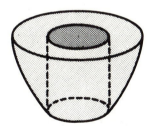

$$V = 2\pi \int_0^2 x \left( 2 - \frac{x^2}{2} \right) dx$$

$$= 2\pi \int_0^2 \left( 2x - \frac{x^3}{2} \right) dx$$

$$= 2\pi \left[ x^2 - \frac{x^4}{8} \right]_0^2 = 2\pi(4 - 2) = 4\pi$$

If a hole drilled in the center with a radius of $x_0$ removes $\frac{1}{4}$ of

this volume, we would have

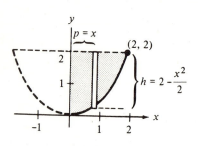

$$\left(\frac{3}{4}\right)V = 2\pi \int_{x_0}^{2} x\left(2 - \frac{x^2}{2}\right)dx$$

$$3\pi = 2\pi\left[x^2 - \frac{x^4}{8}\right]_{x_0}^{2}$$

$$3\pi = 2\pi\left[(4-2) - \left(x_0^2 - \frac{x_0^4}{8}\right)\right]$$

$$3 = 4 - 2x_0^2 + \left(\frac{x_0^4}{4}\right)$$

$$x_0^4 - 8x_0^2 + 4 = 0$$

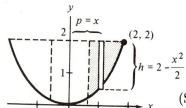

$$x_0^2 = \frac{8 \pm \sqrt{64 - 16}}{2}$$

$$x_0 = \sqrt{4 - 2\sqrt{3}}$$

(Since $0 < x_0 < 2$, we are not interested in $x_0 = \sqrt{4 + 2\sqrt{3}} \approx 2.7$.) Finally, since

$$\text{radius} = x_0 = \sqrt{4 - 2\sqrt{3}} = 0.732$$

we have

$$\text{diameter} = 2x_0 = 2\sqrt{4 - 2\sqrt{3}} \approx 1.464$$

## 6.4   Arc Length and Surfaces of Revolution

**5.** Find the arc length of $y = (x^4/8) + (1/4x^2)$ between $x = 1$ and $x = 2$.

**Solution:**

$$y = \frac{x^4}{8} + \frac{1}{4x^2} \qquad y' = \frac{x^3}{2} - \frac{1}{2x^3}$$

$$s = \int_1^2 \sqrt{1 + (y')^2}\, dx = \int_1^2 \sqrt{1 + \left(\frac{x^3}{2} - \frac{1}{2x^3}\right)^2}\, dx$$

$$= \int_1^2 \sqrt{1 + \frac{x^6}{4} - \frac{1}{2} + \frac{1}{4x^6}}\, dx = \int_1^2 \sqrt{\frac{x^6}{4} + \frac{1}{2} + \frac{1}{4x^6}}\, dx$$

$$= \int_1^2 \sqrt{\left(\frac{x^3}{2} + \frac{1}{2x^3}\right)^2}\, dx$$

(Note that $0 < \dfrac{x^3}{2} + \dfrac{1}{2x^3}$ for $1 \le x \le 2$.)

$$s = \int_1^2 \left( \frac{x^3}{2} + \frac{1}{2x^3} \right) dx = \left[ \frac{x^4}{8} - \frac{1}{4x^2} \right]_1^2$$

$$= \left( \frac{16}{8} - \frac{1}{16} \right) - \left( \frac{1}{8} - \frac{1}{4} \right) = \frac{32 - 1 - 2 + 4}{16} = \frac{33}{16}$$

**17.** Use Simpson's Rule with $n = 4$ to approximate the arc length of $y = x^3$ over the interval $[0, 2]$.

**Solution:**

$$y = x^3 \qquad \frac{dy}{dx} = 3x^2$$

$$s = \int_0^2 \sqrt{1 + \left( \frac{dy}{dx} \right)^2}\, dx$$

$$= \int_0^2 \sqrt{1 + 9x^4}\, dx$$

$$\approx \frac{2}{3(4)} \Big[ \sqrt{1 + 9(0)^4} + 4\sqrt{1 + 9(\tfrac{1}{2})^4} + 2\sqrt{1 + 9(1)^4}$$

$$+ 4\sqrt{1 + 9(\tfrac{3}{2})^4} + \sqrt{1 + 9(2)^4} \Big]$$

$$\approx 8.610$$

**19.** A fleeing object leaves the origin and moves up the $y$-axis. At the same time a pursuer leaves the point $(1, 0)$ and moves always toward the fleeing object. If the pursuer's speed is twice that of the fleeing object, the equation of the path is

$$y = \frac{1}{3}(x^{3/2} - 3x^{1/2} + 2)$$

How far has the fleeing object traveled when it is caught? Show that the pursuer traveled twice as far.

**Solution:**

The $y$-intercept of $y = \frac{1}{3}(x^{3/2} - 3x^{1/2} + 2)$ is $(0, \frac{2}{3})$. Therefore, the fleeing object traveled from $(0,0)$ to $(0, \frac{2}{3})$, a distance of $\frac{2}{3}$.

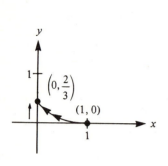

$$y = \frac{1}{3}(x^{3/2} - 3x^{1/2} + 2)$$

$$y' = \frac{1}{3}\left[\frac{3}{2}x^{1/2} - \frac{3}{2}x^{-1/2}\right]$$

$$= \frac{1}{2}\left(\sqrt{x} - \frac{1}{\sqrt{x}}\right) = \frac{x-1}{2\sqrt{x}}$$

$$1 + (y')^2 = 1 + \left(\frac{x-1}{2\sqrt{x}}\right)^2 = \frac{4x + (x^2 - 2x + 1)}{4x}$$

$$= \frac{x^2 + 2x + 1}{4x} = \frac{(x+1)^2}{4x}$$

Therefore, the distance traveled by the pursuer is given by

$$s = \int_0^1 \sqrt{1 + (y')^2}\, dx$$

$$= \frac{1}{2}\int_0^1 \frac{x+1}{\sqrt{x}}\, dx$$

$$= \frac{1}{2}\int_0^1 (x^{1/2} + x^{-1/2})\, dx$$

$$= \frac{1}{2}\left[\frac{2}{3}x^{3/2} + 2x^{1/2}\right]_0^1 = \frac{4}{3}$$

Thus the pursuer traveled a distancce of $\frac{4}{3}$, twice the distance of the fleeing object.

**25.** Find the area of the surface formed by revolving the graph of $y = (x^3/6) + (1/2x), 1 \le x \le 2$, about the $x$-axis.

**Solution:**

$$y = \frac{x^3}{6} + \frac{1}{2x}$$

$$y' = \frac{1}{2}x^2 - \frac{1}{2x^2} = \frac{x^4 - 1}{2x^2}$$

$$1 + (y')^2 = 1 + \left(\frac{x^4 - 1}{2x^2}\right)^2$$

$$= \frac{4x^4 + (x^8 - 2x^4 + 1)}{4x^2}$$

$$= \frac{x^8 + 2x^4 + 1}{4x^4} = \left(\frac{x^4 + 1}{2x^2}\right)^2$$

$$S = 2\pi \int_1^2 y\sqrt{1 + (y')^2}\, dx$$

$$= 2\pi \int_1^2 \left(\frac{x^3}{6} + \frac{1}{2x}\right)\left(\frac{x^4 + 1}{2x^2}\right) dx$$

$$= 2\pi \int_1^2 \left(\frac{x^5}{12} + \frac{x}{3} + \frac{1}{4x^3}\right) dx$$

$$= 2\pi \left[\frac{x^6}{72} + \frac{x^2}{6} - \frac{1}{8x^2}\right]_1^2 = \frac{47\pi}{16}$$

**27.** Find the area of the surface formed by revolving the graph of $y = \sqrt[3]{x} + 2, 1 \le x \le 8$, about the $y$-axis.

**Solution:**

$$y = \sqrt[3]{x} + 2 \qquad y' = \frac{1}{3}x^{-2/3}$$

$$S = 2\pi \int_1^8 x\sqrt{1 + (y')^2}\, dx = 2\pi \int_1^8 x\sqrt{1 + \left(\frac{1}{3x^{2/3}}\right)^2}\, dx$$

$$= 2\pi \int_1^8 x\sqrt{\frac{9x^{4/3} + 1}{9x^{4/3}}}\, dx = 2\pi \int_1^8 \frac{x}{3x^{2/3}}\sqrt{9x^{4/3} + 1}\, dx$$

$$= \frac{2\pi}{3}\left(\frac{1}{12}\right) \int_1^8 (9x^{4/3} + 1)^{1/2}(12x^{1/3})\, dx$$

$$= \frac{\pi}{18}\left[\frac{(9x^{4/3} + 1)^{3/2}}{3/2}\right]_1^8 = \frac{\pi}{27}[145\sqrt{145} - 10\sqrt{10}] \approx 199.48$$

**29.** A right circular cone is generated by revolving the region bounded by $y = hx/r$, $y = h$, and $x = 0$ about the $y$-axis. Verify that the lateral surface area of the cone is $S = \pi r \sqrt{r^2 + h^2}$.

**Solution:**

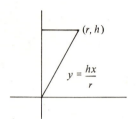

The distance between the $y$-axis and the graph of the line $y = hx/r$ is

$$r(y) = g(y) = \frac{ry}{h}$$

and since $g'(y) = \dfrac{r}{h}$, the surface area is given by

$$S = 2\pi \int_0^h r(y)\sqrt{1 + [g'(y)]^2}\, dy$$

$$= 2\pi \int_0^h \left(\frac{ry}{h}\right)\sqrt{1 + \left(\frac{r}{h}\right)^2}\, dy$$

$$= \frac{2\pi r \sqrt{r^2 + h^2}}{h^2} \int_0^h y\, dy$$

$$= \frac{2\pi r \sqrt{r^2 + h^2}}{h^2} \left[\frac{1}{2}y^2\right]_0^h = \pi r \sqrt{r^2 + h^2}$$

**31.** Find the area of the zone of a sphere formed by revolving the graph of $y = \sqrt{9 - x^2}$, $0 \le x \le 2$, about the $y$-axis.

**Solution:**

$$y = \sqrt{9 - x^2} \qquad y' = \frac{-x}{\sqrt{9 - x^2}}$$

$$S = 2\pi \int_0^2 x\sqrt{1 + \left(\frac{-x}{\sqrt{9 - x^2}}\right)^2}\, dx = 2\pi \int_0^2 x\sqrt{\frac{9}{9 - x^2}}\, dx$$

$$= 2\pi \int_0^2 3x(9 - x^2)^{-1/2}\, dx = -3\pi \int_0^2 (9 - x^2)^{-1/2}(-2x)\, dx$$

$$= -3\pi \left[\frac{(9 - x^2)^{1/2}}{1/2}\right]_0^2 = -6\pi(\sqrt{5} - 3) = 6\pi(3 - \sqrt{5}) \approx 14.40$$

## 6.5   Work

**7.** A force of 60 lb stretches a spring 1 ft. How much work is done in stretching the spring from 9 in to 15 in?

**Solution:**

Let $F(x)$ be the force required to stretch a spring $x$ units. By Hooke's Law we know that $F(x) = kx$. Since 60 lb stretches the spring 1 ft(12 in), we have

$$60 = k(12) \quad \text{or} \quad k = 5$$

Therefore, the work done by stretching the spring from 9 to 15 in is

$$W = \int_9^{15} \underbrace{F(x)}_{\text{(force)}} \underbrace{dx}_{\text{(distance)}} = \int_9^{15} 5x \, dx = \left[\frac{5x^2}{2}\right]_9^{15}$$

$$= \frac{5}{2}(225 - 81) = 360 \ \text{in} \cdot \text{lb}$$

**15.** A hemispherical tank of radius 6 ft is positioned so that its base is circular. How much work is required to fill the tank with water through a hole in the base if the water source is at the base?

**Solution:**

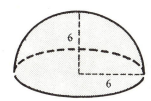

To fill the tank with water from a hole in the bottom, we do not have to move all the water a height of 6 ft. Some of the water must be moved 6 ft, some 5 ft, some 4 ft, and so on. In general, the "disc" of water that must be moved $y$ feet has a volume of

$$\pi x^2 \Delta y = \pi(\sqrt{36 - y^2})^2 \Delta y = \pi(36 - y^2)\Delta y \ \text{cubic feet}$$

The weight of the disc of water is

$$62.4(\pi)(36 - y^2)\Delta y \ \text{pounds}$$

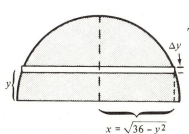

$x = \sqrt{36 - y^2}$

Thus the work done in filling the tank from the bottom is

$$W = \int_0^6 \underbrace{(y)}_{\text{(distance)}} \underbrace{[62.4\pi(36 - y^2)\,dy}_{\text{(force: weight of water)}}$$

$$= 62.4\pi \int_0^6 (36 - y^3)\,dy = 62.4\pi\left[18y^2 - \frac{y^4}{4}\right]_0^6$$

$$= 62.4\pi(648 - 324) = 20,217.6\pi \ \text{ft} \cdot \text{lb}$$

**17.** An open tank has the shape of a right circular cone. If the tank is 8 ft across the top and has a height of 6 ft, how much work is required to empty the tank of water by pumping the water over the top edge?

**Solution:**

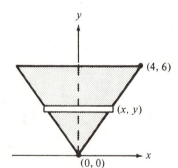

A disc of water at height $y$ has to be moved $(6 - y)$ feet up and has a volume of $\pi x^2 \Delta y$. To find $x$ in terms of $y$, we solve the equation

$$\frac{6 - 0}{4 - 0} = \frac{y - 0}{x - 0}$$

$$x = \frac{2y}{3}$$

Thus the work done in moving the water up over the top of the tank is

$$W = \int_0^6 \underbrace{(6 - y)}_{\text{(distance)}} \underbrace{\left[62.4\pi\left(\frac{2y}{3}\right)^2 dy\right]}_{\text{(force: weight of water)}}$$

$$= \frac{4}{9}(62.4)\pi \int_0^6 (6y^2 - y^3)dy = \frac{83.2}{3}\pi\left[2y^3 - \frac{y^4}{4}\right]_0^6$$

$$= \frac{83.2}{3}\pi(432 - 324) = 2995.2\pi \ \ ft \cdot lb$$

**19.** A cylindrical gasoline tank 3 feet in diameter and 4 feet long is carried on the back of a truck and is used to fuel tractors in the field. The axis of the tank is horizontal. Find the work done to pump the entire contents of the full tank into a tractor if the opening in the tractor tank is 5 ft above the top of the tank in the truck. Assume gasoline weighs 42 pounds per cubic foot.

**Solution:**

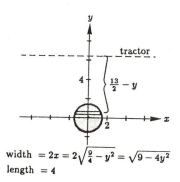

width $= 2x = 2\sqrt{\frac{9}{4} - y^2} = \sqrt{9 - 4y^2}$
length $= 4$

A layer of gasoline at height $y$ (see the accompanying figure) has to be moved $\left(\frac{13}{2} - y\right)$ feet up and has a volume $V = lwh = 4(2x)\Delta y$. To find $x$ in terms of $y$, we solve for $x$ in the equation of the circle representing a cross-section of the tank and obtain

$$x^2 + y^2 = \frac{9}{4}$$

$$x^2 = \frac{9}{4} - y^2$$

$$x = \sqrt{\frac{9}{4} - y^2}$$

Thus the work done is

$$W = \int_{-\frac{1}{5}}^{1.5} \underbrace{\left(\frac{13}{2} - y\right)}_{\text{(distance)}} \underbrace{\left[42(4)(2\sqrt{\frac{9}{4} - y^2}\right) dy\right]}_{\text{(force: weight of water)}}$$

$$= 336\left[\frac{13}{2} \int_{-\frac{1}{5}}^{1.5} \sqrt{\frac{9}{4} - y^2}\, dy - \int_{-\frac{1}{5}}^{1.5} y\sqrt{\frac{9}{4} - y^2}\, dy\right]$$

$$= 336\left[\frac{13}{2} \int_{-1.5}^{1.5} \sqrt{\frac{9}{4} - y^2}\, dy - \int_{-1.5}^{1.5} y\sqrt{\frac{9}{4} - y^2}\, dy\right]$$

The first integral represents the area of a semicircle of radius $3/2$ and the second integral is zero since the integrand is odd and the limits of integration are symmetric to the origin. Therefore,

$$W = 336\left(\frac{13}{2}\right)\left(\frac{1}{2}\right)(\pi)\left(\frac{3}{2}\right)^2 = 2457\pi \text{ ft} \cdot \text{lb}.$$

**31.** A chain 15 ft long and weighing 3lb/ft is suspended vertically from a height of 15 ft. How much work is required to take the bottom of the chain and raise it to the 15-ft level, leaving the chain doubled but still hanging vertically?

**Solution:**

A small piece of chain of length $\Delta y$ at height $y$ must be moved so that it is $y$ feet from the top. Therefore, the distance moved (as seen in the accompanying figure) is $15 - 2y$. (For example, the chain at an initial height of 7.5 is moved 0 ft.) The weight of a piece of chain of length $\Delta y$ is

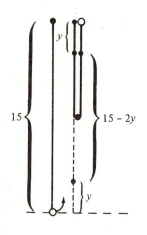

$$\frac{3\,\text{pounds}}{\text{foot}}\Delta y \ \text{feet} = 3\Delta y \ \text{pounds}$$

Finally, since we are only moving chain that has an initial height between 0 and 7.5 ft, we have

$$W = \int_0^{7.5} \underbrace{(15 - 2y)}_{\text{(distance)}}\underbrace{(3\,dy)}_{\text{(force)}}$$

$$= 3\left[15y - y^2\right]_0^{7.5} = 3(112.5 - 56.25) = 168.75 \ \text{ft}\cdot\text{lb}$$

## 6.6   Fluid Pressure and Fluid Force

**7.** Find the force on a vertical side of a tank if the tank is full of water and the side has the shape of a trapezoid, as in the accompanying figure.

**Solution:**

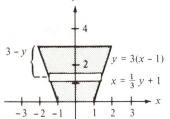

The force against a representative rectangle of length $2x$ is

$$\Delta F = (\text{density})(\text{depth})(\text{area}) = (62.4)(3 - y)(2x\,\Delta y)$$

$$= (62.4)(3 - y)(2)\left(\frac{1}{3}y + 1\right)\Delta y$$

Since $y$ ranges from 0 to 3, the total force is

$$F = \int_0^3 (62.4)(3-y)(2)\left(\frac{1}{3}y+1\right)dy$$

$$= 124.8 \int_0^3 \left(3 - \frac{1}{3}y^2\right)dy$$

$$= 124.8\left[3y - \frac{1}{9}y^3\right]_0^3 = 748.8 \text{ lb}$$

**13.** Find the total force on the vertical plate submerged in water as shown in the accompanying figure.

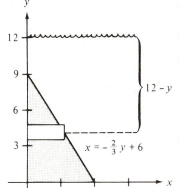

**Solution:**

The force against a representative rectangle of length $x$ is

$$\Delta F = (\text{density})(\text{depth})(\text{area}) = (62.4)(12-y)\left(-\frac{2}{3}y+6\right)\Delta y$$

Since $y$ ranges from 0 to 9, the total force is

$$F = \int_0^9 62.4(12-y)\left(-\frac{2}{3}y+6\right)dy$$

$$= 62.4 \int_0^9 \left(\frac{2}{3}y^2 - 14y + 72\right)dy$$

$$= 62.4\left[\frac{2}{9}y^3 - 7y^2 + 72y\right]_0^9 = 15,163.2 \text{ lb}$$

**19.** A cylindrical gasoline tank is placed so that the axis of the cylinder is horizontal. If the tank is half full, find the force on a circular end of the tank, assuming that the diameter is 3 ft and gasoline has a density of 42 lb/ft³.

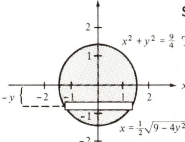

**Solution:**

The force against a representative rectangle of length $2x$ is

$$\Delta F = (\text{density})(\text{depth})(\text{area}) = (42)(-y)(2x\,\Delta y)$$

$$= (42)(-y)(2)\left(\frac{1}{2}\right)(9-4y^2)^{1/2}\Delta y = -42y(9-4y^2)^{1/2}\Delta y$$

Since $y$ varies from $-\frac{3}{2}$ to 0, the total force is

$$F = \int_{-3/2}^{0} (-42)y(9 - 4y^2)^{1/2}\,dy$$

$$= \frac{42}{8} \int_{-3/2}^{0} (9 - 4y^2)^{1/2}(-8y)\,dy$$

$$= \frac{21}{4}\left(\frac{2}{3}\right)\left[(9 - 4y^2)^{3/2}\right]_{-3/2}^{0} = 94.5 \text{ lb}$$

**25.** A swimming pool is 20 ft wide, 40 ft long, 4 ft deep at one end, and 8 ft deep at the other. The bottom is an inclined plane. Find the total force on each of the vertical walls of the pool.

**Solution:**

The force on the rectangular shallow end is given by

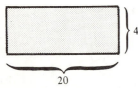

$$F = \int_{0}^{4} 62.4(4 - y)(20)\,dy = 1{,}248 \int_{0}^{4} (4 - y)\,dy$$

$$= 1{,}248\left[4y - \frac{y^2}{2}\right]_{0}^{4} = 1248(8) = 9{,}948 \text{ lb}$$

and for the deep end,

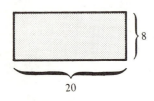

$$F = \int_{0}^{8} 62.4(8 - y)(20)\,dy = 1{,}248 \int_{0}^{8} (8 - y)\,dy$$

$$= 1{,}248\left[8y - \frac{y^2}{2}\right]_{0}^{8} = 1{,}248(32) = 39{,}936 \text{ lb}$$

The force on each side is given by

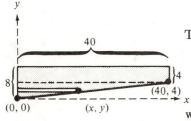

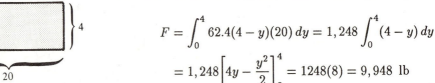

$$F = \int_{0}^{4} 62.4(8 - y)(x)\,dy + \int_{4}^{8} 62.4(8 - y)40\,dy$$

where $x$ and $y$ are related by

$$\frac{4 - 0}{40 - 0} = \frac{y - 0}{x - 0} \quad \text{or} \quad x = 10y$$

Thus

$$F = \int_0^4 62.4(8-y)(10y)\,dy + \int_4^8 62.4(8-y)(40)\,dy$$

$$= 624 \int_0^4 (8y - y^2)\,dy + 2{,}496 \int_4^8 (8-y)\,dy$$

$$= 624 \left[ 4y^2 - \frac{y^3}{3} \right]_0^4 + 2{,}496 \left[ 8y - \frac{y^2}{2} \right]_4^8$$

$$= 624 \left( 64 - \frac{64}{3} \right) + 2{,}496[(64-32)-(32-8)]$$

$$= 624 \left( \frac{128}{3} \right) + 2{,}496(8) = 46{,}592 \text{ lb}$$

## 6.7   Moments, Centers of Mass, and Centroids

**9.** Find the center of mass for the masses located at the given points.

| $m_i$ | 3 | 4 | 2 | 1 | 6 |
|---|---|---|---|---|---|
| $(x_i, y_i)$ | $(-2,-3)$ | $(-1,0)$ | $(7,1)$ | $(0,0)$ | $(-3,0)$ |

**Solution:**

$$\bar{x} = \frac{m_1x_1 + m_2x_2 + m_3x_3 + m_4x_4 + m_5x_5}{m_1 + m_2 + m_3 + m_4 + m_5}$$

$$= \frac{3(-2) + 4(-1) + 2(7) + 1(0) + 6(-3)}{3+4+2+1+6} = -\frac{7}{8}$$

$$\bar{y} = \frac{m_1y_1 + m_2y_2 + m_3y_3 + m_4y_4 + m_5y_5}{m_1 + m_2 + m_3 + m_4 + m_5}$$

$$= \frac{3(-3) + 4(0) + 2(1) + 1(0) + 6(0)}{3+4+2+1+6} = -\frac{7}{16}$$

**11.** A plate of uniform density is formed by a circle and a square as shown in the accompanying figure. Introduce an appropriate rectangular coordinate system and find the coordinates of the center of mass.

**Solution:**

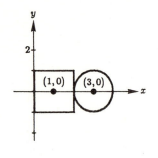

Although a coordinate system may be introduced in many different ways, the one shown in the accompanying figure is a fairly natural choice. Since both the circle and square have a uniform density, their masses are proportional to their areas, $\pi$ and 4, respectively. (For simplicity we assume the density to be 1 unit of mass per 1 unit of area.) Again, because of the uniform density, both the circle and the square have their centers of mass at their geometrical centers, $(3,0)$ and $(1,0)$, respectively. Therefore, we can find the center of mass of the plate by considering a mass of $\pi$ centered at $(3,0)$ and a mass of 4 centered at $(1,0)$. Thus

$$\overline{x} = \frac{\pi(3) + 4(1)}{\pi + 4} = \frac{3\pi + 4}{\pi + 4} \quad \text{and} \quad \overline{y} = \frac{\pi(0) + 4(0)}{\pi + 4} = 0$$

**23.** Find $M_x, M_y$, and $(\overline{x}, \overline{y})$ for the lamina of uniform density $\rho$ bounded by $x = 4 - y^2$ and $x = 0$.

**Solution:**

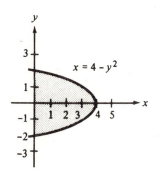

Since the region is symmetric with respect to the $x$-axis, we know that

$$M_x = 0 \text{ and } \overline{y} = \frac{M_x}{m} = 0$$

To find $\overline{x}$, we observe that $x$ is a function of $y$, and we use the formula

$$\overline{x} = \frac{\displaystyle\int_a^b \left[\frac{f(y) + g(y)}{2}\right][f(y) - g(y)]\, dy}{m} = \frac{M_y}{m}$$

where $f(y) = 4 - y^2$, $g(y) = 0$, $a = -2$, $b = 2$.

$$m = \rho \int_{-2}^{2} (4 - y^2)\, dy = \rho \left[ 4y - \frac{y^3}{3} \right]_{-2}^{2} = \frac{32\rho}{3}$$

$$M_y = \frac{\rho}{2} \int_{-2}^{2} (4 - y^2)^2\, dy$$

$$= \frac{\rho}{2} \int_{-2}^{2} (16 - 8y^2 + y^4)\, dy$$

$$= \frac{\rho}{2} \left[ 16y - \frac{8y^3}{3} + \frac{y^5}{5} \right]_{-2}^{2}$$

$$= \frac{\rho}{2} \left[ \left( 32 - \frac{64}{3} + \frac{32}{5} \right) - \left( -32 + \frac{64}{3} - \frac{32}{5} \right) \right]$$

$$= \frac{\rho}{2} \left( \frac{512}{15} \right) = \frac{256\rho}{15}$$

$$\overline{x} = \frac{M_y}{m} = \frac{256\rho/15}{32\rho/3} = \frac{8}{5}$$

**31.** Find the centroid of the region bounded by $y = f(x) = x$ and $y = g(x) = x^2$.

**Solution:**

The two graphs intersect at the points $(0,0)$ and $(1,1)$. The area is given by

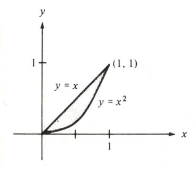

$$A = \int_0^1 [f(x) - g(x)]\, dx$$

$$= \int_0^1 (x - x^2)\, dx = \left[ \frac{1}{2}x^2 - \frac{1}{3}x^3 \right]_0^1 = \frac{1}{6}$$

$$\overline{x} = \frac{\int_0^1 x[f(x) - g(x)]\, dx}{A} = \frac{\int_0^1 x(x - x^2)\, dx}{\frac{1}{6}}$$

$$= 6 \int_0^1 (x^2 - x^3)\, dx = 6 \left[ \frac{1}{3}x^3 - \frac{1}{4}x^4 \right]_0^1 = \frac{1}{2}$$

$$\bar{y} = \frac{\dfrac{1}{2}\displaystyle\int_0^1 \left[\dfrac{f(x)+g(x)}{2}\right][f(x)-g(x)]\,dx}{A}$$

$$= \frac{\dfrac{1}{2}\displaystyle\int_0^1 [(x)^2-(x^2)^2]\,dx}{\dfrac{1}{6}}$$

$$= 3\int_0^1 (x^2-x^4)\,dx = 3\left[\frac{1}{3}x^3-\frac{1}{5}x^5\right]_0^1 = \frac{2}{5}$$

$$(\bar{x},\bar{y}) = \left(\frac{1}{2},\frac{2}{5}\right)$$

**35.** Find the centroid of the triangle with vertices
$(-a,0)$, $(a,0)$, $(b,c)$. Show that it is at the point of intersection
of the medians. (Assume that $-a < b < a$.)

**Solution:**

The equation of the line containing $(-a,0)$ and $(b,c)$ is

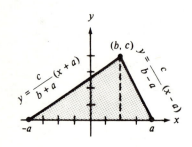

$$y = \left(\frac{c}{b+a}\right)(x+a)$$

The equation of the line containing $(a,0)$, and $(b,c)$ is

$$y = \left(\frac{c}{b-a}\right)(x-a)$$

Since the area of the triangle is

$$A = \left(\frac{1}{2}\right)(2a)(c) = ac$$

we have

$$\bar{x} = \frac{\displaystyle\int_{-a}^{b} x\left(\frac{c}{b+a}\right)(x+a)\,dx + \int_{b}^{a} x\left(\frac{c}{b-a}\right)(x-a)\,dx}{ac}$$

$$= \frac{1}{ac}\left[\frac{c}{b+a}\int_{-a}^{b}(x^2+ax)\,dx + \frac{c}{b-a}\int_{a}^{b}(x^2-ax)\,dx\right]$$

$$= \frac{1}{ac}\left(\frac{c}{b+a}\left[\frac{x^3}{3}+\frac{ax^2}{2}\right]_{-a}^{b} + \frac{c}{b-a}\left[\frac{x^3}{3}-\frac{ax^2}{2}\right]_{b}^{a}\right)$$

$$= \frac{1}{ac}\left[\frac{c}{b+a}\left(\frac{b^3}{3}+\frac{ab^2}{2}+\frac{a^3}{3}-\frac{a^3}{2}\right)\right.$$

$$\left. +\frac{c}{b-a}\left(\frac{a^3}{3}-\frac{a^3}{2}-\frac{b^3}{3}+\frac{ab^2}{2}\right)\right]$$

$$= \frac{2b^3+3ab^2-a^3}{6a(b+a)} + \frac{-2b^3+3ab^2-a^3}{6a(b-a)}$$

$$= \frac{(2b^2+ab-a^2)(a+b)}{6a(b+a)} + \frac{(-2b^2+ab+a^2)(b-a)}{6a(b-a)}$$

$$= \frac{2ab}{6a} = \frac{b}{3}$$

$$\bar{y} = \frac{\displaystyle\frac{1}{2}\int_{-a}^{b}\left[\frac{c}{b+a}(x+a)\right]^2 dx + \frac{1}{2}\int_{b}^{a}\left[\frac{c}{b-a}(x-a)\right]^2 dx}{ac}$$

$$= \frac{1}{2ac}\left\{\frac{c^2}{(b+a)^2}\left[\frac{(x+a)^3}{3}\right]_{-a}^{b} + \frac{c^2}{(b-a)^2}\left[\frac{(x-a)^3}{3}\right]_{b}^{a}\right\}$$

$$= \frac{1}{2ac}\left\{\frac{c^2}{(b+a)^2}\left[\frac{(b+a)^3}{3}\right] - \frac{c^2}{(b-a)^2}\left[\frac{(b-a)^3}{3}\right]\right\}$$

$$= \frac{1}{2ac}\left[\frac{c^2(b+a)}{3} - \frac{c^2(b-a)}{3}\right] = \frac{c}{3}$$

By Exercise 58, Section 1.4, we know that the point $(b/3, c/3)$ is the intersection of the medians of the triangle.

## Review Exercises for Chapter 7

5. Sketch and find the area of the region bounded by the graphs of $x = y^2 - 2y$ and $x = 0$.

**Solution:**

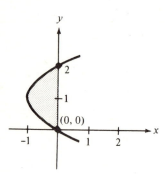

The points of intersection are given by

$$y^2 - 2y = 0$$
$$y(y - 2) = 0 \implies y = 0, \ 2$$

Since $y^2 - 2y \le 0$ for $0 \le y \le 2$, we have

$$A = \int_0^2 [0 - (y^2 - 2y)] \, dy$$

$$= \int_0^2 (-y^2 + 2y) \, dy$$

$$= \left[ \frac{-y^3}{3} + y^2 \right]_0^2 = -\frac{8}{3} + 4 = \frac{4}{3}$$

10. Sketch and find the area of the region bounded by the graphs of $y = x^2 - 4x + 3, y = x^3$, and $x = 0$.

**Solution:**

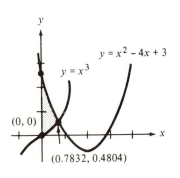

The points of intersection are given by

$$x^3 = x^2 - 4x + 3$$
$$x^3 - x^2 + 4x - 3 = 0$$

Since $x^3 - x^2 + 4x - 3$ does not factor easily, we apply Newton's Method to the function

$$f(x) = x^3 - x^2 + 4x - 3$$

By letting $x_1 = 1$, we have

$$x_2 = x_1 - \frac{f(x_1)}{f'(x_1)} = 0.8000$$

$$x_3 = x_2 - \frac{f(x_2)}{f'(x_2)} = 0.7833$$

$$x_4 = x_3 - \frac{f(x_3)}{f'(x_3)} = 0.7832$$

Since $x^3 \le x^2 - 4x + 3$ for $0 \le x \le 0.7832$, we have

$$A \approx \int_0^{0.7832} (x^2 - 4x + 3 - x^3) \, dx$$

$$= \left[ \frac{x^3}{3} - 2x^2 + 3x - \frac{x^4}{4} \right]_0^{0.7832}$$

$$\approx 0.1601 - 1.2268 + 2.3496 - 0.0941 \approx 1.189$$

**21.** Find the volume of the solid generated by revolving the region bounded by the graphs of $y = -x^2 + 6x - 5$ and $y = 0$ about:
(a)   the $x$-axis.
(b)   the $y$-axis.

**Solution:**

(a)   Disc Method. From the accompany figure we have

$$V = \pi \int_1^5 (-x^2 + 6x - 5)^2 \, dx$$

$$= \pi \int_1^5 (x^4 - 12x^3 + 46x^2 - 60x + 25) \, dx$$

$$= \pi \left[ \frac{x^5}{5} - 3x^4 + \frac{46}{3}x^3 - 30x^2 + 25x \right]_1^5 = \frac{512\pi}{15}$$

(b)   Shell Method. From the accompanying figure we have

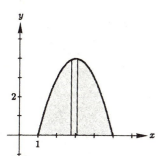

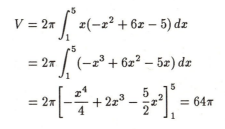

$$V = 2\pi \int_1^5 x(-x^2 + 6x - 5) \, dx$$

$$= 2\pi \int_1^5 (-x^3 + 6x^2 - 5x) \, dx$$

$$= 2\pi \left[ -\frac{x^4}{4} + 2x^3 - \frac{5}{2}x^2 \right]_1^5 = 64\pi$$

**25.** A water well has an 8-in casing (diameter) and is 175 ft deep. If the water is 25 ft from the top of the well, determine the amount of work done in pumping it dry, assuming that no water enters the well while it is being pumped.

**Solution:**

A disk of water at height $y$ has to be moved $(175 - y)$ feet up and has a volume of $\pi(\frac{1}{3})^2 \Delta y$. Thus the work done in moving

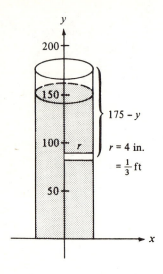

the water up over the top of the well is

$$W = \int_0^{150} \underbrace{(175 - y)}_{\text{(distance)}} \underbrace{\left[62.4\pi \left(\frac{1}{3}\right)^2 dy\right]}_{\text{(force: weight of water)}}$$

$$= \frac{62.4\pi}{9} \int_0^{150} (175 - y) \, dy$$

$$= \frac{62.4\pi}{9} \left[175y - \frac{1}{2}y^2\right]_0^{150}$$

$$= 104,000\pi \text{ ft} \cdot \text{lb} \approx 163.4 \text{ ft} \cdot \text{ton}$$

**30.** Show that the force against any vertical region in a liquid is the product of the density $\rho$ of the liquid, the area of the region, and the depth of the centroid of the region.

**Solution:**

The force is given by

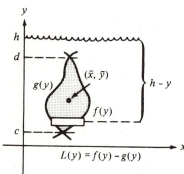

$$F = \int_c^d \rho(h - y)[f(y) - g(y)] \, dy$$

$$= \rho \left\{ h \int_c^d [f(y) - g(y)] \, dy - \int_c^d y[f(y) - g(y)] \, dy \right\}$$

$$= \rho \left\{ \int_c^d [f(y) - g(y)] \, dy \right\} \left\{ h - \frac{\int_c^d y[f(y) - g(y)] \, dy}{\int_c^d [f(y) - g(y)] \, dy} \right\}$$

$$= \rho(\text{area})(h - \bar{y}) = \rho(\text{area})(\text{depth of centroid})$$

**33.** Find the centroid of the region bounded by the graphs of $\sqrt{x} + \sqrt{y} = \sqrt{a}$, $x = 0$, and $y = 0$.

**Solution:**

Solving the equation $\sqrt{x} + \sqrt{y} = \sqrt{a}$ for $y$ we have $y = (\sqrt{a} - \sqrt{x})^2$. Therefore,

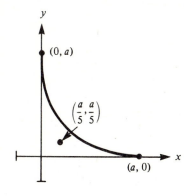

$$A = \int_0^a (\sqrt{a} - \sqrt{x})^2 \, dx = \int_0^a (a - 2\sqrt{a}x^{1/2} + x) \, dx$$

$$= \left[ ax - \frac{4}{3}\sqrt{a}x^{3/2} + \frac{1}{2}x^2 \right]_0^a = \frac{a^2}{6}$$

$$\overline{x} = \frac{1}{(a^2/6)} \int_0^a x(\sqrt{a} - \sqrt{x})^2 \, dx$$

$$= \frac{6}{a^2} \int_0^a (ax - 2\sqrt{a}x^{3/2} + x^2) \, dx$$

$$= \frac{6}{a^2} \left[ \frac{ax^2}{2} - \frac{4}{5}\sqrt{a}x^{5/2} + \frac{1}{3}x^3 \right]_0^a = \frac{1}{5}a$$

By symmetry, we have $\overline{y} = \dfrac{a}{5}$. Therefore, $(\overline{x}, \overline{y}) = \left( \dfrac{a}{5}, \dfrac{a}{5} \right)$.

---

**45.** Find the arc length of the graph of $f(x) = \frac{4}{5}x^{5/4}$ from $x = 0$ to $x = 4$.

**Solution:**

Since $f'(x) = x^{1/4}$, we have

$$s = \int_0^4 \sqrt{1 + [f'(x)]^2} \, dx = \int_0^4 \sqrt{1 + \sqrt{x}} \, dx$$

Let $u = \sqrt{1 + \sqrt{x}}$. Then $u^2 = 1 + \sqrt{x}, x = (u^2 - 1)^2$, and $dx = 2(u^2 - 1)(2u) \, du$. If $x = 0$, then $u = 1$, and if $x = 4$, then $u = \sqrt{3}$. Thus,

$$s = \int_0^4 \sqrt{1 + \sqrt{x}} \, dx = \int_1^{\sqrt{3}} u(2)(u^2 - 1)(2u) \, du$$

$$= 4 \int_0^{\sqrt{3}} (u^4 - u^2) \, du$$

$$= 4 \left[ \frac{u^5}{5} - \frac{u^3}{3} \right]_1^{\sqrt{3}} = \frac{8}{15}(6\sqrt{3} + 1) \approx 6.076$$

# 7 EXPONENTIAL AND LOGARITHMIC FUNCTIONS

## 7.1   Exponential Functions

**9.** Solve for $x$ if $\left(\dfrac{1}{3}\right)^{x-1} = 27$.

**Solution:**

Using the properties of exponents we have

$$\left(\frac{1}{3}\right)^{x-1} = 27$$
$$(3^{-1})^{x-1} = 3^3$$
$$3^{-(x-1)} = 3^3$$
$$-(x-1) = 3$$
$$x - 1 = -3 \quad \Longrightarrow \quad x = -2$$

**27.** Sketch the graph of $y = e^{-x^2}$.

**Solution:**

We begin by making the following observations:

(a)   The graph is symmetric with respect to the $y$-axis since

$$y = e^{-x^2} = e^{-(-x)^2}$$

(b)   The $y$-intercept is $(0, 1)$.

(c)   The $x$-axis is a horizontal asymptote since

$$\lim_{x \to \infty} e^{-x^2} = \lim_{x \to \infty} \frac{1}{e^{x^2}} = 0$$

(d)   The graph lies entirely above the $x$-axis, since for all $x$

$$0 < e^{-x^2}$$

(e)   The graph is decreasing, since $0 < x_1 < x_2$ implies that

$$\frac{1}{e^{x_1^2}} > \frac{1}{e^{x_1^2}}.$$

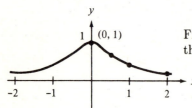

Finally, by plotting a few points, we have the graph shown in the accompanying figure.

| $x$ | 0 | 0.5 | 1 | 2 |
|---|---|---|---|---|
| $y$ | 1 | 0.607 | 0.368 | 0.135 |

**41.** In a group project in learning theory, a mathematical model for the proportion $P$ of correct responses after $n$ trials was found to be

$$P = \frac{0.83}{1 + e^{-0.2n}}$$

(a)   Find the proportion of correct responses after $n = 10$ trials.

(b)   Find the limiting proportion of correct responses as $n \to \infty$.

**Solution:**

(a)
$$P(n) = \frac{0.83}{1 + e^{-0.2n}}$$
$$P(10) = \frac{0.83}{1 + e^{-0.2(10)}}$$
$$= \frac{0.83}{1 + e^{-2}} \approx 0.731$$

(b)
$$\lim_{n \to \infty} \frac{0.83}{1 + e^{-0.2n}} = \frac{0.83}{1 + 0} = 0.83 = 83\%$$

## 7.2   Differentiation and Integration of Exponential Functions

**13.** Find $dy/dx$ given $y = e^{\sqrt{x}}$.

**Solution:**

$$y = e^{\sqrt{x}}$$

$$\frac{dy}{dx} = e^{\sqrt{x}} \frac{d}{dx}[\sqrt{x}]$$

$$= e^{\sqrt{x}}\left(\frac{1}{2\sqrt{x}}\right) = \frac{e^{\sqrt{x}}}{2\sqrt{x}}$$

**19.** Find $dy/dx$ given $y = (e^{-x} + e^{x})^3$.

**Solution:**

$$y = (e^{-x} + e^{x})^3$$

$$\frac{dy}{dx} = 3(e^{-x} + e^{x})^2[e^{-x}(-1) + e^{x}]$$

$$= 3(e^{-x} + e^{x})^2(e^{x} - e^{-x})$$

**33.** Find (if any exist) the extrema and the points of inflection, and sketch the graph of $f(x) = x^2 e^{-x}$.

**Solution:**

$$f(x) = x^2 e^{-x} \qquad \text{Intercepts:} \quad (0,0)$$

$$f'(x) = x^2(-e^{-x}) + e^{-x}(2x) = xe^{-x}(-x + 2)$$

$$f''(x) = (xe^{-x})(-1) + (-x + 2)[x(-e^{-x}) + e^{-x}(1)]$$

$$= e^{-x}[-x + (-x + 2)(-x + 1)] = e^{-x}(x^2 - 4x + 2)$$

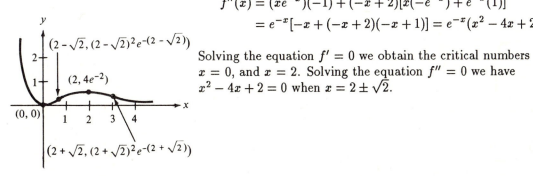

Solving the equation $f' = 0$ we obtain the critical numbers $x = 0$, and $x = 2$. Solving the equation $f'' = 0$ we have $x^2 - 4x + 2 = 0$ when $x = 2 \pm \sqrt{2}$.

| Interval | $f(x)$ | $f'(x)$ | $f''(x)$ | Shape of graph |
|---|---|---|---|---|
| $x$ in $(-\infty, 0)$ | | $-$ | $+$ | decreasing, concave up |
| $x = 0$ | $0$ | $0$ | $+$ | relative minimum |
| $x$ in $(0, 2 - \sqrt{2})$ | | $+$ | $+$ | increasing, concave up |
| $x = 2 - \sqrt{2}$ | $0.191$ | $+$ | $0$ | point of inflection |
| $x$ in $(2 - \sqrt{2}, 2)$ | | $+$ | $-$ | increasing, concave down |
| $x = 2$ | $0.541$ | $0$ | $-$ | relative maximum |
| $x$ in $(2, 2 + \sqrt{2})$ | | $-$ | $-$ | decreasing, concave down |
| $x = 2 + \sqrt{2}$ | $0.384$ | $-$ | $0$ | point of inflection |
| $x$ in $(2 + \sqrt{2}, \infty)$ | | $-$ | $+$ | decreasing, concave up |

**41.** In a group project in learning theory, a mathematical model for the proportion $P$ of correct responses after $n$ trials was found to be

$$P = \frac{0.83}{1 + e^{-0.2n}}.$$

Find the rate at which $P$ is changing after $n = 3$ and $n = 10$ trials.

**Solution:**

$$P(n) = \frac{0.83}{1 + e^{-0.2n}} = 0.83(1 + e^{-0.2n})^{-1}$$

$$P'(n) = 0.83(-1)(1 + e^{-0.2n})^{-2}(e^{-0.2n})(-0.2)$$

$$= \frac{0.166e^{-0.2n}}{(1 + e^{-0.2n})^2}$$

$$P'(3) = \frac{0.166e^{-0.6}}{(1 + e^{-0.6})^2} \approx 0.038$$

$$P'(10) = \frac{0.166e^{-2}}{(1 + e^{-2})^2} \approx 0.017$$

**45.** Evaluate $\int_0^2 (x^2 - 1) e^{x^3 - 3x + 1} \, dx$.

**Solution:**

Letting $u = x^3 - 3x + 1$, we have $du = (3x^2 - 3) \, dx$ or $(x^2 - 1) \, dx = du/3$. Therefore, when $x = 0$, $u = 1$ and when $x = 2$, $u = 3$.

$$\int_0^2 (x^2 - 1) e^{x^3 - 3x + 1} \, dx = \int_0^2 e^{x^3 - 3x + 1} (x^2 - 1) \, dx$$
$$= \int_1^3 e^u \left( \frac{du}{3} \right)$$
$$= \frac{1}{3} [e^u]_1^3 = \frac{e}{3}(e^2 - 1) \approx 5.789$$

**51.** Evaluate $\int_1^3 \frac{e^{3/x}}{x^2} \, dx$.

**Solution:**

Letting $u = 3/x$, we have $du = (-3/x^2) \, dx$. When $x = 1$, $u = 3$ and when $x = 3$, $u = 1$. Therefore,

$$\int_1^3 \frac{e^{3/x}}{x^2} \, dx = -\frac{1}{3} \int_1^3 e^{3/x} \left( -\frac{3}{x^2} \right) dx$$
$$= -\frac{1}{3} \int_3^1 e^u \, du$$
$$= -\frac{1}{3} [e^u]_3^1 = -\frac{1}{3}(e - e^3) = \frac{e}{3}(e^2 - 1)$$

**57.** Evaluate

$$\int \frac{e^x + e^{-x}}{\sqrt{e^x - e^{-x}}} \, dx$$

**Solution:**

Letting $u = e^x - e^{-x}$, we have $du = (e^x + e^{-x}) \, dx$

$$\int \frac{e^x + e^{-x}}{e^x - e^{-x}} \, dx = \int \frac{1}{\sqrt{u}} \, du$$
$$= 2\sqrt{u} + C = 2\sqrt{e^x - e^{-x}} + C$$

**75.** The standard normal probability density function is

$$f(z) = \frac{1}{\sqrt{2\pi}} e^{-z^2/2}.$$

The probability that $z$ is in the interval $[a, b]$ is the area of the region defined by $y = f(z)$, $y = 0$, $z = a$, and $z = b$ and is denoted by $Pr(a \leq z \leq b)$. Estimate $Pr(0 \leq z \leq 1)$ using Simpson's Rule with $n = 6$.

**Solution:**

$$Pr(0 \leq z \leq 1) = \frac{1}{\sqrt{2\pi}} \int_0^1 e^{-z^2/2}\, dz$$

$$\approx \frac{1}{\sqrt{2\pi}} \left[\frac{1}{3(6)}\right] \left[e^0 + 4e^{-(1/6)^2/2} + 2e^{-(1/3)^2/2}\right.$$

$$\left. + 4e^{-(1/2)^2/2} + 2e^{-(2/3)^2/2} + 4e^{-(5/6)^2/2} + e^{-1/2}\right]$$

$$\approx 0.3413$$

## 7.3   Inverse Functions

**5.** (a) Show that $f(x) = \sqrt{x - 4}$ and $g(x) = x^2 + 4$ $(x \geq 0)$ are inverse functions by showing that $f(g(x)) = g(f(x)) = x$, and (b) graph $f$ and $g$ on the same set of coordinate axes.

**Solution:**

$$f(x) = \sqrt{x - 4} \quad \text{and} \quad g(x) = x^2 + 4 \quad (x \geq 0)$$

The composite of $f$ and $g$ is given by

$$f(g(x)) = f(x^2 + 4) = \sqrt{(x^2 + 4) - 4} = \sqrt{x^2} = x$$

The composite of $g$ with $f$ is given by

$$g(f(x)) = g(\sqrt{x - 4}) = (\sqrt{x - 4})^2 + 4 = x - 4 + 4 = x$$

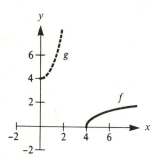

Since $f(g(x)) = g(f(x)) = x$, we conclude that $f$ and $g$ are inverses of each other.

**9.** Find the inverse of $f(x) = 2x - 3$ and then graph both $f$ and $f^{-1}$.

**Solution:**

We begin by observing that $f$ is increasing on its entire domain. To find an equation for the inverse, we let $y = f(x)$ and solve for $x$ in terms of $y$.

$$2x - 3 = y$$

$$x = \frac{y + 3}{2}$$

$$f^{-1}(y) = \frac{y + 3}{2}$$

Finally, using $x$ as the independent variable, we have

$$f^{-1}(x) = \frac{x + 3}{2}$$

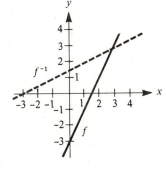

**19.** Find the inverse of $f(x) = x^{2/3}$ $(x \geq 0)$ and then graph both $f$ and $f^{-1}$.

**Solution:**

We begin by observing that $f$ is increasing on $[0, \infty)$. To find an equation for the inverse, we let $y = f(x)$ and solve for $x$ in terms of $y$.

$$x^{2/3} = y \qquad x \geq 0$$

$$x = y^{3/2} \qquad y \geq 0$$

$$f^{-1}(y) = y^{3/2} \qquad y \geq 0$$

Finally, using $x$ as the independent variable, we have

$$f^{-1}(x) = x^{3/2} \qquad x \geq 0$$

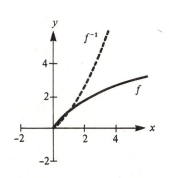

**23.** Use the graph of the function $f$ in the accompanying figure to complete the table and sketch the graph of $f^{-1}$.

| $x$ | 1 | 2 | 3 | 4 |
|---|---|---|---|---|
| $f^{-1}(x)$ |  |  |  |  |

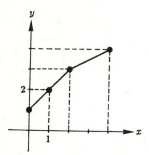

**Solution:**

From the accompanying graph we complete the following table for the given function $f$.

| $x$ | 0 | 1 | 2 | 4 |
|---|---|---|---|---|
| $f(x)$ | 1 | 2 | 3 | 4 |

From the reflective property of inverses, we know the graph of $f$ contains the point $(a, b)$ if and only if the graph of $f^{-1}$ contains the point $(b, a)$. Therefore we obtain the required table by interchanging the $x$ and $y$ coordinates of the preceeding table.

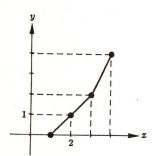

| $x$ | 1 | 2 | 3 | 4 |
|---|---|---|---|---|
| $f^{-1}(x)$ | 0 | 1 | 2 | 4 |

**41.** Show that $f(x) = 4/x^2$ is strictly monotonic on the interval $(0, \infty)$ and therefore has an inverse on that interval.

**Solution:**

Given the function $f(x) = 4/x^2$, we have

$$f'(x) = \frac{-8}{x^3} < 0$$

on the interval $(0, \infty)$. Therefore, $f$ is monotonically decreasing on $(0, \infty)$ and has an inverse.

**45.** Show that the slopes of the graphs of $f(x) = \sqrt{x - 4}$ and $f^{-1}(x) = x^2 + 4$ are reciprocals at the points $(5, 1)$ and $(1, 5)$, respectively.

**Solution:**

For the function $f(x) = \sqrt{x - 4}$, we have

$$f'(x) = \frac{1}{2\sqrt{x - 4}} \qquad \text{and} \qquad f'(5) = \frac{1}{2\sqrt{5 - 4}} = \frac{1}{2}$$

For the function $f^{-1}(x) = x^2 + 4$, we have

$$(f^{-1})'(x) = 2x \qquad \text{and} \qquad (f^{-1})'(1) = 2$$

Therefore, $(f^{-1})'(1) = \dfrac{1}{f'(5)}$.

## 7.4   Logarithmic Functions

**5.** Write the following logarithmic equations as exponential equations:

(a)   $\ln 2 = 0.6931\ldots$ \qquad (b)   $\ln 8.4 = 2.128\ldots$

**Solution:**

By definition,

$$\log_a x = b \qquad \text{if and only if} \qquad a^b = x$$

(a)   By letting $a = e$, $x = 2$, and $b = 0.6931\ldots$, we have

$$\ln 2 = \log_e 2 = 0.6931\ldots \text{ if and only if } e^{0.6931\cdots} = 2$$

(b)   By letting $a = 3$, $x = 8.4$, and $b = 2.128\ldots$, we have

$$\ln 8.4 = \log_e 8.4 = 2.128\ldots \text{ if and only if } e^{2.128\cdots} = 8.4$$

**8.** Find $x$ if (a) $\log_4(1/64) = x$, and (b) $\log_5 25 = x$.

**Solution:**

(a)   $x = \log_4 \dfrac{1}{64} = \log_4 \dfrac{1}{4^3} = \log_4(4^{-3}) = -3$

(b)   $x = \log_5 25 = \log_5(5^2) = 2$

**13.** Find $x$ if (a) $x^2 - x = \log_5 25$ and (b) $3x + 5 = \log_2 64$.

**Solution:**

(a)
$$x^2 - x = \log_5 25$$
$$x^2 - x = \log_5(5^2)$$
$$x^2 - x = 2$$
$$x^2 - x - 2 = 0$$
$$(x - 2)(x + 1) = 0 \quad \implies \quad x = -1,\ 2$$

(b)
$$3x + 5 = \log_2 64$$
$$3x + 5 = \log_2(2^6)$$
$$3x + 5 = 6$$
$$3x = 1 \quad \implies \quad x = \frac{1}{3}$$

**31.** Using properties of logarithms and the fact that $\ln 2 \approx 0.6931$ and $\ln 3 \approx 1.0986$, find the following:

(a)  $\ln 6$             (b)  $\ln \frac{2}{3}$

(c)  $\ln 81$           (d)  $\ln \sqrt{3}$

**Solution:**

(a)  $\ln 6 = \ln (2 \cdot 3)$
$$= \ln 2 + \ln 3 \approx 0.6931 + 1.0986 = 1.7917$$

(b)  $\ln \frac{2}{3} = \ln 2 - \ln 3$
$$\approx 0.6931 - 1.0986 = -0.4055$$

(c)  $\ln 81 = \ln (3^4) = 4 \ln 3 \approx 4(1.0986) = 4.3944$

(d)  $\ln \sqrt{3} = \ln (3^{1/2}) = \frac{1}{2} \ln 3 \approx \frac{1}{2}(1.0986) = 0.5493$

**39.** Use the properties of logarithms to write

$$\ln\left(\frac{x^2-1}{x^3}\right)^3$$

as a sum, difference, or multiple of logarithms.

**Solution:**

$$\ln\left(\frac{x^2-1}{x^3}\right)^3 = 3[\ln(x^2-1) - \ln x^3]$$

$$= 3[\ln[(x+1)(x-1)] - 3\ln x]$$
$$= 3[\ln(x+1) + \ln(x-1) - 3\ln x]$$

**45.** Use the properties of logarithms to write

$$\frac{1}{3}[2\ln(x+3) + \ln x - \ln(x^2-1)]$$

as a single quantity.

**Solution:**

$$\frac{1}{3}[2\ln(x+3) + \ln x - \ln(x^2-1)]$$

$$= \frac{1}{3}[\ln(x+3)^2 + \ln x - \ln(x^2-1)]$$

$$= \frac{1}{3}[\ln x(x+3)^2 - \ln(x^2-1)]$$

$$= \frac{1}{3}\ln\frac{x(x+3)^2}{x^2-1}$$

$$= \ln\left[\frac{x(x+3)^2}{x^2-1}\right]^{1/3} = \ln\sqrt[3]{\frac{x(x+3)^2}{x^2-1}}$$

**61.** A deposit of $1000 is made into a fund with an annual interest rate of 11%. Find the time for the investment to double if the interest is compounded

(a)   annually

(b)   monthly

(c)   daily

(d)   continuously.

**Solution:**

For parts a, b, and c we use the formula

$$A = P\left(1 + \frac{r}{n}\right)^{nt}$$

where $P$ is the principal, $r$ is the annual percentage rate, $n$ is the number of compoundings per year, $t$ is time in years, and $A$ is the amount. To double the investment, we have

$$P\left(1 + \frac{r}{n}\right)^{nt} = 2P$$

$$\left(1 + \frac{r}{n}\right)^{nt} = 2$$

$$\ln\left(1 + \frac{r}{n}\right)^{nt} = \ln 2$$

$$(nt)\ln\left(1 + \frac{r}{n}\right) = \ln 2 \quad \Longrightarrow \quad t = \frac{\ln 2}{n\ln\left(1 + \frac{r}{n}\right)}$$

(a) If the interest is compounded annually, $n = 1$ and we have

$$t = \frac{\ln 2}{(1)\ln\left(1 + \frac{0.11}{1}\right)} \approx 6.642 \text{ years.}$$

(b) If the interest is compounded monthly, $n = 12$ and we have

$$t = \frac{\ln 2}{12\ln\left(1 + \frac{0.11}{12}\right)} \approx 6.330 \text{ years.}$$

(c) If the interest is compounded daily, $n = 365$ and we have

$$t = \frac{\ln 2}{365\ln\left(1 + \frac{0.11}{365}\right)} \approx 6.302 \text{ years.}$$

(d) Since the interest is compounded continuously, use the formula $A = Pe^{rt}$. To double the investment, we have

$$Pe^{rt} = 2P$$
$$e^{rt} = 2$$
$$\ln e^{rt} = \ln 2$$
$$rt = \ln 2 \quad \Longrightarrow \quad t = \frac{\ln 2}{r}.$$

Therefore, the time required to double the investment when $r = 0.11$ is

$$t = \frac{\ln 2}{0.11} \approx 6.301 \text{ years.}$$

## 7.5   Logarithmic Functions and Differentiation

**11.** Find $dy/dx$ given $y = \ln\left(x\sqrt{x^2 - 1}\right)$.

**Solution:**

$$y = \ln\left(x\sqrt{x^2 - 1}\right) = \ln x + \ln\sqrt{x^2 - 1}$$
$$= \ln x + \frac{1}{2}\ln(x^2 - 1)$$
$$\frac{dy}{dx} = \frac{1}{x} + \frac{1}{2}\left(\frac{1}{x^2 - 1}\right)(2x) = \frac{1}{x} + \frac{x}{x^2 - 1} = \frac{2x^2 - 1}{x(x^2 - 1)}$$

**15.** Find $dy/dx$ given $y = (\ln x)/x^2$.

**Solution:**

$$y = \frac{\ln x}{x^2}$$
$$\frac{dy}{dx} = \frac{(x^2)(1/x) - (\ln x)(2x)}{x^4} = \frac{x - 2x\ln x}{x^4} = \frac{1 - 2\ln x}{x^3}$$

**23.** Find $dy/dx$ given $y = -\sqrt{x^2 + 1}/x + \ln(x + \sqrt{x^2 + 1})$.

**Solution:**

$$y = \frac{-\sqrt{x^2 + 1}}{x} + \ln(x + \sqrt{x^2 + 1})$$

$$\frac{dy}{dx} = -\frac{x(\frac{1}{2})(x^2 + 1)^{-1/2}(2x) - (x^2 + 1)^{1/2}(1)}{x^2}$$

$$+ \frac{1}{x + \sqrt{x^2 + 1}}\left[1 + \left(\frac{1}{2}\right)(x^2 + 1)^{-1/2}(2x)\right]$$

$$= -\frac{x^2(x^2 + 1)^{-1/2} - (x^2 + 1)^{1/2}}{x^2}$$

$$+ \left(\frac{1}{x + \sqrt{x^2 + 1}}\right)\left(1 + \frac{x}{\sqrt{x^2 + 1}}\right)$$

$$= \frac{-x^2 + (x^2 + 1)}{x^2\sqrt{x^2 + 1}} + \frac{1}{\sqrt{x^2 + 1}}$$

$$= \frac{1 + x^2}{x^2\sqrt{x^2 + 1}} = \frac{\sqrt{x^2 + 1}}{x^2}$$

**29.** Find $dy/dx$ given $y = x^2 2^x$.

**Solution:**

Using the Product Rule and Part 1 of Theorem 7.11, we have

$$y = x^2 2^x$$

$$\frac{dy}{dx} = x^2[(\ln 2)2^x] + 2^x(2x)$$

$$= x(2^x)(x \ln 2 + 2)$$

**33.** Find $dy/dx$ given $y = \log_2 \left( \dfrac{x^2}{x-1} \right)$.

**Solution:**

Using the properties of logarithms, we have

$$y = \log_2 \left( \frac{x^2}{x-1} \right) = 2 \log_2 x - \log_2 (x-1).$$

To differentiate we use Parts 3 and 4 of Theorem 7.11 and obtain

$$\frac{dy}{dx} = \frac{2}{(\ln 2)x} - \frac{1}{(\ln 2)(x-1)}(1)$$

$$= \frac{1}{\ln 2} \left( \frac{2}{x} - \frac{1}{x-1} \right) = \frac{x-2}{(\ln 2)[x(x-1)]}$$

**41.** Find $dy/dx$ by using logarithmic differentiation given

$$y = \frac{x^2 \sqrt{3x-2}}{(x-1)^2}$$

**Solution:**

We begin by taking the natural logarithm of each member of the equation.

$$y = \frac{x^2 \sqrt{3x-2}}{(x-1)^2}$$

$$\ln y = \ln \frac{x^2 \sqrt{3x-2}}{(x-1)^2} = 2 \ln x + \frac{1}{2} \ln (3x-2) - 2 \ln (x-1)$$

$$\frac{1}{y} \frac{dy}{dx} = \frac{2}{x} + \left( \frac{1}{2} \right) \frac{3}{3x-2} - 2 \frac{1}{x-1}$$

$$\frac{dy}{dx} = y \left[ \frac{3x^2 - 15x + 8}{2x(3x-2)(x-1)} \right]$$

If the solution is desired in terms of $x$ above, we can replace $y$ from the original equation and obtain

$$y' = \frac{3x^3 - 15x^2 + 8x}{2(x-1)^3 \sqrt{3x-2}}$$

49. Show that $y = 2\ln x + 3$ is a solution to the differential equation $x(y'') + y' = 0$.

**Solution:**

$$y = 2(\ln x) + 3$$

$$y' = 2\left(\frac{1}{x}\right)$$

$$y'' = 2\left(\frac{-1}{x^2}\right) = \frac{-2}{x^2}$$

$$x(y'') + y' = x\left(\frac{-2}{x^2}\right) + \left(\frac{2}{x}\right) = \left(\frac{-2}{x}\right) + \left(\frac{2}{x}\right) = 0$$

55. Find any relative extrema and inflection points for $y = (x^2/2) - \ln x$, and sketch the graph of the function.

**Solution:**

$$y = \frac{x^2}{2} - \ln x \qquad \text{Domain:} \quad 0 < x$$

$$y' = x - \frac{1}{x}$$

$$y'' = 1 + \frac{1}{x^2} > 0$$

We first observe that $y' = 0$ when $x = \pm 1$. However, $x = -1$ is not in the domain of the function. Since $y''$ is positive for all $x$ in the domain, the graph is concave up and $(1, \frac{1}{2})$ is a relative minimum point. Furthermore, since $y''$ is never zero, there are no points of inflection. By plotting a few points, we have the graph shown in the accompanying figure.

| $x$ | 0.25 | 0.5 | 1 | 1.5 | 2 | 3 |
|---|---|---|---|---|---|---|
| $y$ | 1.418 | 0.818 | 0.5 | 0.720 | 1.307 | 3.401 |

**61.** Use Newton's Method to approximate, to three decimal places, the value of the $x$-coordinate of the point of intersection of the graphs of the equations $y = \ln x$ and $y = -x$.

**Solution:**

To find the $x$-coordinate of the point of intersection of the graphs of $y = \ln x$ and $y = -x$, we find the zeros of the function $f(x) = \ln x + x$. Therefore, by Newton's Method we have

$$x_{n+1} = x_n - \frac{f(x_n)}{f'(x_n)} = x_n - \frac{\ln x_n + x_n}{(1/x_n) + 1}.$$

From the accompanying figure we choose $x = 0.5$ as the initial estimate

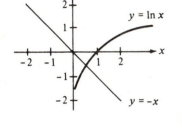

| $n$ | $x_n$ | $f(x_n)$ | $f'(x_n)$ | $f(x_n)/f'(x_n)$ | $x_n - \dfrac{f(x_n)}{f'(x_n)}$ |
|---|---|---|---|---|---|
| 1 | 0.5000 | −0.1931 | 3 | −0.0644 | 0.5644 |
| 2 | 0.5644 | −0.0076 | 2.7718 | −0.0027 | 0.5671 |
| 3 | 0.5671 | 0.0000 | 2.7632 | 0.0000 | 0.5671 |

Therefore, we approximate the zero to be $x = 0.567$.

## 7.6   Logarithmic Functions and Integration

**5.** Evaluate $\displaystyle\int \frac{x}{x^2 + 1}\, dx$.

**Solution:**

Letting $u = x^2 + 1$, we have $du = 2x\, dx$. By multiplying and dividing by 2, we have

$$\int \frac{x}{x^2 + 1}\, dx = \frac{1}{2} \int \frac{2x}{x^2 + 1}\, dx = \frac{1}{2} \int \frac{1}{u}\, du$$

$$= \frac{1}{2} \ln u + C = \frac{1}{2} \ln (x^2 + 1) + C$$

$$= \ln \sqrt{x^2 + 1} + C$$

[Note that absolute value signs around $(x^2 + 1)$ are unnecessary since $x^2 + 1 > 0$ for all $x$.]

**7.** Evaluate $\displaystyle\int \frac{x^2 - 4}{x}\, dx$.

**Solution:**

$$\int \frac{x^2 - 4}{x}\, dx = \int \left( \frac{x^2}{x} - \frac{4}{x} \right) dx$$

$$= \int \left( x - \frac{4}{x} \right) dx = \frac{x^2}{2} - 4\ln|x| + C$$

**11.** Evaluate $\displaystyle\int_1^e \frac{(1 + \ln x)^2}{x}\, dx$.

**Solution:**

Letting $u = 1 + \ln x$, we have $du = (1/x)\, dx$. When $x = 1$, $u = 1$ and when $x = e$, $u = 2$. Therefore,

$$\int_1^e \frac{(1 + \ln x)^2}{x}\, dx = \int_1^e (1 + \ln x)^2 \left( \frac{1}{x} \right) dx$$

$$= \int_1^2 u^2\, du = \left[ \frac{u^3}{3} \right]_1^2 = \frac{7}{3}$$

**15.** Evaluate $\displaystyle\int \frac{1}{\sqrt{x + 1}}\, dx$.

**Solution:**

$$\int \frac{1}{\sqrt{x + 1}}\, dx = \int (x + 1)^{-1/2}(1)\, dx$$

$$= \int u^{-1/2}\, du$$

$$= \frac{u^{1/2}}{\frac{1}{2}} = 2(x + 1)^{1/2} + C = 2\sqrt{x + 1} + C$$

**25.** Evaluate $\displaystyle\int \frac{\sqrt{x}}{1 - x\sqrt{x}}\, dx$.

**Solution:**

Letting $u = 1 - x\sqrt{x} = 1 - x^{3/2}$, we have $du = -\frac{3}{2}\sqrt{x}\, dx$.
Therefore,

$$\int \frac{\sqrt{x}}{1 - x\sqrt{x}}\, dx = \int \frac{\sqrt{x}}{1 - x^{3/2}}\, dx$$

$$= -\frac{2}{3} \int \frac{1}{u}\, du$$

$$== \frac{2}{3} \ln|u| + C = -\frac{2}{3} \ln|1 - x\sqrt{x}| + C$$

**27.** Evaluate $\displaystyle\int \frac{x(x-2)}{(x-1)^3}\, dx$.

**Solution:**

$$\int \frac{x(x-2)}{(x-1)^3}\, dx = \int \frac{x^2 - 2x + 1 - 1}{(x-1)^3}\, dx = \int \frac{(x-1)^2 - 1}{(x-1)^3}\, dx$$

$$= \int \frac{1}{(x-1)}\, dx - \int \frac{1}{(x-1)^3}\, dx$$

$$= \int \frac{1}{(x-1)}\, dx - \int (x-1)^{-3}\, dx$$

$$= \int \frac{1}{u}\, du - \int u^{-3}\, du$$

$$= \ln|u| - \frac{u^{-2}}{-2} + C$$

$$= \ln|x-1| + \frac{1}{2(x-1)^2} + C$$

**33.** Evaluate $\displaystyle\int x\, 5^{x^2}\, dx$.

**Solution:**

Letting $u = x^2$, we have $du = 2x\,dx$. Therefore,

$$\int x\,5^{x^2}\,dx = \frac{1}{2}\int 5^{x^2}(2x)\,dx$$

$$= \frac{1}{2}\int 5^u\,du$$

$$= \frac{1}{2\ln 5}5^u + C = \frac{1}{2\ln 5}5^{x^2} + C$$

**35.** Find the area of the region bounded by the graphs of $y = (x^2 + 4)/x$, $y = 0$, $x = 1$, and $x = 4$.

**Solution:**

$$\text{area} = \int_1^4 \frac{x^2 + 4}{x}\,dx = \int_1^4\left(x + \frac{4}{x}\right)dx$$

$$= \left[\frac{1}{2}x^2 + 4\ln x\right]_1^4$$

$$= 8 + 4\ln 4 - \frac{1}{2}$$

$$= \frac{1}{2}(15 + 16\ln 2) \approx 13.045 \text{ square units}$$

**39.** Find the volume of the solid generated by revolving the region bounded by the graphs of the equations $xy = 1$, $y = 0$, $x = 1$, and $x = 4$ about the line $y = 4$.

**Solution:**

Using the Disc Method with $R(x) = 4$ and $r(x) = 4 - (1/x)$

(see the accompanying figure), we have

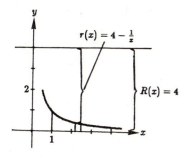

$r(x) = 4 - \frac{1}{x}$

$R(x) = 4$

$$V = \pi \int_a^b \left([R(x)]^2 - [r(x)]^2\right) dx$$

$$= \pi \int_1^4 \left[4^2 - \left(4 - \frac{1}{x}\right)^2\right] dx$$

$$= \pi \int_1^4 \left(\frac{8}{x} - \frac{1}{x^2}\right) dx$$

$$= \pi \left[8 \ln x + \frac{1}{x}\right]_1^4$$

$$= \pi \left[8 \ln 4 - \frac{3}{4}\right] = \frac{\pi}{4}(32 \ln 4 - 3)$$

## 7.7   Growth and Decay

7. Given that an initial investment of \$750 in a savings account in which interest is compounded continuously doubles in $7\frac{3}{4}$ years, find the amount in the account after 10 years and the amount after 25 years.

**Solution:**

Since the interest is compounded coninuously, we use the formula

$$A = Pe^{rt}$$

To determine the rate, we have

$$2(750) = 750e^{r(7.75)}$$

$$2 = e^{7.75r}$$

$$\ln 2 = \ln e^{7.75r} = 7.75r$$

$$r = \frac{\ln 2}{7.75} \approx 0.0894 = 8.94\%$$

The amount after 10 years is given by

$$A = 750e^{\left(\frac{\ln 2}{7.75}\right)10} = \$1834.37$$

The amount after 25 years is given by

$$A = 750e^{\left(\frac{\ln 2}{7.75}\right)25} = \$7016.58$$

**11.** The management at a certain factory has found that a worker can produce at most 30 units in a day. The learning curve for the number of units $N$ produced per day after a new employee has worked $t$ days is given by

$$N(t) = 30(1 - e^{kt})$$

After 20 days on the job, a particular worker produced 19 units.

(a) Find the learning curve for this worker.

(b) How many days should pass before this worker is producing 25 units per day?

**Solution:**

(a) Since $N(20) = 19$, we have

$$19 = 30(1 - e^{20k})$$
$$30e^{20k} = 11$$
$$e^{20k} = \frac{11}{30}$$
$$k = \frac{\ln(11/30)}{20} \approx -0.0502$$

Therefore,
$$N(t) = 30(1 - e^{-0.0502t})$$

(b) To determine the time when the worker will be producing 25 units per day we solve the equation

$$25 = 30(1 - e^{-0.0502t})$$
$$e^{-0.0502t} = \frac{1}{6}$$
$$t = \frac{-\ln 6}{-0.0502} \approx 36 \text{ days}$$

**23.** If the half life of the radioactive isotope $Pu^{230}$ is 24,360 years, find the initial amount of the isotope if after 1000 years there are 2.1 grams. How many grams will there be after 10,000 years?

**Solution:**

Let $y$ represent the mass (in grams) of the isotope after $t$ years. Since the rate of decay is proportional to $y$, we apply the Law of Experimental Decay to conclude that $y = y_0 e^{kt}$ where $y_0$ is the initial amount. Since the half life of $Pu^{230}$ is 24,360 years, we have

$$y_0 e^{24,360k} = \frac{1}{2} y_0$$

$$e^{24,360k} = \frac{1}{2}$$

$$24,360k = \ln \frac{1}{2}$$

$$k = \frac{-\ln 2}{24,360}$$

Therefore, $y = y_0 e^{\left(\frac{-\ln 2}{24,360}\right)t}$. Since $y = 2.1$ when $t = 1000$, we have

$$y_0 e^{\left(\frac{-\ln 2}{24,360}\right)1000} = 2.1$$

$$y_0 = 2.1e^{\frac{1000\ln 2}{24,360}} \approx 2.16 \text{ grams.}$$

After 10,000 years the amount of the isotope remaining will be

$$y = 2.16e^{\left(\frac{-\ln 2}{24,360}\right)10,000} \approx 1.6 \text{ grams.}$$

**31.** Using Newton's Law of Cooling, determine the outdoor temperature if a thermometer is taken from a room where the temperature is 68° to the outdoors, where after $\frac{1}{2}$ min and 1 min, the thermometer reads 53° and 42°, respectively.

**Solution:**

Let $y$ be the temperature of the thermometer and $T$ be the outdoor temperature. From Newton's Law of Cooling, we have

$$\frac{dy}{dt} = k(y - T)$$

$$\left(\frac{1}{y-T}\right)\frac{dy}{dt} = k$$

$$\int\left(\frac{1}{y-T}\right)\frac{dy}{dt}\,dt = \int k\,dt$$

$$\ln(y - T) = kt + C$$

When $t = 0$ min, $y = 68°$. Thus

$$\ln(68 - T) = k(0) + C \qquad \text{or} \qquad C = \ln(68 - T)$$

When $t = 0.5$ min, $y = 53°$. Thus

$$\ln(53 - t) = k(0.5) + \ln(68 - T)$$

$$k = 2[\ln(53 - T) - \ln(68 - T)]$$

$$= \ln\left[\left(\frac{53 - T}{68 - T}\right)^2\right]$$

When $t = 1$ min, $y = 42°$. Thus

$$\ln(42 - T) = k(1) + \ln(68 - T)$$

$$= \ln\left[\left(\frac{53 - T}{68 - T}\right)^2\right] + \ln(68 - T)$$

$$(42 - T)(68 - T) = (53 - T)^2$$

$$2856 - 110T + T^2 = 2809 - 106T + T^2$$

$$-4T = -47$$

$$T = \frac{47}{4} = 11.75°$$

## 7.8   Indeterminate Forms and L'Hôpital's Rule

**3.** Evaluate $\lim\limits_{x \to 0} \dfrac{\sqrt{4 - x^2} - 2}{x}$.

**Solution:**

Since a direct substitution of $x = 0$ leaves us with the indeterminate form $0/0$, we apply L'Hôpital's Rule to obtain

$$\lim_{x \to 0} \frac{\sqrt{4 - x^2} - 2}{x} = \lim_{x \to 0} \frac{(\frac{1}{2})(4 - x^2)^{-1/2}(-2x)}{1}$$

$$= \lim_{x \to 0} \frac{-x}{\sqrt{4 - x^2}} = \frac{0}{2} = 0$$

**7.** Evaluate $\lim\limits_{x \to 0+} \dfrac{e^x - (1 + x)}{x^n}$ where $n = 1, 2, 3 \ldots$.

**Solution:**

Case 1: $n = 1$ (apply L'Hôpital's Rule once).

$$\lim_{x \to 0+} \frac{e^x - (1 + x)}{x} = \lim_{x \to 0+} \frac{e^x - 1}{1} = 0$$

Case 2: $n = 2$ (apply L'Hôpital's Rule twice).

$$\lim_{x \to 0+} \frac{e^x - (1 + x)}{x^2} = \lim_{x \to 0+} \frac{e^x - 1}{2x} = \lim_{x \to 0+} \frac{e^x}{2} = \frac{1}{2}$$

Case 3: $n \geq 3$ (apply L'Hôpital's Rule twice).

$$\lim_{x \to 0+} \frac{e^x - (1 + x)}{x^n} = \lim_{x \to 0+} \frac{e^x - 1}{nx^{n-1}} = \lim_{x \to 0+} \frac{e^2}{n(n - 1)x^{n-2}} = \infty$$

**9.** Evaluate $\lim\limits_{x \to \infty} \dfrac{\ln x}{x}$.

**Solution:**

Since direct substitution leads to the indeterminate form $\infty/\infty$, we use L'Hôpital's Rule.

$$\lim_{x \to \infty} \frac{\ln x}{x} = \lim_{x \to \infty} \frac{1/x}{1} = \lim_{x \to \infty} \frac{1}{x} = 0$$

**15.** Evaluate $\lim\limits_{x \to 0+} x^2 \ln x$.

**Solution:**

Since direct substitution leads to the indeterminate form $0(-\infty)$, we rewrite the limit and use L'Hôpital's Rule.

$$\lim_{x \to 0+} x^2 \ln x = \lim_{x \to 0+} \frac{\ln x}{1/x^2} = \lim_{x \to 0+} \frac{1/x}{-2/x^3} = \lim_{x \to 0+} \frac{-x^2}{2} = 0$$

**17.** Evaluate $\lim\limits_{x \to 2} \left( \dfrac{8}{x^2 - 4} - \dfrac{x}{x - 2} \right)$.

**Solution:**

Since direct substitution yields the indeterminate form $\infty - \infty$, we rewrite the expression.

$$\lim_{x \to 2} \left( \frac{8}{x^2 - 4} - \frac{x}{x - 2} \right) = \lim_{x \to 2} \left[ \frac{8(x - 2) - x(x^2 - 4)}{(x^2 - 4)(x - 2)} \right]$$

$$= \lim_{x \to 2} \left[ \frac{(x - 2)(x - 2)(-x - 4)}{(x - 2)(x + 2)(x - 2)} \right]$$

$$= -\frac{3}{2}$$

Note that L'Hôpital's Rule was not required in evaluating the limit.

**21.** Evaluate $\lim\limits_{x \to 0+} x^{1/x}$.

**Solution:**

$$\lim_{x \to 0+} x^{1/x} = 0^\infty = 0$$

**31.** Use L'Hôpital's Rule to evaluate $\lim\limits_{x \to \infty} \dfrac{(\ln x)^n}{x^m}$, where $0 < n, m$.

**Solution:**

$$\lim_{x \to \infty} \frac{(\ln x)^n}{x^m} = \lim_{x \to \infty} \left[ \frac{n(\ln x)^{n-1}(1/x)}{mx^{m-1}} \right] = \lim_{x \to \infty} \left[ \frac{n(\ln x)^{n-1}}{mx^m} \right]$$

If $n - 1 \leq 0$, this limit is zero. If $n - 1 > 0$, we repeat L'Hôpital's Rule to obtain

$$\lim_{x \to \infty} \left[ \frac{n(n - 1)(\ln x)^{n-2}}{m^2 x^m} \right]$$

Again, if $n - 2 \leq 0$, this limit is zero. If $n - 2 > 0$, repeated applications of L'Hôpital's Rule will eventually yield a form where the numerator approaches a finite number and the denominator approaches infinity. Thus in every case the limit is 0.

**41.** The velocity of an object falling in a resisting medium such as air or water (if the downward direction is positive) is given by

$$v = \frac{32}{k} \left( 1 - e^{-kt} + \frac{v_0 k e^{-kt}}{32} \right)$$

where $v_0$ is the initial velocity, $t$ is the time, and $k$ is the resistance constant of the medium. Use L'Hôpital's Rule to find the formula for the velocity of a falling body in a vacuum by fixing $v_0$ and $t$ and letting $k$ approach zero.

**Solution:**

$$\lim_{k \to 0} \frac{32}{k} \left( 1 - e^{-kt} + \frac{v_0 k e^{-kt}}{32} \right)$$

$$= \lim_{k \to 0} \frac{32(1 - e^{-kt})}{k} + \lim_{k \to 0} \frac{32}{k} \cdot \frac{v_0 k e^{-kt}}{32}$$

$$= \lim_{k \to 0} \frac{32t e^{-kt}}{1} + v_0 = 32t + v_0$$

(Apply L'Hôpital's Rule to the first limit on the right-hand side.)

## Review Exercises for Chapter 7

**11.** Solve for $x$ in the equation $\log_3 x + \log_3 (x-1) - \log_3 (x-2) = 2$.

**Solution:**

$$\log_3 x + \log_3 (x-1) - \log_3 (x-2) = 2$$

$$\log_3 \frac{x(x-1)}{x-2} = 2$$

$$\frac{x(x-1)}{x-2} = 3^2$$

$$x^2 - x = 9x - 18$$

$$x^2 - 10x + 18 = 0$$

$$x = \frac{10 \pm \sqrt{100-72}}{2}$$

$$= 5 \pm \sqrt{7}$$

**17.** Find $dx/dy$ given $y \ln x + y^2 = 0$.

**Solution:**

$$y \ln x + y^2 = 0$$

$$y\left(\frac{1}{x}\right) + (\ln x)y' + 2yy' = 0$$

$$y'(2y + \ln x) = -\frac{y}{x}$$

$$y' = \frac{-y}{x(2y + \ln x)}$$

**21.** Find $dy/dx$ given $y = \frac{1}{b^2}\left[\ln(a+bx) + \frac{a}{a+bx}\right]$.

**Solution:**

$$y = \frac{1}{b^2}\left[\ln(a+bx) + a(a+bx)^{-1}\right]$$

$$\frac{dy}{dx} = \frac{1}{b^2}\left[\left(\frac{1}{a+bx}\right)(b) + a(-1)(a+bx)^{-2}(b)\right]$$

$$= \frac{1}{b^2}\left[\frac{b}{a+bx} - \frac{ab}{(a+bx)^2}\right]$$

$$= \frac{1}{b^2}\left[\frac{b(a+bx) - ab}{(a+bx)^2}\right] = \frac{x}{(a+bx)^2}$$

**33.** Find $dy/dx$ given $ye^x + xe^y = xy$.

**Solution:**

Using the Product Rule on each term and differentiating implicitly, we obtain

$$\left(ye^x + e^x\frac{dy}{dx}\right) + \left(xe^y\frac{dy}{dx} + e^y\right) = \left(x\frac{dy}{dx} + y\right)$$

Solving for $dy/dx$, we have

$$(e^x + xe^y - x)\frac{dy}{dx} = y - e^y - ye^x$$

$$\frac{dy}{dx} = \frac{y - e^y - ye^x}{e^x + xe^y - x}$$

**43.** Evaluate $\displaystyle\int \frac{x^2 + 3}{x}\,dx$.

**Solution:**

We begin by dividing and obtain

$$\int \frac{x^2 + 3}{x}\,dx = \int \left(x + \frac{3}{x}\right)\,dx$$

$$= \frac{1}{2}x^2 + 3\ln|x| + C$$

**51. Evaluate** $\displaystyle\int \frac{e^{4x} - e^{2x} + 1}{e^x}\, dx.$

**Solution:**

$$\int \frac{e^{4x} - e^{2x} + 1}{e^x}\, dx = \int \left(e^{3x} - e^x + e^{-x}\right) dx$$

$$= \frac{e^{3x}}{3} - e^x - e^{-x} + C$$

$$= \frac{e^{4x} - 3e^{2x} - 3}{3e^x} + C$$

**69.** How large a deposit, at 7% interest compounded continuously, must be made to obtain a balance of \$10,000 in 15 years?

**Solution:**

Using the formula for continuous interest, we have

$$A = Pe^{rt}$$

$$10,000 = Pe^{(0.07)(15)}$$

$$P = \frac{10,000}{e^{1.05}}$$

$$\approx \$3499.38$$

**75.** Trucks arrive at a terminal at an average rate of 3 per hour (thus $u = 20$ min is the average time between arrivals). If a truck has just arrived, find the probability that the next arrival will be:

(a) within 10 minutes

(b) within 30 minutes

(c) between 15 and 30 minutes

(d) within 60 minutes.

**Solution:**

$$\int_{t_1}^{t_2} \left(\frac{1}{u}\right) e^{-t/u}\, dt = \int_{t_1}^{t_2} \left(\frac{1}{20}\right) e^{-t/20}\, dt$$

$$= -\int_{t_1}^{t_2} e^{-t/20} \left(\frac{-1}{20}\right) dt$$

$$= \left[-e^{-t/20}\right]_{t_1}^{t_2} = e^{-t_1/20} - e^{-t_2/20}$$

(a) When $t_1 = 0$, $t_2 = 10$,

probability $= e^0 - e^{-1/2} = 1 - e^{-1/2} = 0.3935$

(b) When $t_1 = 0$, $t_2 = 30$,

probability $= e^0 - e^{-3/2} = 1 - e^{-3/2} = 0.7769$

(c) When $t_1 = 15$, $t_2 = 30$,

probability $= e^{-3/4} - e^{-3/2} = 0.2492$

(d) When $t_1 = 0$, $t_2 = 60$,

probability $= e^0 - e^{-3} = 1 - e^{-3} = 0.9502$

**89.** Use L'Hôpital's Rule to evaluate $\lim\limits_{x \to \infty} (\ln x)^{2/x}$.

**Solution:**

We begin by taking the natural logarithm of both sides of the equation

$$y = \lim_{x \to \infty} (\ln x)^{2/x}$$

to obtain

$$\ln y = \ln \left[ \lim_{x \to \infty} (\ln x)^{2/x} \right]$$

$$= \lim_{x \to \infty} \left[ \frac{2}{x} \ln (\ln x) \right]$$

$$= 2 \lim_{x \to \infty} \frac{(1/\ln x)(1/x)}{1} = 0$$

Finally, as $\ln y \longrightarrow 0$, we know that $y \longrightarrow 1$ and conclude that

$$\lim_{x \to \infty} (\ln x)^{2/x} = 1$$

# 8 TRIGONOMETRIC FUNCTIONS AND INVERSE TRIGONOMETRIC FUNCTIONS

## 8.1 Review of Trigonometric Functions

**7.** Express the following angles in degree measure:

  (a)   $3\pi/2$              (b)   $7\pi/6$

  (c)   $-7\pi/12$      (d)   $\pi/9$

**Solution:**

Since $180° = \pi$ radians, we conclude that 1 radian $= 180°/\pi$.

(a)   $\dfrac{3\pi}{2}$ radians $= \left(\dfrac{3\pi}{2}\right)\left(\dfrac{180°}{\pi}\right) = 270°$

(b)   $\dfrac{7\pi}{6}$ radians $= \left(\dfrac{7\pi}{6}\right)\left(\dfrac{180°}{\pi}\right) = 210°$

(c)   $-\dfrac{7\pi}{12}$ radians $= \left(-\dfrac{7\pi}{12}\right)\left(\dfrac{180°}{\pi}\right) = -105°$

(d)   $\dfrac{\pi}{9}$ radians $= \left(\dfrac{\pi}{9}\right)\left(\dfrac{180°}{\pi}\right) = 20°$

**19.** Find $\cot \theta$ given $\cos \theta = \frac{4}{5}$.

**Solution:**

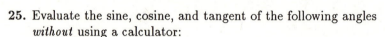

Using the fact that

$$\cos \theta = \frac{4}{5}$$

we construct the accompanying figure and obtain

$$\cot \theta = \frac{4}{y} = \frac{4}{\sqrt{25 - 16}} = \frac{4}{3}.$$

**25.** Evaluate the sine, cosine, and tangent of the following angles *without* using a calculator:

(a)   225°      (b)   −225°      (c)   300°      (d)   330°

**Solution:**

(a) The angle is in quadrant III and the reference angle is 225° − 180° = 45°. Therefore,

$$\sin 225° = -\sin 45° = -\frac{\sqrt{2}}{2}$$

$$\cos 225° = -\cos 45° = -\frac{\sqrt{2}}{2}$$

$$\tan 225° = \tan 45° = 1$$

(b) The angle is in quadrant II and the reference angle is 225° − 180° = 45°. Therefore,

$$\sin (-225°) = \sin 45° = \frac{\sqrt{2}}{2}$$

$$\cos (-225°) = -\cos 45° = -\frac{\sqrt{2}}{2}$$

$$\tan (-225°) = -\tan 45° = -1$$

(c) The angle is in quadrant IV and the reference angle is 360° − 300° = 60°. Therefore,

$$\sin 300° = -\sin 60° = -\frac{\sqrt{3}}{2}$$

$$\cos 300° = \cos 60° = \frac{1}{2}$$

$$\tan 300° = -\tan 60° = -\sqrt{3}$$

(d) The angle is in quadrant IV and the reference angle is
$360° - 330° = 30°$. Therefore,

$$\sin 330° = -\sin 30° = -\frac{1}{2}$$

$$\cos 330° = \cos 30° = \frac{\sqrt{3}}{2}$$

$$\tan 330° = -\tan 30° = -\frac{\sqrt{3}}{3}$$

**37.** Solve the equation $\tan^2 \theta - \tan \theta = 0$ for $\theta$ where $0 \le \theta < 2\pi$.

**Solution:**

$$\tan^2 \theta - \tan \theta = 0$$
$$\tan \theta(\tan \theta - 1) = 0$$

If $\tan \theta = 0$, then $\theta = 0$ or $\theta = \pi$. If $\tan \theta - 1 = 0$, we have
$\tan \theta = 1$ and $\theta = \pi/4$ or $\theta = 5\pi/4$. Thus for $0 \le \theta < 2\pi$, there
are four solutions:

$$\theta = 0, \frac{\pi}{4}, \pi, \frac{5\pi}{4}.$$

**43.** Solve for $y$ and $r$ in the accompanying triangle.

**Solution:**

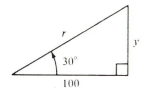

$$\tan 30° = \frac{\sqrt{3}}{3} = \frac{y}{100}$$

Therefore, $y = 100\sqrt{3}/3$.

$$\sec 30° = \frac{2\sqrt{3}}{3} = \frac{r}{100}$$

Therefore, $r = 200\sqrt{3}/3$.

**49.** From a 150-foot observation tower on the coast, a Coast Guard sights a boat in difficulty. The angle of depression to the boat is 4° as shown in the accompanying figure. How far is the boat from the shoreline?

**Solution:**

If $d$ is the required distance, then

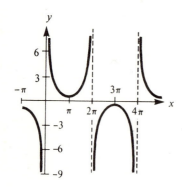

$$\cot 4° = \frac{d}{150}$$
$$d = 150 \cot 4° \approx 2145 \text{ feet}$$

## 8.2   Graphs and Limits of Trigonometric Functions

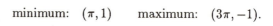

**21.** Sketch the graph of the function $\csc(x/2)$.

**Solution:**

The graph of $y = \csc(x/2)$ has the following characteristics:

$$\text{period:} \ \frac{2\pi}{\frac{1}{2}} = 4\pi$$

$$\text{vertical asymptotes:} \quad x = 2n\pi, n \text{ an integer}$$

Using the basic shape of the graph of the cosecant function, we sketch one period of the function on the interval $[0, 4\pi]$, following the pattern

$$\text{minimum:} \ (\pi, 1) \qquad \text{maximum:} \ (3\pi, -1).$$

**26.** Graph the function $f(x) = \cos(2x - \pi/3)$ through 2 periods.

**Solution:**

$$f(x) = \cos\left(2x - \frac{\pi}{3}\right) = \cos\left[2\left(x - \frac{\pi}{6}\right)\right]$$

Therefore, the graph of $f$ has these characteristics:

amplitude:  1     period:  $\dfrac{2\pi}{2} = \pi$     right shift of  $\dfrac{\pi}{6}$

Using the basic shape of the graph of the cosine function, we sketch two periods of the function on the interval $[\pi/6, 13\pi/6]$, following the pattern

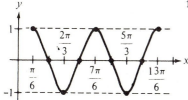

maximum:  $\left(\dfrac{\pi}{6}, 1\right)$     minimum:  $\left(\dfrac{2\pi}{3}, -1\right)$

maximum:  $\left(\dfrac{7\pi}{6}, 1\right)$     minimum:  $\left(\dfrac{5\pi}{3}, -1\right)$

**31.** Use addition of ordinates to sketch the graph of the function $f(x) = \sin x + \sin 2x$.

**Solution:**

To obtain the graph of $f(x) = \sin x + \sin 2x$, we graph $y_1 = \sin x$ and $y_2 = \sin 2x$ on the same set of coordinates axes. Then we use addition of ordinates as demonstrated in the accompanying figure.

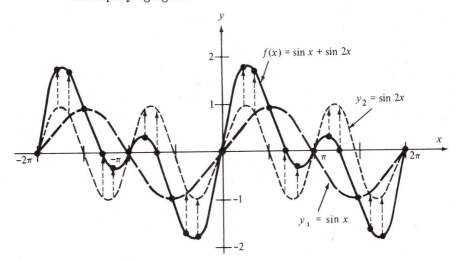

**39.** Find (if it exists) $\lim\limits_{x \to \pi/2} \dfrac{\cos x}{\cot x}$.

**Solution:**

$$\lim_{x \to \pi/2} \frac{\cos x}{\cot x} = \lim_{x \to \pi/2} \frac{\cos x}{(\cos x)/(\sin x)}$$
$$= \lim_{x \to \pi/2} \sin x = 1$$

**49.** Find (if it exists) $\lim\limits_{x \to 0+} \dfrac{2}{\sin x}$.

**Solution:**

Since
$$\lim_{x \to 0+} 2 = 2 \quad \text{and} \quad \lim_{x \to 0+} \sin x = 0^+,$$
we conclude that
$$\lim_{x \to 0+} \frac{2}{\sin x} = \infty.$$

## 8.3   Derivatives of Trigonometric Functions

**3.** Differentiate $y = \dfrac{1}{x} - 3\sin x$.

**Solution:**

$$y = \frac{1}{x} - 3\sin x = x^{-1} - 3\sin x$$
$$y' = -x^{-2} - 3\cos x$$
$$= -\frac{1}{x^2} - 3\cos x$$

**9.** Differentiate $f(t) = \dfrac{\cos t}{t}$.

**Solution:**

$$f(t) = \frac{\cos t}{t}$$
$$f'(t) = \frac{t(-\sin t) - (\cos t)(1)}{t^2} = \frac{-(t\sin t + \cos t)}{t^2}$$

**15.** Differentiate $f(\theta) = -\csc\theta - \sin\theta$.

**Solution:**

$$f(\theta) = -\csc\theta - \sin\theta$$

$$f'(\theta) = -(-\csc\theta\cot\theta) - \cos\theta = \frac{1}{\sin\theta}\left(\frac{\cos\theta}{\sin\theta}\right) - \cos\theta$$

$$= \cos\theta\left(\frac{1}{\sin^2\theta} - 1\right) = \frac{\cos\theta}{\sin^2\theta}(1 - \sin^2\theta) = \cos\theta\cot^2\theta$$

**21.** Differentiate $f(x) = \dfrac{1 + \csc x}{1 - \csc x}$.

**Solution:**

$$f(x) = \frac{1 + \csc x}{1 - \csc x}$$

$$f'(x) = \frac{(1 - \csc x)(-\csc x\cot x) - (1 + \csc x)(\csc x\cot x)}{(1 - \csc x)^2}$$

$$= -\frac{2\csc x\cot x}{(1 - \csc x)^2}$$

**31.** Find $dy/dx$ given $y = \frac{1}{4}\sin^2 2x$.

**Solution:**

$$y = \frac{1}{4}(\sin 2x)^2$$

$$\frac{dy}{dx} = \frac{1}{4}(2)(\sin 2x)(\cos 2x)(2)$$

$$= \frac{1}{2}(2\sin 2x\cos 2x) = \frac{1}{2}\sin 4x$$

**35.** Find $dy/dx$ given $y = \sec^3 2x$.

**Solution:**

$$y = (\sec 2x)^3$$

$$dy/dx = 3(\sec 2x)^2[(\sec 2x\tan 2x)(2)] = 6\sec^3 2x\tan 2x$$

**37.** Find $dy/dx$ given $\ln|\csc x - \cot x|$.

**Solution:**

$$y = \ln|\csc x - \cot x|$$

$$\frac{dy}{dx} = \frac{-\csc x \cot x + \csc^2 x}{\csc x - \cot x}$$

$$= \frac{\csc x(-\cot x + \csc x)}{\csc x - \cot x} = \csc x$$

**47.** Given $\tan(x + y) = x$, find $dy/dx$ by implicit differentiation and evaluate the derivative at $(0, 0)$.

**Solution:**

$$\tan(x + y) = x$$

$$[\sec^2(x + y)](1 + y') = 1$$

$$[\sec^2(x + y)]y' = 1 - \sec^2(x + y)$$

$$y' = \frac{1 - \sec^2(x + y)}{\sec^2(x + y)}$$

$$= \frac{-\tan^2(x + y)}{\tan^2(x + y) + 1} = \frac{-x^2}{x^2 + 1}$$

At $(0, 0)$, $\dfrac{dy}{dx} = \dfrac{0}{1} = 0$.

**51.** Show that $y = 2\sin x + 3\cos x$ satisfies the differential equation $y'' + y = 0$.

**Solution:**

$$y = 2\sin x + 3\cos x$$

$$y' = 2\cos x - 3\sin x$$

$$y'' = -2\sin x - 3\cos x = -y$$

Therefore, $y'' + y = -y + y = 0$.

**59.** Use L'Hôpital's Rule to evaluate $\lim\limits_{x \to 0} \dfrac{x - \tan x}{x - \sin x}$.

**Solution:**

Direct substitution of $x = 0$ gives the indeterminate form $0/0$. Thus by L'Hôpital's Rule, we have

$$\lim_{x \to 0} \frac{x - \tan x}{x - \sin x} = \lim_{x \to 0} \frac{1 - \sec^2 x}{1 - \cos x}.$$

Direct substitution of $x = 0$ still yields the indeterminate form $0/0$ and we use L'Hôpital's Rule again to obtain

$$
\begin{aligned}
\lim_{x \to 0} \frac{x - \tan x}{x - \sin x} &= \lim_{x \to 0} \frac{1 - \sec^2 x}{1 - \cos x} \\
&= \lim_{x \to 0} \frac{-2 \sec x (\sec x \tan x)}{\sin x} \\
&= \lim_{x \to 0} \frac{-2 \sec^2 x (\sin x / \cos x)}{\sin x} \\
&= \lim_{x \to 0} \frac{-2 \sec^2 x}{\cos x} = -2.
\end{aligned}
$$

**63.** Use L'Hôpital's Rule to evaluate $\lim\limits_{x \to \infty} x \sin \dfrac{1}{x}$.

**Solution:**

Since direct substitution yields an indeterminate form we use L'Hôspital Rule.

$$
\begin{aligned}
\lim_{x \to \infty} x \sin \frac{1}{x} &= \lim_{x \to \infty} \frac{\sin(1/x)}{1/x} \\
&= \lim_{x \to \infty} \frac{(-1/x^2) \cos(1/x)}{-1/x^2} \\
&= \lim_{x \to \infty} \cos \frac{1}{x} = 1
\end{aligned}
$$

**65.** Sketch the graph of $f(x) = 2\sin x + \sin 2x$ on the interval $[0, 2\pi]$, and identify all relative extrema and points of inflection.

**Solution:**

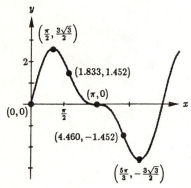

$$f(x) = 2\sin x + \sin 2x$$
$$f'(x) = 2\cos x + 2\cos 2x$$
$$= 2[\cos x + 2\cos^2 x - 1] = 2(2\cos x - 1)(\cos x + 1)$$
$$f'(x) = 0 \quad \text{when} \quad x = \frac{\pi}{3}, \ \pi, \ \frac{5\pi}{3}$$
$$f''(x) = -2\sin x - 4\sin 2x$$
$$= -2(\sin x + 4\sin x \cos x) = -2\sin x(1 + 4\cos x)$$
$$f''(x) = 0 \quad \text{when} \quad x = 0, \ \pi, \ \arccos\left(-\tfrac{1}{4}\right), \ 2\pi - \arccos\left(\tfrac{1}{4}\right)$$

Since $f''(\frac{\pi}{3}) < 0$, there is a relative maximum at $(\pi/3, 3\sqrt{3}/2)$, and since $f''(\frac{5\pi}{3}) > 0$, there is a relative minimum at $(5\pi/3, -3\sqrt{3}/2)$. The inflection points are $(0,0), (\pi, 0), (1.823, 1.452)$, and $(4.460, -1.452)$.

**72.** The general equation giving the height of an ocsillating object attached to a spring is

$$y = A\sin\sqrt{\frac{k}{m}}t + B\cos\sqrt{\frac{k}{m}}t$$

where $k$ is the spring constant and $m$ is the mass of the object. Show that the maximum displacement of the object is $\sqrt{A^2 + B^2}$. Show that the frequency (number of oscillations per second) is $(1/2\pi)\sqrt{k/m}$. How is the frequency changed if the stiffness $k$ of the spring is increased? How is the frequency changed if the mass $m$ of the object is increased?

**Solution:**

$$y = A\sin\sqrt{\frac{k}{m}}t + B\cos\sqrt{\frac{k}{m}}t$$
$$y' = A\sqrt{\frac{k}{m}}\cos\sqrt{\frac{k}{m}}t - B\sqrt{\frac{k}{m}}\sin\sqrt{\frac{k}{m}}t$$

Therefore, $y' = 0$ if

$$A\sqrt{\frac{k}{m}}\cos\sqrt{\frac{k}{m}}t = B\sqrt{\frac{k}{m}}\sin\sqrt{\frac{k}{m}}t$$

$$\frac{A}{B} = \frac{\sin(\sqrt{k/m})t}{\cos(\sqrt{k/m})t}$$

$$\tan\sqrt{\frac{k}{m}}t = \frac{A}{B}$$

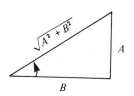

From the accompanying figure we observe that

$$\sin\sqrt{\frac{k}{m}}t = \frac{A}{\sqrt{A^2 + B^2}} \qquad \text{and} \qquad \cos\sqrt{\frac{k}{m}}t = \frac{B}{\sqrt{A^2 + B^2}}$$

Thus when $y' = 0$,

$$y = A\left[\frac{A}{\sqrt{A^2 + B^2}}\right] + B\left[\frac{B}{\sqrt{A^2 + B^2}}\right] = \sqrt{A^2 + B^2}$$

Since the frequency is the reciprocal of the period, we have

$$\text{period} = \frac{2\pi}{\sqrt{k/m}}, \qquad \text{frequency} = \frac{\sqrt{k/m}}{2\pi} = \frac{1}{2\pi}\sqrt{\frac{k}{m}}$$

Therefore, the frequency is proportional to the square root of $k$ and inversely proportional to the square root of the mass.

**73.** A component is designed to slide a block of steel of weight $W$ across a table and into a chute as shown in the accompanying figure. The motion of the block is resisted by a frictional force proportional to its net weight. (let $k$ by the constant of proportionality.) Find the minimum force $F$ needed to slide the block and find the corresponding value of $\theta$.

**Solution:**

The force in the direction of motion is $F\cos\theta$ and the force tending to lift the block is $F\sin\theta$. Therefore, the net weight of the block is $W - F\sin\theta$, and

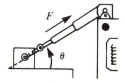

$$F\cos\theta = k(W - F\sin\theta)$$
$$F\cos\theta + kF\sin\theta = kW$$
$$F = \frac{kW}{\cos\theta + k\sin\theta}$$
$$= kW(\cos\theta + k\sin\theta)^{-1}$$

Differentiating and letting $dF/d\theta = 0$, we obtain

$$\frac{dF}{d\theta} = -kW(\cos\theta + k\sin\theta)^{-2}(-\sin\theta + k\cos\theta)$$

$$= \frac{-kW(k\cos\theta - \sin\theta)}{(\cos\theta + k\sin\theta)^2} = 0$$

$$\sin\theta = k\cos\theta$$

$$\tan\theta = k$$

Therefore, $F$ is minimum when $\theta = \arctan k$ and the minimum force is

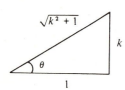

$$F = \frac{kW}{\cos\theta + k\sin\theta}$$

$$= \frac{kW}{(1/\sqrt{k^2+1}) + k(k/\sqrt{k^2+1})}$$

$$= \frac{kW}{\sqrt{k^2+1}}$$

**79.** A wheel of radius 1 ft revolves at a rate of 10 rev/s. A dot is painted at a point $P$ on the rim of the wheel as shown in the accompanying figure. Find the rate of horizontal movement of the dot for the following angles:
(a)  $\theta = 0°$      (b)  $\theta = 30°$      (c)  $\theta = 60°$

**Solution:**

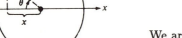

From the figure, we have

$$x = \cos\theta$$

We are given

$$\frac{d\theta}{dt} = 10 \text{ rev/sec} = 10(-2\pi) = -20\pi \text{ rad/sec}$$

Therefore,

$$\frac{dx}{dt} = -\sin\theta \frac{d\theta}{dt}$$

$$= -\sin\theta(-20\pi) = 20\pi\sin\theta$$

(a)   When $\theta = 0°$, $\dfrac{dx}{dt} = 0$ ft/sec.

(b)   When $\theta = 30°$, $\dfrac{dx}{dt} = 20\pi\left(\dfrac{1}{2}\right) = 10\pi$ ft/sec.

(c)   When $\theta = 60°$, $\dfrac{dx}{dt} = 20\pi\left(\dfrac{\sqrt{3}}{2}\right) = 10\pi\sqrt{3}$ ft/sec.

## 8.4   Integrals of Trigonometric Functions

**9.** Evaluate $\int x \cos x^2 \, dx$.

**Solution:**

If we let $u = x^2$, then $du = 2x \, dx$ and $\dfrac{du}{2} = x \, dx$. Therefore,

$$\int x \cos x^2 \, dx = \int \cos x^2 (x) \, dx$$

$$= \int \cos u \, \frac{du}{2}$$

$$= \frac{1}{2} \int \cos u \, du = \frac{1}{2} \sin u + C = \frac{1}{2} \sin x^2 + C$$

**15.** Evaluate $\int \cot^2 x \, dx$.

**Solution:**

$$\int \cot^2 x \, dx = \int (\csc^2 x - 1) \, dx = -\cot x - x + C$$

**25.** Evaluate $\displaystyle\int \frac{\sec x \tan x}{\sec x - 1} \, dx$.

**Solution:**

Letting $u = \sec x - 1$, we have $du = \sec x \tan x \, dx$. Therefore,

$$\int \frac{\sec x \tan x}{\sec x - 1} \, dx = \int \frac{1}{u} \, du$$

$$= \ln |u| + C = \ln |\sec x - 1| + C$$

**33.** Evaluate $\int e^{-x} \tan (e^{-x}) \, dx$.

**Solution:**

Letting $u = e^{-x}$, we have $du = -e^{-x} \, dx$. Therefore,

$$\int e^{-x} \tan (e^{-x}) \, dx = -\int \tan u \, du$$

$$= \ln |\cos u| + C = \ln |\cos (e^{-x})| + C$$

**39.** Evaluate the definite integral $\int_{\pi/2}^{2\pi/3} \sec^2{(x/2)}\,dx$.

**Solution:**

Let $u = \dfrac{x}{2}$. Then $du = \dfrac{1}{2}\,dx$ and $dx = 2\,du$. Also, if $x = \dfrac{\pi}{2}$,

then $u = \dfrac{\pi}{4}$, and if $x = \dfrac{2\pi}{3}$, then $u = \dfrac{\pi}{3}$. Thus,

$$\int_{\pi/2}^{2\pi/3} \sec^2\left(\frac{x}{2}\right)\,dx = \int_{\pi/4}^{\pi/3} \sec^2{u}(2)\,du$$

$$= 2\left[\tan u\right]_{\pi/4}^{\pi/3}$$

$$= 2\left[\tan\frac{\pi}{3} - \tan\frac{\pi}{4}\right] = 2(\sqrt{3} - 1)$$

**47.** Determine the area of the region having the given boundaries in the accompanying figure.

**Solution:**

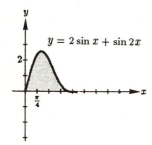

From the accompanying figure we see that the area of the region is given by the definite integral $\int_0^\pi (2\sin x + \sin 2x)\,dx$.

$$\text{area} = \int_0^\pi (2\sin x + \sin 2x)\,dx$$

$$= 2\int_0^\pi \sin x\,dx + \frac{1}{2}\int_0^\pi \sin 2x(2)\,dx$$

$$= \left[-2\cos x - \frac{1}{2}\cos 2x\right]_0^\pi = 4$$

**51.** Find the volume of the solid generated by revolving the region bounded by the graphs of $y = \sqrt{\sin x}$, $y = 0$, $x = 0$, and $x = \pi$ about the $x$-axis.

**Solution:**

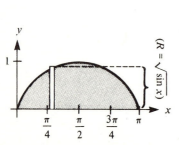

Use the Disc Method for the volume of a solid of revolution, we have

$$V = \pi\int_a^b (R^2 - r^2)\,dx = \pi\int_0^\pi [(\sqrt{\sin x})^2 - 0^2]\,dx$$

$$= \pi\int_0^\pi \sin x\,dx = \pi\left[-\cos x\right]_0^\pi = 2\pi$$

**67.** Approximate $\int_0^{\sqrt{\pi/2}} \cos x^2 \, dx$ using (a) the Trapezoidal Rule and (b) Simpson's Rule with $n = 4$.

**Solution:**

(a)   Trapezoidal Rule with $n = 4$

$$\int_0^{\sqrt{\pi/2}} \cos x^2 \, dx$$

$$\approx \frac{\sqrt{\pi/2}}{2(4)} \left[ \cos(0) + 2 \cos \left( \frac{\sqrt{\pi/2}}{4} \right)^2 + 2 \cos \left( \frac{2\sqrt{\pi/2}}{4} \right)^2 \right.$$

$$\left. + 2 \cos \left( \frac{3\sqrt{\pi/2}}{4} \right)^2 + \cos \left( \frac{4\sqrt{\pi/2}}{4} \right)^2 \right]$$

$$\approx 0.957$$

(b)   Simpson's Rule with $n = 4$

$$\int_0^{\sqrt{\pi/2}} \cos x^2 \, dx$$

$$\approx \frac{\sqrt{\pi/2}}{3(4)} \left[ \cos(0) + 4 \cos \left( \frac{\sqrt{\pi/2}}{4} \right)^2 + 2 \cos \left( \frac{2\sqrt{\pi/2}}{4} \right)^2 \right.$$

$$\left. + 4 \cos \left( \frac{3\sqrt{\pi/2}}{4} \right)^2 + \cos \left( \frac{4\sqrt{\pi/2}}{4} \right)^2 \right]$$

$$\approx 0.978$$

## 8.5   Inverse Trigonometric Functions and Differentiation

**3.** Evaluate $\arccos \left( \frac{1}{2} \right)$

**Solution:**

Since $y = \arccos \left( \frac{1}{2} \right)$ if and only if $\cos y = \frac{1}{2}$, we choose $y = \pi/3$ in the interval $[0, \pi]$. Thus

$$\arccos \left( \frac{1}{2} \right) = \frac{\pi}{3}.$$

**13.** Evaluate (a)  $\sin\left[\arctan\left(\frac{3}{4}\right)\right]$  and (b)  $\sec\left[\arcsin\left(\frac{4}{5}\right)\right]$  without the use of a calculator.

**Solution:**

(a) We begin by sketching a triangle to represent $\theta$, "the angle whose tangent is $\frac{3}{4}$". Then

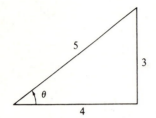

$$\theta = \arctan\left(\frac{3}{4}\right)$$

and

$$\sin\left[\arctan\left(\frac{3}{4}\right)\right] = \sin\theta = \frac{3}{5}.$$

(b) We begin by sketching a triangle to represent $\theta$, "the angle whose sine is $\frac{4}{5}$". Then

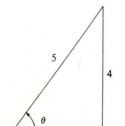

$$\theta = \arcsin\left(\frac{4}{5}\right)$$

and

$$\sec\left[\arcsin\left(\frac{4}{5}\right)\right] = \sec\theta = \frac{5}{3}.$$

**21.** Write an algebraic expression that is equivalent to the expression  $\sin\left(\text{arcsec } x\right)$.

**Solution:**

We begin by sketching a triangle to represent $\theta$, "the angle whose secant is $x$". Then

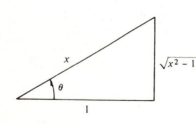

$$\theta = \text{arcsec } x$$

and

$$\sin\left(\text{arcsec } x\right) = \sin\theta = \frac{\sqrt{x^2 - 1}}{x}.$$

**25.** Write an algebraic expression equivalent to the expression

$$\csc\left(\arctan\frac{x}{\sqrt{2}}\right).$$

**Solution:**

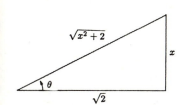

·We begin by sketching a triangle to represent $\theta$, "the angle whose tangent is $x/\sqrt{2}$." Then

$$\theta = \arctan\frac{x}{\sqrt{2}}$$

and

$$\csc\left(\arctan\frac{x}{\sqrt{2}}\right) = \csc\theta = \frac{\sqrt{x^2+2}}{x}$$

**37.** Solve $\arcsin\sqrt{2x} = \arccos\sqrt{x}$ for $x$.

**Solution:**

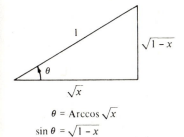

$\theta = \text{Arccos }\sqrt{x}$

$\sin\theta = \sqrt{1-x}$

Taking the sine of each member of the equation, we have

$$\sin(\arcsin\sqrt{2x}) = \sin(\arccos\sqrt{x})$$
$$\sqrt{2x} = \sqrt{1-x}$$
$$2x = 1-x$$
$$3x = 1 \quad \Longrightarrow \quad x = \frac{1}{3}$$

**41.** Find the derivative of $f(x) = 2\arcsin(x-1)$.

**Solution:**

$$f(x) = 2\arcsin(x-1)$$
$$f'(x) = 2\frac{1}{\sqrt{1-(x-1)^2}}(1) = \frac{2}{\sqrt{2x-x^2}}$$

**47.** Find the derivative of $f(x) = \arccos(1/x)$.

**Solution:**

$$f(x) = \arccos\left(\frac{1}{x}\right)$$

$$f'(x) = \frac{-1}{\sqrt{1 - (1/x)^2}}\left(\frac{-1}{x^2}\right) = \frac{1}{x^2\sqrt{(x^2 - 1)/x^2}}$$

$$= \frac{1}{(x^2/|x|)/\sqrt{x^2 - 1}} = \frac{1}{|x|\sqrt{x^2 - 1}}$$

Note: We could have arrived at the same result more quickly had we recognized that $\arccos(1/x) = \operatorname{arcsec} x$. Then

$$f(x) = \arccos\left(\frac{1}{x}\right) = \operatorname{arcsec} x$$

$$f'(x) = \frac{1}{|x|\sqrt{x^2 - 1}}$$

**51.** Find the derivative of $h(t) = \sin(\arccos t)$.

**Solution:**

From the accompanying figure we observe that

$$h(t) = \sin(\arccos t) = \sqrt{1 - t^2}$$

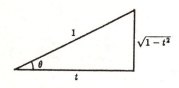

Therefore,

$$h'(t) = \frac{1}{2}(1 - t^2)^{-1/2}(-2t) = \frac{-t}{\sqrt{1 - t^2}}$$

**55.** Find the derivative of $f(x) = \frac{1}{2}\left[\frac{1}{2}\ln\left(\frac{1 + x}{1 - x}\right) + \arctan x\right]$.

**Solution:**

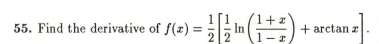

$$f(x) = \frac{1}{2}\left[\frac{1}{2}\ln\left(\frac{1 + x}{1 - x}\right) + \arctan x\right]$$

$$= \frac{1}{2}\left[\frac{1}{2}\ln(1 + x) - \frac{1}{2}\ln(1 - x) + \arctan x\right]$$

$$f'(x) = \frac{1}{2}\left[\frac{1}{2(1 + x)} - \frac{-1}{2(1 - x)} + \frac{1}{1 + x^2}\right]$$

$$= \frac{1}{2}\left[\frac{1 - x + 1 + x}{2(1 - x^2)} + \frac{1}{1 + x^2}\right]$$

$$= \frac{1}{2}\left[\frac{1}{1 - x^2} + \frac{1}{1 + x^2}\right] = \frac{1}{2}\left[\frac{1 + x^2 + 1 - x^2}{1 - x^4}\right] = \frac{1}{1 - x^4}$$

**57.** Find the derivative of $f(x) = x \arcsin x + \sqrt{1-x^2}$.

**Solution:**

$$f(x) = x \arcsin x + \sqrt{1-x^2}$$

$$f'(x) = x\left(\frac{1}{\sqrt{1-x^2}}\right) + \arcsin x + \frac{1}{2}(1-x^2)^{-1/2}(-2x)$$

$$= \frac{x}{\sqrt{1-x^2}} + \arcsin x + \frac{-x}{\sqrt{1-x^2}} = \arcsin x$$

**65.** A small boat is being pulled toward a dock that is 10 feet above the level of the water. The rope is being pulled in at a rate of 1.5 feet per second. Find the rate at which the angle the rope makes with the horizontal is changing when there is 20 feet of rope out.

**Solution:**

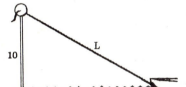

From the accompanying figure we ovserve that

$$\sin \theta = \frac{10}{L} \qquad \text{or} \qquad \theta = \arcsin \frac{10}{L}$$

Differentiating with respect to $t$, we have

$$\frac{d\theta}{dt} = \frac{1}{\sqrt{1-(100/L^2)}}\left(-\frac{10}{L^2}\right)\frac{dL}{dt} = \frac{-10}{L\sqrt{L^2-100}}\frac{dL}{dt}$$

When $dL/dt = -1.5$ feet per second and $L = 20$ feet we have

$$\frac{d\theta}{dt} = \frac{-10}{20\sqrt{20^2-100}}\left(-\frac{3}{2}\right) = \frac{\sqrt{3}}{40} \text{ rad/sec}$$

## 8.6  Inverse Trigonometric Functions: Integration and Completing the Square

**1.** Evaluate $\displaystyle\int_0^{1/6} \frac{1}{\sqrt{1-9x^2}}\,dx$.

**Solution:**

If we let $a = 1$ and $u = 3x$, then $du = 3\,dx$. Thus

$$\int_0^{1/6} \frac{1}{\sqrt{1-9x^2}}\,dx = \frac{1}{3}\int_0^{1/6}\left(\frac{1}{\sqrt{1-(3x)^2}}\right)(3)\,dx$$

$$= \frac{1}{3}\Big[\arcsin(3x)\Big]_0^{1/6}$$

$$= \frac{1}{3}\left[\arcsin\left(\frac{1}{2}\right) - \arcsin 0\right] = \frac{1}{3}\left(\frac{\pi}{6}\right) = \frac{\pi}{18}$$

**7.** Evaluate $\displaystyle\int \frac{x^3}{x^2+1}\,dx$.

**Solution:**

Since the degree of the numerator is greater than the degree of the denominator, we divide to obtain

$$\frac{x^3}{x^2+1} = x - \frac{x}{x^2+1}$$

Therefore,

$$\int \frac{x^3}{x^2+1}\,dx = \int\left(x - \frac{x}{x^2+1}\right)dx$$

$$= \int x\,dx - \frac{1}{2}\int \frac{2x}{x^2+1}\,dx$$

$$= \frac{x^2}{2} - \frac{1}{2}\ln(x^2+1) + C$$

$$= \frac{1}{2}x^2 - \frac{1}{2}\ln(x^2+1) + C$$

**15.** Evaluate $\displaystyle\int_0^{1/\sqrt{2}} \frac{\arcsin x}{\sqrt{1-x^2}}\,dx$.

**Solution:**

If we let $u = \arcsin x$, then $du = \dfrac{1}{\sqrt{1-x^2}}\,dx$. Thus

$$\int_0^{1/\sqrt{2}} \frac{\arcsin x}{\sqrt{1-x^2}}\,dx = \int_0^{1/\sqrt{2}} (\arcsin x)^1 \frac{1}{\sqrt{1-x^2}}\,dx$$

$$= \left[\frac{(\arcsin x)^2}{2}\right]_0^{1/\sqrt{2}}$$

$$= \frac{1}{2}\left\{\left[\arcsin\left(\frac{1}{\sqrt{2}}\right)\right]^2 - [\arcsin(0)]^2\right\}$$

$$= \frac{1}{2}\left[\left(\frac{\pi}{4}\right)^2 - (0)^2\right] = \frac{\pi^2}{32} \approx 0.308$$

**23.** Evaluate $\displaystyle\int \frac{1}{\sqrt{x}(1+x)}\,dx$.

**Solution:**

If we let $a = 1$ and $u = \sqrt{x}$, then $du = (1/2\sqrt{x})\,dx$. Thus

$$\int \frac{1}{\sqrt{x}(1+x)}\,dx = 2\int \frac{1}{2\sqrt{x}[1+(\sqrt{x})^2]}\,dx$$

$$= 2\int \frac{1/2\sqrt{x}}{1+(\sqrt{x})^2}\,dx$$

$$= 2\int \frac{du}{1+u^2} = 2\arctan\sqrt{x} + C$$

**29.** Evaluate $\displaystyle\int \frac{2x}{x^2 + 6x + 13}\,dx$.

**Solution:**

$$\int \frac{2x}{x^2 + 6x + 13}\,dx = \int \frac{(2x+6) - 6}{x^2 + 6x + 13}\,dx$$

$$= \int \frac{2x + 6}{x^2 + 6x + 13}\,dx - \int \frac{6}{x^2 + 6x + 13}\,dx$$

$$= \int \frac{2x + 6}{x^2 + 6x + 13} \, dx - \int \frac{6}{(x^2 + 6x + 9) + 4} \, dx$$

$$= \int \frac{2x + 6}{x^2 + 6x + 13} \, dx - 6 \int \frac{1}{(x + 3)^2 + 2^2} \, dx$$

$$= \ln (x^2 + 6x + 13) - 3 \arctan \left( \frac{x + 3}{2} \right) + C$$

**31.** Evaluate $\displaystyle\int \frac{1}{\sqrt{-x^2 - 4x}} \, dx$.

**Solution:**

$$\int \frac{1}{\sqrt{-x^2 - 4x}} \, dx = \int \frac{1}{\sqrt{-(x^2 + 4x)}} \, dx$$

$$= \int \frac{1}{\sqrt{4 - (x^2 + 4x + 4)}} \, dx$$

$$= \int \frac{1}{\sqrt{2^2 - (x + 2)^2}} \, dx$$

$$= \arcsin \left( \frac{x + 2}{2} \right) + C$$

**35.** Evaluate $\displaystyle\int_2^3 \frac{2x - 3}{\sqrt{4x - x^2}} \, dx$.

**Solution:**

$$\int_2^3 \frac{2x - 3}{\sqrt{4x - x^2}} \, dx = \int_2^3 \frac{(2x - 4) + 1}{\sqrt{4x - x^2}} \, dx$$

$$= \int_2^3 \frac{2x - 4}{\sqrt{4x - x^2}} \, dx + \int_2^3 \frac{1}{\sqrt{4x - x^2}} \, dx$$

$$= -\int_2^3 (4x - x^2)^{-1/2}(4 - 2x) \, dx$$

$$+ \int_2^3 \frac{1}{\sqrt{4 - (x^2 - 4x + 4)}} \, dx$$

$$= -\int_2^3 (4x - x^2)^{-1/2}(4 - 2x) \, dx$$

$$+ \int_2^3 \frac{1}{\sqrt{2^2 - (x - 2)^2}} \, dx$$

$$= \left[ -\frac{(4x - x^2)^{1/2}}{\frac{1}{2}} + \arcsin\left(\frac{x - 2}{2}\right) \right]_2^3$$

$$= -2\sqrt{3} + \frac{\pi}{6} - (-4 + 0) = 4 - 2\sqrt{3} + \frac{\pi}{6} \approx 1.059$$

**43.** Evaluate $\int \sqrt{e^t - 3} \, dt$.

**Solution:**

Let $u = \sqrt{e^t - 3}$. Then $e^t = u^2 + 3, t = \ln(u^2 + 3)$, and $dt = 2u/(u^2 + 3) \, du$. Therefore,

$$\int \sqrt{e^t - 3} \, dt = \int \frac{2u^2}{u^2 + 3} \, du$$

Now since the numerator and denominator are of equal degree, we divide to obtain

$$2\int \frac{u^2}{u^2 + 3} \, du = 2\int \left[1 - \frac{3}{u^2 + 3}\right] du$$

$$= 2\left[\int du - 3\int \frac{1}{(\sqrt{3})^2 + u^2} \, du\right]$$

$$= 2\left[u - 3\left(\frac{1}{\sqrt{3}}\right) \arctan\left(\frac{u}{\sqrt{3}}\right)\right] + C$$

$$= 2\sqrt{e^t - 3} - 2\sqrt{3} \arctan\left(\frac{\sqrt{e^t - 3}}{\sqrt{3}}\right) + C$$

**47.** Find the area of the region bounded by the graphs of

$$y = \frac{1}{x^2 - 2x + 5}, \ y = 0, \ x = 1, \ \text{and} \ x = 3$$

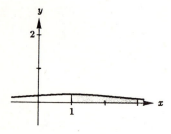

**Solution:**

$$A = \int_1^3 \frac{1}{x^2 - 2x + 5} \, dx$$

$$= \int_1^3 \frac{1}{(x^2 - 2x + 1) + 5 - 1} \, dx$$

$$= \int_1^3 \frac{1}{(x - 1)^2 + 2^2} \, dx$$

$$= \frac{1}{2} \arctan \frac{x - 1}{2} \Big]_1^3$$

$$= \frac{1}{2} \left[ \arctan 1 - \arctan 0 \right] = \frac{\pi}{8}$$

## 8.7   Hyperbolic Functions

**3.** Evaluate (a) csch (ln 2) and (b) coth (ln 5). If the function value is not a rational number, give the answer to three decimal place accuracy.

**Solution:**

(a)   $\text{csch} (\ln 2) = \dfrac{1}{\sinh (\ln 2)} = \dfrac{2}{e^{\ln 2} - e^{-\ln 2}} = \dfrac{2}{2 - \left(\frac{1}{2}\right)} = \dfrac{4}{3}$

(b)   $\coth (\ln 5) = \dfrac{\cosh (\ln 5)}{\sinh (\ln 5)} = \dfrac{e^{\ln 5} + e^{-\ln 5}}{e^{\ln 5} - e^{-\ln 5}} = \dfrac{5 + \left(\frac{1}{5}\right)}{5 - \left(\frac{1}{5}\right)} = \dfrac{13}{12}$

**17.** If $y = \ln[\tanh (x/2)]$, find $y'$ and simplify.

**Solution:**

$$y = \ln\left[\tanh\left(\frac{x}{2}\right)\right]$$

$$y' = \frac{1}{\tanh(x/2)}\left[\frac{1}{2}\text{sech}^2\left(\frac{x}{2}\right)\right] = \frac{1}{2}\left[\frac{\cosh(x/2)}{\sinh(x/2)}\right]\left[\frac{1}{\cosh^2(x/2)}\right]$$

$$= \frac{1}{2\sinh(x/2)\cosh(x/2)} = \frac{1}{\sinh x} = \text{csch } x$$

**23.** If $y = x^{\cosh x}$, find $y'$ and simplify.

**Solution:**

Using logarithmic differentiation, we have

$$y = x^{\cosh x}$$

$$\ln y = \ln(x^{\cosh x}) = (\cosh x)(\ln x)$$

$$\frac{y'}{y} = (\cosh x)\left(\frac{1}{x}\right) + (\ln x)(\sinh x)$$

$$y' = y\left[\frac{\cosh x}{x} + (\ln x)(\sinh x)\right]$$

$$= \frac{y}{x}[\cosh x + x(\sinh x)\ln x]$$

**33.** If $y = \sinh^{-1}(2x) - \sqrt{1 + 4x^2}$, find $y'$ and simplify.

**Solution:**

$$y = 2x\sinh^{-1}(2x) - \sqrt{1 + 4x^2}$$

$$y' = 2x\left[\frac{1}{\sqrt{1 + (2x)^2}}\right](2) + 2\sinh^{-1}(2x) - \frac{1}{2}(1 + 4x^2)^{-1/2}(8x)$$

$$= \frac{4x}{\sqrt{1 + 4x}} + 2\sinh^{-1}(2x) - \frac{4x}{\sqrt{1 + 4x^2}} = 2\sinh^{-1}(2x)$$

**43.** Evaluate $\displaystyle\int \frac{\text{csch}(1/x)\coth(1/x)}{x^2}\, dx$

**Solution:**

If we let $u = \dfrac{1}{x}$, then $du = -\dfrac{1}{x^2}\, dx$ and we have

$$\int \frac{\text{csch}(1/x)\coth(1/x)}{x^2}\, dx = -\int \text{csch}\frac{1}{x}\coth\frac{1}{x}\left(-\frac{1}{x^2}\right) dx$$

$$= -\int \text{csch}\, u \coth u\, du$$

$$= \text{csch}\, u + C$$

$$= \text{csch}\frac{1}{x} + C$$

**51.** Evaluate $\displaystyle\int \frac{1}{\sqrt{1 + e^{2x}}}\, dx.$

**Solution:**

If we let $u = e^x$, then $du = e^x\, dx$ and we have

$$\int \frac{1}{\sqrt{1 + e^{2x}}}\, dx = \int \frac{e^x}{e^x\sqrt{1 + (e^x)^2}}\, dx$$

$$= \int \frac{du}{u\sqrt{1 + u^2}}$$

$$= -\ln\left(\frac{1 + \sqrt{1 + e^{2x}}}{e^x}\right) + C = -\text{csch}^{-1}(e^x) + C$$

**53.** Evaluate $\displaystyle\int \frac{1}{\sqrt{x}\sqrt{1 + x}}\, dx.$

**Solution:**

If we let $u = \sqrt{x}$, then $du = \dfrac{1}{2\sqrt{x}}\, dx$ and we have

$$\int \frac{1}{\sqrt{x}\sqrt{1 + x}}\, dx = 2\int \frac{1}{\sqrt{1 + (\sqrt{x})^2}}\left(\frac{1}{2\sqrt{x}}\right) dx$$

$$= 2\int \frac{du}{\sqrt{1 + u^2}}$$

$$= 2\sinh^{-1}\sqrt{x} + C$$

$$= 2\ln(\sqrt{x} + \sqrt{1 + x}) + C$$

**67.** Find $dy/dx$ for the tractrix $y = a \operatorname{sech}^{-1}(x/a) - \sqrt{a^2 - x^2}$.

**Solution:**

Note that the domain of this function restricts $x$ to the interval $(0, a)$. (Assume $0 < a$.)

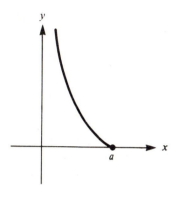

$$y = a \operatorname{sech}^{-1}\left(\frac{x}{a}\right) - \sqrt{a^2 - x^2}$$

$$\frac{dy}{dx} = a\left[\frac{-1}{|x/a|\sqrt{1 - (x/a)^2}}\right]\left(\frac{1}{a}\right) - \frac{1}{2}(a^2 - x^2)^{-1/2}(-2x)$$

$$= \frac{-1}{|x/a|\sqrt{(a^2 - x^2)/a^2}} + \frac{x}{\sqrt{a^2 - x^2}}$$

$$= \frac{-a^2}{x\sqrt{a^2 - x^2}} + \frac{x}{\sqrt{a^2 - x^2}}$$

(Note that we may drop the absolute value signs since $0 < x$.)

$$\frac{dy}{dx} = \frac{-a^2 + x^2}{x\sqrt{a^2 - x^2}} = \frac{-(a^2 - x^2)}{x\sqrt{a^2 - x^2}} = \frac{-\sqrt{a^2 - x^2}}{x}$$

## Review Exercises for Chapter 8

**1.** Find $dy/dx$ if $y = \dfrac{\sin x}{x^2}$.

**Solution:**

$$y = \frac{\sin x}{x^2}$$

$$\frac{dy}{dx} = \frac{x^2(\cos x) - (\sin x)(2x)}{x^4}$$

$$= \frac{x(x \cos x - 2\sin x)}{x^4} = \frac{x \cos x - 2\sin x}{x^3}$$

**19.** Find $dy/dx$ if $y = \left(\dfrac{x^2 + 1}{2}\right) \arctan x$.

**Solution:**

$$y = \frac{1}{2}(x^2 + 1) \arctan x$$

$$\frac{dy}{dx} = \frac{1}{2}(x^2 + 1)\left(\frac{1}{1 + x^2}\right) + \frac{1}{2}\arctan x(2x) = \frac{1}{2} + x\arctan x$$

**23.** Find $dy/dx$ if $\cos x^2 = xe^y$.

**Solution:**

Using implicit differentiation, we have

$$\cos x^2 = xe^y$$

$$-\sin(x^2)(2x) = x(e^y)\frac{dy}{dx} + e^y$$

$$-xe^y\frac{dy}{dx} = 2x\sin x^2 + e^y$$

$$\frac{dy}{dx} = \frac{2x\sin x^2 + e^y}{-xe^y}.$$

From the original equation we have $xe^y = \cos x^2$. Therefore,

$$\frac{dy}{dx} = \frac{2x\sin x^2}{-\cos x^2} + \frac{e^y}{-xe^y} = -2x\tan x^2 - \frac{1}{x}.$$

**31.** Evaluate $\displaystyle\int \frac{\cos x}{1 + \sin^2 x}\, dx$.

**Solution:**

If we let $u = \sin x$ then $du = \cos x\, dx$. Thus

$$\int \frac{\cos x}{1 + \sin^2 x}\, dx = \int \frac{\cos x}{1 + (\sin x)^2}\, dx$$

$$= \int \frac{1}{1 + u^2}\, du = \arctan(\sin x) + C$$

**39.** Evaluate $\displaystyle\int \frac{1}{e^{2x} + e^{-2x}}\, dx$.

**Solution:**

If we let $u = e^{2x}$, then $du = 2e^{2x}\, dx$. Thus

$$\int \frac{1}{e^{2x} + e^{-2x}}\, dx = \int \left(\frac{1}{e^{2x} + e^{-2x}}\right)\left(\frac{e^{2x}}{e^{2x}}\right) dx$$

$$= \int \frac{e^{2x}}{e^{4x} + 1}\, dx$$

$$= \frac{1}{2} \int \left[\frac{1}{1 + (e^{2x})^2}\right](2e^{2x})\, dx$$

$$= \frac{1}{2} \int \frac{1}{1 + u^2}\, du = \frac{1}{2} \arctan(e^{2x}) + C$$

**49.** Evaluate $\displaystyle\int \frac{4 - x}{\sqrt{4 - x^2}}\, dx$.

**Solution:**

$$\int \frac{4 - x}{\sqrt{4 - x^2}}\, dx$$

$$= \int \frac{4}{\sqrt{4 - x^2}}\, dx - \int \frac{x}{\sqrt{4 - x^2}}\, dx$$

$$= 4 \int \frac{1}{\sqrt{2^2 - x^2}}\, dx - \left(-\frac{1}{2}\right) \int (4 - x^2)^{-1/2}(-2x)\, dx$$

$$= 4 \arcsin \frac{x}{2} + \frac{1}{2}\left[\frac{(4 - x^2)^{1/2}}{\frac{1}{2}}\right] + C$$

$$= 4 \arcsin \frac{x}{2} + \sqrt{4 - x^2} + C$$

**57.** A hallway of width 6 ft meets a hallway of width 9 ft at right angles. Find the length of the longest pipe that can be carried horizontally around this corner. (Hint: If $L$ is the length of the pipe, show that

$$L = 6 \cos \theta + 9 \csc\left(\frac{\pi}{2} - \theta\right)$$

where $\theta$ is the angle between the pipe and the wall of the narrower hallway.)

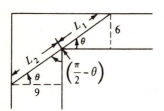

**Solution:**

From the accompanying figure we observe that

$$\csc\theta = \frac{L_1}{6} \quad \text{or} \quad L_1 = 6\csc\theta$$

$$\csc\left(\frac{\pi}{2} - \theta\right) = \frac{L_2}{9} \quad \text{or} \quad L_2 = 9\csc\left(\frac{\pi}{2} - \theta\right)$$

Therefore, the length of the pipe is given by

$$L = L_1 + L_2 = 6\csc\theta + 9\csc\left(\frac{\pi}{2} - \theta\right)$$

$$= 6\csc\theta + 9\sec\theta$$

Note that $\csc[(\pi/2) - \theta] = \sec\theta$. To maximize $L$, we set $dL/d\theta = 0$ as follows.

$$\frac{dL}{d\theta} = -6\csc\theta\cot\theta + 9\sec\theta\tan\theta = 0$$

$$9\sec\theta\tan\theta = 6\csc\theta\cot\theta$$

$$\frac{\sec\theta\tan\theta}{\csc\theta\cot\theta} = \frac{6}{9}$$

$$\tan^3\theta = \frac{2}{3}$$

$$\tan\theta = \frac{2^{1/3}}{3^{1/3}}$$

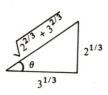

From the accompanying figure we observe that

$$\csc\theta = \frac{\sqrt{2^{2/3} + 3^{2/3}}}{2^{1/3}} \quad \text{and} \quad \sec\theta = \frac{\sqrt{2^{2/3} + 3^{2/3}}}{3^{1/3}}$$

$$L = (6)\left(\frac{\sqrt{2^{2/3} + 3^{2/3}}}{2^{1/3}}\right) + (9)\left(\frac{\sqrt{2^{2/3} + 3^{2/3}}}{3^{1/3}}\right)$$

$$= 3\sqrt{2^{2/3} + 3^{2/3}}(2^{2/3} + 3^{2/3}) = 3(2^{2/3} + 3^{2/3})^{3/2}$$

**65.** Find the volume of the solid generated by revolving the plane region bounded by the graphs of $y = 1/(x^4 + 1)$, $y = 0$, $x = 0$, and $x = 1$ about the $y$-axis.

**Solution:**

We use the Shell Method with $p(x) = x$ and $h(x) = 1/(x^4 + 1)$. (See the accompanying figure.)

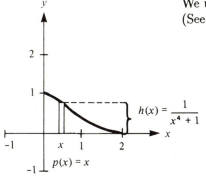

$$V = 2\pi \int_a^b p(x)h(x)\, dx$$

$$= 2\pi \int_0^1 (x)\, \frac{1}{x^4 + 1}\, dx$$

$$= \pi \int_0^1 \frac{1}{x^4 + 1}(2x)\, dx \qquad (u = x^2)$$

$$= \pi\left[\arctan x^2\right]_0^1 = \frac{\pi^2}{4}$$

# 9 INTEGRATION TECHNIQUES, AND IMPROPER INTEGRALS

## 9.1 Basic Integration Formulas

**7.** Evaluate $\displaystyle\int \frac{t^2 - 3}{-t^3 + 9t + 1}\, dt$.

**Solution:**

If we let $u = -t^3 + 9t + 1$, then $du = (-3t^2 + 9)\, dt = -3(t^2 - 3)\, dt$. Thus,

$$\int \frac{t^2 - 3}{-t^3 + 9t + 1}\, dt = -\frac{1}{3}\int \frac{1}{-t^3 + 9t + 1}[-3(t^2 - 3)]\, dt$$

$$= -\frac{1}{3}\int \frac{1}{u}\, du$$

$$= -\frac{1}{3}\ln |-t^3 + 9t + 1| + C$$

**13.** Evaluate $\int t \sin t^2 \, dt$.

**Solution:**

If we let $u = t^2$, then $du = 2t\,dt$. Thus,

$$
\begin{aligned}
\int t \sin t^2 \, dt &= \frac{1}{2} \int \sin(t^2)(2t) \, dt \\
&= \frac{1}{2} \int \sin u \, du \\
&= \frac{1}{2}(-\cos t^2) + C \\
&= -\frac{1}{2} \cos t^2 + C
\end{aligned}
$$

**23.** Evaluate $\displaystyle \int \frac{1}{1 - \cos x} \, dx$.

**Solution:**

$$
\begin{aligned}
\int \frac{1}{1 - \cos x} \, dx &= \int \left( \frac{1}{1 - \cos x} \right) \left( \frac{1 + \cos x}{1 + \cos x} \right) \, dx \\
&= \int \frac{1 + \cos x}{1 - \cos^2 x} \, dx \\
&= \int \frac{1 + \cos x}{\sin^2 x} \, dx \\
&= \int \frac{1}{\sin^2 x} \, dx + \int \frac{\cos x}{\sin^2 x} \, dx \\
&= \int \csc^2 x \, dx + \int \overbrace{(\sin x)^{-2}}^{u^{-2}} \overbrace{\cos x \, dx}^{du} \\
&= -\cot x + \frac{(\sin x)^{-1}}{-1} + C \\
&= -\cot x - \csc x + C
\end{aligned}
$$

**25.** Evaluate $\displaystyle \int \frac{2t - 1}{t^2 + 4} \, dt$.

**Solution:**

$$
\begin{aligned}
\int \frac{2t - 1}{t^2 + 4} \, dt &= \int \frac{2t}{t^2 + 4} \, dt - \int \frac{1}{t^2 + 4} \, dt \\
&= \ln(t^2 + 4) - \frac{1}{2} \arctan \frac{t}{2} + C
\end{aligned}
$$

**31.** Evaluate $\displaystyle\int \frac{-1}{\sqrt{1-(2t-1)^2}}\, dt$.

**Solution:**

If we let $u = 2t - 1$, then $du = 2\, dt$. Thus,

$$\int \frac{-1}{\sqrt{1-(2t-1)^2}}\, dt = \frac{-1}{2}\int \frac{2}{\sqrt{1-(2t-1)^2}}\, dt$$

$$= \frac{-1}{2}\int \frac{1}{\sqrt{a^2-u^2}}\, du$$

$$= -\frac{1}{2}\arcsin(2t-1) + C$$

**37.** Evaluate $\displaystyle\int \frac{\sec^2 x}{4+\tan^2 x}\, dx$

**Solution:**

If we let $u = \tan x$, then $du = \sec^2 x\, dx$ and we have

$$\int \frac{\sec^2 x}{4+\tan^2 x}\, dx = \int \frac{1}{a^2+u^2}\, du$$

$$= \frac{1}{2}\arctan \frac{\tan x}{2} + C$$

**47.** Evaluate $\displaystyle\int \frac{3}{\sqrt{6x-x^2}}\, dx$.

**Solution:**

By completing the square we have

$$\int \frac{3}{\sqrt{6x-x^2}}\, dx = 3\int \frac{1}{\sqrt{9-(9-6x+x^2)}}\, dx$$

$$= 3\int \frac{1}{\sqrt{9-(x-3)^2}}\, dx$$

$$= 3\arcsin\left(\frac{x-3}{3}\right) + C$$

**53.** Evaluate $\displaystyle\int_1^e \frac{1 - \ln x}{x}\, dx$.

**Solution:**

If we let $u = 1 - \ln x$, then $du = \dfrac{1}{x}\, dx$ and we have

$$\int_1^e \frac{1 - \ln x}{x}\, dx = -\int_1^e (1 - \ln x)^1 \left(\frac{-1}{x}\right)\, dx$$

$$= -\left[\frac{(1 - \ln x)^2}{2}\right]_1^e$$

$$= -\frac{1}{2}[(1 - \ln e)^2 - (1 - \ln 1)^2]$$

$$= -\frac{1}{2}[(0)^2 - 1] = \frac{1}{2}$$

**59.** Evaluate $\displaystyle\int_1^2 \frac{1}{2x\sqrt{4x^2 - 1}}\, dx$.

**Solution:**

$$\int_1^2 \frac{1}{2x\sqrt{4x^2 - 1}}\, dx = \frac{1}{2}\int_1^2 \frac{2}{2x\sqrt{(2x)^2 - 1}}\, dx$$

$$= \frac{1}{2}\left[\operatorname{arcsec} 2x\right]_1^2$$

$$= \frac{1}{2}[\operatorname{arcsec} 4 - \operatorname{arcsec} 2]$$

$$= \frac{1}{2}\left[\operatorname{arcsec} 4 - \frac{\pi}{3}\right] \approx 0.135$$

**65.** The region bounded by $y = e^{-x^2}$, $y = 0$, and $x = b$ is revolved around the $y$-axis. Find $b$ so that the volume of the genereated solid is $\frac{4}{3}$ cubic units.

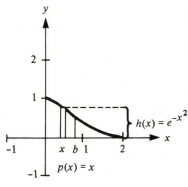

$h(x) = e^{-x^2}$

$p(x) = x$

**Solution:**

By using the Shell Method we have

$$V = 2\pi \int_0^b x e^{-x^2}\, dx$$

$$= -\pi \int_0^b e^{-x^2}(-2x\, dx) \qquad (\text{Let } u = -x^2, du = -2x\, dx)$$

$$= \left[-\pi e^{-x^2}\right]_0^b = \pi(1 - e^{-b^2}) = \frac{4}{3}$$

Solving for $b$ we have

$$e^{-b^2} = \frac{3\pi - 4}{3\pi}$$

$$-b^2 = \ln\left(\frac{3\pi - 4}{3\pi}\right),$$

$$b = \sqrt{\ln\left(\frac{3\pi - 4}{3\pi}\right)^{-1}} = \sqrt{\ln\left(\frac{3\pi}{3\pi - 4}\right)} \approx 0.743$$

## 9.2   Integration by Parts

**5.** Evaluate $\int x e^{-2x}\, dx$.

**Solution:**

Letting $u = x$ and $dv = e^{-2x}\, dx$, we have

$$dv = e^{-2x}\, dx \quad \Longrightarrow \quad v = \int e^{-2x}\, dx = -\frac{1}{2}e^{-2x}$$

$$u = x \qquad\qquad \Longrightarrow \quad du = dx$$

Therefore, we have

$$\int u\, dv = uv - \int v\, du$$

$$\int x e^{-2x}\, dx = x\left(-\frac{1}{2}e^{-2x}\right) - \int -\frac{1}{2}e^{-2x}\, dx$$

$$= -\frac{x}{2}e^{-2x} - \frac{1}{4}e^{-2x} + C$$

$$= -\frac{1}{4e^{2x}}(2x + 1) + C$$

9. Evaluate $\int x^3 \ln x \, dx$.

**Solution:**

Letting $u = \ln x$ and $dv = x^3 \, dx$, we have

$$dv = x^3 \, dx \quad \Longrightarrow \quad v = \int x^3 \, dx = \frac{x^4}{4}$$

$$u = \ln x \quad \Longrightarrow \quad du = \frac{1}{x} \, dx$$

Therefore, we have

$$\int u \, dv = uv - \int v \, du$$

$$\int x^3 \ln x \, dx = (\ln x)\left(\frac{x^4}{4}\right) - \int \left(\frac{x^4}{4}\right)\left(\frac{1}{x}\right) dx$$

$$= \frac{1}{4} x^4 \ln x - \frac{1}{16} x^4 + C = \frac{x^4}{16}(4 \ln x - 1) + C$$

21. Evaluate $\int x\sqrt{x-1} \, dx$.

**Solution:**

Letting $u = x$ and $dv = \sqrt{x-1} \, dx$, we have

$$dv = \sqrt{x-1} \, dx \quad \Longrightarrow \quad v = \int \sqrt{x-1} \, dx = \frac{2}{3}(x-1)^{3/2}$$

$$u = x \quad \Longrightarrow \quad du = dx$$

Therefore,

$$\int u \, dv = uv - \int v \, du$$

$$\int x\sqrt{x-1} \, dx = \frac{2}{3}x(x-1)^{3/2} - \int \frac{2}{3}(x-1)^{3/2} \, dx$$

$$= \frac{2}{3}x(x-1)^{3/2} - \frac{4}{15}(x-1)^{5/2} + C$$

$$= \frac{2}{15}(x-1)^{3/2}[5x - 2(x-1)] + C$$

$$= \frac{2}{15}(x-1)^{3/2}(3x+2) + C$$

**31.** Evaluate $\int \arctan x \, dx$.

**Solution:**

Let

$$dv = dx \qquad \Longrightarrow \qquad v = \int dx = x$$

$$u = \arctan x \quad \Longrightarrow \quad du = \frac{1}{1 + x^2} \, dx$$

Therefore, we have

$$\int \arctan x \, dx = x \arctan x - \int \frac{x}{1 + x^2} \, dx$$

$$= x \arctan x - \frac{1}{2} \ln(1 + x^2) + C$$

**33.** Evaluate $\int e^{2x} \sin x \, dx$.

**Solution:**

Let

$$dv = \sin x \, dx \quad \Longrightarrow \quad v = \int \sin x \, dx = - \cos x$$

$$u = e^{2x} \qquad \Longrightarrow \quad du = 2e^{2x} \, dx$$

$$\int e^{2x} \sin x \, dx = -e^{2x} \cos x + \int \cos x (2e^{2x}) \, dx$$

Using integration by parts again, we let

$$dv = \cos x \, dx \quad \Longrightarrow \quad v = \int \cos x \, dx = \sin x$$

$$u = 2e^{2x} \qquad \Longrightarrow \quad du = 4e^{2x} \, dx$$

$$\int e^{2x} \sin x \, dx = -e^{2x} \cos x + 2e^{2x} \sin x - \int 4e^{2x} \sin x \, dx$$

By adding the right integral to both sides, we have

$$5 \int e^{2x} \sin x \, dx = -e^{2x} \cos x + 2e^{2x} \sin x + C_1$$

Finally, dividing both sides by 5, we conclude that

$$\int e^{2x} \sin x \, dx = \frac{e^{2x}}{5} (2 \sin x - \cos x) + C$$

**36.** Evaluate $\int_0^1 x \arcsin x^2 \, dx$.

**Solution:**

Let

$$dv = x \, dx \qquad \Longrightarrow \qquad v = \int x \, dx = \frac{x^2}{2}$$

$$u = \arcsin x^2 \qquad \Longrightarrow \qquad du = \frac{2x}{\sqrt{1 - x^4}} \, dx$$

$$\int x \arcsin x^2 \, dx = \frac{x^2 \arcsin x^2}{2} - \int \frac{x^2}{2} \left( \frac{2x}{\sqrt{1 - x^4}} \right) dx$$

$$= \frac{x^2 \arcsin x^2}{2} - \int \frac{x^3}{\sqrt{1 - x^4}} \, dx$$

$$= \frac{x^2 \arcsin x^2}{2} + \frac{1}{4} \int (1 - x^4)^{-1/2}(-4x^3) \, dx$$

$$= \frac{x^2 \arcsin x^2}{2} + \frac{1}{4} \left[ \frac{(1 - x^4)^{1/2}}{\frac{1}{2}} \right] + C$$

$$= \frac{1}{2}[x^2 \arcsin x^2 + \sqrt{1 - x^4}] + C$$

Finally,

$$\int_0^1 x \arcsin x^2 \, dx = \frac{1}{2}[x^2 \arcsin x^2 + \sqrt{1 - x^4}]_0^1$$

$$= \frac{1}{2}\left( \frac{\pi}{2} + 0 - 0 - 1 \right) = \frac{1}{2}\left( \frac{\pi}{2} - 1 \right)$$

**39.** Evaluate $\int_0^{\pi/2} x \cos x \, dx$.

**Solution:**

Let

$$dv = \cos \, dx \qquad \Longrightarrow \qquad v = \int \cos \, dx = \sin x$$

$$u = x \qquad \Longrightarrow \qquad du = dx$$

$$\int x \cos x \, dx = x \sin x - \int \sin x \, dx$$

$$= x \sin x + \cos x + C$$

Finally,

$$\int_0^{\pi/2} x \cos x \, dx = [x \sin x + \cos x]_0^{\pi/2}$$

$$= \frac{\pi}{2} - 1$$

**41.** Use the tabular method for repeated application of Integration by Parts to evaluate

$$\int x^2 e^{2x}\, dx.$$

**Solution:**

We begin by letting $u = x^2$ and $dv = v'\, dx = e^{2x}\, dx$. We next create a table consisting of three columns as follows:

| Alternate Signs | $u$ and its Derivatives | $v'$ and its Antiderivatives |
|:---:|:---:|:---:|
| $+$ | $x^2$ | $e^{2x}$ |
| $-$ | $2x$ | $\frac{1}{2}e^{2x}$ |
| $+$ | $2$ | $\frac{1}{4}e^{2x}$ |
| $-$ | $0$ | $\frac{1}{8}e^{2x}$ |

Finally, the solution is given by multiplying the signed products of the diagonal entries of the table to obtain

$$\int x^2 e^{2x}\, dx = \frac{1}{2}x^2 e^{2x} - 2x\left(\frac{1}{4}\right)e^{2x} + 2\left(\frac{1}{8}\right)e^{2x} + C$$

$$= \frac{e^{2x}}{4}(2x^2 - 2x + 1) + C$$

**53.** Use integration by parts to verify the formula

$$\int x^n \ln x\, dx = \frac{x^{n+1}}{(n+1)^2}[-1 + (n+1)\ln x] + C$$

**Solution:**

Let

$$dv = x^n\, dx \quad \Longrightarrow \quad v = \int x^n\, dx = \frac{x^{n+1}}{n+1}$$

$$u = \ln x \quad \Longrightarrow \quad du = \frac{1}{x}\, dx$$

$$\int x^n \ln x\, dx = \frac{x^{n+1}}{n+1}\ln x - \int \frac{x^{n+1}}{n+1}\left(\frac{1}{x}\right)dx$$

$$= \frac{x^{n+1}}{n+1}\ln x - \frac{1}{n+1}\int x^n\, dx$$

$$= \frac{x^{n+1}}{n+1}\ln x - \frac{x^{n+1}}{(n+1)^2} + C$$

$$= \frac{x^{n+1}}{(n+1)^2}[-1 + (n+1)\ln x] + C$$

**65.** Given the region bounded by the graphs of $y = \ln x$, $y = 0$, and $x = e$, find

(a) the area of the region.

(b) the volume of the solid generated by revolving the region about the $x$-axis.

(c) the volume of the solid generated by revolving the region about the $y$-axis.

(d) the centroid of the region

**Solution:**

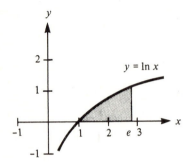

(a) From the accompanying figure we have

$$A = \int_1^e \ln x \, dx$$

$$dv = dx \quad\Longrightarrow\quad v = x$$
$$u = \ln x \quad\Longrightarrow\quad du = \frac{1}{x} \, dx$$

Therefore,

$$A = \int_1^e \ln x \, dx = x \ln x - \int dx$$
$$= \big[x \ln x - x\big]_1^e = 1$$

(b) Using the Disk Method we have

$$V = \pi \int_1^e (\ln x)^2 \, dx$$

Let

$$dv = dx \quad\Longrightarrow\quad v = x$$
$$u = (\ln x)^2 \quad\Longrightarrow\quad du = \frac{2 \ln x}{x} \, dx$$

Therefore,

$$\int (\ln x)^2 \, dx = x(\ln x)^2 - 2 \int \ln x \, dx$$
$$= x(\ln x)^2 - 2x(\ln x - 1) \qquad \text{From part (a)}$$

Finally,

$$V = \pi \int_1^e (\ln x)^2 \, dx = \pi \big[x(\ln x)^2 - 2x(\ln x - 1)\big]_1^e$$
$$= \pi(e - 2) \approx 2.257$$

(c) Using the Shell Method we have

$$V = 2\pi \int_1^e x \ln x \, dx$$

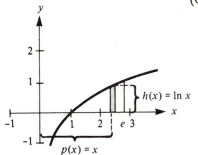

Let

$$dv = x \, dx \quad \Longrightarrow \quad v = \frac{x^2}{2}$$

$$u = \ln x \quad \Longrightarrow \quad du = \frac{1}{x} \, dx$$

Therefore,

$$\int x \ln x \, dx = \frac{x^2}{2} \ln x - \frac{1}{2} \int x \, dx = \frac{x^2}{2} \ln x - \frac{x^2}{4}$$

Finally,

$$V = 2\pi \int_1^e x \ln x \, dx$$

$$= 2\pi \left[ \frac{x^2}{4}(2\ln x - 1) \right]_1^e = \frac{\pi}{2}(e^2 + 1) \approx 13.177$$

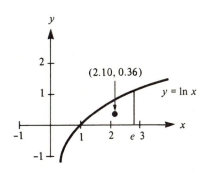

(d) $\bar{x} = \dfrac{1}{A} \displaystyle\int_1^e x \ln x \, dx$

$$= \left[ \frac{x^2}{4}(2\ln x - 1) \right]_1^e \qquad \text{From part (c)}$$

$$= \frac{1}{4}(e^2 + 1) \approx 2.097$$

$$\bar{y} = \frac{1}{2A} \int_1^e (\ln x)^2 \, dx$$

$$= \frac{1}{2} \left[ x(\ln x)^2 - 2x(\ln x - 1) \right]_1^e \qquad \text{From part (b)}$$

$$= \frac{e - 2}{2} \approx 0.359$$

## 9.3  Trigonometric Integrals

**5.** Evaluate $\int \sin^5 x \cos^2 x \, dx$.

**Solution:**

$$
\begin{aligned}
\int \sin^5 x \cos^2 x \, dx &= \int \sin x (\sin^2 x)^2 \cos^2 x \, dx \\
&= \int \sin x (1 - \cos^2 x)^2 \cos^2 x \, dx \\
&= -\int (\cos^2 x - 2\cos^4 x + \cos^6 x)(-\sin x) \, dx \\
&= -\frac{1}{3} \cos^3 x + \frac{2}{5} \cos^5 x - \frac{1}{7} \cos^7 x + C
\end{aligned}
$$

**9.** Evaluate $\int \sin^4 \pi x \, dx$.

**Solution:**

$$
\begin{aligned}
\int \sin^4 \pi x \, dx &= \int (\sin^2 \pi x)^2 \, dx \\
&= \int \left( \frac{1 - \cos 2\pi x}{2} \right)^2 dx \\
&= \frac{1}{4} \int (1 - 2\cos 2\pi x + \cos^2 2\pi x) \, dx \\
&= \frac{1}{4} \int \left( 1 - 2\cos 2\pi x + \frac{1 + \cos 4\pi x}{2} \right) dx \\
&= \frac{1}{8} \int (3 - 4\cos 2\pi x + \cos 4\pi x) \, dx \\
&= \frac{1}{8} \left( 3x - \frac{2}{\pi} \sin 2\pi x + \frac{1}{4\pi} \sin 4\pi x \right) + C \\
&= \frac{1}{32\pi} (12\pi x - 8\sin 2\pi x + \sin 4\pi x) + C
\end{aligned}
$$

**23.** Evaluate $\int \tan^5 \frac{x}{4}\, dx$.

**Solution:**

$$\int \tan^5 \frac{x}{4}\, dx = \int \tan^2 \frac{x}{4} \tan^3 \frac{x}{4}\, dx$$

$$= \int \left(\sec^2 \frac{x}{4} - 1\right) \tan^3 \frac{x}{4}\, dx$$

$$= \int \tan^3 \frac{x}{4} \sec^2 \frac{x}{4}\, dx - \int \tan^3 \frac{x}{4}\, dx$$

$$= \tan^4 \frac{x}{4} - \int \tan^2 \frac{x}{4} \tan \frac{x}{4}\, dx$$

$$= \tan^4 \frac{x}{4} - \int \left(\sec^2 \frac{x}{4} - 1\right) \tan \frac{x}{4}\, dx$$

$$= \tan^4 \frac{x}{4} - \int \tan \frac{x}{4} \sec^2 \frac{x}{4}\, dx + \int \tan \frac{x}{4}\, dx$$

$$= \tan^4 \frac{x}{4} - 2 \tan^2 \frac{x}{4} - 4 \ln \left|\cos \frac{x}{4}\right| + C$$

**31.** Evaluate $\int \sec^6 4x \tan 4x\, dx$.

**Solution:**

$$\int \sec^6 4x \tan 4x\, dx$$

$$= \int (\sec^2 4x)(\sec^2 4x)^2 \tan 4x\, dx$$

$$= \frac{1}{4} \int \tan 4x (\tan^2 4x + 1)^2 (4 \sec^2 4x)\, dx$$

$$= \frac{1}{4} \int (\tan^5 4x + 2 \tan^3 4x + \tan 4x)(4 \sec^2 4x)\, dx$$

$$= \frac{1}{4} \left[\frac{\tan^6 4x}{6} + \frac{\tan^4 4x}{2} + \frac{\tan^2 4x}{2}\right] + C$$

$$= \frac{1}{24} (\tan^2 4x)(\tan^4 4x + 3 \tan^2 4x + 3) + C$$

or

$$\int \sec^6 4x \tan 4x \, dx = \frac{1}{4} \int \sec^5 4x (4 \sec 4x \tan 4x) \, dx$$
$$= \frac{1}{24} \sec^6 4x + C_1$$

(See Exercise 61 for a comparison of the two methods.)

**43.** Evaluate $\int \sin 3x \cos 2x \, dx$.

**Solution:**

Using the trigonometric identity

$$\sin u \cos v = \frac{1}{2}[\sin(u+v) + \sin(u-v)]$$

we have

$$\int \sin 3x \cos 2x \, dx = \int \frac{1}{2}(\sin 5x + \sin x) \, dx$$
$$= \frac{1}{2}\left(\frac{1}{5}\right) \int \sin 5x(5) \, dx + \frac{1}{2} \int \sin x \, dx$$
$$= -\frac{1}{10} \cos 5x - \frac{1}{2} \cos x + C$$
$$= -\frac{1}{10}(\cos 5x + 5 \cos x) + C$$

**53.** Evaluate $\int_0^{\pi/4} \tan^3 x \, dx$.

**Solution:**

$$\int_0^{\pi/4} \tan^3 x \, dx = \int_0^{\pi/4} (\sec^2 x - 1)(\tan x) \, dx$$
$$= \int_0^{\pi/4} \tan x \sec^2 x \, dx - \int_0^{\pi/4} \tan x \, dx$$
$$= \left[\frac{1}{2} \tan^2 x + \ln|\cos x|\right]_0^{\pi/4} = \frac{1}{2}(1 - \ln 2)$$

**61.** Evaluate $\int \sec^4 3x \tan^3 3x \, dx$ in two ways and show that the results differ only by a constant.

**Solution:**

Method 1

Since there is an odd power of the tangent, we write

$$\int \sec^4 3x \tan^3 3x \, dx = \int \sec^3 3x \tan^2 3x (\sec 3x \tan 3x) \, dx$$

$$= \int \sec^3 3x (\sec^2 3x - 1) \sec 3x \tan 3x \, dx$$

$$= \frac{1}{3} \int (\sec^5 3x - \sec^3 3x)(3 \sec 3x \tan 3x \, dx)$$

$$= \frac{1}{3} \left( \frac{1}{6} \sec^6 3x - \frac{1}{4} \sec^4 3x \right) + C$$

Method 2

Since the power of the secant is even, we write

$$\int \sec^4 3x \tan^3 3x \, dx = \int \sec^2 3x \tan^3 3x (\sec^2 3x) \, dx$$

$$= \int (1 + \tan^2 3x) \tan^3 3x (\sec^2 3x) \, dx$$

$$= \frac{1}{3} \int (\tan^3 3x + \tan^5 3x)(3 \sec^2 3x) \, dx$$

$$= \frac{1}{3} \left( \frac{\tan^4 3x}{4} + \frac{\tan^6 3x}{6} \right) + C$$

Comparing the results of the two methods, we have

$$\frac{1}{3} \left[ \frac{1}{4} \tan^4 3x + \frac{1}{6} \tan^6 3x \right] + C$$

$$= \frac{1}{3} \left[ \frac{1}{4} (\sec^2 3x - 1)^2 + \frac{1}{6} (\sec^2 3x - 1)^3 \right] + C$$

$$= \frac{1}{3} \left[ \frac{1}{4} (\sec^4 3x - 2 \sec^2 3x + 1) \right.$$

$$\left. + \frac{1}{6} (\sec^6 3x - 3 \sec^4 3x + 3 \sec^2 3x - 1) \right] + C$$

$$= \frac{1}{3} \left[ \sec^6 3x - \frac{1}{4} \sec^4 3x + \frac{1}{4} - \frac{1}{6} \right] + C$$

Therefore, the results differ only by the constant $\frac{1}{3}(\frac{1}{4} - \frac{1}{6})$.

**67.** Find (a) the volume of the solid generated by revolving the region bounded by the graphs of $y = \sin x$, $y = 0$, $x = 0$, and $x = \pi$ about the $x$-axis and (b) find the centroid of the region.

**Solution:**

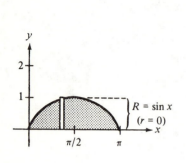

(a)

$$V = \pi \int_0^\pi R^2 \, dx = \pi \int_0^\pi \sin^2 x \, dx$$

$$= \pi \int_0^\pi \frac{1 - \cos 2x}{2} \, dx$$

$$= \frac{\pi}{2} \int_0^\pi (1 - \cos 2x) \, dx$$

$$= \frac{\pi}{2} \left[ x - \frac{1}{2} \sin 2x \right]_0^\pi$$

$$= \frac{\pi}{2} \left[ \pi - \frac{1}{2}(0) - 0 \right] = \frac{\pi^2}{2}$$

(b) By symmetry $\bar{x} = \dfrac{\pi}{2}$.

The area of the region is given by

$$A = \int_0^\pi \sin x \, dx = \left[ -\cos x \right]_0^\pi = 2$$

Therefore,

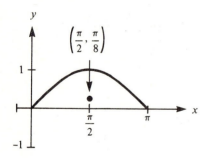

$$\bar{y} = \frac{1}{2A} \int_0^\pi \sin^2 x \, dx$$

$$= \frac{1}{8} \int_0^\pi (1 - \cos 2x) \, dx = \left[ \frac{1}{8} \left( x - \frac{1}{2} \sin 2x \right) \right]_0^\pi = \frac{\pi}{8}$$

Therefore, $(\bar{x}, \bar{y}) = (\pi/2, \pi/8)$.

**77.** Use integration by parts to verify the formula

$$\int \cos^m x \sin^n x \, dx$$

$$= -\frac{\cos^{m+1} x \sin^{n-1} x}{m+n} + \frac{n-1}{m+n} \int \cos^m x \sin^{n-2} x \, dx$$

**Solution:**

Let $dv = \cos^m x \sin x \, dx$ and $u = \sin^{n-1} x$. Then
$v = (-\cos^{m+1} x)/(m+1)$ and $du = (n-1)\sin^{n-2} x(\cos x) \, dx$.
Therefore,

$$\int \cos^m x \sin^n x \, dx$$

$$= -\frac{\sin^{n-1} x \cos^{m+1} x}{m+1} + \frac{n-1}{m+1} \int \sin^{n-2} x \cos^{m+2} x \, dx$$

$$= -\frac{\sin^{n-1} x \cos^{m+1} x}{m+1} + \frac{n-1}{m+1} \int \sin^{n-2} x \cos^m x(1 - \sin^2 x) \, dx$$

$$= -\frac{\sin^{n-1} x \cos^{m+1} x}{m+1} + \frac{n-1}{m+1} \int \sin^{n-2} x \cos^m x \, dx$$

$$\qquad - \frac{n-1}{m+1} \int \sin^n x \cos^m x \, dx$$

Now observe that the last integral is a multiple of the original.
Adding yields

$$\frac{m+n}{m+1} \int \cos^m x \sin^n x \, dx$$

$$= -\frac{\sin^{n-1} x \cos^{m+1} x}{m+1} + \frac{n-1}{m+1} \int \cos^m x \sin^{n-2} x \, dx$$

$$\int \cos^m x \sin^n x \, dx$$

$$= -\frac{\cos^{m+1} x \sin^{n-1} x}{m+n} + \frac{n-1}{m+n} \int \cos^m x \sin^{n-2} x \, dx$$

## 9.4   Trigonometric Substitution

**3.** Evaluate $\displaystyle\int \frac{\sqrt{25 - x^2}}{x}\, dx$.

**Solution:**

Let $x = 5\sin\theta$. Then $\sqrt{25 - x^2} = 5\cos\theta$ and $dx = 5\cos\theta\, d\theta$. Thus,

$$\int \frac{\sqrt{25 - x^2}}{x}\, dx$$

$$= \int \frac{5\cos\theta}{5\sin\theta} 5\cos\theta\, d\theta = 5\int \frac{\cos^2\theta}{\sin\theta}\, d\theta$$

$$= 5\int \frac{1 - \sin^2\theta}{\sin\theta}\, d\theta = 5\int (\csc\theta - \sin\theta)\, d\theta$$

$$= 5(\ln|\csc\theta - \cot\theta| + \cos\theta) + C$$

$$= 5\left(\ln\left|\frac{5}{x} - \frac{\sqrt{25 - x^2}}{x}\right| + \frac{\sqrt{25 - x^2}}{5}\right) + C$$

$$= 5\ln\left|\frac{5 - \sqrt{25 - x^2}}{x}\right| + \sqrt{25 - x^2} + C$$

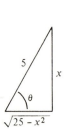

**7.** Evaluate $\int x^3\sqrt{x^2 - 4}\, dx$.

**Solution:**

Let $x = 2\sec\theta$. Then $\sqrt{x^2 - 4} = 2\tan\theta$ and $dx = 2\sec\theta\tan\theta\, d\theta$. Thus,

$$\int x^3\sqrt{x^2 - 4}\, dx$$

$$= \int (2\sec\theta)^3(2\tan\theta)(2\sec\theta\tan\theta)\, d\theta$$

$$= 32\int \sec^4\theta \tan^2\theta\, d\theta$$

$$= 32\int \sec^2\theta \tan^2\theta \sec^2\theta\, d\theta$$

$$= 32\int (\tan^2\theta + 1)\tan^2\theta \sec^2\theta\, d\theta$$

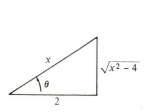

$$= 32 \int (\tan^4 \theta + \tan^2 \theta) \sec^2 \theta \, d\theta$$

$$= 32 \left[ \frac{1}{5} \tan^5 \theta + \frac{1}{3} \tan^3 \theta \right] + C$$

$$= 32 \left[ \frac{1}{5} \left( \frac{\sqrt{x^2 - 4}}{2} \right)^5 + \frac{1}{3} \left( \frac{\sqrt{x^2 - 4}}{2} \right)^3 \right] + C$$

$$= \frac{1}{15} (x^2 - 4)^{3/2} (3x^2 + 8) + C$$

**11.** Evaluate $\int \frac{1}{(1 + x^2)^2} \, dx$.

**Solution:**

Let $x = \tan \theta$. Then $1 + x^2 = \sec^2 \theta$ and $dx = \sec^2 \theta \, d\theta$. Thus,

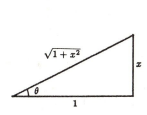

$$\int \frac{1}{(1 + x^2)^2} \, dx = \int \frac{1}{\sec^4 \theta} (\sec^2 \theta) \, d\theta$$

$$= \int \cos^2 \theta \, d\theta$$

$$= \frac{1}{2} \int (1 + \cos 2\theta) \, d\theta$$

$$= \frac{1}{2} \left( \theta + \frac{1}{2} \sin 2\theta \right) + C$$

$$= \frac{1}{2} (\theta + \sin \theta \cos \theta) + C$$

$$= \frac{1}{2} \left( \arctan x + \frac{x}{\sqrt{1 + x^2}} \cdot \frac{1}{\sqrt{1 + x^2}} \right) + C$$

$$= \frac{1}{2} \left( \arctan x + \frac{x}{1 + x^2} \right) + C$$

**23.** Evaluate $\int \frac{1}{x \sqrt{4x^2 + 9}} \, dx$.

**Solution:**

Let $2x = 3\tan\theta$. Then $x = \frac{3}{2}\tan\theta$, $\sqrt{4x^2 + 9} = 3\sec\theta$, and $dx = \frac{3}{2}\sec^2\theta\,d\theta$.

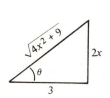

$$\int \frac{1}{x\sqrt{4x^2 + 9}}\,dx$$

$$= \int \frac{\frac{3}{2}\sec^2\theta\,d\theta}{\frac{3}{2}\tan\theta(3\sec\theta)} = \frac{1}{3}\int \csc\theta\,d\theta$$

$$= \frac{1}{3}\ln|\csc\theta - \cot\theta| + C$$

$$= \frac{1}{3}\ln\left|\frac{\sqrt{4x^2 + 9} - 3}{2x}\right| + C$$

$$= -\frac{1}{3}\ln\left|\frac{2x}{\sqrt{4x^2 + 9} - 3}\right| + C$$

$$= -\frac{1}{3}\ln\left|\left(\frac{2x}{\sqrt{4x^2 + 9} - 3}\right)\left(\frac{\sqrt{4x^2 + 9} + 3}{\sqrt{4x^2 + 9} + 3}\right)\right| + C$$

$$= -\frac{1}{3}\ln\left|\frac{3 + \sqrt{4x^2 + 9}}{2x}\right| + C$$

**37.** Evaluate $\displaystyle\int \frac{1}{4 + 4x^2 + x^4}\,dx$.

**Solution:**

$$\int \frac{1}{4 + 4x^2 + x^4}\,dx = \int \frac{1}{(2 + x^2)^2}\,dx$$

Let $x = \sqrt{2}\tan\theta$. Then $2 + x^2 = 2\sec^2\theta$ and $dx = \sqrt{2}\sec^2\theta\,d\theta$. Thus,

$$\int \frac{1}{(2 + x^2)^2}\,dx = \int \frac{\sqrt{2}\sec^2\theta\,d\theta}{(2\sec^2\theta)^2} = \frac{\sqrt{2}}{4}\int \cos^2\theta\,d\theta$$

$$= \frac{\sqrt{2}}{4}\int \frac{1 + \cos 2\theta}{2}\,d\theta = \frac{\sqrt{2}}{8}\left(\theta + \frac{1}{2}\sin 2\theta\right) + C$$

$$= \frac{\sqrt{2}}{8}(\theta + \sin\theta\cos\theta) + C$$

$$= \frac{\sqrt{2}}{8}\left[\arctan\frac{x}{\sqrt{2}} + \left(\frac{x}{\sqrt{x^2 + 2}}\right)\left(\frac{\sqrt{2}}{\sqrt{x^2 + 2}}\right)\right] + C$$

$$= \frac{1}{4}\left[\frac{x}{x^2 + 2} + \frac{1}{\sqrt{2}}\arctan\frac{x}{\sqrt{2}}\right] + C$$

**43.** Evaluate $\int \text{arcsec}\, 2x\, dx$.

**Solution:**

Let

$$dv = dx \qquad \Longrightarrow \qquad v = x$$

$$u = \text{arcsec}\, x \qquad \Longrightarrow \qquad du = \frac{1}{x\sqrt{4x^2 - 1}}\, dx$$

Therefore,

$$\int \text{arcsec}\, 2x\, dx$$

$$= uv - \int v\, du$$

$$= x\, \text{arcsec}\, 2x - \int x\left(\frac{1}{x\sqrt{4x^2 - 1}}\right) dx$$

$$= x\, \text{arcsec}\, 2x - \frac{1}{2}\int \frac{2}{\sqrt{(2x)^2 - 1^2}}\, dx \qquad \text{Let}\ 2x = \sec\theta$$

$$= x\, \text{arcsec}\, 2x - \int \frac{(1/2)\sec\theta\tan\theta\, d\theta}{\tan\theta}$$

$$= x\, \text{arcsec}\, 2x - \frac{1}{2}\int \sec\theta\, d\theta$$

$$= x\, \text{arcsec}\, 2x = \frac{1}{2}\ln|\sec\theta + \tan\theta| + C$$

$$= x\, \text{arcsec}\, 2x - \frac{1}{2}\ln|2x + \sqrt{4x^2 - 1}| + C$$

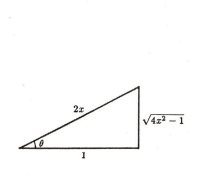

**53.** The region bounded by the circle $(x - 3)^2 + y^2 = 1$ is revolved about the $y$-axis. The resulting doughnut-shaped solid is called a torus. Find the volume of this solid.

**Solution:**

Shell Method

$$V = 2\pi \int_2^4 x\left[2\sqrt{1 - (x - 3)^2}\right] dx = 4\pi \int_2^4 x\sqrt{1 - (x - 3)^2}\, dx$$

Let $x - 3 = \sin\theta$. Then $\sqrt{1 - (x - 3)^2} = \cos\theta$ and $dx = \cos\theta\, d\theta$. Also when $x = 2, \sin\theta = -1$ and $\theta = -\pi/2$.

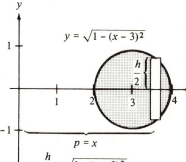

When $x = 4$, $\sin \theta = 1$ and $\theta = \pi/2$. Therefore,

$$V = 4\pi \int_{-\pi/2}^{\pi/2} (3 + \sin \theta)(\cos \theta)(\cos \theta)\, d\theta$$

$$= 4\pi \left[ \int_{-\pi/2}^{\pi/2} 3\cos^2 \theta\, d\theta + \int_{-\pi/2}^{\pi/2} \cos^2 \theta \sin \theta\, d\theta \right]$$

$$= 4\pi \left[ \int_{-\pi/2}^{\pi/2} \frac{3}{2}(1 + \cos 2\theta)\, d\theta + \int_{-\pi/2}^{\pi/2} \cos^2 \theta \sin \theta\, d\theta \right]$$

$$= 4\pi \left[ \frac{3}{2}(\theta + \frac{1}{2}\sin 2\theta) - \frac{1}{3}\cos^3 \theta \right]_{-\pi/2}^{\pi/2} = 6\pi^2$$

59. Find the surface area of the solid generated by revolving the region bounded by $y = x^2$, $y = 0$, $x = 0$, and $x = \sqrt{2}$ about the $x$-axis.

**Solution:**

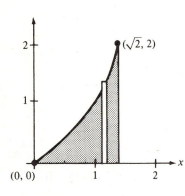

$$S = 2\pi \int_a^b y\sqrt{1 + (y')^2}\, dx = 2\pi \int_0^{\sqrt{2}} x^2 \sqrt{1 + 4x^2}\, dx$$

Using trigonometric substitution, we let $2x = \tan \theta$. Then $\sqrt{1 + 4x^2} = \sec \theta$, $x^2 = \frac{1}{4}\tan^2 \theta$, and $dx = \frac{1}{2}\sec^2 \theta\, d\theta$. Therefore,

$$\int x^2 \sqrt{1 + 4x^2}\, dx = \int \frac{\tan^2 \theta}{4}(\sec \theta)\left(\frac{1}{2}\sec^2 \theta\right) d\theta$$

$$= \frac{1}{8} \int \sec^3 \theta \tan^2 \theta\, d\theta$$

$$= \frac{1}{8} \int \sec^3 \theta(\sec^2 \theta - 1)\, d\theta$$

$$= \frac{1}{8} \left[ \int \sec^5 \theta\, d\theta - \int \sec^3 \theta\, d\theta \right]$$

Now using integration by parts on $\int \sec^5 \theta\, d\theta$, we let

$$dv = \sec^2 \theta\, d\theta \quad \Longrightarrow \quad v = \tan \theta$$

$$u = \sec^3 \theta \quad \Longrightarrow \quad du = 3\sec^3 \theta \tan \theta\, d\theta \ = 3\sec^3 \theta \tan \theta\, d\theta$$

Thus,

$$\int \sec^5 \theta \, d\theta = \sec^3 \theta \tan \theta - 3 \int \sec^3 \theta \tan^2 \theta \, d\theta$$

$$= \sec^3 \theta \tan \theta - 3 \int \sec^3 \theta (\sec^2 \theta - 1) \, d\theta$$

$$= \sec^3 \theta \tan \theta - 3 \int \sec^5 \theta \, d\theta + 3 \int \sec^3 \theta \, d\theta$$

$$4 \int \sec^5 \theta \, d\theta = \sec^3 \theta \tan \theta + 3 \int \sec^3 \theta \, d\theta$$

$$\int \sec^5 \theta \, d\theta = \frac{1}{4} \left( \sec^3 \tan \theta + 3 \int \sec^3 \theta \, d\theta \right)$$

Thus,

$$\int x^2 \sqrt{1 + 4x^2} \, dx$$

$$= \frac{1}{8} \left[ \frac{1}{4} \sec^3 \theta \tan \theta + \frac{3}{4} \int \sec^3 \theta \, d\theta - \int \sec^3 \theta \, d\theta \right]$$

$$= \frac{1}{32} \left( \sec^3 \theta \tan \theta - \int \sec^3 \theta \, d\theta \right)$$

Using the formula from Section 9.2 for $\int \sec^3 \theta \, d\theta$, we have

$$\int x^2 \sqrt{1 + 4x^2} \, dx$$

$$= \frac{1}{32} \left[ \sec^3 \theta \tan \theta - \frac{1}{2} (\sec \theta \tan \theta + \ln |\sec \theta + \tan \theta|) \right] + C$$

Finally, when $x = 0$, $\theta = 0$, and when $x = \sqrt{2}$, $\theta = \arctan 2\sqrt{2}$. Therefore,

$$S = 2\pi \int_0^{\sqrt{2}} x^2 \sqrt{1 + 4x^2} \, dx$$

$$= \frac{\pi}{16} \left[ \sec^3 \theta \tan \theta - \frac{1}{2} (\sec \theta \tan \theta + \ln |\sec \theta + \tan \theta|) \right]_0^{\arctan 2\sqrt{2}}$$

$$= \frac{\pi}{16} \left[ 51\sqrt{2} - \frac{1}{2} \ln \left( 2\sqrt{2} + 3 \right) \right]$$

$$= \frac{\pi}{32} [102\sqrt{2} - \ln \left( 2\sqrt{2} + 3 \right) \approx 13.989$$

## 9.5    Partial Fractions

**7.** Evaluate $\displaystyle\int \frac{x^2 + 12x + 12}{x^3 - 4x}\, dx$.

**Solution:**

$$\frac{x^2 + 12x + 12}{x^3 - 4x} = \frac{x^2 + 12x + 12}{x(x-2)(x+2)} = \frac{A}{x} + \frac{B}{x-2} + \frac{C}{x+2}$$

Multiplying by $(x)(x-2)(x+2)$, we have

$$x^2 + 12x + 12 = A(x-2)(x+2) + B(x)(x+2) + C(x)(x-2)$$

If $x = 0$, then $12 = A(-2)(2)$    or    $A = -3$.

If $x = 2$, then $4 + 24 + 12 = B(2)(4)$    or    $B = 5$

If $x = -2$, then $4 - 24 + 12 = C(-2)(-4)$    or    $C = -1$

Thus,

$$\int \frac{x^2 + 12x + 12}{x^3 - 4x}\, dx = \int \left(\frac{-3}{x} + \frac{5}{x-2} + \frac{-1}{x+2}\right) dx$$

$$= -3\ln|x| + 5\ln|x-2| - \ln|x+2| + C$$

$$= \ln\left|\frac{(x-2)^5}{x^3(x+2)}\right| + C$$

**13.** Evaluate $\displaystyle\int \frac{x^4}{(x-1)^3}\, dx$.

**Solution:**

Division yields

$$\frac{x^4}{(x-1)^3} = x + 3 + \frac{6x^2 - 8x + 3}{(x-1)^3}$$

Furthermore, by partial fractions we have

$$\frac{6x^2 - 8x + 3}{(x-1)^3} = \frac{A}{x-1} + \frac{B}{(x-1)^2} + \frac{C}{(x-1)^3}$$

Multiplying by $(x - 1)^3$, we have

$$6x^2 - 8x + 3 = A(x - 1)^2 + B(x - 1) + C$$
$$= A(x^2 - 2x + 1) + B(x - 1) + C$$
$$= Ax^2 + (B - 2A)x + (A - B + C)$$

Now by equating coefficients of like terms, we have the three equations

$$6 = A, \quad -8 = B - 2A, \quad 3 = A - B + C$$

Solving these equations, we have

$$A = 6, \quad B = 4, \quad C = 1$$

Thus,

$$\int \frac{x^4}{(x - 1)^3} \, dx$$
$$= \int \left[ x + 3 + \frac{6}{x - 1} + \frac{4}{(x - 1)^2} + \frac{1}{(x - 1)^3} \right] dx$$
$$= \frac{x^2}{2} + 3x + 6 \ln |x - 1| - \frac{4}{x - 1} - \frac{1}{2(x - 1)^2} + C$$

**25.** Evaluate $\displaystyle\int \frac{x^2 + 5}{x^3 - x^2 + x + 3} \, dx$.

**Solution:**

Since $x^3 - x^2 + x + 3 = (x + 1)(x^2 - 2x + 3)$, we have

$$\int \frac{x^2 + 5}{x^3 - x^2 + x + 3} = \frac{A}{x + 1} + \frac{Bx + C}{x^2 - 2x + 3}$$

Multiplying by $(x + 1)(x^2 - 2x + 3)$ yields

$$x^2 + 5 = A(x^2 - 2x + 3) + (Bx + C)(x + 1)$$

If $x = -1$, then $1 + 5 = A(1 + 2 + 3)$ or $A = 1$.

Therefore,

$$x^2 + 5 = x^2 - 2x + 3 + Bx^2 + Bx + Cx + C$$
$$2x + 2 = Bx^2 + (B + C)x + C$$

Equating coefficients, we have $B = 0, B + C = 2$, and $C = 2$.
Finally, we have

$$\int \frac{x^2}{x^3 - x^2 + x + 3}\, dx$$

$$= \int \left( \frac{1}{x + 1} + \frac{2}{x^2 - 2x + 3} \right) dx$$

$$= \int \frac{1}{x + 1}\, dx + 2 \int \frac{1}{(x - 1)^2 + 2}\, dx$$

$$= \ln |x + 1| + 2 \left( \frac{1}{\sqrt{2}} \right) \arctan \left( \frac{x - 1}{\sqrt{2}} \right) + C$$

$$= \ln |x + 1| + \sqrt{2} \arctan \left( \frac{x - 1}{\sqrt{2}} \right) + C$$

**33.** Evaluate $\displaystyle\int_1^2 \frac{x + 1}{x(x^2 + 1)}\, dx$.

**Solution:**

$$\frac{x + 1}{x(x^2 + 1)} = \frac{A}{x} + \frac{Bx + C}{x^2 + 1}$$

Multiplying by $(x)(x^2 + 1)$ yields

$$x + 1 = A(x^2 + 1) + (Bx + C)(x) = Ax^2 + A + Bx^2 + Cx$$

$$= (A + B)x^2 + Cx + A$$

By equating coefficients, we have $A + B = 0$, $C = 1$, and $A = 1$.
Thus, $B = -1$, and we get

$$\int_1^2 \frac{x + 1}{x(x^2 + 1)}\, dx$$

$$= \int_1^2 \left( \frac{1}{x} + \frac{-x + 1}{x^2 + 1} \right) dx$$

$$= \int_1^2 \frac{1}{x}\, dx - \frac{1}{2} \int_1^2 \frac{2x}{x^2 + 1}\, dx + \int_1^2 \frac{1}{x^2 + 1}\, dx$$

$$= \left[ \ln |x| - \frac{1}{2} \ln |x^2 + 1| + \arctan x \right]_1^2$$

$$= \ln 2 - \frac{1}{2} \ln 5 + \arctan 2 - \ln 1 + \frac{1}{2} \ln 2 - \arctan 1$$

$$= \frac{3}{2} \ln 2 - \frac{1}{2} \ln 5 + \arctan 2 - \frac{\pi}{4}$$

$$= \frac{1}{2} \ln \frac{8}{5} + \arctan 2 - \frac{\pi}{4} \approx 0.557$$

**41.** Evaluate $\int \dfrac{e^x}{(e^x - 1)(e^x + 4)}\, dx$ by letting $u = e^x$.

**Solution:**

If $u = e^x$, then $du = e^x\, dx$, and

$$\int \frac{e^x}{(e^x - 1)(e^x + 4)}\, dx = \int \frac{du}{(u - 1)(u + 4)}$$

By partial fractions,

$$\frac{1}{(u - 1)(u + 4)} = \frac{A}{u - 1} + \frac{B}{u + 4}$$

Multiplying by $(u - 1)(u + 4)$ yields

$$1 = A(u + 4) + B(u - 1)$$

If $u = 1$, then $1 = A(5)$    or    $A = \dfrac{1}{5}$.

If $u = -4$, then $1 = B(-5)$    or    $B = -\dfrac{1}{5}$.

Therefore,

$$\int \frac{du}{(u - 1)(u + 4)} = \frac{1}{5} \int \left( \frac{1}{u - 1} - \frac{1}{u + 4} \right) du$$
$$= \frac{1}{5}(\ln |u - 1| - \ln |u + 4|) + C$$
$$= \frac{1}{5} \ln \left| \frac{u - 1}{u + 4} \right| + C$$
$$= \frac{1}{5} \ln \left| \frac{e^x - 1}{e^x + 4} \right| + C$$

**51.** A single infected individual enters a community of $n$ individuals susceptible to the disease. Let $x$ be the number of newly infected individuals after time $t$. The common Epidemic Model assumes that the disease spreads at a rate proportional to the product of the total number infected and the number of susceptible not yet infected. Thus, $dx/dt = k(x+1)(n-x)$, and we obtain

$$\int \frac{1}{(x+1)(n-x)}\, dx = \int k\, dt$$

Solve for $x$ as a function of $t$.

**Solution:**

$$\frac{1}{(x+1)(n-x)} = \frac{A}{x+1} + \frac{B}{n-x}$$

Multiplying by $(x+1)(n-x)$, we have

$$1 = A(n-x) + B(x+1)$$

If $x = -1$, then $1 = A(n+1)$   or   $A = \dfrac{1}{n+1}$.

If $x = n$, then $1 = B(n+1)$   or   $B = \dfrac{1}{n+1}$.

Therefore,

$$\int \frac{1}{(x+1)(n-x)}\, dx = \int k\, dt$$

$$\frac{1}{n+1}\int \left( \frac{1}{x+1} + \frac{1}{n-x} \right) dx = \int k\, dt$$

$$\frac{1}{n+1}[\ln(x+1) - \ln(n-x)] = kt + C$$

$$\left( \frac{1}{n+1} \right)\ln\left( \frac{x+1}{n-x} \right) = kt + C$$

When $t = 0$, $x = 0$. Thus,

$$\left( \frac{1}{n+1} \right)\ln\left( \frac{1}{n} \right) = C$$

Therefore,

$$\ln\left( \frac{x+1}{n-x} \right) = k(n+1)t + \ln\left( \frac{1}{n} \right)$$

$$\frac{x+1}{n-x} = \frac{1}{n}e^{k(n+1)t}$$

$$x = \frac{n[e^{(n+1)kt} - 1]}{e^{(n+1)kt} + n}$$

## 9.6   Integration by Tables and Other Techniques

**7.** Use the integration table at the end of the text to evaluate

$$\int \frac{1}{x^2\sqrt{x^2 - 4}}\, dx$$

**Solution:**

Consider Formula 35, where $u = x$ and $a = 2$. Then

$$\int \frac{1}{x^2\sqrt{x^2 - 4}}\, dx = \int \frac{du}{u^2\sqrt{u^2 - a^2}}$$

$$= \frac{\sqrt{u^2 - a^2}}{a^2 u} + C = \frac{\sqrt{x^2 - 4}}{4x} + C$$

**11.** Use the integration table at the end of the text to evaluate

$$\int \frac{2x}{(1 - 3x)^2}\, dx$$

**Solution:**

Consider Formula 4, where $u = x$, $a = 1$, $b = -3$, and $a + bu = 1 - 3x$. Then

$$\int \frac{2x}{(1 - 3x)^2}\, dx = 2\int \frac{x}{(1 - 3x)^2}\, dx = 2\int \frac{u}{(a + bu)^2}\, du$$

$$= 2\left(\frac{1}{b^2}\right)\left(\frac{1}{a + bu} + \ln|a + bu|\right) + C$$

$$= \frac{2}{9}\left[\frac{1}{1 - 3x} + \ln|1 - 3x|\right] + C$$

**23.** Use the integration table at the end of the text to evaluate

$$\int \frac{1}{1 + e^{2x}} \, dx$$

**Solution:**

Consider Formula 84, where $u = 2x$ and $du = 2 \, dx$. Then

$$\int \frac{1}{1 + e^{2x}} \, dx = \frac{1}{2} \int \frac{2}{1 + e^{2x}} \, dx$$

$$= \frac{1}{2}[2x - \ln(1 + e^{2x})] + C = x - \frac{1}{2} \ln(1 + e^{2x}) + C$$

**33.** Use the integration table at the end of the text to evaluate

$$\int \frac{1}{\sqrt{x}(1 - \cos\sqrt{x})} \, dx$$

**Solution:**

Consider Formula 57, with $u = \sqrt{x}$ and $du = (1/2\sqrt{x}) \, dx$. Then

$$\int \frac{1}{\sqrt{x}(1 - \cos\sqrt{x})} \, dx = 2 \int \frac{(1/2\sqrt{x})}{1 - \cos\sqrt{x}} \, dx$$

$$= 2(-\cot\sqrt{x} - \csc\sqrt{x}) + C$$
$$= -2(\cot\sqrt{x} + \csc\sqrt{x}) + C$$

**39.** Use the integration table at the end of the text to evaluate

$$\int \frac{\ln x}{x(3 + 2\ln x)} \, dx$$

**Solution:**

Consider Formula 3, where $u = \ln x$, $a = 3$, $b = 2$, and $du = (1/x) \, dx$. Then

$$\int \frac{\ln x}{x(3 + 2\ln x)} \, dx = \int \frac{(\ln x)(1/x)}{3 + 2\ln x} \, dx$$

$$= \frac{1}{4}[2\ln|x| - 3\ln(3 + 2\ln|x|)] + C$$

**47.** Use the integration table at the end of the text to evaluate

$$\int \frac{x^3}{\sqrt{4 - x^2}}\, dx$$

**Solution:**

Consider Formula 21, with $u = x^2$, $a = 4$, $b = -1$, $du = 2x\, dx$, and $\sqrt{a + bu} = \sqrt{4 - x^2}$. Then

$$\int \frac{x^2}{\sqrt{4 - x^2}}\, dx = \frac{1}{2} \int \frac{x^2(2x)}{\sqrt{4 - x^2}}\, dx = \frac{1}{2} \int \frac{u}{\sqrt{a + bu}}\, du$$

$$= \frac{1}{2}\left(\frac{(-2)(8 + x^2)}{3}\right)\sqrt{4 - x^2} + C$$

$$= -\left(\frac{x^2 + 8}{3}\right)\sqrt{4 - x^2} + C$$

**53.** Verify the formula

$$\int \frac{u^2}{(a + bu)^2}\, du = \frac{1}{b^3}\left(bu - \frac{a^2}{a + bu} - 2a \ln |a + bu|\right) + C$$

by the method of partial fractions.

**Solution:**

Since the numerator and denominator are of the same degree, we begin by dividing

$$\frac{u^2}{(a + bu)^2} = \frac{1}{b^2} - \frac{(2a/b)u + (a^2/b^2)}{(a + bu)^2}$$

Now we use partial fractions to obtain

$$\frac{(2a/b)u + (a^2/b^2)}{(a + bu)^2} = \frac{A}{a + bu} + \frac{B}{(a + bu)^2}$$

Multiplying by $(a + bu)^2$, we have

$$\left(\frac{2a}{b}\right)u + \frac{a^2}{b^2} = A(a + bu) + B = bAu + (aA + B)$$

Equating the coefficients of like terms, we have

$$bA = \frac{2a}{b} \quad \text{and} \quad A = \frac{2a}{b^2}$$

$$aA + B = \frac{a^2}{b^2} \quad \text{and} \quad B = \frac{a^2}{b^2} - a\left(\frac{2a}{b^2}\right) = -\frac{a^2}{b^2}$$

Therefore,

$$\int \frac{u^2}{(a+bu)^2} \, du$$

$$= \frac{1}{b^2} \int du - \frac{2a}{b^2}\left(\frac{1}{b}\right) \int \frac{b}{a+bu} \, du + \frac{a^2}{b^2}\left(\frac{1}{b}\right) \int \frac{b}{(a+bu)^2} \, du$$

$$= \left(\frac{1}{b^2}\right)u - \frac{2a}{b^3}(\ln|a+bu|) - \frac{a^2}{b^3}\left(\frac{1}{a+bu}\right) + C$$

$$= \frac{1}{b^3}\left[bu - \frac{a^2}{a+bu} - 2a\ln|a+bu|\right] + C$$

**61.** Evaluate $\displaystyle\int_0^{\pi/2} \frac{1}{1+\sin\theta+\cos\theta} \, d\theta$.

**Solution:**

Let $u = \dfrac{\sin\theta}{1+\cos\theta}$. Then $\cos\theta = \dfrac{1-u^2}{1+u^2}$, $\sin\theta = \dfrac{2u}{1+u^2}$, and $d\theta = \dfrac{2\,du}{1+u^2}$. Furthermore, when $\theta = \pi/2$, $u = 1$, and when $\theta = 0$, $u = 0$.

$$\int_0^{\pi/2} \frac{1}{1+\sin\theta+\cos\theta} \, d\theta$$

$$= \int_0^1 \frac{[2/(1+u^2)]\,du}{1+[(2u)/(1+u^2)]+[(1-u^2)/(1+u^2)]}$$

$$= \int_0^1 \frac{1}{u+1} \, du = \left[\ln|u+1|\right]_0^1 = \ln 2$$

## 9.7   Improper Integrals

**3.** Determine the divergence or convergence of

$$\int_0^2 \frac{1}{(x-1)^{2/3}}\,dx$$

and evaluate it if it converges.

**Solution:**

$$\int_0^2 \frac{1}{(x-1)^{2/3}}\,dx$$

$$= \lim_{b\to1^-} \int_0^b (x-1)^{-2/3}\,dx + \lim_{c\to1^+} \int_c^2 (x-1)^{-2/3}\,dx$$

$$= \lim_{b\to1^-} \left[3(x-1)^{1/3}\right]_0^b + \lim_{c\to1^+} \left[3(x-1)^{1/3}\right]_c^2$$

$$= \lim_{b\to1^-} \left[3(b-1)^{1/3} - (-3)\right] + \lim_{c\to1^+} \left[3 - 3(c-1)^{1/3}\right] = 6$$

**13.** Determine the divergence or convergence of

$$\int_0^\infty e^{-x} \cos x\,dx$$

and evaluate it if it converges.

**Solution:**

Since

$$\int e^{-x} \cos x\,dx = \frac{1}{2}e^{-x}(-\cos x + \sin x) + C$$

we have

$$\int_0^\infty e^{-x} \cos x\,dx = \lim_{b\to\infty} \int_0^b e^{-x} \cos x\,dx$$

$$= \lim_{b\to\infty} \frac{1}{2}\left[\frac{(-\cos x + \sin x)}{e^x}\right]_0^b$$

$$= \frac{1}{2}[0 - (-1)] = \frac{1}{2}$$

**23.** Determine the divergence or convergence of

$$\int_0^8 \frac{1}{\sqrt[3]{8-x}}\, dx$$

and evaluate it if it converges.

**Solution:**

$$\int_0^8 \frac{1}{\sqrt[3]{8-x}}\, dx = \lim_{b \to 8^-} \int_0^b \frac{1}{\sqrt[3]{8-x}}\, dx$$

$$= \lim_{b \to 8^-}\left[ -\frac{(8-x)^{2/3}}{\frac{2}{3}} \right]_0^b$$

$$= -\frac{3}{2}(0) + \frac{3}{2}(4) = 6$$

**29.** Determine the divergence or convergence of

$$\int_2^4 \frac{1}{\sqrt{x^2-4}}\, dx$$

and evaluate it if it converges.

**Solution:**

$$\int_2^4 \frac{1}{\sqrt{x^2-4}}\, dx = \lim_{a \to 2^+} \int_a^4 \frac{1}{\sqrt{x^2-4}}\, dx$$

$$= \lim_{a \to 2^+}\left[ \ln|x + \sqrt{x^2-4}| \right]_a^4$$

$$= \ln(4+\sqrt{12}) - \ln(2+0)$$

$$= \ln\left(\frac{4+2\sqrt{3}}{2}\right) = \ln(2+\sqrt{3})$$

**35.** Use mathematical induction to verify that the following integral converges for any positive integer $n$.

$$\int_0^\infty x^n e^{-x}\, dx$$

**Solution:**

When $n = 1$, the integral converges, since

$$\int_0^\infty x e^{-x}\, dx = \lim_{b \to \infty} \int_0^b x e^{-x}\, dx$$

$$= \lim_{b \to \infty} \left[ -e^{-x}(x+1) \right]_0^b \quad \text{(Integration by parts)}$$

$$= \lim_{b \to \infty} \left[ -e^{-b}(b+1) + 1 \right]$$

$$= 0 + 1 = 1 \quad \text{(L'Hôpital's Rule)}$$

Now assume that the integral converges for $n = k$ and verify that it converges for $n = k + 1$.

$$\int_0^\infty x^{k+1} e^{-x}\, dx$$

$$= \lim_{b \to \infty} \int_0^b x^{k+1} e^{-x}\, dx$$

$$= \lim_{b \to \infty} \left[ -x^{k+1} e^{-x} - \frac{k+1}{-1} \int_0^b x^k e^{-x}\, dx \right]_0^\infty \quad \text{(Integration by parts)}$$

$$= 0 \div (k+1) \int_0^\infty x^k e^{-x}\, dx \quad \text{(L'Hôpital's Rule)}$$

$$= (k+1) \int_0^\infty x^k e^{-x}\, dx$$

Therefore, we have shown the integral for $n = k + 1$ converges if the integral for $n = k$ converges. Combining this with the results for $n = 1$, we conclude by mathematical induction that the integral converges for any positive integer $n$.

**45.** Use the result of Exercise 36 to determine if $\int_0^\infty e^{-x^2}\,dx$ converges.

**Solution:**

On $[1, \infty)$ we have

$$x \le x^2$$

$$e^x \le e^{x^2}$$

$$\frac{1}{e^x} \ge \frac{1}{e^{x^2}}$$

$$e^{-x} \ge e^{-x^2}$$

Therefore, by Exercise 36 we have

$$\int_0^\infty e^{-x^2}\,dx = \int_0^1 e^{-x^2}\,dx + \int_1^\infty e^{-x^2}\,dx$$

$$< \int_0^1 e^{-x^2}\,dx + \int_1^\infty e^{-x}\,dx$$

$$= \int_0^1 e^{-x^2}\,dx + \left[ -e^{-x} \right]_1^\infty = \int_0^1 e^{-x^2}\,dx + e^{-1}$$

Therefore, the integral converges.

**49.** Sketch the graph of the hypocycloid of four cusps, $x^{2/3} + y^{2/3} = 1$, and find its perimeter.

**Solution:**

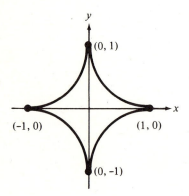

By symmetry the perimeter of this figure is four times the length of the arc in the first quadrant. Thus,

$$s = 4 \int_0^1 \sqrt{1 + (y')^2}\,dx$$

By implicit differentiation we have

$$x^{2/3} + y^{2/3} = 1$$

$$\frac{2}{3}x^{-1/3} + \frac{2}{3}y^{-1/3}y' = 0$$

$$y' = -\frac{y^{1/3}}{x^{1/3}}$$

Therefore,

$$s = 4 \int_0^1 \sqrt{1 + \left(-\frac{y^{1/3}}{x^{1/3}}\right)^2} \, dx = 4 \int_0^1 \sqrt{\frac{x^{2/3} + y^{2/3}}{x^{2/3}}} \, dx$$

$$= 4 \lim_{b \to 0^+} \int_b^1 \sqrt{\frac{1}{x^{2/3}}} \, dx$$

$$= 4 \lim_{b \to 0^+} \int_b^1 x^{-1/3} \, dx = 4 \lim_{b \to 0^+} \left[\frac{x^{2/3}}{2/3}\right]_b^1 = 4\left(\frac{3}{2}\right) = 6$$

## Review Exercises for Chapter 8

3. Evaluate $\displaystyle\int \frac{1}{1 - \sin \theta} \, d\theta$.

**Solution:**

$$\int \frac{1}{1 - \sin \theta} \, d\theta = \int \frac{1}{1 - \sin \theta}\left(\frac{1 + \sin \theta}{1 + \sin \theta}\right) d\theta$$

$$= \int \frac{1 + \sin \theta}{1 - \sin^2 \theta} \, d\theta = \int \frac{1 + \sin \theta}{\cos^2 \theta} \, d\theta$$

$$= \int \left(\frac{1}{\cos^2 \theta} + \frac{\sin \theta}{\cos \theta \cos \theta}\right) d\theta$$

$$= \int (\sec^2 \theta + \sec \theta \tan \theta) \, d\theta$$

$$= \tan \theta + \sec \theta + C$$

7. Evaluate $\displaystyle\int \frac{\ln (2x)}{x^2} \, dx$.

**Solution:**

Using integration by parts, let

$$dv = \frac{1}{x^2} \, dx \quad \Longrightarrow \quad v = -\frac{1}{x}$$

$$u = \ln (2x) \quad \Longrightarrow \quad du = \frac{1}{x}$$

$$\int \frac{\ln(2x)}{x^2}\,dx = -\frac{\ln(2x)}{x} - \int \frac{1}{x}\left(-\frac{1}{x}\right)dx$$

$$= -\frac{\ln(2x)}{x} + \int x^{-2}\,dx$$

$$= -\frac{\ln(2x)}{x} - \frac{1}{x} + C = -\frac{1}{x}(1 + \ln 2x) + C$$

**13.** Evaluate $\displaystyle\int \sec^4 \frac{x}{2}\,dx$.

**Solution:**

$$\int \sec^4 \frac{x}{2}\,dx = \int \sec^2 \frac{x}{2} \sec^2 \frac{x}{2}\,dx$$

$$= \int \left[\tan^2 \frac{x}{2} + 1\right] \sec^2 \frac{x}{2}\,dx$$

$$= 2\int \tan^2 \frac{x}{2} \sec^2 \frac{x}{2}\left(\frac{1}{2}\right)dx + 2\int \sec^2 \frac{x}{2}\left(\frac{1}{2}\right)dx$$

$$= \frac{2}{3}\tan^3 \frac{x}{2} + 2\tan \frac{x}{2} + C$$

$$= \frac{2}{3}\left[\tan^3 \frac{x}{2} + 3\tan \frac{x}{2}\right] + C$$

**17.** Evaluate $\displaystyle\int \frac{x^2 + 2x}{x^3 - x^2 + x - 1}\,dx$.

**Solution:**

Since $x^3 - x^2 + x - 1 = (x - 1)(x^2 + 1)$, we have

$$\frac{x^2 + 2x}{(x - 1)(x^2 + 1)} = \frac{A}{x - 1} + \frac{Bx + C}{x^2 + 1}$$

Multiplying by $(x - 1)(x^2 + 1)$ yields

$$x^2 + 2x = A(x^2 + 1) + (Bx + C)(x - 1)$$

If $x = 1$, then $1 + 2 = A(2) + (B + C)(0)$    or    $A = \frac{3}{2}$.

Furthermore,

$$x^2 + 2x = Ax^2 + A + Bx^2 - Bx + Cx - C$$
$$= (A + B)x^2 + (C - B)x + (A - C) \quad .$$

Now by equating coefficients, we have

$$1 = A + B = \frac{3}{2} + B \qquad \text{or} \qquad B = -\frac{1}{2}$$

$$2 = C - B = C + \frac{1}{2} \qquad \text{or} \qquad C = \frac{3}{2}$$

Therefore,

$$\int \frac{x^2 + 2x}{(x - 1)(x^2 + 1)} \, dx$$

$$= \int \frac{\frac{3}{2}}{x - 1} + \frac{(-x/2) + (\frac{3}{2})}{x^2 + 1} \, dx$$

$$= \frac{3}{2} \int \frac{1}{x - 1} \, dx - \frac{1}{2} \int \frac{x}{x^2 + 1} \, dx + \frac{3}{2} \int \frac{1}{x^2 + 1} \, dx$$

$$= \frac{3}{2} \ln |x - 1| - \frac{1}{4} \ln (x^2 + 1) + \frac{3}{2} \arctan x + C$$

$$= \frac{1}{4} [6 \ln |x - 1| - \ln (x^2 + 1) + 6 \arctan x] + C$$

**31. Evaluate** $\displaystyle \int \frac{x^{1/4}}{1 + x^{1/2}} \, dx$.

**Solution:**

Let $u = x^{1/4}$. Then $u^4 = x$, $u^2 = x^{1/2}$, $dx = 4u^3 \, du$, and

$$\int \frac{x^{1/4}}{1 + x^{1/2}} \, dx = \int \frac{u}{1 + u^2} (4u^3 \, du) = 4 \int \frac{u^4}{1 + u^2} \, du$$

$$= 4 \int \left( u^2 - 1 + \frac{1}{1 + u^2} \right) du$$

$$= 4 \left( \frac{u^3}{3} - u + \arctan u \right) + C$$

$$= \frac{4}{3} (x^{3/4} - 3x^{1/4} + 3 \arctan x^{1/4}) + C$$

**43.** Integrate $\int \dfrac{x^3}{\sqrt{4+x^2}}\, dx$ by the following methods:

(a) Trigonometric substitution

(b) Substitution: $u^2 = 4 + x^2$

(c) By parts: $dv = (x/\sqrt{4+x^2})\, dx$

**Solution:**

(a) Let $x = 2\tan\theta$. Then $\sqrt{4+x^2} = 2\sec\theta$ and $dx = 2\sec^2\theta\, d\theta$. Thus,

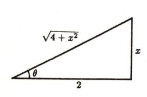

$$\int \frac{x^3}{\sqrt{4+x^2}}\, dx = \int \frac{(2\tan\theta)^3}{2\sec\theta}(2\sec^2\theta\, d\theta)$$

$$= 8\int \tan^3\theta \sec\theta\, d\theta$$

$$= 8\int (\sec^2\theta - 1)\tan\theta \sec\theta\, d\theta$$

$$= 8\left(\frac{1}{3}\sec^3\theta - \sec\theta\right) + C$$

$$= \frac{8}{3}\sec\theta(\sec^2\theta - 3) + C$$

$$= \frac{\sqrt{4+x^2}}{3}(x^2 - 8) + C$$

(b) Let $u^2 = 4 + x^2$. Thus, $2u\, du = 2x\, dx$ and we have

$$\int \frac{x^3}{\sqrt{4+x^2}}\, dx = \int \frac{x^2(x\, dx)}{\sqrt{4+x^2}}$$

$$= \int \frac{(u^2 - 4)(u\, du)}{u}$$

$$= \int (u^2 - 4)\, du$$

$$= \frac{1}{3}u^3 - 4u + C$$

$$= \frac{u}{3}(u^2 - 12) + C = \frac{\sqrt{4+x^2}}{3}(x^2 - 8) + C$$

(c)    Let

$$dv = \frac{x}{\sqrt{4+x^2}}\, dx \quad \Longrightarrow \quad v = \int \frac{x}{\sqrt{4+x^2}}\, dx = \sqrt{4+x^2}$$

$$u = x^2 \quad\quad\quad\quad\quad \Longrightarrow \quad du = 2x\, dx$$

$$\int u\,dv = uv - \int v\,du$$

$$\int \frac{x^3}{\sqrt{4+x^2}}\,dx = x^2\sqrt{4+x^2} - \int 2x\sqrt{4+x^2}\,dx$$

$$= x^2\sqrt{4+x^2} - \frac{2}{3}(4+x^2)^{3/2} + C$$

$$= \frac{\sqrt{4+x^2}}{3}(x^2 - 8) + C$$

# 10 INFINITE SERIES

## 10.1  Sequences

**11.** Write an expression for the $n$th term of the sequence
$\{-1, 2, 7, 14, 23, \dots\}$.

**Solution:**

Compare the terms of the sequence $\{-1, 2, 7, 14, 23, \dots\}$ with the sequence of squares

$$\{1^2, 2^2, 3^2, 4^2, 5^2, \dots\} = \{1, 4, 9, 16, 25, \dots\}$$

Observe that each term of the given sequence is two less than the sequence of squares. Thus we write the $n$th term of the given sequence as $a_n = n^2 - 2$.

**15.** Write an expression for the $n$th term of the sequence

$$\left\{2, -1, \frac{1}{2}, \frac{-1}{2}, \frac{1}{8}, \cdots\right\}$$

**Solution:**

First observe that the denominators are powers of 2, where for $n = 4$ the denominator is $2^2$. This implies that the $n$th term has a denominator $2^{n-2}$. Note further that the signs alternate starting with a positive sign, and thus the $n$th term is

$$a_n = (-1)^{n-1}\left(\frac{1}{2^{n-2}}\right) = \frac{(-1)^{n-1}}{2^{n-2}}$$

**21.** Write an expression for the $n$th term of the sequence

$$\left\{1, \frac{1}{1\cdot 3}, \frac{1}{1\cdot 3\cdot 5}, \frac{1}{1\cdot 3\cdot 5\cdot 7}, \cdots\right\}$$

**Solution:**

The denominator of the $n$th term is the product of the first $n$ positive odd integers.  Therefore

$$a_n = \frac{1}{1\cdot 3\cdot 5\cdots (2n-1)}$$

A second form of the $n$th term is obtained when we multiply the numerator and denominator by the $n$ missing even integers.  Then

$$\begin{aligned}
a_n &= \frac{1}{1\cdot 3\cdot 5\cdots (2n-1)}\\
&= \frac{2\cdot 4\cdot 6\cdot 8\cdots (2n)}{1\cdot 2\cdot 3\cdot 4\cdot 5\cdots (2n-1)(2n)}\\
&= \frac{2^n(1\cdot 2\cdot 3\cdot 4\cdots n)}{1\cdot 2\cdot 3\cdot 4\cdot 5\cdots (2n-1)(2n)} = \frac{2^n\, n!}{(2n)!}
\end{aligned}$$

**27.** Given $a_n = (-1)^n[n/(n+1)]$, determine if the sequence $\{a_n\}$ converges or diverges.  If it converges, find its limit.

**Solution:**

Since

$$\lim_{n\to\infty} a_n = \lim_{n\to\infty} (-1)^n\left(\frac{n}{n+1}\right) = \pm 1$$

the limit does not exist and the sequence $\{a_n\}$ diverges.

**39.** Given $a_n = f^{(n-1)}(2)$, $f(x) = \ln x$, determine if the sequence $\{a_n\}$ converges or diverges. If it converges, find its limit.

**Solution:**

$$f(x) = \ln x \qquad\qquad a_1 = f(2) = \ln 2$$
$$f'(x) = \frac{1}{x} \qquad\qquad a_2 = f'(2) = \frac{1}{2}$$
$$f''(x) = \frac{-1}{x^2} \qquad\qquad a_3 = f''(2) = \frac{-1}{2^2}$$
$$f'''(x) = \frac{2}{x^3} \qquad\qquad a_4 = f'''(2) = \frac{2}{2^3}$$
$$f^{(4)}(x) = \frac{-6}{x^4} \qquad\qquad a_5 = f^{(4)}(2) = \frac{-6}{2^4}$$
$$f^{(5)}(x) = \frac{24}{x^5} \qquad\qquad a_6 = f^{(5)}(2) = \frac{24}{2^5}$$
$$. \qquad\qquad .$$
$$. \qquad\qquad .$$
$$. \qquad\qquad .$$
$$f^{(n-1)}(x) = \frac{(-1)^n(n-2)!}{x^{n-1}} \qquad a_n = f^{(n-1)}(2) = \frac{(-1)^n(n-2)!}{2^{n-1}}$$

Since

$$\lim_{n\to\infty} \frac{(-1)^n(n-2)!}{2^{n-1}}$$

does not exist, the sequence diverges.

**43.** Given $a_n = [1 + (k/n)]^n$, determine if the sequence $\{a_n\}$ converges or diverges. If it converges, find its limit.

**Solution:**

$$\lim_{n\to\infty} a_n = \lim_{n\to\infty}\left(1 + \frac{k}{n}\right)^n = \lim_{n\to\infty}\left[\left(1 + \frac{k}{n}\right)^{n/k}\right]^k$$

Now we let $u = k/n$. Then as $n$ approaches infinity, $u$ approaches zero, and

$$\lim_{n\to\infty} a_n = \lim_{u\to 0}[(1+u)^{1/u}]^k = e^k$$

Therefore the sequence converges to $e^k$.

**51.** Determine if the sequence $a_n = (-1)^n \left(\dfrac{1}{n}\right)$ is monotonic.

**Solution:**

Writing out the first few terms of the sequence we have

$$a_1 = -1, \; a_2 = \frac{1}{2}, \; a_3 = -\frac{1}{3}, \; a_4 = \frac{1}{4}, \cdots$$

Because of the alternating signs, we observe that the terms are neither nondecreasing nor nonincreasing. Therefore, the sequence is *not* monotonic.

**63.** A government program that currently costs taxpayers $2.5 billion per year is to be cut back by 20% per year.

(a) Write an expression for the amount budgeted for this program after $n$ years.

(b) Compute the budgets for the first 4 years.

(c) Determine the convergence or divergence of the sequence of reduced budgets. If the sequence converges, find its limit.

**Solution:**

(a) If we let $A_n$ be the amount budgeted after $n$ years, we have

$$A_1 = 2.5 - 0.2(2.5) = 0.8(2.5) \text{ billion}$$
$$A_2 = A_1 - 0.2A_1 = 0.8A_1 = 0.8^2(2.5) \text{ billion}$$
$$A_3 = A_2 - 0.2A_2 = 0.8A_2 = 0.8^2(2.5) \text{ billion}$$

$$\cdot$$
$$\cdot$$
$$\cdot$$

$$A_n = 2.5(0.8)^n \text{ billion}$$

(b)

$$A_1 = 2.5(0.8) = \$2 \text{ billion}$$
$$A_2 = 2.5(0.8)^2 = \$1.6 \text{ billion}$$
$$A_3 = 2.5(0.8)^3 = \$1.28 \text{ billion}$$
$$A_4 = 2.5(0.8)^4 = \$1.024 \text{ billion}$$

(c) $\lim\limits_{n\to\infty} 2.5(0.8)^n = 0$ and therefore the sequence converges.

## 10.2   Series and Convergence

**3.** Find the first five terms of the sequence of partial sums for the series

$$3 - \frac{9}{2} + \frac{27}{4} - \frac{81}{8} + \frac{243}{16} - \cdots$$

**Solution:**

$$S_1 = 3$$
$$S_2 = 3 - \frac{9}{2} = -\frac{3}{2} = -1.5$$
$$S_3 = 3 - \frac{9}{2} + \frac{27}{4} = \frac{21}{4} = 5.25$$
$$S_4 = 3 - \frac{9}{2} + \frac{27}{4} - \frac{81}{8} = -\frac{39}{8} = -4.875$$
$$S_5 = 3 - \frac{9}{2} + \frac{27}{4} - \frac{81}{8} + \frac{243}{16} = \frac{165}{16} = 10.3125$$

**13.** Verify that the series $\displaystyle\sum_{n=0}^{\infty} 3\left(\frac{3}{2}\right)^n$ diverges.

**Solution:**

Writing out a few terms of the series we obtain

$$\sum_{n=0}^{\infty} 3\left(\frac{3}{2}\right)^n = 3 + \frac{9}{2} + \frac{27}{4} + \frac{81}{8} + \cdots$$

The series is geometric with common ration $r = \frac{3}{2}$. Since $|r| \geq 1$, the series diverges.

**17.** Verify that the infinite series $\displaystyle\sum_{n=1}^{\infty} \frac{2^n + 1}{2^{n+1}}$ diverges.

**Solution:**

$$\lim_{n\to\infty} a_n = \lim_{n\to\infty} \frac{2^n - 1}{2^{n+1}} = \lim_{n\to\infty} \frac{1 - (1/2^n)}{2} = \frac{1}{2} \neq 0$$

Therefore, by Theorem 10.9 the series diverges.

**19.** Verify that the infinite series $2 + \frac{3}{2} + \frac{9}{8} + \frac{27}{32} + \frac{81}{128} + \cdots$ converges.

**Solution:**

$$2 + \frac{3}{2} + \frac{9}{8} + \frac{27}{32} + \frac{81}{128} + \cdots = \sum_{n=1}^{\infty} 2\left(\frac{3}{4}\right)^{n-1}$$

Therefore we have a geometric series with $a = 2$ and $r = \frac{3}{4}$. Since $|r| < 1$, the series converges by Theorem 10.6.

**23.** Verify that the infinite series $\displaystyle\sum_{n=1}^{\infty} \frac{1}{n(n+1)}$ converges.

**Solution:**

Using partial fractions we can write

$$a_n = \frac{1}{n(n+1)} = \frac{1}{n} - \frac{1}{n+1}.$$

Therefore,

$$\sum_{n=1}^{\infty} \frac{1}{n(n+1)} = \sum_{n=1}^{\infty} \left(\frac{1}{n} - \frac{1}{n+1}\right)$$

$$= \left(1 - \frac{1}{2}\right) + \left(\frac{1}{2} - \frac{1}{3}\right) + \left(\frac{1}{3} - \frac{1}{4}\right) + \cdots$$

where each term after the first term of 1 cancels. Thus, the series converges to 1.

**29.** Find the sum of the series $1 + 0.1 + 0.01 + 0.001 + \cdots$.

**Solution:**

The series

$$1 + 0.1 + 0.01 + 0.001 + \cdots = \sum_{n=0}^{\infty} (0.1)^n$$

is geometric with $a = 1$ and $r = 0.1$. Therefore, the sum is

$$S = \frac{a}{1 - r} = \frac{1}{1 - 0.1} = \frac{10}{9}$$

**37.** Find the sum of the series $\displaystyle\sum_{n=1}^{\infty} \frac{4}{n(n+2)}$.

**Solution:**

Using partial fractions we have

$$\frac{4}{n(n+2)} = \frac{A}{n} + \frac{B}{n+2}$$

Multiplying by $n(n+2)$, we obtain

$$4 = A(n+2) + Bn$$

If $n = 0$, then $4 = 2A$ or $A = 2$. If $n = -2$, then $4 = -2B$ or $B = -2$. Thus

$$\sum_{n=1}^{\infty} \frac{4}{n(n+2)} = \sum_{n=1}^{\infty} \left[ \frac{2}{n} - \frac{2}{n+2} \right]$$

$$= \left[ 2 - \frac{2}{3} \right] + \left[ \frac{2}{2} - \frac{2}{4} \right] + \left[ \frac{2}{3} - \frac{2}{5} \right] + \left[ \frac{2}{4} - \frac{2}{6} \right]$$

$$+ \left[ \frac{2}{5} - \frac{2}{7} \right] + \cdots = 2 + 1 = 3$$

**49.** Determine the convergence or divergence of the series

$$\sum_{n=1}^{\infty} \frac{3n-1}{2n+1}$$

**Solution:**

Since

$$\lim_{n\to\infty} a_n = \lim_{n\to\infty} \frac{3n-1}{2n+1} = \frac{3}{2} \neq 0$$

we conclude by Theorem 10.9 that the given series diverges.

**63.** A deposit of \$50 is made at the beginning of each month in an account that pays 12% compounded monthly. Find the balance after ten years by using the formula

$$\sum_{i=0}^{n-1} ar^i = \frac{a(1 - r^n)}{1 - r}$$

for the $n$th partial sum of a geometric series.

**Solution:**

The balance $A$ in the account at the end of the 10 years is given by

$$A = 50\left[1 + \frac{0.12}{12}\right] + 50\left[1 + \frac{0.12}{12}\right]^2$$

$$+ \cdots + 50\left[1 + \frac{0.12}{12}\right]^{120}$$

$$A + 50 = 50 + 50\left[1 + \frac{0.12}{12}\right] + 50\left[1 + \frac{0.12}{12}\right]^2$$

$$+ \cdots + 50\left[1 + \frac{0.12}{12}\right]^{120}$$

$$= \sum_{i=0}^{120} 50\left[1 + \frac{0.12}{12}\right]^i = \frac{50[1 - (1.01)^{121}]}{1 - 1.01} = 11,666.95$$

Therefore, $A = 11,666.95 - 50 = \$11,616.95$.

## 10.3   The Integral Test and $p$-Series

**7.** Determine the convergence of divergence of the $p$-series

$$\frac{\ln 2}{2} + \frac{\ln 3}{3} + \frac{\ln 4}{4} + \frac{\ln 5}{5} + \cdots$$

by the Integral Test.

**Solution:**

First observe that the $n$th term of the series is

$$a_n = \frac{\ln n}{n}$$

Now, since

$$f(x) = \frac{\ln x}{x}$$

satisfies the conditions for the Integral Test, we have

$$\int_1^\infty \frac{\ln x}{x} \, dx = \int_1^\infty (\ln x) \frac{1}{x} \, dx = \left[ \frac{(\ln x)^2}{2} \right]_1^\infty = \infty$$

Therefore the given series diverges.

**17.** Determine the convergence or divergence of the $p$-series

$$1 + \frac{1}{2\sqrt{2}} + \frac{1}{3\sqrt{3}} + \frac{1}{4\sqrt{4}} + \frac{1}{5\sqrt{5}} + \cdots$$

**Solution:**

Since

$$1 + \frac{1}{2\sqrt{2}} + \frac{1}{3\sqrt{3}} + \frac{1}{4\sqrt{4}} + \frac{1}{5\sqrt{5}} + \cdots = \sum_{n=1}^\infty \frac{1}{n\sqrt{n}}$$

$$= \sum_{n=1}^\infty \frac{1}{n^{3/2}}$$

we have a $p$-series with $p = \frac{3}{2} > 1$. Therefore by Theorem 10.11 the series converges.

**29.** Determine the convergence or divergence of the series

$$\sum_{n=1}^\infty \left( 1 + \frac{1}{n} \right)^n$$

**Solution:**

From the definition of the irrational number $e$ we have

$$\lim_{n \to \infty} a_n = \lim_{n \to \infty} \left( 1 + \frac{1}{n} \right)^n = e \neq 0$$

Thus, by Theorem 10.9, the series diverges.

**33.** Find the values of $p$ for which the series

$$\sum_{n=2}^{\infty} \frac{1}{n(\ln n)^p}$$

converges.

**Solution:**

Since the conditions of the Theorem 10.10 are met, we use the Integral Test to find the values of $p$ for which the series converges. If $u = \ln x$ then $du = (1/x)\,dx$. Thus

$$\int_{2}^{\infty} \frac{1}{x(\ln x)^p}\,dx = \lim_{b \to \infty} \int_{2}^{b} (\ln x)^{-p} \frac{1}{x}\,dx$$

$$= \lim_{b \to \infty} \left[ \frac{(\ln x)^{-p+1}}{-p+1} \right]_{2}^{b}$$

$$= \lim_{b \to \infty} \left[ \left( \frac{1}{1-p} \right) \frac{1}{(\ln b)^{p-1}} - \left( \frac{1}{1-p} \right) \frac{1}{(\ln 2)^{p-1}} \right]$$

$$= 0 + \frac{1}{(p-1)(\ln 2)^{p-1}} \quad \text{for} \quad p-1 > 0.$$

Therefore, the given series converges if $p > 1$

**41.** Approximate the sum of the series $\displaystyle\sum_{n=1}^{\infty} ne^{-n^2}$ using the first four terms. Include an estimate of the maximum error for your approximation.

**Solution:**

Since $f(x) = xe^{-x^2}$ satisfies the conditions for the Intergral Test, we have

$$\int_{1}^{\infty} xe^{-x^2}\,dx = -\frac{1}{2} \int_{1}^{\infty} e^{-x^2}(-2x)\,dx$$

$$= -\frac{1}{2} \lim_{b \to \infty} \left[ e^{-x^2} \right]_{1}^{b}$$

$$= -\frac{1}{2} \lim_{b \to \infty} \left[ e^{-b^2} - e^{-1} \right] = \frac{1}{2e}$$

Thus, the series converges and has a sum $S$. Furthermore, since

$$R_4 \leq \int_{4}^{\infty} xe^{-x^2}\,dx = \frac{1}{2e^{16}} \approx 5.6 \times 10^{-8}$$

and $S_4 = e^{-1} + 2e^{-2} + 3e^{-3} + 4e^{-4} \approx 0.40488$, we conclude that $S = 0.4049$.

## 10.4   Comparisons of Series

**3.** Use the Direct Comparison Test to determine the convergence or divergence of the series

$$\sum_{n=2}^{\infty} \frac{1}{n-1}$$

**Solution:**

This series resembles

$$\sum_{n=2}^{\infty} \frac{1}{n}, \qquad \text{a divergent } p\text{-series}$$

Term-by-term comparison yields

$$a_n = \frac{1}{n} < \frac{1}{n-1} = b_n$$

Thus, it follows by the Direct Comparison Test that the series diverges.

**9.** Use the Direct Comparison Test to determine the convergence or divergence of the series

$$\sum_{n=0}^{\infty} \frac{1}{n!}$$

**Solution:**

We compare the given series to the convergent $p$-series $\sum_{n=1}^{\infty} \frac{1}{n^2}$.

For $n > 3$ we have $n^2 < n!$ and therefore $\frac{1}{n!} < \frac{1}{n^2}$. Therefore, by the Direct Comparison Test, the given series converges.

**13.** Use the Limit Comparison Test to determine the convergence or divergence of the series

$$\sum_{n=1}^{\infty} \frac{n}{n^2 + 1}$$

**Solution:**

This series can be compared with the series

$$\sum_{n=1}^{\infty} \frac{1}{n}$$

since the degree of the denominator is only one greater than the degree of the numerator. Furthermore, since

$$\lim_{n \to \infty} \frac{n/(n^2 + 1)}{1/n} = \lim_{n \to \infty} \frac{n^2}{n^2 + 1} = 1$$

and since the series

$$\sum_{n=1}^{\infty} \frac{1}{n}$$

diverges, we conclude that the given series also diverges by the Limit Comparison Test.

**21.** Use the Limit Comparison Test to determine the convergence or divergence of the series

$$\sum_{n=1}^{\infty} \frac{1}{n\sqrt{n^2 + 1}}$$

**Solution:**

Disregarding all but the highest power of $n$, we compare the series to

$$\sum_{n=1}^{\infty} \frac{1}{n\sqrt{n^2}} = \sum_{n=1}^{\infty} \frac{1}{n^2}, \qquad \text{a convergent } p\text{-series}$$

Since

$$\lim_{n \to \infty} \frac{a_n}{b_n} = \lim_{n \to \infty} \left( \frac{1}{n\sqrt{n^2 + 1}} \right) \left( \frac{n^2}{1} \right) = \lim_{n \to \infty} \frac{1}{\sqrt{1 + (1/n)}} = 1$$

we conclude by the Limit Comparison Test that the given series converges.

**31.** Use one of the tests studied thus far to determine the convergence or dievergence of the series

$$\sum_{n=1}^{\infty} \frac{n}{2n+3}$$

**Solution:**

Since

$$\lim_{n\to\infty} a_n = \lim_{n\to\infty} \frac{n}{2n+3} = \frac{1}{2} \neq 0,$$

we conclude that the series diverges by the $n$th-Term Divergence Test.

**37.** Use the Polynomial Test of Exercise 36 to determine the convergence or divergence of the series

$$\frac{1}{2} + \frac{2}{5} + \frac{3}{10} + \frac{4}{17} + \frac{5}{26} + \cdots$$

**Solution:**

The $n$th term of the series is

$$a_n = \frac{n}{n^2+1}$$

and the series can be written

$$\sum_{n=1}^{\infty} \frac{n}{n^2+1} = \sum_{n=1}^{\infty} \frac{P(n)}{Q(n)}$$

In this case $P(n)$ has degree $k = 1$ and $Q(n)$ has degree $k = 2$. Hence, we conclude from the Polynomial Test that the given series diverges.

## 10.5    Alternating Series

**3.** Use the Alternating Series Test to determine the convergence or divergence of

$$\sum_{n=1}^{\infty} \frac{(-1)^{n+1}}{2n-1}$$

**Solution:**

We observe that

$$\lim_{n \to \infty} \frac{1}{2n-1} = 0$$

and

$$a_{n+1} = \frac{1}{2(n+1)-1} = \frac{1}{2n+1} < \frac{1}{2n-1} = a_n$$

for all $n$. Therefore, by the Alternating Series Test the given series converges.

**9.** Use the Alternating Series Test to determine the convergence or divergence of

$$\sum_{n=1}^{\infty} \frac{(-1)^{n+1}(n+1)}{\ln(n+1)}$$

**Solution:**

Since

$$\lim_{n \to \infty} a_n = \lim_{n \to \infty} \frac{n+1}{\ln(n+1)} \qquad \text{(L'Hôpital's Rule)}$$
$$= \infty \neq 0$$

the series diverges.

**21.** Use the Alternating Series Test to determine the convergence or divergence of

$$\sum_{n=1}^{\infty} \frac{2(-1)^{n+1}}{e^n + e^{-n}}$$

**Solution:**

We begin by using differentiaiton to establishh that $a_{n+1} \leq a_n$.

$$f(x) = \frac{2}{e^x + e^{-x}}$$

$$f'(x) = \frac{-2(e^x - e^{-x})}{(e^x + e^{-x})^2}$$

We observe that $f'(x)$ is negative for $x > 0$. Hence, $f$ is a decreasing function and it follows that $a_{n+1} \leq a_n$ for $n \geq 1$. Since

$$\lim_{n \to \infty} \frac{2}{e^n + e^{-n}} = 0,$$

the series converges by the Alternating Series Test.

**25.** Examine

$$\sum_{n=1}^{\infty} \frac{(-1)^{n+1}}{\sqrt{n}}$$

for conditional convergence or absolute convergence.

**Solution:**

In this case, the Alternating Series Test indicates that the given series converges. However, the series

$$\sum_{n=1}^{\infty} \left| \frac{(-1)^{n+1}}{\sqrt{n}} \right| = \sum_{n=1}^{\infty} \frac{1}{\sqrt{n}}$$

is a divergent $p$-series with $p = \frac{1}{2}$. Therefore, we conclude that the given series is conditionally convergent.

**29.** Examine

$$\sum_{n=2}^{\infty} \frac{(-1)^n n}{n^3 - 1}$$

for conditional convergence or absolute convergence.

**Solution:**

In this case, the Alternating Series Test indicates that the given series converges. Also, the series

$$\sum_{n=2}^{\infty} \left| \frac{(-1)^{n+1} n}{n^3 - 1} \right| = \sum_{n=2}^{\infty} \frac{n}{n^3 - 1}$$

converges absolutely by the Limit Comparison Test with the series $\sum_{n=2}^{\infty} \frac{1}{n^2}$. Therefore, we conclude that the given series is absolutely convergent.

**41.** Approximate the sum of the series

$$\sum_{n=0}^{\infty} \frac{(-1)^n}{n!}$$

with an error less that 0.001. (The actual sum is $1/e$.)

**Solution:**

The series is alternating, and since

$$\lim_{n \to \infty} \frac{1}{n!} = 0 \quad \text{and} \quad \frac{1}{(n+1)!} < \frac{1}{n!}$$

the given series converges. By Theorem 10.15 the error, $R_N$, after $N$ terms is such that $|R_N| \le a_{N+1}$. Therefore, to ensure an error less than 0.001, we choose $N$ sufficiently large so that

$$\frac{1}{(N+1)!} \le 0.001$$

Since $1/6! = 1/720 = 0.0013888$ and $1/7! = 1/5040 < 0.001$, we choose $N = 6$ and obtain the approximation

$$\sum_{n=0}^{6} \frac{(-1)^n}{n!} = 1 - 1 + \frac{1}{2} - \frac{1}{6} + \frac{1}{24} - \frac{1}{120} + \frac{1}{720} \approx 0.368$$

which is within 0.001 of the actual sum.

**47.** Find the number of terms necessary to approximate the sum of the series

$$\sum_{n=0}^{\infty} \frac{(-1)^n}{2n+1}$$

with an error less than 0.001.

**Solution:**

By Theorem 10.15 the error, $R_N$, after $N$ terms is such that $|R_N| \le a_{N+1}$. Therefore to insure an error less than 0.001, we choose $N$ sufficiently large so that

$$\frac{1}{2(N+1)+1} \le 0.001$$

$$\frac{1}{2N+3} \le \frac{1}{1000}$$

$$1000 \le 2N+3$$

$$498.5 \le N$$

Therefore, we choose $N = 499$.

## 10.6   The Ratio and Root Tests

**3.** Use the Ratio Test to test for convergence or divergence of

$$\sum_{n=0}^{\infty} \frac{3^n}{n!}$$

**Solution:**

Since

$$\lim_{n\to\infty} \left| \frac{a_{n+1}}{a_n} \right| = \lim_{n\to\infty} \left| \frac{3^{n+1}/(n+1)!}{3^n/n!} \right|$$

$$= \lim_{n\to\infty} \left| \frac{3^{n+1}}{1\cdot 2\cdot 3\cdot 4\cdots n\cdot(n+1)} \cdot \frac{1\cdot 2\cdot 3\cdot 4\cdots n}{3^n} \right|$$

$$= \lim_{n\to\infty} \left( \frac{3}{n+1} \right) = 0 < 1$$

we conclude by the Ratio Test that the series

$$\sum_{n=0}^{\infty} \frac{3^n}{n!}$$

converges.

**17.** Use the Ratio Test to test for convergence or divergence of

$$\sum_{n=0}^{\infty} \frac{4n}{3^n + 1}$$

**Solution:**

Since

$$\lim_{n \to \infty} \left| \frac{a_{n+1}}{a_n} \right| = \lim_{n \to \infty} \left[ \frac{4^{n+1}}{3^{n+1} + 1} \cdot \frac{3^n + 1}{4^n} \right]$$

$$= \lim_{n \to \infty} \left[ \frac{4(3^n + 1)}{3^{n+1} + 1} \right]$$

$$= 4 \lim_{n \to \infty} \left[ \frac{1 + (1/3^n)}{3 + (1/3^n)} \right] = \frac{4}{3} > 1$$

we conclude by the Ratio Test that the given series diverges

**19.** Use the Ratio Test to test for convergence or divergence of

$$\sum_{n=0}^{\infty} \frac{(-1)^{n+1} n!}{1 \cdot 3 \cdot 5 \cdots (2n + 1)}$$

**Solution:**

Since

$$\lim_{n \to \infty} \left| \frac{a_{n+1}}{a_n} \right|$$

$$= \lim_{n \to \infty} \left[ \frac{(n + 1)!}{1 \cdot 3 \cdot 5 \cdots (2n + 1)(2n + 3)} \cdot \frac{1 \cdot 3 \cdot 5 \cdots (2n + 1)}{n!} \right]$$

$$= \lim_{n \to \infty} \frac{(n + 1)}{(2n + 3)} = \frac{1}{2}$$

we conclude by the Ratio Test that the given series converges.

**25.** Use the Root Test to test for convergence or divergence of

$$\sum_{n=1}^{\infty} (2\sqrt[n]{n} + 1)^n$$

**Solution:**

Using the Root Test, we have

$$\lim_{n\to\infty} \sqrt[n]{a_n} = \lim_{n\to\infty} \sqrt[n]{(2\sqrt[n]{n}+1)^n} = \lim_{n\to\infty} (2\sqrt[n]{n}+1)$$

Now we let $y = \sqrt[n]{n} = n^{1/n}$. Then $\ln y = (1/n)\ln n$ and

$$\lim_{n\to\infty} \left(\frac{\ln n}{n}\right) = \lim_{n\to\infty} \left(\frac{1/n}{1}\right) \qquad \text{(L'Hôpital's Rule)}$$
$$= 0$$

Therefore, as $n$ approaches infinity, $\ln y$ approaches zero and $y = \sqrt[n]{n}$ approaches one. Thus

$$\lim_{n\to\infty} (2\sqrt[n]{n}+1) = 2+1 = 3$$

and the series diverges.

**39.** Test for convergence or divergence using any appropriate test from this chapter for the series

$$\sum_{n=1}^{\infty} \frac{10n+3}{n2^n}$$

Identify the test used.

**Solution:**

We compare the given series to the convergent geometric series

$$\sum_{n=0}^{\infty} \frac{1}{2^n}$$

Since
$$\lim_{n\to\infty} \frac{(10n+3)/n2^n}{1/2^n} = \lim_{n\to\infty} \frac{10n+3}{n} = 10$$

we conclude by the Limit Comparison Test that the given series also converges.

**43.** Test for convergence or divergence using any appropriate test from this chapter for the series

$$\sum_{n=1}^{\infty} \frac{\cos(n)}{2^n}$$

Identify the test used.

**Solution:**

This is not an alternating series; hence we use the Comparison Test and observe that

$$\left| \frac{\cos(n)}{2^n} \right| \le \frac{1}{2^n}$$

Now since the series

$$\sum_{n=0}^{\infty} \frac{1}{2^n}$$

is a convergent geometric series, we conclude by the Comparison Test that

$$\sum_{n=1}^{\infty} \left| \frac{\cos(n)}{2^n} \right|$$

also converges. Finally, by Theorem 10.16, since

$$\sum_{n=1}^{\infty} \left| \frac{\cos(n)}{2^n} \right|$$

converges, then

$$\sum_{n=1}^{\infty} \frac{\cos(n)}{2^n}$$

also converges.

**49.** Test for convergence or divergence using any appropriate test from this chapter for the series

$$\sum_{n=1}^{\infty} \frac{(-3)^n}{3 \cdot 5 \cdot 7 \cdots (2n+1)}$$

Identify the test used.

**Solution:**

We test for convergence or divergence by using the Ratio Test and obtain

$$\lim_{n \to \infty} \left| \frac{a_{n+1}}{a_n} \right|$$

$$= \lim_{n \to \infty} \left| \frac{(-3)^{n+1}}{3 \cdot 5 \cdot 7 \cdots (2n+1)[2(n+1)+1]} \cdot \frac{3 \cdot 5 \cdot 7 \cdots (2n+1)}{(-3)^n} \right|$$

$$= \lim_{n \to \infty} \left| \frac{-3}{2(n+1)+1} \right| = 0 < 1$$

Therefore, by the Ratio Test the given series converges.

## 10.7  Taylor Polynomials and Approximations

**5.** Find the Maclaurin polynomial of degree 5 for $f(x) = \sin x$.

**Solution:**

We have

$$
\begin{array}{ll}
f(x) = \sin x & f(0) = 0 \\
f'(x) = \cos x & f'(0) = 1 \\
f''(x) = -\sin x & f''(0) = 0 \\
f'''(x) = -\cos x & f'''(0) = -1 \\
f^{(4)}(x) = \sin x & f^{(4)}(0) = 0 \\
f^{(5)}(x) = \cos x & f^{(5)}(0) = 1
\end{array}
$$

Therefore, the expansion yields

$$P_5(x) = f(0) + f'(0)x + \frac{f''(0)}{2!}x^2 + \frac{f'''(0)}{3!} + \frac{f^{(4)}(0)}{4!}x^4 + \frac{f^{(5)}(0)}{5!}x^5$$

$$= x - \frac{x^3}{6} + \frac{x^5}{120}$$

**9.** Find the Maclaurin polynomial of degree 4 for $f(x) = \dfrac{1}{x+1}$.

**Solution:**

We have

$$
\begin{aligned}
f(x) &= \frac{1}{x+1} & f(0) &= 1 \\
f'(x) &= \frac{-1}{(x+1)^2} & f'(0) &= -1 \\
f''(x) &= \frac{2}{(x+1)^3} & f''(0) &= 2 \\
f'''(x) &= \frac{-6}{(x+1)^4} & f'''(0) &= -6 \\
f^{(4)}(x) &= \frac{24}{(x+1)^5} & f^{(4)}(0) &= 24
\end{aligned}
$$

Therefore, the expansion yields

$$P_4(x) = f(0) + f'(0)x + \frac{f''(0)}{2!}x^2 + \frac{f'''(0)}{3!}x^3 + \frac{f^{(4)}(0)}{4!}x^4$$
$$= 1 - x + x^2 - x^3 + x^4$$

**11.** Find the Maclaurin polynomial of degree 3 for $f(x) = \tan x$.

**Solution:**

We have

$$
\begin{aligned}
f(x) &= \tan x & f(0) &= 0 \\
f'(x) &= \sec^2 x & f'(0) &= 1 \\
f''(x) &= 2\sec^2 x \tan x & f''(0) &= 0 \\
f'''(x) &= 2\sec^4 x + 4\sec^2 x \tan^2 x & f'''(0) &= 0
\end{aligned}
$$

Therefore, the expansion yields

$$P_3(x) = f(0) + f'(0)x + \frac{f''(0)}{2!}x^2 + \frac{f'''(0)}{3!}x^3$$
$$= x + \frac{1}{3}x^3$$

**17.** Find the Taylor polynomial of degree 4, centered at $c = 1$ for $f(x) = \ln x$.

**Solution:**

We have

$$
\begin{aligned}
f(x) &= \ln x & f(1) &= 0 \\
f'(x) &= \frac{1}{x} & f'(1) &= 1 \\
f''(x) &= -\frac{1}{x^2} & f''(1) &= -1 \\
f'''(x) &= \frac{2}{x^3} & f'''(1) &= 2 \\
f^{(4)}(x) &= -\frac{6}{x^4} & f^{(4)}(1) &= -6
\end{aligned}
$$

Therefore, the expansion yields

$$
\begin{aligned}
P_4(x) &= f(1) + f'(1)(x-1) + \frac{f''(1)}{2!}(x-1)^2 \\
&\quad + \frac{f'''(1)}{3!}(x-1)^3 + \frac{f^{(4)}(1)}{4!}(x-1)^4 \\
&= (x-1) - \frac{1}{2}(x-1)^2 + \frac{1}{3}(x-1)^3 - \frac{1}{4}(x-1)^4
\end{aligned}
$$

**23.** Approximate $f(7\pi/8)$ for $f(x) = x^2 \cos x$ by using the Taylor polynomial of degree 2, centered at $c = \pi$.

**Solution:**

We have

$$
\begin{aligned}
f(x) &= x^2 \cos x & f(\pi) &= -\pi^2 \\
f'(x) &= -x^2 \sin x + 2x \cos x & f'(\pi) &= -2\pi \\
f''(x) &= -x^2 \cos x - 4x \sin x + 2 \cos x & f''(\pi) &= \pi^2 - 2
\end{aligned}
$$

Therefore, the expansion yields

$$
\begin{aligned}
P_2(x) &= f(\pi) + f'(\pi)(x - \pi) + \frac{f''(\pi)}{2!}(x - \pi)^2 \\
&= -\pi^2 - 2\pi(x - \pi) + \frac{\pi^2 - 2}{2}(x - \pi)^2
\end{aligned}
$$

Now we have

$$f\left(\frac{7\pi}{8}\right) = \left(\frac{7\pi}{8}\right)^2 \cos\left(\frac{7\pi}{8}\right)$$

$$\approx -\pi^2 - 2\pi\left(\frac{7\pi}{8} - \pi\right) + \frac{\pi^2 - 2}{2}\left(\frac{7\pi}{8} - \pi\right)^2$$

$$\approx -6.7954$$

(The actual functional value accurate to four decimal places is $-6.9812$.)

27. For what values of $x < 0$ can $e^x$ be replaced by the polynomial $1 + x + (x^2/2!) + (x^3/3!)$ if the error cannot exceed 0.001?

**Solution:**

By Theorem 10.19

$$e^x = 1 + x + \frac{x^2}{2!} + \frac{x^3}{3!} + R_3$$

where

$$R_3 = \frac{f^{(4)}(z)}{4!} x^4 = \frac{e^z}{4!} x^4$$

For $z < 0$ we have

$$R_3 = \frac{e^z x^4}{4!} < \frac{x^4}{4!}$$

and we wish to find $x < 0$ such that

$$\frac{x^4}{4!} < 0.001$$

$$x^4 < 24(0.001) = 0.024$$

$$|x| < (0.024)^{1/4} \approx 0.3936$$

Therefore, for values of $x$ such that $-0.39 < x < 0$, we have

$$e^x \approx 1 + x + \frac{x^2}{2!} + \frac{x^3}{3!}.$$

**31.** (a)   Find the Taylor polynomial $P(x)$ of degree 3 for

$$f(x) = \arcsin x.$$

(b)   Complete the accompanying table for $f(x)$ and $P_3(x)$.

| $x$ | $-1$ | $-0.75$ | $-0.50$ | $-0.25$ | 0 | 0.25 | 0.50 | 0.75 | 1 |
|-----|------|---------|---------|---------|---|------|------|------|---|
| $f(x)$ | | | | | | | | | |
| $P_3(x)$ | | | | | | | | | |

(c)   Sketch the graphs of $f(x)$ and $P_3(x)$.

**Solution:**

(a)

$$f(x) = \arcsin x \qquad\qquad f(0) = 0$$

$$f'(x) = \frac{1}{\sqrt{1-x^2}} \qquad\qquad f'(0) = 1$$

$$f''(x) = \frac{x}{(1-x^2)^{3/2}} \qquad\qquad f''(0) = 0$$

$$f'''(x) = \frac{2x^2 + 1}{(1-x^2)^{5/2}} \qquad\qquad f'''(0) = 1$$

$$P_3(x) = f(0) + f'(0)x + \frac{f''(0)}{2!}x^2 + \frac{f'''(0)}{3!}x^3 = x + \frac{x^3}{6}$$

(b)

| $x$ | $-1.0$ | $-0.75$ | $-0.50$ | $-0.25$ | 0 | 0.25 | 0.50 | 0.75 | 1.0 |
|-----|--------|---------|---------|---------|---|------|------|------|-----|
| $f(x)$ | $-1.571$ | $-0.848$ | $-0.524$ | $-0.253$ | 0 | 0.253 | 0.524 | 0.848 | 1.571 |
| $P_3(x)$ | $-1.167$ | $-0.820$ | $-0.521$ | $-0.253$ | 0 | 0.253 | 0.521 | 0.820 | 1.167 |

## 10.8   Power Series

**9.** Find the interval of convergence for the power series

$$\sum_{n=1}^{\infty} \frac{(-1)^n x^n}{n}$$

Include a check for convergence at the endpoints of the interval.

**Solution:**

Since

$$\lim_{n \to \infty} \left| \frac{u_{n+1}}{u_n} \right| = \lim_{n \to \infty} \left[ \frac{1/(n+1)}{1/n} \right] = \lim_{n \to \infty} \left[ \frac{1}{n+1} \cdot \frac{n}{1} \right]$$

$$= \lim_{n \to \infty} \frac{n}{n+1} = 1$$

we conclude by the Ratio Test that the radius of convergence is 1 and that the interval of convergence includes $-1 < x < 1$. When $x = -1$, we have the harmonic series

$$\sum_{n=1}^{\infty} \frac{1}{n}$$

which diverges. When $x = 1$, we have the alternating series

$$\sum_{n=1}^{\infty} \frac{(-1)^n}{n}$$

which converges. Thus the interval of convergence is $-1 < x \leq 1$.

13. Find the interval of convergence for the power series

$$\sum_{n=0}^{\infty} (2n)! \left( \frac{x}{2} \right)^n$$

Include a check for convergence at the endpoints of the interval.

**Solution:**

Since

$$\lim_{n \to \infty} \left| \frac{u_{n+1}}{u_n} \right| = \lim_{n \to \infty} \left| \frac{(2n+2)!/2^{n+1}}{(2n)!/2^n} \right|$$

$$= \lim_{n \to \infty} \frac{(2n+2)(2n+1)}{2} = \infty$$

Therefore, by the Ratio Test we conclude that the series converges only when $x = 0$.

**17.** Find the interval of convergence for the power series

$$\sum_{n=1}^{\infty} \frac{(-1)^{n+1}(x-5)^n}{n5^n}$$

Include a check for convergence at the endpoints of the interval.

**Solution:**

Since

$$\lim_{n\to\infty}\left|\frac{u_{n+1}}{u_n}\right| = \lim_{n\to\infty}\frac{n5^n}{(n+1)5^{n+1}} = \lim_{n\to\infty}\frac{1}{5}\left(\frac{n}{n+1}\right) = \frac{1}{5}$$

the radius of convergence is $R = 5$, and since the series is centered at $x = 5$, the series will converge in the interval $(0, 10)$. Furthermore, when $x = 0$ we have the series

$$\sum_{n=1}^{\infty}\frac{(-1)^{n+1}(-1)^n}{n} = \sum_{n=1}^{\infty}\frac{(-1)^{2n+1}}{n} = -\sum_{n=1}^{\infty}\frac{1}{n}$$

which diverges $(p = 1)$. When $x = 10$, we have the series

$$\sum_{n=1}^{\infty}\frac{(-1)^{n+1}}{n}$$

which converges by the Alternating Series Test. Hence the interval of convergence of the given series is $0 < x \le 10$.

**27.** Find the interval of convergence for the power series

$$\sum_{n=1}^{\infty}\frac{k(k+1)(k+2)\cdots(k+n-1)x^n}{n!} \qquad (k \ge 1)$$

Include a check for convergence at the endpoints of the interval.

**Solution:**

Since

$$\lim_{n\to\infty}\left|\frac{u_{n+1}}{u_n}\right|$$

$$= \lim_{n\to\infty}\left[\frac{k(k+1)\cdots(k+n-1)(k+n)}{(n+1)!}\cdot\frac{n!}{k(k+1)\cdots(k+n-1)}\right]$$

$$= \lim_{n\to\infty}\frac{k+n}{n+1} = 1$$

the radius of convergence is $R = 1$. Since the series is centered at $x = 0$, it will converge on the interval $(-1, 1)$. To test for convergence at the endpoints, we note that for $k \geq 1$, we have

$$\lim_{n \to \infty} a_n = \lim_{n \to \infty} \left[ \left( \frac{k}{1} \right) \left( \frac{k+1}{2} \right) \left( \frac{k+2}{3} \right) \cdots \left( \frac{k+n-1}{n} \right) \right] \neq 0$$

Thus for $x = \pm 1$ the series diverges, and we conclude that the interval of convergence is $-1 < x < 1$.

31. Find the interval of convergence (check for convergence at the endpoints) of (a) $f(x)$, (b) $f'(x)$, and (d) $\int f(x)\, dx$ if

$$f(x) = \sum_{n=0}^{\infty} \left( \frac{x}{2} \right)^n$$

**Solution:**

(a) The given series is geometric with $r = x/2$ and converges is

$$\left| \frac{x}{2} \right| < 1 \qquad \text{or} \qquad -2 < x < 2$$

(b) $\quad f'(x) = \sum_{n=1}^{\infty} n \left( \frac{x}{2} \right)^{n-1} \left( \frac{1}{2} \right) = \sum_{n=1}^{\infty} \left( \frac{n}{2} \right) \left( \frac{x}{2} \right)^{n-1}$

Therefore the series for $f'(x)$ diverges for $x = \pm 2$, and its interval of convergence is $-2 < x < 2$. (c)

$$f''(x) = \sum_{n=2}^{\infty} \left( \frac{n}{2} \right) (n-1) \left( \frac{x}{2} \right)^{n-2} \left( \frac{1}{2} \right)$$

$$= \sum_{n=2}^{\infty} \frac{n(n-1)}{4} \left( \frac{x}{2} \right)^{n-2}$$

Therefore the series for $f''(x)$ diverges for $x = \pm 2$, and its interval of convergence is $-2 < x < 2$.

(d) The series for $\int f(x)\, dx$ is

$$\sum_{n=0}^{\infty} \frac{2}{n+1} \left( \frac{x}{2} \right)^{n+1}$$

and converges (Alternating Series Test) for $x = -2$ and diverges (Limit Comparison Test with $\sum_{n=1}^{\infty} (1/n)$ for $x = 2$. Therefore, its interval of convergence is $-2 \leq x < 2$.

## 10.9   Representation of Functions by Power Series

**3.** Find a power series centered at $c = 5$ for $f(x) = 1/(2-x)$, and find the interval of convergence.

**Solution:**

Writing $f(x)$ in the form $a/(1-r)$, we have

$$\frac{1}{2-x} = \frac{1}{-3-x+5} = \frac{-\frac{1}{3}}{1-\frac{x-5}{-3}}$$

which implies that $a = -1/3$ and $r = (x-5)/(-3)$. Therefore,

$$\frac{1}{2-x} = -\frac{1}{3}\sum_{n=0}^{\infty}\left(\frac{x-5}{-3}\right)^n = \sum_{n=0}^{\infty}\frac{(x-5)^n}{(-3)^{n+1}}$$

Since

$$\lim_{n\to\infty}\left|\frac{u_{n+1}}{u_n}\right| = \lim_{n\to\infty}\left|\frac{1}{(-3)^{n+2}}\cdot\frac{(-3)^{n+1}}{1}\right| = \frac{1}{3},$$

the radius of convergence is $R = 3$. Since the series is centered at $c = 5$, it converges in the interval $(2,8)$. Finally, since the series diverges at both endpoints, the interval of convergence is $2 < x < 8$.

**11.** Find a power series centerd at $c = 0$ for $f(x) = 3x/(x^2+x-2)$, and find the interval of convergence.

**Solution:**

Using the method for finding partial fractions (Section 8.5), we have

$$\frac{3x}{x^2+x-2} = \frac{2}{x+2} + \frac{1}{x-1} = \frac{1}{1-\frac{-x}{2}} - \frac{1}{1-x}$$

Therefore,

$$\frac{1}{1-\frac{-x}{2}} - \frac{1}{1-x} = \sum_{n=0}^{\infty}\left(\frac{x}{-2}\right)^n - \sum_{n=0}^{\infty}x^n$$

$$= \sum_{n=0}^{\infty}\left[\frac{x^n}{(-2)^n} - x^n\right] = \sum_{n=0}^{\infty}\left[\frac{1}{(-2)^n} - 1\right]x^n$$

Now since

$$\lim_{n\to\infty}\left|\frac{u_{n+1}}{u_n}\right| = \lim_{n\to\infty}\left|\frac{[1/(-2)^{n+1}]-1}{[1/(-2)^n]-1}\right| = 1$$

we have $R = 1$. Finally, when $x = \pm 1$, we have

$$\lim_{n\to\infty} a_n = \lim_{n\to\infty}\left[\frac{1}{(-2)^n}-1\right] \neq 0$$

Thus the series diverges when $x = \pm 1$ and the interval of convergence is $-1 < x < 1$.

**13.** Find a power series centered at $c = 0$ for $f(x) = 2/(1 - x^2)$, and find the interval of convergence.

**Solution:**

Letting $u = x^2$, we have

$$\frac{2}{1-x^2} = \frac{2}{1-u}$$

$$= 2\sum_{n=0}^{\infty} u^n$$

$$= 2\sum_{n=0}^{\infty}(x^2)^n = 2\sum_{n=0}^{\infty} x^{2n}$$

The series will converge if $x^2 < 1$ or $-1 < x < 1$.

**17.** Use the power series

$$\frac{1}{x+1} = \sum_{n=0}^{\infty}(-1)^n x^n$$

to determine a power series representation about $c = 0$ for $f(x) = \ln(x+1)$. Specify the interval of convergence.

**Solution:**

Since $\qquad \displaystyle\int \frac{1}{x+1}\,dx = \ln(x+1) + C$

we can integrate the power series for $1/(x+1)$ to obtain the series for $\ln(x+1)$ as follows:

$$\frac{1}{x+1} = \sum_{n=0}^{\infty}(-1)^n x^n$$

$$\ln(x+1) = \sum_{n=0}^{\infty} \frac{(-1)^n x^{n+1}}{n+1} + C$$

By substituting $x = 0$ on both sides of this equation, we determine that $C = 0$. Furthermore, since

$$\lim_{n \to \infty} \left| \frac{u_{n+1}}{u_n} \right| = \lim_{n \to \infty} \left[ \frac{1}{n+2} \cdot \frac{n+1}{1} \right] = 1$$

we conclude that $R = 1$ and the series converges in the interval $(-1, 1)$. At $x = -1$ we have the divergent series

$$\sum_{n=0}^{\infty} \frac{(-1)^{2n+1}}{n+1} = -\sum_{n=0}^{\infty} \frac{1}{n+1}$$

At $x = 1$ we have the convergent alternating series

$$\sum_{n=0}^{\infty} \frac{(-1)^n}{n+1}$$

Therefore,

$$\ln(x+1) = \sum_{n=0}^{\infty} \frac{(-1)^n x^{n+1}}{n+1}, \quad \text{for} \quad -1 < x \leq 1.$$

**27.** Use the series for the function $f(x) = \arctan x$ to approximate (with $R_N \leq 0.001$)

$$\int_0^{1/2} \frac{\arctan x^2}{x} \, dx$$

**Solution:**

From Example 5 we have

$$\arctan x = \sum_{n=0}^{\infty} \frac{(-1)^n x^{2n+1}}{2n+1}$$

Therefore

$$\arctan x^2 = \sum_{n=0}^{\infty} \frac{(-1)^n (x^2)^{(2n+1)}}{2n+1}$$

$$= \sum_{n=0}^{\infty} \frac{(-1)^n (x^{(4n+2)})}{2n+1}$$

and

$$\frac{\arctan x^2}{x} = \sum_{n=0}^{\infty} \frac{(-1)^n x^{4n+1}}{2n+1}$$

Thus

$$\int_0^{1/2} \frac{\arctan x^2}{x}\,dx = \left[\sum_{n=0}^{\infty} \frac{(-1)^n x^{4n+2}}{(2n+1)(4n+2)}\right]_0^{1/2}$$

$$= \sum_{n=0}^{\infty} \frac{(-1)^n (1/2)^{4n+2}}{(2n+1)(4n+2)}$$

$$\approx \left(\frac{1}{2}\right)\left(\frac{1}{2}\right)^2 \approx \frac{1}{8}$$

Note that the second term of the series is $-(\frac{1}{18})(\frac{1}{2})^6 < 0.001$. Therefore, by Theorem 10.15 one term is sufficient to approximate the integral.

## 10.10   Taylor and Maclaurin Series

7. Find the Taylor Series centered at $c = 0$ for the function $f(x) = \sin 2x$.

**Solution:**

Since

$$
\begin{array}{ll}
f(x) = \sin 2x & f(0) = 0 \\
f'(x) = 2\cos 2x & f'(0) = 2 \\
f''(x) = -4\sin 2x & f''(0) = 0 \\
f'''(x) = -8\cos 2x & f'''(0) = -8 = -2^3 \\
f^{(4)}(x) = 16\sin 2x & f^{(4)}(0) = 0 \\
f^{(5)}(x) = 32\cos 2x & f^{(5)}(0) = 32 = 2^5
\end{array}
$$

we can see that the signs alternate and that $|f^{(n)}(0)| = 2^n$ if $n$ is odd. Therefore the Taylor Series is

$$\sin 2x$$
$$= f(0) + f'(0)x + \frac{f''(0)x^2}{2!} + \frac{f'''(0)x^3}{3!} + \frac{f^{(4)}(0)x^4}{4!} + \cdots$$
$$= \frac{2x}{1!} - \frac{2^3x^3}{3!} + \frac{2^5x^5}{5!} - \cdots + \frac{(-1)^n(2x)^{2n+1}}{(2n+1)!} + \cdots$$
$$= \sum_{n=0}^{\infty} \frac{(-1)^n(2x)^{2n+1}}{(2n+1)!}$$

Note that we could have arrived at the same result by substituting $2x$ into the series for $\sin x$ as follows:

$$\sin x = x - \frac{x^3}{3!} + \frac{x^5}{5!} - \frac{x^7}{7!} + \cdots$$
$$\sin (2x) = (2x) - \frac{(2x)^3}{3!} + \frac{(2x)^5}{5!} - \frac{(2x)^7}{7!} + \cdots$$
$$= \sum_{n=0}^{\infty} \frac{(-1)^n(2x)^{2n+1}}{(2n+1)!}$$

**13.** Use the binomial series to find the power series centered at $c = 0$ for the function $f(x) = 1/\sqrt{4 + x^2}$.

**Solution:**

Consider $f$ in the form

$$f(x) = \frac{1}{\sqrt{4 + x^2}} = \frac{1}{2\sqrt{1 + (x/2)^2}} = \frac{1}{2}\left[1 + \left(\frac{x}{2}\right)^2\right]^{-1/2}$$

which is similar to the binommial form $(1 + x)^{-k}$. Since

$$(1 + x)^{-k} = 1 - kx + \frac{k(k+1)x^2}{2!} - \frac{k(k+1)(k+2)x^3}{3!} + \cdots$$

we have, for $k = \frac{1}{2}$,

$$(1 + x)^{-1/2} = 1 - \frac{1}{2} + \frac{(\frac{1}{2})(\frac{3}{2})x^2}{2!} - \frac{(\frac{1}{2})(\frac{3}{2})(\frac{5}{2})x^3}{3!} + \cdots$$
$$= 1 - \frac{x}{2} + \frac{1 \cdot 3x^2}{2^2 2!} - \frac{1 \cdot 3 \cdot 5x^3}{2^3 3!} + \cdots$$
$$= 1 + \sum_{n=1}^{\infty} \frac{(-1)^{n+1} 1 \cdot 3 \cdot 5 \cdots (2n-1)x^n}{2^n n!}$$

Now by substituting $(x/2)^2$ for $x$, we obtain

$$f(x) = \frac{1}{\sqrt{4+x^2}} = \frac{1}{2}\left[1 + \left(\frac{x}{2}\right)^2\right]^{-1/2}$$

$$= \frac{1}{2}\left[1 + \sum_{n=1}^{\infty} \frac{(-1)^{n+1} 1 \cdot 3 \cdot 5 \cdots (2n-1)(x/2)^{2n}}{2^n n!}\right]$$

$$= \frac{1}{2} + \sum_{n=1}^{\infty} \frac{(-1)^{n+1} 1 \cdot 3 \cdot 5 \cdots (2n-1) x^{2n}}{2^{3n+1} n!}$$

$$= \frac{1}{2}\left[1 + \sum_{n=1}^{\infty} \frac{(-1)^{n+1} 1 \cdot 3 \cdot 5 \cdots (2n-1) x^{2n}}{2^{3n} n!}\right]$$

**17.** Find the power series for $f(x) = e^{-x^2/2}$ by using the series for $e^x$

**Solution:**

Since $\qquad e^x = 1 + x + \frac{x^2}{2!} + \frac{x^3}{3!} + \frac{x^4}{4!} + \frac{x^5}{5!} + \cdots$

we can substitute $(-x^2/2)$ for $x$ and obtain the series

$$e^{-x^2/2} = 1 - \frac{x^2}{2} + \frac{(-x^2/2)^2}{2!} + \frac{(-x^2/2)^3}{3!} + \frac{(-x^2/2)^4}{4!} + \cdots$$

$$= 1 - \frac{x^2}{2} + \frac{x^4}{2^2 2!} - \frac{x^6}{2^3 3!} + \frac{x^8}{2^4 4!} - \cdots$$

$$= \sum_{n=0}^{\infty} \frac{(-1)^n x^{2n}}{2^n n!}$$

**23.** Find the power series for $f(x) = (\sin x)/x$ by using the series for $\sin x$.

**Solution:**

Since $\qquad \sin x = x - \frac{x^3}{3!} + \frac{x^5}{5!} - \frac{x^7}{7!} + \frac{x^9}{9!} - \cdots$

we can divide by $x$ to obtain

$$\frac{\sin x}{x} = 1 - \frac{x^2}{3!} + \frac{x^4}{5!} - \frac{x^6}{7!} + \frac{x^8}{9!} - \cdots = \sum_{n=0}^{\infty} \frac{(-1)^n x^{2n}}{(2n+1)!}$$

**27.** Use the power series for $e^x$ to show that

$$g(x) = \tfrac{1}{2}(e^{ix} - e^{ix}) = \sin x.$$

**Solution:**

We first observe the following powers of $i$:

$$
\begin{array}{ll}
i = \sqrt{-1} & i^5 = i^4 \cdot i = 1 \\
i^2 = -1 & i^6 = i^4 \cdot i^2 = -1 \\
i^3 = i^2 \cdot i = -i & i^7 = i^4 \cdot i^3 = -i \\
i^4 = i^2 \cdot i^2 = 1 & i^8 = 1^4 \cdot i^4 = 1
\end{array}
$$

Since $\qquad e^x = 1 + x + \dfrac{x^2}{2!} + \dfrac{x^3}{3!} + \dfrac{x^4}{4!} + \dfrac{x^5}{5!} + \cdots$

we can substitute $(ix)$ and $(-ix)$ for $x$ and obtain the series for $e^{ix}$ and $e^{-ix}$.

$$e^{ix} = 1 + (ix) + \frac{(ix)^2}{2!} + \frac{(ix)^3}{3!} + \frac{(ix)^4}{4!} + \frac{(ix)^5}{5!} + \cdots$$

$$= 1 + ix - \frac{x^2}{2!} - \frac{ix}{3!} + \frac{x^4}{4!} + \frac{ix^5}{5!} - \cdots$$

$$e^{-ix} = 1 + (-ix) + \frac{(-ix)^2}{2!} + \frac{(-ix)^3}{3!} + \frac{(-ix)^4}{4!} + \frac{(-ix)^5}{5!} + \cdots$$

$$= 1 - ix - \frac{x^2}{2!} + \frac{ix^3}{3!} + \frac{x^4}{4!} - \frac{ix^5}{5!} - \cdots$$

Therefore, subtracting the series for $e^{ix}$ and $e^{-ix}$ we have

$$e^{ix} - e^{-ix} = 2ix - \frac{2ix^3}{3!} + \frac{2ix^5}{5!} - .$$

and

$$\frac{e^{ix} - e^{-ix}}{2i} = x - \frac{x^3}{3!} + \frac{x^5}{5!} - \cdots$$

$$= \sum_{n=0}^{\infty} \frac{(-1)^n x^{2n+1}}{(2n+1)!} = \sin x$$

**31.** Use power series to approximate

$$\int_0^{\pi/2} \frac{\sin x}{x}\, dx$$

accurate to four decimal places.

**Solution:**

From Exercise 23 we have

$$\frac{\sin x}{x} = \sum_{n=0}^{\infty} \frac{(-1)^n x^{2n}}{(2n+1)!}$$

Therefore

$$\int_0^{\pi/2} \frac{\sin x}{x}\, dx = \left[ \sum_{n=0}^{\infty} \frac{(-1)^n x^{2n+1}}{(2n+1)(2n+1)!} \right]_0^{\pi/2}$$

$$= \sum_{n=0}^{\infty} \frac{(-1)^n (\pi/2)^{2n+1}}{(2n+1)(2n+1)!}$$

$$\approx \frac{\pi}{2} - \frac{\pi^3}{3 \cdot 3! \cdot 2^3} + \frac{\pi^5}{5 \cdot 5! \cdot 2^5} + \frac{\pi^7}{7 \cdot 7! \cdot 2^7} - \cdots$$

$$= 1.3708$$

# Review Exercises for Chapter 10

**7.** Determine the convergence or divergence of $\{a_n\}$ when $a_n = \sqrt{n+1} - \sqrt{n}$.

**Solution:**

$$\lim_{n \to \infty} a_n = \lim_{n \to \infty} [\sqrt{n+1} - \sqrt{n}]$$

$$= \lim_{n \to \infty} \left[ (\sqrt{n+1} - \sqrt{n}) \frac{\sqrt{n+1} + \sqrt{n}}{\sqrt{n+1} + \sqrt{n}} \right]$$

$$= \lim_{n \to \infty} \frac{1}{\sqrt{n+1} + \sqrt{n}} = 0$$

Therefore the sequence converges to 0.

**13.** Find the first five terms of the sequence of partial sums for the series

$$\sum_{n=1}^{\infty} \frac{(-1)^{n+1}}{(2n)!}$$

**Solution:**

Since

$$\sum_{n=1}^{\infty} \frac{(-1)^{n+1}}{(2n)!} = \frac{1}{2!} - \frac{1}{4!} + \frac{1}{6!} - \frac{1}{8!} + \frac{1}{10!} - \cdots$$

$$= \frac{1}{2} - \frac{1}{24} + \frac{1}{720} - \frac{1}{40,320} + \frac{1}{3,628,800} - \cdots$$

we have

$$S_1 = \frac{1}{2} = 0.5$$

$$S_2 = \frac{1}{2} - \frac{1}{24} \approx 0.45833$$

$$S_3 = \frac{1}{2} - \frac{1}{24} + \frac{1}{720} \approx 0.45972$$

$$S_4 = \frac{1}{2} - \frac{1}{24} + \frac{1}{720} - \frac{1}{40,320} \approx 0.45970$$

$$S_5 = \frac{1}{2} - \frac{1}{24} + \frac{1}{720} - \frac{1}{40,320} + \frac{1}{3,628,800} \approx 0.45970$$

**17.** Find the sum of the infinite series

$$\sum_{n=0}^{\infty} \left( \frac{1}{2^n} - \frac{1}{3^n} \right)$$

**Solution:**

We begin by writing the given series as the difference of two geometric series

$$\sum_{n=0}^{\infty} \left( \frac{1}{2^n} - \frac{1}{3^n} \right) = \sum_{n=0}^{\infty} \left( \frac{1}{2} \right)^n - \sum_{n=0}^{\infty} \left( \frac{1}{3} \right)^n$$

Since these two geometric series have the values $a = 1$ and $r = \frac{1}{2}$ and $\frac{1}{3}$, respectively, we conclude that the sum of the given series is

$$\sum_{n=0}^{\infty} \left( \frac{1}{2^n} - \frac{1}{3^n} \right) = \frac{1}{1 - (\frac{1}{2})} - \frac{1}{1 - (\frac{1}{3})} = 2 - \frac{3}{2} = \frac{1}{2}$$

**25.** Determine the convergence or divergence of the series

$$\sum_{n=1}^{\infty} \frac{1}{(n^3 + 2n)^{1/3}}$$

**Solution:**

Since the denominator is the cube root of a third-degree polynomial, we compare the given series to the divergent harmonic series

$$\sum_{n=1}^{\infty} \frac{1}{n}$$

Thus

$$\lim_{n \to \infty} \frac{[1/(n^3 + 2n)^{1/3}]}{1/n} = \lim_{n \to \infty} \frac{n}{(n^3 + 2n)^{1/3}}$$

$$= \lim_{n \to \infty} \frac{1}{[1 + (2/n^2)]^{1/3}} = 1$$

and we conclude by the Limit Comparison Test that the given series diverges.

**35.** Find the interval of convergence of the power series

$$\sum_{n=0}^{\infty} n!(x - 2)^n$$

**Solution:**

Since

$$\lim_{n \to \infty} \left| \frac{u_{n+1}}{u_n} \right| = \lim_{n \to \infty} \frac{(n + 1)!}{n!} = \lim_{n \to \infty} (n + 1) = \infty$$

we conclude that the radius of convergence is $R = 0$. Therefore the given series will converge only at $x = 2$.

39. Find the power series for $f(x) = 3^x$ centered at $c = 0$.

**Solution:**

Since
$$3^x = e^{\ln(3^x)} = e^{x(\ln 3)}$$
we can substitute $x(\ln 3)$ for the $x$ in the series

$$e^x = 1 + x + \frac{x^2}{2!} + \frac{x^3}{3!} + \frac{x^4}{4!} + \cdots$$

to obtain

$$3^x = e^{x(\ln 3)}$$
$$= 1 + (\ln 3)x + \frac{(\ln 3)^2 x^2}{2!} + \frac{(\ln 3)^3 x^3}{3!} + \frac{(\ln 3)^4 x^4}{4!} + \cdots$$
$$= \sum_{n=0}^{\infty} \frac{(x \ln 3)^n}{n!}$$

# 11  CONICS

## 11.1  Parabolas

9. Find the vertex, focus, and directrix of the parabola given by $y^2 = -6x$, and sketch its graph.

**Solution:**

We begin by writing the equation of this parabola in standard form.

$$(y - k)^2 = 4p(x - h)$$
$$y^2 = -6x$$
$$(y - 0)^2 = 4\left(-\frac{3}{2}\right)(x - 0)$$

Thus $k = 0, h = 0$, and $p = -\frac{3}{2}$. We conclude that

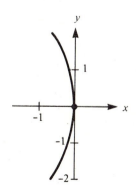

$$\text{vertex, } (h, k): \qquad (0, 0)$$
$$\text{focus, } (h + p, k): \qquad \left(-\frac{3}{2}, 0\right)$$
$$\text{directrix, } (x = h - p): \qquad x = \frac{3}{2}$$

**17.** Find the vertex, focus, and directrix of the parabola given by $y = \frac{1}{4}(x^2 - 2x + 5)$, and sketch its graph.

**Solution:**

We begin by writing the equation of this parabola in standard form.

$$(x - h)^2 = 4p(y - k)$$

$$\frac{1}{4}(x^2 - 2x + 5) = y$$

$$x^2 - 2x + 5 = 4y$$

$$x^2 - 2x + 1 = 4y - 4$$

$$(x - 1)^2 = 4(1)(y - 1)$$

Thus $h = 1, k = 1$, and $p = 1$. We conclude that

$$\begin{array}{lll} \text{vertex, } (h, k): & (1, 1) \\ \text{focus, } (h, k + p): & (1, 2) \\ \text{directrix, } (y = k - p): & y = 0 \end{array}$$

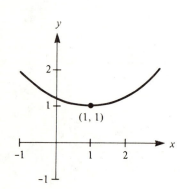

**23.** Find the vertex, focus, and directrix of the parabola given by $y^2 - 4y - 4x = 0$, and sketch its graph.

**Solution:**

We begin by writing the equation of this parabola in standard form.

$$(y - k)^2 = 4p(x - h)$$

$$y^2 - 4y = 4x$$

$$y^2 - 4y + 4 = 4x + 4$$

$$(y - 2)^2 = 4(1)(x + 1)$$

Thus $h = -1, k = 2$, and $p = 1$. We conclude that

$$\begin{array}{lll} \text{vertex, } (h, k): & (-1, 2) \\ \text{focus, } (h + p, k): & (0, 2) \\ \text{directrix, } (x = h - p): & x = -2 \end{array}$$

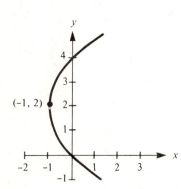

**29.** Find the equation of a parabola with vertex at $(3, 2)$ and focus at $(1, 2)$.

**Solution:**

Since the vertex and focus lie on a horizontal line, the axis of this parabola must be horizontal and its standard form would be

$$(y - k)^2 = 4p(x - h)$$

Since the vertex lies at $(3, 2)$, we have $h = 3$ and $k = 2$. Furthermore, the *directed distance* from the focus to the vertex is

$$p = 1 - 3 = -2$$

Thus the standard equation is

$$(y - 2)^2 = 4(-2)(x - 3)$$
$$y^2 - 4y + 4 = -8x + 24$$
$$y^2 - 4y + 8x - 20 = 0$$

**35.** Find an equation for the parabola whose axis is parallel to the $y$-axis and passes through the points $(0, 3), (3, 4), (4, 11)$.

**Solution:**

Since the axis of the parabola is vertical, the standard form is

$$(x - h)^2 = 4p(y - k)$$

A more convenient form for this equation is

$$y = ax^2 + bx + c$$

Now substituting the values of the given coordinates into this equation, we obtain three equations:

(i)  $\qquad\qquad 3 = a(0)^2 + b(0) + c$

(ii)  $\qquad\qquad 4 = a(3)^2 + b(3) + c$

(iii)  $\qquad\qquad 11 = a(4)^2 + b(4) + c$

Simplification yields

(i)  $\qquad\qquad 3 = c$

(ii)  $\qquad\qquad 4 = 9a + 3b + c \;\rightarrow\; 1 = 9a + 3b$

(iii)  $\qquad\qquad 11 = 16a + 4b + c \;\rightarrow\; 8 = 16a + 4b$

Solving (ii) and (iii) for $a$ and $b$ gives us

(ii) $\qquad 9a + 3b = 1 \;\rightarrow\; 9a + 3b = 1$

(iii) $\qquad 4a + b = 2 \;\rightarrow\; 12a + 3b = 6$

$$-3a = -5$$
$$a = \frac{5}{3}$$
$$b = -\frac{14}{3}$$

Finally, we have

$$y = ax^2 + bx + c$$
$$y = \frac{5}{3}x^2 - \frac{14}{3}x + 3$$
$$3y = 5x^2 - 14x + 9$$
$$5x^2 - 14x - 3y + 9 = 0$$

45. Find the coordinates of the centroid of the region bounded by the graphs of $y^2 = 4x$ and $x = 1$.

**Solution:**

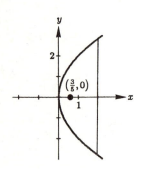

From the accompanying figure we observe that $\bar{y} = 0$, the area of the region is given by

$$A = \int_{-2}^{2} \left(1 - \frac{y^2}{4}\right) dy$$
$$= 2 \int_{0}^{2} \left(1 - \frac{y^2}{4}\right) dy$$
$$= 2\left[y - \frac{y^3}{12}\right]_{0}^{2} = \frac{8}{3}$$

Therefore, the $x$-coordinate of the centroid is given by

$$\bar{x} = \frac{1}{A} \int_{-2}^{2} \left(\frac{1 + \frac{y^2}{4}}{2}\right)\left(1 - \frac{y^2}{4}\right) dy$$
$$= \frac{3}{8}\left(\frac{1}{2}\right) \int_{-2}^{2} \left(1 - \frac{y^4}{16}\right) dy$$
$$= \frac{3}{8} \int_{0}^{2} \left(1 - \frac{y^4}{16}\right) dy$$
$$= \frac{3}{8}\left[y - \frac{y^5}{80}\right]_{0}^{2} = \frac{3}{5}$$

Therefore the coordinates of the centroid are $\left(\frac{3}{5}, 0\right)$.

**53.** A cable of a parabolic suspension bridge is suspended between two towers that are 400 feet apart and 50 feet above the roadway as shown in the accompanying figure. The cable touches the roadway midway between the towers. Find the length of the cable.

**Solution:**

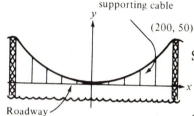

Since the axis of the parabola is vertical, the standard form is

$$(x - h)^2 = 4p(y - k)$$

Since the vertex is $(0,0)$, $h = 0$ and $k = 0$. Therefore

$$x^2 = 4py$$

Since the parabola passes throught the point $(200, 50)$ we have

$$200^2 = 4p(50) \qquad \text{or} \qquad 200 = p$$

Therefore

$$x^2 = 4(200)y \qquad \text{or} \qquad y = \frac{1}{800}x^2$$

$$y' = \frac{1}{400}x$$

$$s = \int_{-200}^{200} \sqrt{1 + (y')^2}\, dx = 2 \int_{0}^{200} \sqrt{1 + \left(\frac{x}{400}\right)^2}\, dx$$

$$= \frac{1}{200} \int_{0}^{200} \sqrt{400^2 + x^2}\, dx$$

$$= \frac{1}{400}\left[x\sqrt{400^2 + x^2} + 400^2 \ln\left(x + \sqrt{400^2 + x^2}\right)\right]_{0}^{200}$$

$$= 100\left[\sqrt{5} + 4\ln\left(\frac{1 + \sqrt{5}}{2}\right)\right] \approx 416.1\,\text{ft}$$

## 11.2   Ellipses

**13.** Find the center, foci, vertices, eccentricity, and sketch the graph of the ellipse given by $x^2 + 4y^2 = 4$.

**Solution:**

In standard form we have

$$x^2 + 4y^2 = 4$$

$$\frac{x^2}{4} + y^2 = 1$$

$$\frac{(x-0)^2}{2^2} + \frac{(y-0)^2}{1^2} = 1$$

$$\frac{(x-h)^2}{a^2} + \frac{(y-k)^2}{b^2} = 1$$

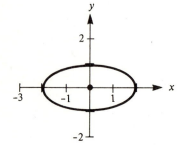

Thus $h = 0$, $k = 0$, $a = 2$, $b = 1$, and $c = \sqrt{2^2 - 1^2} = \sqrt{3}$. We conclude that

$$
\begin{aligned}
\text{center, } (h,k): & \quad (0,0) \\
\text{foci, } (h \pm c, k): & \quad (\pm\sqrt{3}, 0) \\
\text{vertices, } (h \pm a, k): & \quad (\pm 2, 0)
\end{aligned}
$$

$$e = \frac{c}{a} = \frac{\sqrt{3}}{2}$$

**21.** Find the center, foci, vertices, eccentricity, and sketch the graph of the ellipse given by $9x^2 + 4y^2 + 36x - 24y + 36 = 0$.

**Solution:**

In standard form we have

$$9x^2 + 4y^2 + 36x - 24y + 36 = 0$$

$$9x^2 + 36x + 4y^2 - 24y = -36$$

$$9(x^2 + 4x + 4) + 4(y^2 - 6y + 9) = -36 + 36 + 36$$

$$9(x+2)^2 + 4(y-3)^2 = 36$$

$$\frac{(x+2)^2}{4} + \frac{(y-3)^2}{9} = 1$$

$$\frac{(x+2)^2}{2^2} + \frac{(y-3)^2}{3^2} = 1$$

$$\frac{(x-h)^2}{b^2} + \frac{(y-k)^2}{a^2} = 1$$

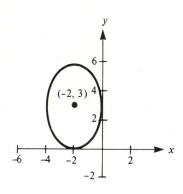

Thus $h = -2$, $k = 3$, $a = 3$, $b = 2$, and $c = \sqrt{3^2 - 2^2} = \sqrt{5}$. We conclude that

$$\text{center, } (h, k): \qquad (-2, 3)$$
$$\text{foci, } (h, k \pm c): \qquad (-2, 3 \pm \sqrt{5})$$
$$\text{vertices, } (h, k \pm a): \qquad (-2, 3 \pm 3)$$

$$e = \frac{c}{a} = \frac{\sqrt{5}}{3}$$

**25.** Find the center, foci, vertices, eccentricity, and sketch the graph of the ellipse given by $12x^2 + 20y^2 - 12x + 40y - 37 = 0$.

**Solution:**

In standard form we have

$$12x^2 + 20y^2 - 12x + 40y - 37 = 0$$
$$12(x^2 - x + \tfrac{1}{4}) + 20(y^2 + 2y + 1) = 37 + 3 + 20$$
$$12(x - \tfrac{1}{2})^2 + 20(y + 1)^2 = 60$$
$$\frac{(x - \tfrac{1}{2})^2}{5} + \frac{(y + 1)^2}{3} = 1$$

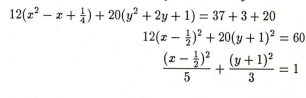

Thus $h = \tfrac{1}{2}, a; k = -1$, $a = \sqrt{5}$, $b = \sqrt{3}$, and $c = \sqrt{5 - 3} = \sqrt{2}$. We conclude that

$$\text{center, } (h, k): \qquad (\tfrac{1}{2}, -1)$$
$$\text{foci, } (h \pm c, k): \qquad (\tfrac{1}{2} \pm \sqrt{2}, -1)$$
$$\text{vertices, } (h \pm a, k): \qquad (\tfrac{1}{2} \pm \sqrt{5}, -1)$$

$$e = \frac{c}{a} = \frac{\sqrt{2}}{\sqrt{5}} = \frac{\sqrt{10}}{5}$$

**29.** Find an equation for the ellipse with vertices $(5,0)$ and $(-5,0)$ and eccentricity $\frac{3}{5}$.

**Solution:**

Since the vertices lie on the $x$-axis, the standard form for the ellipse is

$$\frac{(x-h)^2}{a^2} + \frac{(y-k)^2}{b^2} = 1$$

Since the center is the midpoint of the line segment connecting the vertices, we have $(h,k) = (0,0)$. Furthermore, since $a$ is the distance from the center to the vertices, we have $a = 5$. Also,

$$e = \frac{c}{a} = \frac{c}{5} = \frac{3}{5}$$

Therefore $c = 3$ and $b^2 = a^2 - c^2 = 25 - 9 = 16$. Finally, the equation is

$$\frac{x^2}{25} + \frac{y^2}{16} = 1$$

**31.** Find an equation of the ellipse with vertices $(3,1)$ and $(3,9)$, and minor axis of length 6.

**Solution:**

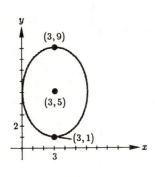

Since the vertices lie on a vertical line (see the accompanying figure), the standard form of the equation of the ellipse is

$$\frac{(x-h)^2}{b^2} + \frac{(y-k)^2}{a^2} = 1$$

Since the center of the ellipse is the midpoint of the line segment connecting the vertices, we have

$$(h,k) = \left(\frac{3+3}{2}, \frac{1+9}{2}\right) = (3,5)$$

The length of the major axis is $2a = 8$ from which we obtain $a = 4$. Furthermore, since the length of the minor axis is 6, we have $2b = 6$ or $b = 3$. Finally, the equation is

$$\frac{(x-3)^2}{9} + \frac{(y-5)^2}{16} = 1$$

**41.** A particle is traveling clockwise on the elliptical orbit given by

$$\frac{x^2}{10^2} + \frac{y^2}{5^2} = 1$$

The particle leaves the orbit at the point $(-8, 3)$ and travels in a straight line tangent to the ellipse. At which point will the particle cross the $y$-axis?

**Solution:**

To find the slope of the tangent line at $(-8, 3)$, we differentiate implicitly as follows:

$$\frac{x^2}{100} + \frac{y^2}{25} = 1$$

$$\frac{x}{50} + \frac{2yy'}{25} = 0$$

$$x + 4yy' = 0$$

$$y' = \frac{-x}{4y}$$

Thus at $(-8, 3)$ the slope is

$$m = \frac{-(-8)}{4(3)} = \frac{8}{12} = \frac{2}{3}$$

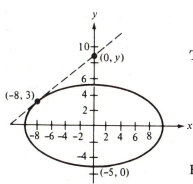

The equation of the tangent line is

$$(y - 3) = \frac{2}{3}[x - (-8)]$$

$$3y - 9 = 2x + 16$$

$$3y = 2x + 25$$

Finally, when $x = 0$,

$$3y = 25 \qquad \text{or} \qquad y = \frac{25}{3}$$

and the required point is $(0, 25/3)$.

**43.** Use Simpson's Rule with $n = 8$ to approximate the elliptic integral representing the circumference of the ellipse

$$\frac{x^2}{9} + \frac{y^2}{16} = 1$$

**Solution:**

We begin by observing that the circumferences of the ellipses

$$\frac{x^2}{9} + \frac{y^2}{16} = 1 \qquad \text{and} \qquad \frac{x^2}{16} + \frac{y^2}{9} = 1$$

are the same, and therefore, we use the formula of Example 5. Since $a^2 = 16$ and $b^2 = 9$, we have $c^2 = 7$ and $e^2 = 7/16$. Hence the circumference is given by

$$C = 4(4) \int_0^{\pi/2} \sqrt{1 - \frac{7\sin^2\theta}{16}} \, d\theta$$

Applying Simpson's Rule with $n = 8$, the endpoints of the subintervals are $\frac{n\pi}{16}$, where $n = 0, 1, 2, \ldots, 8$ and

$$C \approx 16\left(\frac{\pi}{6}\right)\left(\frac{1}{8}\right)[1 + 4(0.9916) + 2(0.9674) + 4(0.9300)$$
$$+ 2(0.8839) + 4(0.8359) + 2(0.7916) + 4(0.7610) + 0.7500]$$
$$\approx 22.10$$

**47.** Given the region bounded by the graph of

$$\frac{x^2}{4} + \frac{y^2}{1} = 1$$

Find (a) its area, (b) the volume and surface area of the solid generated by revolving the region about its major axis (prolate spheriod), and (c) the volume and surface area of the solid generated by revolving the region about its minor axis (oblate spheriod).

**Solution:**

(a)  Using the symmetry of the region we have

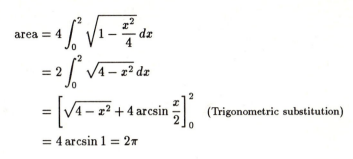

$$\text{area} = 4 \int_0^2 \sqrt{1 - \frac{x^2}{4}}\, dx$$

$$= 2 \int_0^2 \sqrt{4 - x^2}\, dx$$

$$= \left[ \sqrt{4 - x^2} + 4 \arcsin \frac{x}{2} \right]_0^2 \quad \text{(Trigonometric substitution)}$$

$$= 4 \arcsin 1 = 2\pi$$

(b)  Using the disc method and the symmetry of the region, we have

$$V = 2\pi \int_0^2 \left( \sqrt{1 - \frac{x^2}{4}} \right)^2 dx$$

$$= \frac{\pi}{2} \int_0^2 (4 - x^2)\, dx = \frac{\pi}{2} \left[ 4x - \frac{1}{3}x^3 \right]_0^2 = \frac{8\pi}{3}$$

Since $y = \sqrt{1 - \frac{x^2}{4}} = \frac{1}{2}\sqrt{4 - x^2}$, we obtain $y' = \frac{x}{2\sqrt{4 - x^2}}$.

Therefore,

$$\sqrt{1 + \left( \frac{dy}{dx} \right)^2} = \sqrt{1 + \frac{x^2}{4(4 - x^2)}}$$

$$= \frac{16 - 3x^2}{2\sqrt{4 - x^2}} = \frac{\sqrt{16 - 3x^2}}{4y}$$

and

$$S = 2(2\pi) \int_0^2 y \sqrt{1 + \left( \frac{dy}{dx} \right)^2}\, dx$$

$$= 4\pi \int_0^2 y \frac{\sqrt{16 - 3x^2}}{4y}\, dx$$

$$= \pi \int_0^2 \sqrt{4^2 - (\sqrt{3}x)^2}\, dx \quad \text{(Trigonometric substitution)}$$

$$= \frac{\pi}{2\sqrt{3}} \left[ \sqrt{3}x \sqrt{16 - 3x^2} + 16 \arcsin \left( \frac{\sqrt{3}x}{4} \right) \right]_0^2$$

$$= \frac{2\pi}{9} (9 + 4\sqrt{3}\pi) \approx 21.48$$

(c) Using the Shell Method and the symmetry of the region we have

$$V = 2(2\pi) \int_0^2 x \left( \frac{1}{2} \sqrt{4 - x^2} \right) dx$$

$$= 2\pi \left( -\frac{1}{2} \right) \int_0^2 (4 - x^2)^{1/2} (-2x) \, dx$$

$$= -\frac{2\pi}{3} \left[ (4 - x^2)^{3/2} \right]_0^2 = \frac{16\pi}{3}$$

Since $x = 2\sqrt{1 - y^2}$, we obtain $\dfrac{dx}{dy} = \dfrac{-2y}{\sqrt{1 - y^2}}$.

Therefore,

$$\sqrt{1 + \left( \frac{dx}{dy} \right)^2} = \sqrt{1 + \frac{4y^2}{1 - y^2}}$$

$$= \frac{\sqrt{1 + 3y^2}}{\sqrt{1 - y^2}} = \frac{2\sqrt{1 + 3y^2}}{x}$$

and

$$S = 2(2\pi) \int_0^1 x \sqrt{1 + \left( \frac{dx}{dy} \right)^2} \, dy$$

$$= 4\pi \int_0^1 x \, \frac{2\sqrt{1 + 3y^2}}{x} \, dy$$

$$= 8\pi \int_0^1 \sqrt{1 + 3y^2} \, dy \quad \text{(Trigonometric substitution)}$$

$$= \frac{8\pi}{2\sqrt{3}} \left[ \sqrt{3}y \sqrt{1 + 3y^2} + \ln \left| \sqrt{3}y + \sqrt{1 + 3y^2} \right| \right]_0^1$$

$$= \frac{4\pi}{3} [6 + \sqrt{3} \ln(2 + \sqrt{3})] \approx 34.69$$

**49.** Find the dimensions of the rectangle of maximum area that can be inscribed in the ellipse

$$\frac{x^2}{a^2} + \frac{y^2}{b^2} = 1$$

**Solution:**

Let $(x, y)$ be a vertex of the rectangle located on the ellipse in Quadrant I (see the accompanying figure). Solving the equation of the ellipse for $y$, we obtain

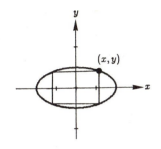

$$\frac{y^2}{b^2} = 1 - \frac{x^2}{a^2} = \frac{a^2 - x^2}{a^2}$$

$$y^2 = \frac{b^2}{a^2}(a^2 - x^2)$$

$$y = \pm \frac{b}{a}\sqrt{a^2 - x^2}$$

Therefore, the dimensions of the rectangle are:

$$\text{length} = 2x$$

$$\text{width} = 2y = \frac{2b}{a}\sqrt{a^2 - x^2}$$

The area of the rectangle is given by

$$A = lw = 2x\left[\frac{2b}{a}\sqrt{a^2 - x^2}\right] = \frac{4b}{a}(x\sqrt{a^2 - x^2})$$

and

$$\frac{dA}{dx} = \frac{4b}{a}\left[\frac{-x^2}{\sqrt{a^2 - x^2}} + \sqrt{a^2 - x^2}\right]$$

$$= \frac{4b}{a}\left[\frac{a^2 - 2x^2}{\sqrt{a^2 - x^2}}\right]$$

Thus $\dfrac{dA}{dx} = 0$ when $x = \dfrac{a}{\sqrt{2}}$ and the dimensions of the rectangle of maximum area are

$$\text{length} = 2x = \sqrt{2}a$$

$$\text{width} = \frac{2b}{a}\sqrt{a^2 - x^2} = \frac{2b}{a}\sqrt{a^2 - \frac{a^2}{2}} = \sqrt{2}b$$

## 11.3  Hyperbolas

**9.** Find the center, vertices, and foci of the hyperbola given by $y^2 - (x^2/4) = 1$ and sketch its graph, using asypmtotes as an aid.

**Solution:**

Writing the equation in standard form, we have

$$y^2 - \frac{x^2}{4} = 1$$

$$\frac{(y-0)^2}{1^2} - \frac{(x-0)^2}{2^2} = 1$$

$$\frac{(y-k)^2}{a^2} - \frac{(x-h)^2}{b^2} = 1$$

Thus $h = 0$, $k = 0$, $a = 1$, $b = 2$, and $c = \sqrt{1^2 + 2^2} = \sqrt{5}$. We conclude that

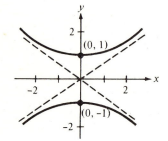

$$\begin{aligned}
\text{center, } (h,k): & \quad (0,0) \\
\text{vertices, } (h, k \pm a): & \quad (0, \pm 1) \\
\text{foci, } (h, k \pm c): & \quad (0, \pm\sqrt{5})
\end{aligned}$$

Finally, the asymptotes are given by

$$y = k \pm \frac{a}{b}(x - h) = \pm\frac{x}{2}$$

**17.** Find the center, vertices, and foci of the hyperbola given by $[(x-1)^2/4] - [(y+2)^2/1] = 1$ and sketch its graph, using asymptotes as an aid.

**Solution:**

$$\frac{(x-1)^2}{4} - \frac{(y+2)^2}{1} = 1$$

$$\frac{(x-h)^2}{a^2} - \frac{(y+k)^2}{b^2} = 1$$

Thus $h = 1$, $k = -2$, $a = 2$, $b = 1$, and $c = \sqrt{4+1} = \sqrt{5}$. We conclude that

$$
\begin{aligned}
\text{center, } (h, k): \quad & (1, -2) \\
\text{vertices, } (h \pm a, k): \quad & (-1, -2), (3, -2) \\
\text{foci, } (h \pm c, k): \quad & (1 \pm \sqrt{5}, -2)
\end{aligned}
$$

Finally, the asymptotes are given by

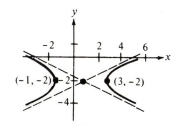

$$
y = k \pm \frac{b}{a}(x - h) = -2 \pm \frac{1}{2}(x - 1)
$$

$$
y = \frac{1}{2}x - \frac{5}{2} \quad \text{and} \quad y = -\frac{1}{2}x - \frac{3}{2}
$$

**21.** Find the center, vertices, and foci of the hyperbola given by $9x^2 - y^2 - 36x - 6y + 18 = 0$ and sketch its graph, using asymptotes as an aid.

**Solution:**

Writing the equation in standard form, we have

$$
\begin{aligned}
9x^2 - y^2 - 36x - 6y + 18 &= 0 \\
9x^2 - 36x - (y^2 + 6y) &= -18 \\
9(x^2 - 4x + 4) - (y^2 + 6y + 9) &= -18 + 36 - 9 \\
9(x - 2)^2 - (y + 3)^2 &= 9 \\
\frac{(x - 2)^2}{1^2} - \frac{(y + 3)^2}{3^2} &= 1
\end{aligned}
$$

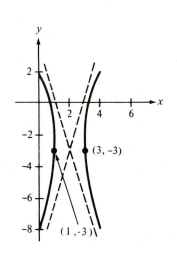

Thus $h = 2$, $k = -3$, $a = 1$, $b = 3$, and $c = \sqrt{1^2 + 3^2} = \sqrt{10}$. We conclude that

$$
\begin{aligned}
\text{center, } (h, k): \quad & (2, -3) \\
\text{vertices, } (h \pm a, k): \quad & (1, -3) \quad \text{and} \quad (3, -3) \\
\text{foci, } (h \pm c, k): \quad & (2 \pm \sqrt{10}, -3)
\end{aligned}
$$

Finally, the asymptotes are given by

$$
y = k \pm \frac{b}{a}(x - h) = -3 \pm \frac{3}{1}(x - 2)
$$

$$
y = 3x - 9 \quad \text{and} \quad y = -3x + 3
$$

**29.** Find an equation for the hyperbola with vertices $(-1, 0)$ and $(1, 0)$ and whose asymptotes are given by $y = \pm 3x$.

**Solution:**

Since the vertices lie on a horizontal line, the standard form is

$$\frac{(x-h)^2}{a^2} - \frac{(y-k)^2}{b^2} = 1$$

The center of the hyperbola lies at the midpoint of the line segment connecting the vertices. Thus

$$(h, k) = \left(\frac{1 + (-1)}{2}, \frac{0 + 0}{2}\right) = (0, 0)$$

and we have $h = 0$ and $k = 0$. Since the asymptotes are of the form

$$y = k \pm \frac{b}{a}(x - h) = \pm 3x$$

we have

$$\pm \frac{b}{a} = \pm 3 \qquad \text{or} \qquad b = 3a$$

Finally, since $a = 1$, we have $b = 3$ and the equation is

$$\frac{x^2}{1^2} - \frac{y^2}{3^2} = 1 \qquad \text{or} \qquad x^2 - \frac{y^2}{9} = 1$$

**35.** Find an equation of the hyperbola such that for any point on the hyperbola, the difference of its distances form the points $(2, 2)$ and $(10, 2)$ is 6.

**Solution:**

From the accompanying figure and the distance formula we have

$$d_1 - d_2 = 6$$

$$\sqrt{(x-2)^2 + (y_2 - 2)^2} - \sqrt{(x-10)^2 + (y-2)^2} = 6.$$

We now isolate the radicals one at a time, square each member of the resulting equation, and simplify.

$$\sqrt{(x-2)^2 + (y-2)^2} = 6 + \sqrt{(x-10)^2 + (y-2)^2}$$

$$(x-2)^2 + (y-2)^2 = 36 + 12\sqrt{(x-10)^2 + (y-2)^2}$$
$$+ (x-10)^2 + (y-2)^2$$

$$(x-2)^2 - (x-10)^2 - 36 = 12\sqrt{(x-10)^2 + (y-2)^2}$$

$$4x - 33 = 3\sqrt{(x-10)^2 + (y-2)^2}$$

$$16x^2 - 264x + 1089 = 9(x^2 - 20x + 100 + y^2 - 4y + 4)$$

$$7x^2 - 9y^2 - 84x + 36y + 153 = 0$$

$$7(x^2 - 12x) - 9(y^2 - 4y) = -153$$

$$7(x-6)^2 - 9(y-2)^2 = 63$$

$$\frac{(x-6)^2}{9} - \frac{(y-2)^2}{7} = 1$$

**39.** Find the volume and the surface area of the solid generated by revolving the region bounded by the graphs of $x^2 - y^2 = 1$, $y = 0$, and $x = 2$ about the $x$-axis.

**Solution:**

Using the Disc Method to find the volume, we have

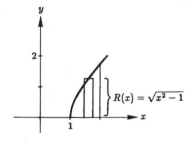

$$V = \pi \int_1^2 \left(\sqrt{x^2 - 1}\right)^2 dx$$

$$= \pi \int_1^2 (x^2 - 1)\, dx$$

$$= \pi \left(\frac{x^3}{3} - x\right)\Big]_1^2 = \pi\left[\left(\frac{8}{3} - 2\right) - \left(\frac{1}{3} - 1\right)\right] = \frac{4\pi}{3}$$

To find the surface area we begin by differentiating the equation for the hyperbola implicitly to obtain $y'$.

$$x^2 - y^2 = 1$$

$$2x - 2y\, y' = 0 \implies y' = \frac{x}{y}$$

Therefore,

$$\sqrt{1 + (y')^2} = \sqrt{1 + \left(\frac{x}{y}\right)^2}$$

$$= \sqrt{\frac{y^2 + x^2}{y^2}}$$

$$= \frac{\sqrt{(x^2 - 1) + x^2}}{y} \qquad \text{(equation of the hyperbola)}$$

$$= \frac{\sqrt{2x^2 - 1}}{y}.$$

Thus, the surface area is

$$S = 2\pi \int_1^2 y\sqrt{1 + (y')^2}\, dx$$

$$= 2\pi \int_1^2 y\left(\frac{\sqrt{2x^2 - 1}}{y}\right) dx$$

$$= \sqrt{2}\pi \int_1^2 \sqrt{(\sqrt{2}x)^2 - 1}\sqrt{2}\, dx$$

$$= \frac{\sqrt{2}\pi}{2}\left[\sqrt{2}x\sqrt{2x^2 - 1} - \ln\left|\sqrt{2}x + \sqrt{2x^2 - 1}\right|\right]_1^2$$

$$= \pi(2\sqrt{7} - 1) + \frac{\sqrt{2}\pi}{2}\ln\left(\frac{\sqrt{2} + 1}{2\sqrt{2} + \sqrt{7}}\right) \approx 11.66$$

**43.** Show that an equation of the tangent line to

$$\frac{x^2}{a^2} - \frac{y^2}{b^2} = 1$$

at the point $(x_0, y_0)$ is

$$\frac{x_0}{a^2}x - \frac{y_0}{b^2}y = 1$$

**Solution:**

To find the slope of the tangent line at the point $(x_0, y_0)$, we differentiate the equation for the hyperbola implicitly.

$$\frac{x^2}{a^2} - \frac{y^2}{b^2} = 1$$

$$\frac{2x}{a^2} - \frac{2y\, y'}{b^2} = 0 \implies y' = \frac{b^2 x}{a^2 y}$$

Therefore, an equation of the tangent line is

$$y - y_0 = \frac{b^2 x_0}{a^2 y_0}(x - x_0)$$
$$a^2 y_0 y - a^2 y_0^2 = b^2 x_0 x - b^2 x_0^2$$
$$b^2 x_0^2 - a^2 y_0^2 = b^2 x_0 x - a^2 y_0 y$$
$$a^2 b^2 = b^2 x_0 x - a^2 y_0 y \qquad \text{(equation of the hyperbola)}$$
$$1 = \frac{x_0 x}{a^2} - \frac{y_0 y}{b^2}$$

## 11.4   Rotation and the General Second-Degree Equation

**3.** Rotate the axes to eliminate the $xy$ term in the equation $9x^2 + 24xy + 16y^2 + 90x - 130y = 0$. Sketch the graph of the resulting equation, showing both sets of axes.

**Solution:**

From the equations

$$9x^2 + 24xy + 16y^2 + 90x - 130y = 0$$
$$Ax^2 + Bxy + Cy^2 + Dx + Ey + F = 0$$

we have $A = 9$, $B = 24$, $C = 16$, $D = 90$, $E = -130$, and $F = 0$. Thus

$$\cot 2\theta = \frac{A - C}{B} = \frac{9 - 16}{24} = \frac{-7}{24} \qquad \theta \approx 53.13°$$

From the identity

$$\cot 2\theta = \frac{\cot^2 \theta - 1}{2 \cot \theta}$$

we have

$$\frac{\cot^2 \theta - 1}{2 \cot \theta} = \frac{-7}{24}$$
$$24 \cot^2 \theta - 24 = -14 \cot \theta$$
$$12 \cot^2 \theta + 7 \cot \theta - 12 = 0$$
$$(4 \cot \theta - 3)(3 \cot \theta + 4) = 0$$
$$\cot \theta = \frac{3}{4} \qquad \text{or} \qquad -\frac{4}{3}$$

Since $0 < \theta < 90°$, we choose $\cot\theta = \frac{3}{4}$ and conclude that $\sin\theta = \frac{4}{5}$ and $\cos\theta = \frac{3}{5}$. Therefore, using the equations

$$x = x'\cos\theta - y'\sin\theta = \frac{3}{5}x' - \frac{4}{5}y'$$

$$y = x'\sin\theta + y'\cos\theta = \frac{4}{5}x' + \frac{3}{5}y'$$

and

$$9x^2 + 24xy + 16y^2 + 90x - 130y = 0$$

we have

$$9\left(\frac{3}{5}x' - \frac{4}{5}y'\right)^2 + 24\left(\frac{3}{5}x' - \frac{4}{5}y'\right)\left(\frac{4}{5}x' + \frac{3}{5}y'\right)$$

$$+ 16\left(\frac{4}{5}x' + \frac{3}{5}y'\right)^2 + 90\left(\frac{3}{5}x' - \frac{4}{5}y'\right) - 130\left(\frac{4}{5}x' + \frac{3}{5}y'\right) = 0$$

After expanding and combining like terms we have

$$25(x')^2 - 50x' - 150y' = 0$$

$$(x')^2 - 2x' - 6y' = 0$$

$$(x' - 1)^2 = 4\left(\frac{3}{2}\right)\left(y' + \frac{1}{6}\right)$$

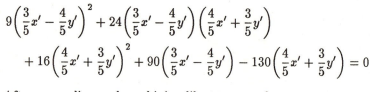

5. Rotate the axes to eliminate the $xy$ term in the equation $x^2 - 10xy + y^2 + 1 = 0$. Sketch the graph of the resulting equation, showing both sets of axes.

**Solution:**

From the equations

$$x^2 - 10xy + y^2 + 1 = 0$$

$$Ax^2 + Bxy + Cy^2 + Dx + Ey + F = 0$$

we have $A = 1$, $B = -10$, $C = 1$, $D = 0$, $E = 0$, and $F = 1$. Thus

$$\cot 2\theta = \frac{1-1}{-10} = 0$$

and $\theta = 45°$. Now, since $\sin\theta = \sqrt{2}/2$ and $\cos\theta = \sqrt{2}/2$, we have

$$x = x'\cos\theta - y'\sin\theta = \frac{\sqrt{2}}{2}x' - \frac{\sqrt{2}}{2}y'$$

$$y = x'\sin\theta + y'\cos\theta = \frac{\sqrt{2}}{2}x' + \frac{\sqrt{2}}{2}y'$$

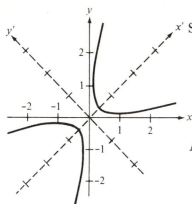

x' Substitution into $x^2 - 10xy + y^2 + 1 = 0$ yields

$$\left(\frac{\sqrt{2}}{2}x' - \frac{\sqrt{2}}{2}y'\right)^2 - 10\left(\frac{\sqrt{2}}{2}x' - \frac{\sqrt{2}}{2}y'\right)\left(\frac{\sqrt{2}}{2}x' + \frac{\sqrt{2}}{2}y'\right)$$
$$+ \left(\frac{\sqrt{2}}{2}x' + \frac{\sqrt{2}}{2}y'\right)^2 + 1 = 0$$

After expanding and combining like terms we have

$$-4(x')^2 + 6(y')^2 + 1 = 0 \qquad \text{or} \qquad \frac{(x')^2}{\frac{1}{4}} - \frac{(y')^2}{\frac{1}{6}} = 1$$

**9.** Rotate the axes to eliminate the $xy$ term in the equation $5x^2 - 2xy + 5y^2 - 12 = 0$. Sketch the graph of the resulting equation, showing both sets of axes.

**Solution:**

From the equations

$$5x^2 - 2xy + 5y^2 - 12 = 0$$
$$Ax^2 + Bxy + Cy^2 + Dx + Ey + F = 0$$

we have $A = 5, B = -2, C = 5, D = 0, E = 0,$ and $F = -12$. Thus

$$\cot 2\theta = \frac{A - C}{B} = 0 \qquad \text{or} \qquad 2\theta = \frac{\pi}{2} \quad \text{and} \quad \theta = \frac{\pi}{4}$$

Therefore, $\sin \theta = \cos \theta = \frac{\sqrt{2}}{2}$ and

$$x = x' \cos \theta - y' \sin \theta = \frac{\sqrt{2}}{2}x' - \frac{\sqrt{2}}{2}y'$$
$$y = x' \sin \theta + y' \cos \theta = \frac{\sqrt{2}}{2}x' + \frac{\sqrt{2}}{2}y'$$

Substituting into $5x^2 - 2xy + 5y^2 - 12 = 0$ yields

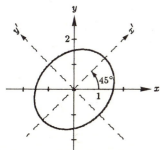

$$5\left(\frac{\sqrt{2}}{2}x' - \frac{\sqrt{2}}{2}y'\right)^2 - 2\left(\frac{\sqrt{2}}{2}x' - \frac{\sqrt{2}}{2}y'\right)\left(\frac{\sqrt{2}}{2}x' + \frac{\sqrt{2}}{2}y'\right)$$
$$+ 5\left(\frac{\sqrt{2}}{2}x' + \frac{\sqrt{2}}{2}y'\right)^2 - 12 = 0$$

After expanding and combining like terms we have

$$4(x')^2 + 6(y')^2 - 12 = 0$$

$$\frac{(x')^2}{3} + \frac{(y')^2}{2} = 1$$

## Review Exercises for Chapter 11

**15.** Analyze the equation $3x^2 + 2y^2 - 12x + 12y + 29 = 0$ and sketch its graph.

**Solution:**

In standard form the equation is

$$3x^2 + 2y^2 - 12x + 12y + 29 = 0$$
$$3(x^2 - 4x) + 2(y^2 + 6y) = -29$$
$$3(x^2 - 4x + 4) + 2(y^2 + 6y + 9) = -29 + 12 + 18$$
$$3(x - 2)^2 + 2(y + 3)^2 = 1$$
$$\frac{(x - 2)^2}{(1/\sqrt{3})^2} + \frac{(y + 3)^2}{(1/\sqrt{2})^2} = 1$$

Therefore the graph is an ellipse where $h = 2, k = -3, a = 1/\sqrt{2}, b = 1/\sqrt{3}$, and $c = \sqrt{(1/2) - (1/3)} = 1/\sqrt{6}$.

We conclude that

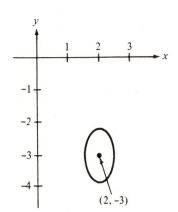

(2, –3)

$$\text{center, } (h, k): \quad (2, -3)$$

$$\text{vertices, } (h, k \pm a): \quad \left(2, -3 \pm \frac{\sqrt{2}}{2}\right)$$

$$\text{foci, } (h, k \pm c): \quad \left(2, -3 \pm \frac{\sqrt{6}}{6}\right)$$

**27.** Find an equation of the parabola whose vertex is at $(0,0)$ and whose focus is at $(1,1)$.

**Solution:**

Since the vertex and focus lie on the line $y = x$, we let $\theta = 45°$. Then in the $x'y'$-coordinate system, the vertex lies at $(0,0)$ and the focus is at $(\sqrt{2}, 0)$. Now $p$, the directed distance from the vertex to the focus, is $p = \sqrt{2}$, and we have

$$(y' - 0)^2 = 4\sqrt{2}(x' - 0) \qquad \text{or} \qquad (y')^2 = 4\sqrt{2}x'$$

Since $\theta = 45°$, we have

$$x' = x\cos\theta + y\sin\theta = \frac{1}{\sqrt{2}}(x+y)$$

$$y' = -x\sin\theta + y\cos\theta = \frac{1}{\sqrt{2}}(-x+y)$$

and

$$\left[\frac{1}{\sqrt{2}}(-x+y)\right]^2 = 4\sqrt{2}\left(\frac{1}{\sqrt{2}}\right)(x+y)$$

$$\frac{x^2 - 2xy + y^2}{2} = 4x + 4y$$

$$x^2 - 2xy + y^2 = 8x + 8y$$

$$x^2 - 2xy + y^2 - 8x - 8y = 0$$

**31.** Find an equation of the hyperbola with foci $(\pm4, 0)$ where the absolute value of the difference of the distances from a point on the hyperbola to the foci is 4.

**Solution:**

We first observe that the center of the hyperbola is at the origin and therefore $(h, k) = (0, 0)$. Since the foci are $(\pm4, 0)$ we know that the transverse axis is horizontal and $c = 4$. Also, from the definition of a hyperbola we know that the absolute value of the difference of the distances from a point on the hyperbola to the foci is $2a$. Thus $2a = 4$ or $a = 2$. Since $b^2 = c^2 - a^2$, we have $b^2 = 16 - 4 = 12$. Therefore, an equation of the hyperbola is

$$\frac{x^2}{4} - \frac{y^2}{12} = 1$$

**38.** The ellipse $(x^2/a^2) + (y^2/b^2) = 1$ is revolved about its minor axis to form an oblate spheriod. Show that the volume of the spheroid is $\frac{4}{3}\pi a^2 b$ and its surface area is $2\pi a^2 + \pi(b^2/e) \ln\left[(1+e)/(1-e)\right]$.

**Solution:**

Solving for $x$ as a function of $y$, we have

$$x = \frac{a}{b}\sqrt{b^2 - y^2}$$

and

$$\frac{dx}{dy} = \left(\frac{a}{b}\right)\frac{-y}{\sqrt{b^2 - y^2}}$$

Revolving about the $y-$axis and using the disc method, we have

$$V = \pi \int_{-b}^{b} \left(\frac{a}{b}\sqrt{b^2 - y^2}\right)^2 dy = \frac{\pi a^2}{b^2}\int_{-b}^{b}(b^2 - y^2)\,dy$$

$$= \frac{\pi a^2}{b^2}\left[b^2 y - \frac{y^3}{3}\right]_{-b}^{b}$$

$$= \frac{\pi a^2}{b^2}\left(b^3 - \frac{b^3}{3} + b^3 - \frac{b^3}{3}\right) = \frac{4\pi a^2 b}{3}$$

$$S = 4\pi \int_{0}^{b} x\sqrt{1 + \left(\frac{dx}{dy}\right)^2}\,dy$$

$$= 4\pi \int_{0}^{b} \frac{a}{b}\sqrt{b^2 - y^2}\sqrt{1 + \frac{a^2 y^2}{b^2(b^2 - y^2)}}\,dy$$

$$= \frac{4\pi a}{b^2}\int_{0}^{b}\sqrt{b^4 + (a^2 - b^2)y^2}\,dy = \frac{4\pi a}{b^2}\int_{0}^{b}\sqrt{b^4 + c^2 y^2}\,dy$$

$$= \frac{4\pi a}{b^2}\left[\frac{1}{2c}\left(cy\sqrt{b^4 + c^2 y^2} + b^4 \ln\left|cy + \sqrt{b^4 + c^2 y^2}\right|\right)\right]_{0}^{b}$$

$$= \frac{2\pi a}{b^2 c}\left[bc\sqrt{b^4 + c^2 b^2} + b^4 \ln\left|bc + \sqrt{b^4 + b^2 c^2}\right| - b^4 \ln b^2\right]$$

$$= \frac{2\pi a}{b^2 c}\left[bc\sqrt{b^4 + (a^2 - b^2)b^2} + b^4 \ln\left(\frac{bc + \sqrt{b^4 + (a^2 - b^2)b^2}}{b^2}\right)\right]$$

$$= \frac{2\pi a}{b^2 c}\left[ab^2 c + b^4 \ln\left(\frac{cb + ab}{b^2}\right)\right] = 2\pi a^2 + \frac{2\pi ab^2}{c}\ln\left(\frac{a+c}{b}\right)$$

$$= 2\pi a^2 + \frac{\pi b^2}{c/a}\ln\frac{(a+c)^2}{b^2}$$

Note that

$$\frac{(a+c)^2}{a^2-c^2} = \frac{(a+c)(a+c)}{(a+c)(a-c)} = \frac{a+c}{a-c} = \frac{1-(c/a)}{1-(c/a)} = \frac{1+e}{1-e}$$

Thus

$$S = 2\pi a^2 + \pi\frac{b^2}{e}\ln\left(\frac{1+e}{1-e}\right)$$

43. Consider a fire truck with a water tank 16 feet long whose vertical cross section are ellipses as described by the equation

$$\frac{x^2}{16} + \frac{y^2}{9} = 1$$

Find the depth of water in the tank if it is $\frac{3}{4}$ full (by volume) and the truck is on level ground.

**Solution:**

The truck will be carrying $\frac{3}{4}$ of its total capacity when the water covers $\frac{3}{4}$ of the area of a cross section of the tank. One-half of this area will be below the major axis and $\frac{1}{4}$ above the major axis of the ellipse

$$\frac{x^2}{16} + \frac{y^2}{9} = 1$$

The total area is given by

$$A = \pi a b = \pi(4)(3) = 12\pi$$

$$\frac{1}{4}A = 3\pi$$

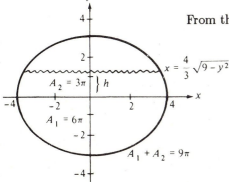

From the accompanying figure we have

$$\int_0^h 2\left(\frac{4}{3}\sqrt{9-y^2}\right)dy = 3\pi$$

$$\int_0^y \sqrt{3^2-y^2}\,dy = \frac{9\pi}{8}$$

$$\frac{1}{2}\left[y\sqrt{9-y^2} + 9\arcsin\frac{y}{3}\right]_0^h = \frac{9\pi}{8}$$

$$h\sqrt{9-h^2} + 9\arcsin\frac{h}{3} = \frac{9\pi}{4}$$

Find $h$ such that

$$f(h) = \sqrt{9 - h^2} + \arcsin\left(\frac{h}{3}\right) - \frac{9\pi}{4} = 0$$

By Newton's Method, with an initial estimate of $h = 1$, we have

$$h_{n+1} = h_n - \frac{f(h_n)}{f'(h_n)}$$

$$= h_n - \frac{h_n\sqrt{9 - h_n^2} + 9\arcsin(h_n/3) - (9\pi/4)}{2\sqrt{9 - h_n^2}}$$

| $n$ | 1 | 2 | 3 | 4 |
|-----|-----|-----|-----|-----|
| $h_n$ | 1.0000 | 1.2089 | 1.2119 | 1.2119 |

We conclude that $h \approx 1.212$ and therefore the total height of the water in the tank is $3 + 1.212 = 4.212$ ft.

# 12 PLANE CURVES, PARAMETRIC EQUATIONS, AND POLAR COORDINATES

## 12.1 Plane Curves and Parametric Equations

**7.** Sketch the curve represented by the parametric equations $x = t^3$, $y = t^2/2$ and write the corresponding rectangular equation by eliminating the parameter.

**Solution:**

Since $x = t^3$ and $y = t^2/2$, we have

$$t = x^{1/3} \quad \text{or} \quad y = \frac{1}{2}(x^{1/3})^2 = \frac{1}{2}x^{2/3}$$

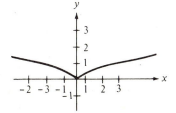

**19.** Sketch the curve represented by the parametric equations $x = \cos\theta$, $y = 2\sin^2\theta$ and write the corresponding rectangular equation by eliminating the parameter.

**Solution:**

We first observe that $-1 \le x \le 1$ and $y \ge 0$.

$$x = \cos\theta \qquad y = 2\sin^2\theta$$
$$x^2 = \cos^2\theta \qquad \frac{y}{2} = \sin^2\theta$$

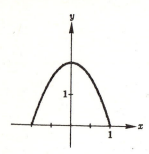

Therefore

$$x^2 + \frac{y}{2} = \cos^2 \theta + \sin^2 \theta = 1$$

$$y = 2 - 2x^2, \qquad -1 \le x \le 1$$

**25.** Sketch the curve represented by the parametric equations $x = 4\sec\theta$, $y = 3\tan\theta$ and write the corresponding rectangular equation by eliminating the parameter.

**Solution:**

Since the parametric equations involve secants and tangents, we consider the identity $\sec^2\theta - \tan^2\theta = 1$. Therefore we write

$$\frac{x}{4} = \sec\theta \quad \text{and} \quad \frac{y}{3} = \tan\theta$$

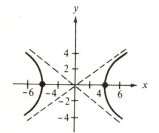

or

$$\frac{x^2}{16} - \frac{y^2}{9} = \sec^2\theta - \tan^2\theta = 1$$

The graph of this equation is a hyperbola centered at the origin with vertices $(\pm 4, 0)$.

**37.** Sketch the curve (a prolate cycloid) represented by $x = \theta - \frac{3}{2}\sin\theta$, $y = 1 - \frac{3}{2}\cos\theta$.

**Solution:**

Using the point-by-point plotting method, we have the following table:

| $\theta$ | 0 | $\dfrac{\pi}{6}$ | $\dfrac{\pi}{4}$ | $\dfrac{\pi}{3}$ | $\dfrac{\pi}{2}$ | $\dfrac{2\pi}{3}$ | $\pi$ | $\dfrac{3\pi}{2}$ | $\dfrac{5\pi}{3}$ | $\dfrac{11\pi}{6}$ |
|---|---|---|---|---|---|---|---|---|---|---|
| $x$ | 0 | $-0.23$ | $-0.28$ | $-0.25$ | 0.07 | 0.795 | 3.14 | 6.21 | 6.54 | 6.51 |
| $y$ | $-1/2$ | $-0.30$ | $-0.06$ | 0.25 | 1.00 | 1.75 | 2.50 | 1.00 | 0.25 | $-0.30$ |

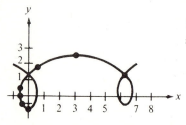

Using the values in the table, we make the accompanying sketch.

**43.** Eliminate the parameter from the equations $x = h + a\sec\theta$ and $y = k + b\tan\theta$ to obtain the standard form of the equation of a hyperbola.

**Solution:**

$$x = h + a\sec\theta \qquad\qquad y = k + b\tan\theta$$

$$\frac{x - h}{a} = \sec\theta \qquad\qquad \frac{y - k}{b} = \tan\theta$$

$$\frac{(x - h)^2}{a^2} = \sec^2\theta \qquad\qquad \frac{(y - k)^2}{b^2} = \tan^2\theta$$

$$\frac{(x - h)^2}{a^2} - \frac{(y - k)^2}{b^2} = \sec^2\theta - \tan^2\theta = 1$$

$$\frac{(x - h)^2}{a^2} - \frac{(y - k)^2}{b^2} = 1$$

**51.** Find a set of parametric equation for the hyperbola with vertices at $(\pm 4, 0)$, and foci at $(\pm 5, 0)$.

**Solution:**

The midpoint of the line segment joining the vertices is the center of the hyperbola. Therefore, $(h, k) = (0, 0)$. The transverse axis of the hyperbola is horizontal with $a = 4$ and $c = 5$. Since $b^2 = c^2 - a^2$, we have $b = 3$. Using the result of Exercise 43 a set of parametric equations are

$$x = h + a\sec\theta, \quad y = k + b\tan\theta$$

or

$$x = 4\sec\theta, \quad y = 3\tan\theta$$

## 12.2    Parametric Equations and Calculus

1. For the parametric equations $x = 2t$, $y = 3t - 1$, find $dy/dx$ and $d^2y/dx^2$ and evaluate there two derivatives when $t = 3$.

**Solution:**

Since $x = 2t$ and $y = 3t - 1$, we have

$$\frac{dy}{dx} = \frac{dy/dt}{dx/dt} = \frac{3}{2} \quad \text{for all values of } t$$

and

$$\frac{d^2y}{dx^2} = \frac{d[dy/dx]/dt}{dx/dt} = \frac{0}{2} = 0 \quad \text{for all values of } t.$$

7. For the parametric equations $x = 2 + \sec\theta$ and $y = 1 + 2\tan\theta$, find $dy/dx$ and $d^2y/dx^2$ and evaluate these two derivatives when $\theta = \pi/6$.

**Solution:**

Since $x = 2 + \sec\theta$ and $y = 1 + 2\tan\theta$, we have

$$\frac{dy}{dx} = \frac{dy/d\theta}{dx/d\theta} = \frac{2\sec^2\theta}{\sec\theta\tan\theta} = \frac{2\sec\theta}{\tan\theta} = 2\csc\theta$$

$$\frac{d^2y}{dx^2} = \frac{d[dy/dx]/d\theta}{dx/d\theta} = \frac{-2\csc\theta\cot\theta}{\sec\theta\tan\theta} = -2\cot^3\theta$$

At $\theta = \pi/6$,

$$\frac{dy}{dx} = 2\csc\frac{\pi}{6} = 2(2) = 4$$

$$\frac{d^2y}{dx^2} = -2\cot^3\frac{\pi}{6} = -2(\sqrt{3})^3 = -6\sqrt{3}$$

15. Find an equation of the tangent line to the graph of
$x = 2\cot\theta$, $y = 2\sin^2\theta$ at $\theta = \dfrac{\pi}{4}$.

**Solution:**

$$\frac{dy}{dx} = \frac{dy/d\theta}{dx/d\theta} = \frac{4\sin\theta\cos\theta}{-2\csc^2\theta} = -2\sin^3\theta\cos\theta$$

At $\theta = \dfrac{\pi}{4}$,

$$\frac{dy}{dx} = -2\sin^3\frac{\pi}{4}\cos\frac{\pi}{4} = -2\left(\frac{\sqrt{2}}{2}\right)^3\left(\frac{\sqrt{2}}{2}\right) = -\frac{1}{2}$$

and

$$x = 2\cot\frac{\pi}{4} = 2$$
$$y = 2\sin^2\frac{\pi}{4} = 1$$

Therefore the equation of the tangent line is

$$y - 1 = -\frac{1}{2}(x - 2)$$
$$2y - 2 = -x + 2$$
$$x + 2y - 4 = 0$$

19. Find all points (if any) of horizontal and vertical tangency on the graph of $x = 1 - t$, $y = t^3 - 3t$.

**Solution:**

Since $\dfrac{dy}{dx} = \dfrac{dy/dt}{dx/dt} = \dfrac{3t^2 - 3}{-1} = 3 - 3t^2$
the horizontal tangents occur when

$$3 - 3t^2 = 0 \qquad \text{or} \qquad t = \pm 1$$

The corresponding points are $(0, -2)$, and $(2, 2)$. Since $dy/dx$ is never undefined, there are no points of vertical tangency.

**29.** Find the length of the arc given by $x = e^{-t} \cos t$, $y = e^{-t} \sin t$ for $0 \le t \le \pi/2$.

**Solution:**

$$x = e^{-t} \cos t$$

$$\frac{dx}{dt} = -e^{-t}(\sin t + \cos t)$$

$$\left(\frac{dx}{dt}\right)^2 = e^{-2t}(\sin^2 t + 2 \sin t \cos t + \cos^2 t)$$

$$= e^{-2t}(1 + \sin 2t)$$

$$y = e^{-t} \sin t$$

$$\frac{dy}{dt} = e^{-t}(\cos t - \sin t)$$

$$\left(\frac{dy}{dt}\right)^2 = e^{-2t}(\cos^2 t - 2 \sin t \cos t + \sin^2 t)$$

$$= e^{-2t}(1 - \sin 2t)$$

$$\left(\frac{dx}{dt}\right)^2 + \left(\frac{dy}{dt}\right)^2 = 2e^{-2t}$$

Therefore,

$$s = \int_0^{\pi/2} \sqrt{\left(\frac{dx}{dt}\right)^2 + \left(\frac{dy}{dt}\right)^2}\, dt = \int_0^{\pi/2} \sqrt{2e^{-2t}}\, dt$$

$$= -\sqrt{2} \int_0^{\pi/2} e^{-t}(-1)\, dt = -\sqrt{2}\left[e^{-t}\right]_0^{\pi/2} = \sqrt{2}(1 - e^{-\pi/2}) \approx 1.12$$

**35.** Find the perimeter of the hypocycloid $x = a \cos^3 \theta$, $y = a \sin^3 \theta$.

**Solution:**

Since $dx/d\theta = -3a \cos^2 \theta \sin \theta$ and $dy/d\theta = 3a \sin^2 \theta \cos \theta$, then from the accompanying figure, we find the length of the

hypocycloid to be

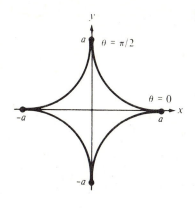

$$s = 4 \int_0^{\pi/2} \sqrt{\left(\frac{dx}{d\theta}\right)^2 + \left(\frac{dy}{d\theta}\right)^2}\, d\theta$$

$$= 4 \int_0^{\pi/2} \sqrt{9a^2 \cos^4\theta \sin^2\theta + 9a^2 \sin^4\theta \cos^2\theta}\, d\theta$$

$$= 4 \int_0^{\pi/2} \sqrt{9a^2 \sin^2\theta \cos^2\theta (\cos^2\theta + \sin^2\theta)}\, d\theta$$

$$= 4 \int_0^{\pi/2} 3a \sin\theta \cos\theta\, d\theta = 12a \left[\frac{\sin^2\theta}{2}\right]_0^{\pi/2} = 6a$$

**45.** A portion of a sphere is removed by a circular cone with its vertex at the center of the sphere. Find the surface area removed from the sphere if the vertex of the cone forms an angle of $2\theta$.

**Solution:**

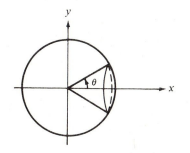

We represent the circle by

$$x = f(\phi) = r \cos\phi \qquad\qquad y = g(\phi) = r \sin\phi$$
$$\frac{dx}{d\phi} = -r \sin\phi \qquad\qquad \frac{dy}{d\phi} = r \cos\phi$$

From the integrals of Theorem 12.3 and the accompanying figure, we have

$$S = 2\pi \int_0^\theta g(\phi) \sqrt{\left(\frac{dx}{d\phi}\right)^2 + \left(\frac{dy}{d\phi}\right)^2}\, d\phi$$

$$= 2\pi \int_0^\theta r \sin\phi \sqrt{r^2 \sin^2\phi + r^2 \cos^2\phi}\, d\phi$$

$$= 2\pi r^2 \int_0^\theta \sin\phi\, d\phi = -2\pi r^2 \left[\cos\phi\right]_0^\theta = 2\pi r^2 (1 - \cos\theta)$$

**49.** Sketch the graph of the cissoid $x = 2\sin^2\theta$, $y = 2\sin^2\theta\tan\theta$ and find the area of the region enclosed between the curve and its asymptote in the first quadrant.

**Solution:**

Using values from the following table, we make the accompanying sketch

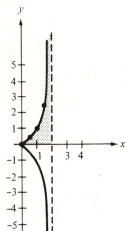

| $\theta$ | 0 | $\dfrac{\pi}{6}$ | $\dfrac{\pi}{4}$ | $\dfrac{\pi}{3}$ | $\dfrac{\pi}{2}$ |
|---|---|---|---|---|---|
| $x$ | 0 | $\dfrac{1}{2}$ | 1 | $\dfrac{3}{2}$ | 2 |
| $y$ | 0 | $\dfrac{1}{2\sqrt{3}}$ | 1 | $\dfrac{3\sqrt{3}}{2}$ | Undefined |

From Exercise 46, we have the following convergent improper integral

$$A = \int_0^2 y\,dx = \int_0^{\pi/2} y\frac{dx}{d\theta}\,d\theta$$

$$= \int_0^{\pi/2} 2\sin^2\theta\tan\theta(4\sin\theta\cos\theta)\,d\theta$$

$$= 8\int_0^{\pi/2} \sin^4\theta\,d\theta$$

$$= 8\int_0^{\pi/2} \left(\frac{1-\cos2\theta}{2}\right)^2 d\theta$$

$$= 2\int_0^{\pi/2} (1 - 2\cos2\theta + \cos^2 2\theta)\,d\theta$$

$$= 2\int_0^{\pi/2} \left(1 - 2\cos2\theta + \frac{1+\cos4\theta}{2}\right) d\theta$$

$$= 2\int_0^{\pi/2} \left(\frac{3}{2} - 2\cos2\theta + \frac{1}{2}\cos4\theta\right) d\theta$$

$$= 2\left[\frac{3}{2}\theta - \sin2\theta + \frac{1}{8}\sin4\theta\right]_0^{\pi/2} = 2\left(\frac{3\pi}{4}\right) = \frac{3\pi}{2}$$

## 12.3   Polar Coordinates and Polar Graphs

**7.** Plot the point $(\sqrt{2}, 2.36)$ in polar coordinates and find its corresponding rectangular coordinates.

**Solution:**

Using the conversion from polar to rectangular coordinates, we have

$$x = r\cos\theta = \sqrt{2}\cos(2.36) \approx -1.004$$
$$y = r\sin\theta = \sqrt{2}\sin(2.36) \approx 0.996$$

Therefore the rectangular coordinates are $(-1.004, 0.996)$.

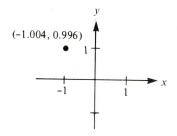

(−1.004, 0.996)

**11.** Find two sets of polar coordinates for the point $(-3, 4)$ given in rectangular coordinates, using $0 \le \theta < 2\pi$.

**Solution:**

First,

$$r = \pm\sqrt{x^2 + y^2} = \pm\sqrt{(-3)^2 + 4^2} = \pm 5$$
$$\tan\theta = -\frac{4}{3} \quad \text{and} \quad \arctan\left(-\frac{4}{3}\right) \approx 0.9273$$

Since $(-3, 4)$ lies in the second quadrant, we let $\theta = \pi - 0.9273 = 2.214$. Thus one polar representation is

$$(5, 2.214) \quad (r > 0, \; 0 \le \theta < 2\pi)$$

To obtain the second representation, we increase $\theta$ by $\pi$ radians to obtain

$$(-5, 5.356) \quad (r < 0, \; 0 \le \theta < 2\pi).$$

19. Find a polar equation of the graph having the rectangular equation $x^2 + y^2 - 2ax = 0$.

**Solution:**

Since   $x = r \cos \theta,\ y = r \sin \theta$, and $x^2 + y^2 = r^2$, we have

$$x^2 + y^2 - 2ax = 0$$
$$r^2 - 2ar \cos \theta = 0$$
$$r(r - 2a \cos \theta) = 0$$

we disregard the solution $r = 0$ since its graph is only the pole and the pole is included in the graph of $r - 2a \cos \theta$. The required solution is

$$r - 2a \cos \theta = 0$$
$$r = 2a \cos \theta$$

25. Find a polar equation of the graph having the rectangular equation $3x - y + 2 = 0$.

**Solution:**

Since $x = r \cos \theta$ and $y = r \sin \theta$, we have

$$3x - y + 2 = 0$$
$$3(r \cos \theta) - r \sin \theta + 2 = 0$$
$$r(3 \cos \theta - \sin \theta) = -2$$
$$r = \frac{-2}{3 \cos \theta - \sin \theta}$$

31. Find a polar equation of the graph having the rectangular equation $x^2 - 4ay - 4a^2 = 0$.

**Solution:**

Since $x = r \cos \theta$ and $y = r \sin \theta$, we have

$$x^2 - 4ay - 4a^2 = 0$$
$$r^2 \cos^2 \theta - 4ar \sin \theta - 4a^2 = 0$$
$$r^2(1 - \sin^2 \theta) - 4ar \sin \theta - 4a^2 = 0$$
$$r^2 = r^2 \sin \theta + 4ar \sin \theta + 4a^2$$

$$r^2 = (r\sin\theta + 2a)^2$$
$$r = \pm(r\sin\theta + 2a)$$
$$r = \frac{2a}{1 - \sin\theta}$$
or $\quad r = \dfrac{-2a}{1 + \sin\theta}$

**41.** Find a rectangular equation having the polar equation

$$r = \frac{6}{2 - 3\sin\theta}$$

**Solution:**

Since $r^2 = x^2 + y^2$ and $y = r\sin\theta$ we have

$$r = \frac{6}{2 - 3\sin\theta}$$
$$2r - 3r\sin\theta = 6$$
$$2\sqrt{x^2 + y^2} - 3y = 6$$
$$2\sqrt{x^2 + y^2} = 3y + 6$$
$$4(x^2 + y^2) = 9y^2 + 36y + 36$$
$$4x^2 - 5y^2 - 36y - 36 = 0$$

**49.** Sketch the graph of the polar equation $r = \sin\theta$, and indicate any symmetry possessed by the graph.

**Solution:**

Since the sine is an odd function, we replace $(r, \theta)$ by $(-r, -\theta)$.

$$-r = \sin(-\theta) = -\sin\theta \implies r = \sin\theta$$

Since the substitution produced an equivalent equation, the curve is symmetric with respect to the line $\theta = \pi/2$. This means that we need only use $\theta$ values from the first and fourth quadrants as shown in the following table.

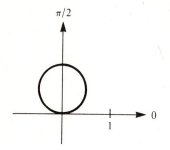

| $\theta$ | $\dfrac{-\pi}{2}$ | $\dfrac{-\pi}{3}$ | $\dfrac{-\pi}{4}$ | $\dfrac{-\pi}{6}$ | $0$ | $\dfrac{\pi}{6}$ | $\dfrac{\pi}{4}$ | $\dfrac{\pi}{3}$ | $\dfrac{\pi}{2}$ |
|---|---|---|---|---|---|---|---|---|---|
| $r$ | $-1$ | $\dfrac{-\sqrt{3}}{2}$ | $\dfrac{-\sqrt{2}}{2}$ | $\dfrac{-1}{2}$ | $0$ | $\dfrac{1}{2}$ | $\dfrac{\sqrt{2}}{2}$ | $\dfrac{\sqrt{3}}{2}$ | $1$ |

Plotting these points, we obtain the graph of the function.

## 12.4    Tangent Lines and Curve Sketching in Polar Coordinates

**7.** Find the slope of the graph of $r = 3(1 - \cos\theta)$ at $\theta = \pi/2$.

**Solution:**

Since $f(\theta) = 3(1 - \cos\theta)$, we have $f'(\theta) = 3\sin\theta$ and

$$
\begin{aligned}
\frac{dy}{dx} &= \frac{f'(\theta)\sin\theta + f(\theta)\cos\theta}{f'(\theta)\cos(\theta) - f(\theta)\sin(\theta)} \\[2mm]
&= \frac{(3\sin\theta)\sin\theta + 3(1 - \cos\theta)\cos\theta}{(3\sin\theta)\cos\theta - 3(1 - \cos\theta)\sin\theta} \\[2mm]
&= \frac{\sin^2\theta + \cos\theta - \cos^2\theta}{\sin\theta(2\cos\theta - 1)} \\[2mm]
&= \frac{(1 + 2\cos\theta)(1 - \cos\theta)}{\sin\theta(2\cos\theta - 1)}
\end{aligned}
$$

When $\theta = \pi/2$, the slope is

$$
\frac{dy}{dx} = \frac{(1 + 0)(1 - 0)}{1(0 - 1)} = -1
$$

**11.** Find the slope of the graph of $r = \theta$ at $\theta = \pi$.

**Solution:**

Since $f(\theta) = \theta$, we have $f'(\theta) = 1$ and

$$\frac{dy}{dx} = \frac{f'(\theta)\sin\theta + f(\theta)\cos\theta}{f'(\theta)\cos\theta - f(\theta)\sin\theta} = \frac{\sin\theta + \theta\cos\theta}{\cos\theta - \theta\sin\theta}$$

When $\theta = \pi$, the slope is

$$\frac{dy}{dx} = \frac{0 + \pi(-1)}{(-1) - \pi(0)} = \frac{-\pi}{-1} = \pi$$

**17.** Find the horizontal and vertical tangents to the polar curve $r = 1 + \sin\theta$.

**Solution:**

Since $f(\theta) = 1 + \sin\theta$ and $f'(\theta) = \cos\theta$, we have

$$\begin{aligned}
\frac{dy}{dx} &= \frac{f'(\theta)\sin\theta + f(\theta)\cos\theta}{f'(\theta)\cos\theta - f(\theta)\sin\theta} \\
&= \frac{\cos\theta\sin\theta + (1 + \sin\theta)\cos\theta}{\cos^2\theta - (1 + \sin\theta)\sin\theta} \\
&= \frac{\cos\theta(1 + 2\sin\theta)}{1 - \sin\theta - 2\sin^2\theta} \\
&= \frac{\cos\theta(1 + 2\sin\theta)}{(1 - 2\sin\theta)(1 + \sin\theta)}
\end{aligned}$$

Since $dy/dx = 0$ when $\cos\theta(1 + 2\sin\theta) = 0$ or $\theta = \pi/2,\ 7\pi/6,\ 11\pi/6$, we have horizontal tangents at the points

$$\left(2, \frac{\pi}{2}\right), \left(\frac{1}{2}, \frac{7\pi}{6}\right), \quad \text{and} \quad \left(\frac{1}{2}, \frac{11\pi}{6}\right)$$

Since $dy/dx$ is undefined when $(1 - 2\sin\theta)(1 + \sin\theta) = 0$ or $\theta = \pi/6,\ 5\pi/6,\ 3\pi/2$, we have vertical tangents at the points

$$\left(\frac{3}{2}, \frac{\pi}{6}\right), \left(\frac{3}{2}, \frac{5\pi}{6}\right), \quad \text{and} \quad \left(0, \frac{3\pi}{2}\right)$$

(Note that L'Hopital's Rule must be used to show that $dy/dx$ is undefined at $\theta = 3\pi/2$.)

**29.** Sketch and identify the graph of $r = 2 + 3\sin\theta$. Find the tangents at the pole.

**Solution:**

We first observe that the given equation has the form

$$r = a + b\sin\theta = 2 + 3\sin\theta$$

with $b > a$, which means its graph is a *limacon* with two loops. Furtheremore, since $r$ is a function of $\sin\theta$, the graph has vertical axis symmetry. The relative extrema of $r$ are $(-1, 3\pi/2)$ and $(5, \pi/2)$. The tangents at the pole are $\theta = \arcsin\left(-\frac{2}{3}\right)$ and $\theta = \pi + \arcsin\left(\frac{2}{3}\right)$ since $r = 0$ and $dr/d\theta \neq 0$ at these values. Using this information and the points in the table, we make the accompanying sketch.

arcsin $\left(-\frac{2}{3}\right)$

$\pi + \arcsin\left(\frac{2}{3}\right)$

| $\theta$ | 0 | $\dfrac{\pi}{6}$ | $\dfrac{\pi}{2}$ | $-\pi$ | $\dfrac{3\pi}{2}$ | $\dfrac{-\pi}{6}$ |
|---|---|---|---|---|---|---|
| $r$ | 2 | $\dfrac{7}{2}$ | 5 | 2 | $-1$ | $\dfrac{1}{2}$ |

**37.** Sketch and identify the graph of $r = 2\cos 3\theta$. Find the tangents at the pole.

**Solution:**

The equation has the form

$$r = a\cos(n\theta) = 2\cos 3\theta$$

where $n$ is odd. This means the graph is a *rose curve* with $n = 3$ petals. The curve has polar axis symmetry. The relative extrema of $r$ are $(2, 0)$, $(-2, \pi/3)$, and $(2, 2\pi/3)$. The tangents at the pole are $\theta = \pi/6$, $\theta = \pi/2$, and $\theta = 5\pi/6$, since $r = 0$ and $dr/d\theta \neq 0$ at these values. Using this information and the points in the table, we make the acccompanying sketch.

| $\theta$ | 0 | $\dfrac{\pi}{12}$ | $\dfrac{\pi}{6}$ | $\dfrac{\pi}{3}$ | $\dfrac{\pi}{2}$ | $\dfrac{2\pi}{3}$ |
|---|---|---|---|---|---|---|
| $r$ | 2 | $\sqrt{2}$ | 0 | $-2$ | 0 | 2 |

**47.** Sketch and identify the graph of $r^2 = 4\sin 2\theta$. Find the tangents at the pole.

**Solution:**

The equation has the form

$$r^2 = a^2 \sin 2\theta = 4\sin 2\theta$$

and therefore is a *lemniscate* with symmetry with respect to the pole. The tangents at the pole are $\theta = 0$ and $\theta = \pi/2$ since $r = 0$ and $dr/d\theta \neq 0$ at these values. The relative extrema of $r$ are $(2, \pi/4)$ and $(2, 5\pi/4)$. Using this information and the points in the table, we make the accompanying sketch

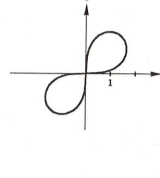

| $\theta$ | 0 | $\dfrac{\pi}{4}$ | $\dfrac{\pi}{2}$ | $\dfrac{5\pi}{4}$ | $\pi$ |
|---|---|---|---|---|---|
| $r$ | 0 | $\pm 2$ | 0 | $\pm 2$ | 0 |

**55.** Sketch the graph of $r = 2 - \sec\theta$ and show that $x = -1$ is a vertical asymptote of the graph.

**Solution:**

First, we write

$$r = 2 - \sec\theta = 2 - \frac{1}{\cos\theta}$$

and note that the graph will have polar axis symmetry. The tangents at the pole are $\theta = \pi/3$ and $\theta = -\pi/3$. Furthermore,

$$r \to -\infty \quad \text{as} \quad \theta \to \frac{\pi^-}{2}$$

and

$$r \to \infty \quad \text{as} \quad \theta \to \frac{\pi^+}{2}$$

To see that the graph has a vertical asymptote at $x = -1$, we write

$$r = 2 - \frac{1}{\cos\theta} = 2 - \frac{r}{r\cos\theta} = 2 - \frac{r}{x}$$
$$rx = 2x - r$$
$$r(1 + x) = 2x$$
$$r = \frac{2x}{1 + x}$$

Thus $r \to \pm\infty$ as $x \to -1$, as shown in the accompanying figure.

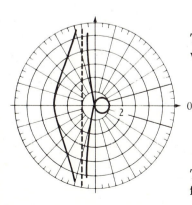

## 12.5   Area and Arc Length in Polar Coordinates

5. Find the points of intersection of the graphs of $r = 4 - 5\sin\theta$ and $r = 3\sin\theta$.

**Solution:**

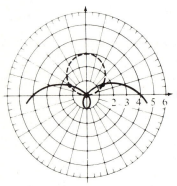

From Section 12.4, we know the graph of $r = 4 - 5\sin\theta$ is a limacon and the graph of $r = 3\sin\theta$ is a circle (see figure). Solving the two equations simultaneously, we have

$$4 - 5\sin\theta = 3\sin\theta$$
$$8\sin\theta = 4$$
$$\sin\theta = \frac{1}{2}$$
$$\theta = \frac{\pi}{6}, \frac{5\pi}{6} \qquad (0 \le \theta < 2\pi)$$

From these values we obtain the points $(3/2, \pi/6)$ and $(3/2, 5\pi/6)$. To test for additional points of intersection, we replace $r$ by $-r$ and $\theta$ by $\pi + \theta$ in $r = 4 - 5\sin\theta$ to obtain

$$-r = 4 - 5\sin(\pi + \theta) = 4 + 5\sin\theta$$

Solving this equation simultaneously with $r = 3\sin\theta$, we have

$$-4 - 5\sin\theta = 3\sin\theta$$
$$8\sin\theta = -4$$
$$\sin\theta = -\frac{1}{2}$$
$$\theta = \frac{7\pi}{6}, \frac{11\pi}{6}$$

which yields the points $(-3/2, 7\pi/6/)$ and $(-3/2, 11\pi/6)$. However, these two points coincide with the previous two points. Finally, we observe that both curves pass through the pole. Hence there are three points of intersection, $(3/2, \pi/6)$, $(3/2, 5\pi/6)$, and $(0,0)$ as seen in the accompanying figure.

**9.** Find the points of intersection of the graphs of $r = 4\sin 2\theta$ and $r = 2$.

**Solution:**

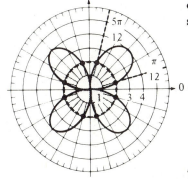

From Section 12.4, we know the graph of $r = 4\sin 2\theta$ is a rose curve with 4 petals and is symmetric to the polar axis, the vertical axis and the pole. Also, the graph of $r = 2$ is a circle of radius 2 centered at the pole. Solving the two equations simultaniously, we have

$$4\sin 2\theta = 2$$
$$\sin 2\theta = \frac{1}{2}$$
$$2\theta = \frac{\pi}{6}, \ \frac{5\pi}{6}$$
$$\theta = \frac{\pi}{12}, \ \frac{5\pi}{12}$$

Therefore, the points of intersection for one petal are $(2, \pi/12)$ and $(2, 5\pi/12)$. By symmetry, the other points of intersection are $(2, 7\pi/12)$, $(2, 11\pi/12)$, $(2, 13\pi/12)$, $(2, 17\pi/12)$, $(2, 19\pi/12)$, and $(2, 23\pi/12)$.

**19.** Find the area of the region within the inner loop of the graph of $r = 1 + 2\cos\theta$.

**Solution:**

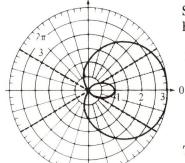

Solving the equation $r = 0$ to find the tangents at the pole we have

$$1 + 2\cos\theta = 0$$
$$\cos\theta = -\frac{1}{2}$$
$$\theta = \frac{2\pi}{3} \quad \text{or} \quad \theta = \frac{4\pi}{3}$$

Therefore, the lower half of the inner loop is generated when $\theta$ is in the interval $\dfrac{2\pi}{3} \le \theta \le \pi$. From the symmetry of the graph in the accompanying figure, the area of the inner loop is given by

$$2\int_{2\pi/3}^{\pi}\frac{1}{2}r^2\,d\theta$$

$$=\int_{2\pi/3}^{\pi}(1+2\cos\theta)^2 d\theta$$

$$=\int_{2\pi/3}^{\pi}(1+4\cos\theta+4\cos^2\theta)\,d\theta$$

$$=\int_{2\pi/3}^{\pi}\left(1+4\cos\theta+4\frac{1+\cos 2\theta}{2}\right)d\theta$$

$$=\int_{2\pi/3}^{\pi}(3+4\cos\theta+2\cos 2\theta)\,d\theta$$

$$=\left[3\theta+4\sin\theta+\sin 2\theta\right]_{2\pi/3}^{\pi}$$

$$=\pi-\frac{3\sqrt{3}}{2}=\frac{2\pi-3\sqrt{3}}{2}$$

**21.** Find the area of the region between the loops of the graph of $r=1+2\cos\theta$.

**Solution:**

From the symmetry of the graph given in the accompanying figure, the area of the region inside the outer loop is

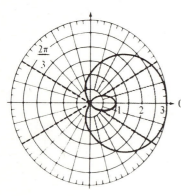

$$2\int_{0}^{2\pi/3}\frac{1}{2}r^2 d\theta=\int_{0}^{2\pi/3}(1+2\cos\theta)^2 d\theta$$

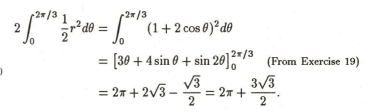

$$=\left[3\theta+4\sin\theta+\sin 2\theta\right]_{0}^{2\pi/3}\quad\text{(From Exercise 19)}$$

$$=2\pi+2\sqrt{3}-\frac{\sqrt{3}}{2}=2\pi+\frac{3\sqrt{3}}{2}.$$

From Exercise 19, we see that the area of the inner loop is given by $\pi-\dfrac{3\sqrt{3}}{2}$. Finally, the area of the region between the two loops is

$$A=\left(2\pi+\frac{3\sqrt{3}}{2}\right)-\left(\pi-\frac{3\sqrt{3}}{2}\right)=\pi+3\sqrt{3}$$

**23.** Find the area of the region common to the interiors of the graphs of $r = 4\sin 2\theta$ and $r = 2$.

**Solution:**

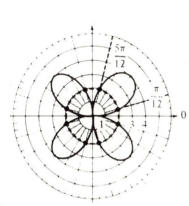

From the accompanying sketch we see that we need only consider the region in one petal common to both curves and multiply the result by four. From Exercise 9 the points of intersection on the first petal occur when $\theta = \pi/12$ and $5\pi/12$. There are three subregions within one petal:
(a)  for $0 \le \theta \le \pi/12$,    $r = 4\sin 2\theta$
(b)  for $\pi/12 \le \theta \le 5\pi/12$,    $r = 2$
(c)  for $5\pi/12 \le \theta \le \pi/2$,    $r = 4\sin 2\theta$

Therefore the area withing one petal is

$$A = \int_0^{\pi/12} \frac{1}{2}(4\sin 2\theta)^2 \, d\theta + \int_{\pi/12}^{5\pi/12} \frac{1}{2}(2)^2 \, d\theta$$

$$+ \int_{5\pi/12}^{\pi/2} \frac{1}{2}(4\sin 2\theta)^2 \, d\theta$$

By the symmetry of the petal, the first and third integrals are equal. Thus

$$A = \int_0^{\pi/12} \frac{1}{2}(4\sin 2\theta)^2 \, d\theta + \int_{\pi/12}^{5\pi/12} \frac{1}{2}(2)^2 \, d\theta$$

$$= 16 \int_0^{\pi/12} \sin^2 2\theta \, d\theta + 2 \int_{\pi/12}^{5\pi/12} d\theta$$

$$= 8 \int_0^{\pi/12} (1 - \cos 4\theta) \, d\theta + 2 \int_{\pi/12}^{5\pi/12} d\theta$$

$$= 8 \left[ \theta - \frac{1}{4}\sin 4\theta \right]_0^{\pi/12} + \left[ 2\theta \right]_{\pi/12}^{5\pi/12}$$

$$= \frac{2\pi}{3} - \frac{2\sqrt{3}}{2} + \frac{2\pi}{3} = \frac{4\pi}{3} - \sqrt{3}$$

Finally, multiplying by 4, we obtain the total area of

$$\frac{4}{3}(4\pi - 3\sqrt{3}).$$

**39.** Find the length of the graph of $r = 1/\theta$ over the interval $\pi \leq \theta \leq 2\pi$.

**Solution:**

$$
s = \int_\alpha^\beta \sqrt{[f(\theta)]^2 + [f'(\theta)]^2}\, d\theta
$$

$$
= \int_\pi^{2\pi} \sqrt{\left(\frac{1}{\theta}\right)^2 + \left(\frac{-1}{\theta^2}\right)^2}\, d\theta
$$

$$
= \int_\pi^{2\pi} \frac{1}{\theta^2}\sqrt{\theta^2 + 1}\, d\theta
$$

$$
= \left[ -\frac{\sqrt{\theta^2 + 1}}{\theta} + \ln\left|\theta + \sqrt{\theta^2 + 1}\right| \right]_\pi^{2\pi}
$$

$$
= \frac{2\sqrt{\pi^2 + 1} - \sqrt{4\pi^2 + 1}}{2\pi} + \ln\left|\frac{2\pi + \sqrt{4\pi^2 + 1}}{\pi + \sqrt{\pi^2 + 1}}\right| \approx 0.7112
$$

**43.** Find the area of the surface formed by revolving the curve $r = e^{a\theta}$ over the interval $0 \leq \theta \leq \pi/2$ and the line $\theta = \pi/2$.

**Solution:**

$$
S = 2\pi \int_\alpha^\beta f(\theta)\cos\theta\sqrt{[f(\theta)]^2 + [f'(\theta)]^2}\, d\theta
$$

$$
= 2\pi \int_0^{\pi/2} e^{a\theta}(\cos\theta)\sqrt{(e^{a\theta})^2 + (ae^{a\theta})^2}\, d\theta
$$

$$
= 2\pi \int_0^{\pi/2} \cos\theta\, e^{a\theta}\sqrt{e^{2a\theta}(1 + a^2)}\, d\theta
$$

$$
= 2\pi\sqrt{1 + a^2} \int_0^{\pi/2} \cos\theta\, e^{2a\theta}\, d\theta
$$

$$
= 2\pi\sqrt{1 + a^2}\left[\frac{e^{2a\theta}}{4a^2 + 1}(2a\cos\theta + \sin\theta)\right]_0^{\pi/2} \qquad \text{(Integration by Parts)}
$$

$$
= \frac{2\pi\sqrt{1 + a^2}}{4a^2 + 1}(e^{\pi a} - 2a)
$$

## 12.6   Polar Equations for Conics and Kepler's Laws

**5.** Sketch the graph of the equation $r = 2/(2 - \cos\theta)$ and identify the curve.

**Solution:**

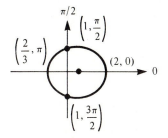

To determine the type of conic, we rewrite the equation as

$$r = \frac{2}{2 - \cos\theta} = \frac{1}{1 - \frac{1}{2}\cos\theta} = \frac{ed}{1 - e\cos\theta}$$

From this form we conclude that the graph is an ellipse with $e = 1/2$. We sketch the upper half of the ellipse by plotting the points in the accompanying table. Then, using symmetry with respect to the polar axis we sketch the lower half.

| $\theta$ | 0 | $\dfrac{\pi}{6}$ | $\dfrac{\pi}{4}$ | $\dfrac{\pi}{3}$ | $\dfrac{\pi}{2}$ | $\dfrac{2\pi}{3}$ | $\dfrac{3\pi}{4}$ | $\dfrac{5\pi}{6}$ | $\pi$ |
|---|---|---|---|---|---|---|---|---|---|
| $r$ | 2 | 1.76 | 1.55 | 1.33 | 1 | 0.80 | 0.74 | 0.70 | 0.67 |

**9.** Sketch the graph of the equation $r = -1/(1 - \sin\theta)$ and identify the curve.

**Solution:**

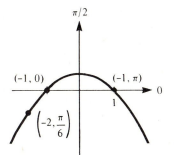

From the form of the equation we have

$$r = \frac{-1}{1 - \sin\theta} = \frac{ed}{1 - e\sin\theta}$$

We conclude that the graph of the equation is a parabola $(e = 1)$. We sketch the left half of the parabola by plotting points in the accompanying table. Then using symmetry with respect to $\theta = \pi/2$, we sketch the right half.

| $\theta$ | $-\dfrac{\pi}{2}$ | $-\dfrac{\pi}{3}$ | $-\dfrac{\pi}{4}$ | $-\dfrac{\pi}{6}$ | 0 | $\dfrac{\pi}{6}$ | $\dfrac{\pi}{4}$ | $\dfrac{\pi}{3}$ | $\dfrac{\pi}{2}$ |
|---|---|---|---|---|---|---|---|---|---|
| $r$ | $-0.50$ | $-0.54$ | $-0.59$ | $-0.67$ | $-1$ | $-2$ | $-3.41$ | $-7.46$ | Undefined |

**13.** Sketch the graph of the equation $r = 3/(2 - 6\cos\theta)$ and identify the curve.

**Solution:**

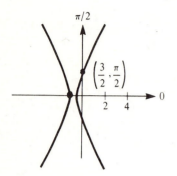

To determine the type of conic, we rewrite the equation as

$$r = \frac{3}{2 - 6\cos\theta} = \frac{3/2}{1 - 3\cos\theta} = \frac{ed}{1 - e\cos\theta}$$

From this form we conclude that the graph is a hyperbola with $e = 3$. We sketch the upper half of the right branch and the lower half of the left branch of the hyperbola by plotting points in the accompanying table. Then, using symmetry with respect to the polar axis, we sketch the other half of the graph.

| $\theta$ | $0$ | $\dfrac{\pi}{6}$ | $\dfrac{\pi}{4}$ | $\dfrac{\pi}{3}$ | $\arccos\dfrac{1}{3}$ | $\dfrac{\pi}{2}$ | $\dfrac{2\pi}{3}$ | $\dfrac{3\pi}{4}$ | $\dfrac{5\pi}{6}$ | $\pi$ |
|---|---|---|---|---|---|---|---|---|---|---|
| $r$ | $-0.75$ | $-0.94$ | $-1.34$ | $-3$ | Undefined | $1.5$ | $0.60$ | $0.48$ | $0.41$ | $0.375$ |

**17.** Find a polar equation for the ellipse with focus at the pole, eccentricity $e = \frac{1}{2}$ and directrix $y = 1$.

**Solution:**

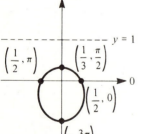

Since the directrix is horizontal and above the pole (see the accompanying figure), we choose an equation of the form

$$r = \frac{ed}{1 + e\sin\theta}$$

Moreover, since the eccentricity of the ellipse is $\frac{1}{2}$ and the directed distance from the focus to the directrix is $d = 1$, we have the equation

$$r = \frac{\frac{1}{2}}{1 + \frac{1}{2}\sin\theta} = \frac{1}{2 + \sin\theta}$$

**23.** Find a polar equation of the parabola with focus at the pole and vertex at $(1, -\pi/2)$.

**Solution:**

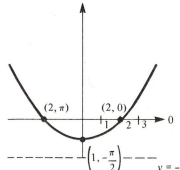

Since the directrix is horizontal and below the pole (see the accompanying figure), we choose an equation of the form

$$r = \frac{ed}{1 - e\sin\theta}$$

Moreover, since the eccentricity of a parabola is $e = 1$ and the distance from the focus to the directrix is $d = 2$, we have the equation

$$r = \frac{2}{1 - \sin\theta}$$

**27.** Find an equation of the hyperbola with focus at the pole, vertex at $(2, 3\pi/2)$ and directrix $y = -3$.

**Solution:**

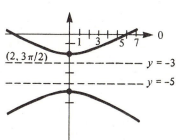

Since the directrix is horizontal and below the pole we choose an equation of the form

$$r = \frac{ed}{1 - e\sin\theta}$$

A hyperbola is the locus of a point that moves so that its distance from the focus is in constant ratio to its distance from the directrix. Therefore, when $\theta = 3\pi/2$ in the accompanying figure we have

$$e = \frac{2}{1} = 2$$

Moreover, since the distance between the pole and the directrix is $d = 3$, we have the equation

$$r = \frac{2(3)}{1 - 2\sin\theta} = \frac{6}{1 - 2\sin\theta}$$

**39.** Use Simpson's Rule with $n = 6$ to find the area of the region bounded by the graph of the polar equation

$$r = \frac{3}{2 - \cos \theta}$$

**Solution:**

From the symmetry of the graph in the accompanying figure, the area of the region is given by

$$2 \int_0^\pi \frac{1}{2} r^2 d\theta = \int_0^\pi \left( \frac{3}{2 - \cos \theta} \right)^2 d\theta$$

$$= \frac{\pi}{3(6)} \left[ \left( \frac{3}{2 - \cos \theta} \right)^2 + 4 \left( \frac{3}{2 - \cos \frac{\pi}{6}} \right)^2 + 2 \left( \frac{3}{2 - \cos \frac{\pi}{3}} \right)^2 \right.$$

$$+ 4 \left( \frac{3}{2 - \cos \frac{\pi}{2}} \right)^2 + 2 \left( \frac{3}{2 - \cos \frac{2\pi}{3}} \right)^2 + 4 \left( \frac{3}{2 - \cos \frac{5\pi}{6}} \right)^2$$

$$\left. + \left( \frac{3}{2 - \cos \pi} \right)^2 \right] \approx 10.87$$

**47.** Find the angle $\psi$ for the graph of $r = 6/(1 - \cos \theta)$ at $\theta = 2\pi/3$.

**Solution:**

Since

$$f(\theta) = r = \frac{6}{1 - \cos \theta}$$

we get

$$f'(\theta) = \frac{-6 \sin \theta}{(1 - \cos \theta)^2}$$

Now from the definition of the angle between the radial line and the tangent line (see Exercise 42), we have

$$\tan \psi = \left| \frac{f(\theta)}{f'(\theta)} \right| = \left| \frac{6/(1 - \cos \theta)}{-6 \sin \theta/(1 - \cos \theta)^2} \right| = \left| \frac{1 - \cos \theta}{- \sin \theta} \right|$$

At $\theta = 2\pi/3$,

$$\tan \psi = \left| \frac{1 - (-\frac{1}{2})}{-\sqrt{3}/2} \right| = \left| \frac{\frac{3}{2}}{-\sqrt{3}/2} \right| = \sqrt{3}$$

Therefore we conclude that $\psi = \pi/3$.

## Review Exercises for Chapter 12

3. (a) Find $dy/dx$ and all points of horizontal tangency, (b) eliminate the parameter if possible, and (c) sketch the curve represented by the parametric equations $x = 3 + 2\cos\theta$ and $y = 2 + 5\sin\theta$.

**Solution:**

(a) Since $x = 3 + 2\cos\theta$ and $y = 2 + 5\sin\theta$, we have

$$\frac{dy}{dx} = \frac{dy/d\theta}{dx/d\theta} = \frac{5\cos\theta}{-2\sin\theta} = -\frac{5}{2}\cot\theta$$

The points of horizontal tangency occur at $\theta = \pi/2$ and $\theta = 3\pi/2$, and the points are $(3, 7)$ and $(3, -3)$.

(b) To eliminate the parameter, we consider the identity $\sin^2\theta + \cos^2\theta = 1$, and write

$$\frac{x - 3}{2} = \cos\theta, \qquad \frac{y - 2}{5} = \sin\theta$$

By squaring and adding these equations, we obtain

$$\frac{(x - 3)^2}{4} + \frac{(y - 2)^2}{25} = 1$$

which is an equation for the ellipse centered at $(3, 2)$ with vertices at $(3, -3)$ and $(3, 7)$.

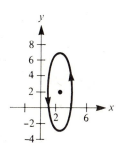

15. Find the length of the involute $x = r(\cos\theta + \theta\sin\theta)$, $y = r(\sin\theta - \theta\cos\theta)$, when $0 \le \theta \le \pi$. (See Exercise 14.)

**Solution:**

$$x = r(\cos\theta + \theta\sin\theta) \qquad y = r(\sin\theta - \theta\cos\theta)$$
$$\frac{dx}{d\theta} = r\theta\cos\theta \qquad\qquad \frac{dy}{d\theta} = r\theta\sin\theta$$

$$s = \int_0^\pi \sqrt{\left(\frac{dx}{d\theta}\right)^2 + \left(\frac{dy}{d\theta}\right)^2}\, d\theta$$

$$= \int_0^\pi \sqrt{(r\theta\cos\theta)^2 + (r\theta\sin\theta)^2}\, d\theta$$

$$= r\int_0^\pi \theta\, d\theta = r\left[\frac{\theta^2}{2}\right]_0^\pi = \frac{1}{2}\pi^2 r$$

**31.** Sketch the graph of $r = 4 \cos 2\theta \sec \theta$.

**Solution:**

The graph is symmetric with respect to the polar axis, and the tangents at the pole are $\theta = \pi/4$ and $3\pi/4$. Note also that

$$r \to -\infty \qquad \text{as} \qquad \theta \to \frac{\pi}{2}^{-}$$

and

$$r \to -\infty \qquad \text{as} \qquad \theta \to -\frac{\pi}{2}^{-}$$

With this information and the following table we make the accompanying sketch

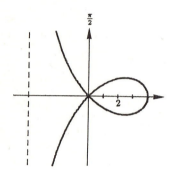

| $\theta$ | 0 | $\dfrac{\pi}{6}$ | $\dfrac{\pi}{4}$ | $\dfrac{\pi}{3}$ | $\dfrac{\pi}{2}$ |
|---|---|---|---|---|---|
| $r$ | 4 | 2.31 | 0 | $-4$ | Undefined |

**37.** Find a rectangular equation that has the same graph as $r = 4 \cos 2\theta \sec \theta$.

**Solution:**

First, we replace $\cos 2\theta$ by $2 \cos^2 \theta - 1$ and $\sec \theta$ by $1/\cos \theta$ to obtain

$$r = 4 \cos 2\theta \sec \theta = 4(2 \cos^2 \theta - 1)\frac{1}{\cos \theta}$$

$$r \cos \theta = 8 \cos^2 \theta - 4$$

Now since $x = r \cos \theta$ and $r^2 = x^2 + y^2$, we have

$$x = 8\left(\frac{x^2}{r^2}\right) - 4 = \frac{8x^2 - 4(x^2 + y^2)}{x^2 + y^2}$$

$$x^3 + xy^2 = 4x^2 - 4y^2$$

$$(4 + x)y^2 = (x^2)(4 - x)$$

$$y^2 = x^2\left(\frac{4 - x}{4 + x}\right)$$

**41.** Find a polar equation for the rectangular equation

$$x^2 + y^2 = a^2 \left( \arctan \frac{y}{x} \right)^2$$

**Solution:**

Since $r^2 = x^2 + y^2$ and $\theta = \arctan \dfrac{y}{x}$, we have

$$x^2 + y^2 = a^2 \left( \arctan \frac{y}{x} \right)^2$$
$$r^2 = a^2 \theta^2$$

**49.** Find the tangent lines at the pole and all points of vertical and horizontal tangency for the graph of $r = 1 - 2\cos\theta$. Sketch the graph of the equation.

**Solution:**

The graph has polar axis symmetry and the tangents at the pole are $\theta = \pi/3$ and $\theta = -\pi/3$. To find the points of vertical or horizontal tangency, we note that $f'(\theta) = 2\sin\theta$ and find $dy/dx$ as follows:

$$\frac{dy}{dx} = \frac{f'(\theta)\sin\theta + f(\theta)\cos\theta}{f'(\theta)\cos\theta - f(\theta)\sin\theta}$$
$$= \frac{2\sin^2\theta + (1 - 2\cos\theta)\cos\theta}{2\sin\theta\cos\theta - (1 - 2\cos\theta)\sin\theta}$$
$$= \frac{2\sin^2\theta + \cos\theta - 2\cos^2\theta}{4\sin\theta\cos\theta - \sin\theta}$$
$$= \frac{2(1 - \cos^2\theta) + \cos\theta - 2\cos^2\theta}{\sin\theta(4\cos\theta - 1)}$$
$$= \frac{2 + \cos\theta - 4\cos^2\theta}{\sin\theta(4\cos\theta - 1)}$$

The graph has horizontal tangents when $dy/dx = 0$, and this occurs when

$$-4\cos^2\theta + \cos\theta + 2 = 0$$
$$\cos\theta = \frac{-1 \pm \sqrt{1 + 32}}{-8} = \frac{1 \mp \sqrt{33}}{8}$$

When $\cos\theta = (1 \mp \sqrt{33})/8$,

$$r = 1 - 2\left(\frac{1 \mp \sqrt{33}}{8}\right) = \frac{3 \pm \sqrt{33}}{4}$$

Therefore the points of horizontal tangency are

$$\left(\frac{3 - \sqrt{33}}{4}, \arccos\left[\frac{1 + \sqrt{33}}{8}\right]\right) \approx (-0.686, 0.568)$$

$$\left(\frac{3 - \sqrt{33}}{4}, -\arccos\left[\frac{1 + \sqrt{33}}{8}\right]\right) \approx (-0.686, -0.568)$$

$$\left(\frac{3 + \sqrt{33}}{4}, \arccos\left[\frac{1 - \sqrt{33}}{8}\right]\right) \approx (2.186, 2.206)$$

$$\left(\frac{3 + \sqrt{33}}{4}, -\arccos\left[\frac{1 - \sqrt{33}}{8}\right]\right) \approx (2.186, -2.206)$$

The graph has vertical tangents when

$$\sin\theta(4\cos\theta - 1) = 0$$

$$\theta = 0, \ \pi, \ \text{ or } \ \pm\arccos\frac{1}{4}$$

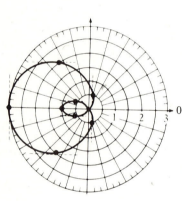

When $\cos\theta = \frac{1}{4}$, $r = 1 - 2(\frac{1}{4}) = \frac{1}{2}$. Thus the points of vertical tangency are

$$(-1, 0), \ (3, \pi), \ \text{ and } \ \left(\frac{1}{2}, \pm\arccos\frac{1}{4}\right) \approx (0.5, \pm1.318)$$

as shown in the accompanying figure.

59. Find the area of the region common to the interiors of $r = 4\cos\theta$ and $r = 2$.

**Solution:**

To find the point of intersection of the two graphs we solve the two equations simultaneously to obtain

$$4\cos\theta = 2$$
$$\cos\theta = \frac{1}{2}$$
$$\theta = \frac{\pi}{3}$$

From the accompanying figure we see that the area is given by

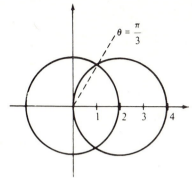

$$A = 2\left[\frac{1}{2}\int_0^{\pi/3} 2^2\, d\theta + \frac{1}{2}\int_{\pi/3}^{\pi/2}(4\cos\theta)^2\, d\theta\right]$$

$$= \int_0^{\pi/3} 4\, d\theta + 16\int_{\pi/3}^{\pi/2}\frac{1+\cos 2\theta}{2}\, d\theta$$

$$= \left[4\theta\right]_0^{\pi/3} + 8\left[\theta + \frac{1}{2}\sin 2\theta\right]_{\pi/3}^{\pi/2}$$

$$= \frac{8\pi - 6\sqrt{3}}{3}$$

**61.** Find the perimeter of a cardiod $r = a(1 - \cos\theta)$.

**Solution:**

$$r = a(1 - \cos\theta) \qquad \text{and} \qquad \frac{dr}{d\theta} = a\sin\theta$$

Due to the symmetry with respect to the polar axis, we have

$$s = 2\int_0^\pi \sqrt{r^2 + \left(\frac{dr}{d\theta}\right)^2}\, d\theta$$

$$= 2\int_0^\pi \sqrt{a^2(1-\cos\theta)^2 + a^2\sin^2\theta}\, d\theta$$

$$= 2\sqrt{2}a\int_0^\pi \sqrt{1 - \cos\theta}\, d\theta$$

$$= 2\sqrt{2}a\int_0^\pi \sqrt{1 - \cos\theta}\frac{\sqrt{1+\cos\theta}}{\sqrt{1+\cos\theta}}\, d\theta$$

$$= 2\sqrt{2}a\int_0^\pi \frac{\sin\theta}{\sqrt{1+\cos\theta}}\, d\theta$$

$$= -4\sqrt{2}a\left[(1+\cos\theta)^{1/2}\right]_0^\pi = 8a$$

# 13 VECTORS AND CURVES IN THE PLANE

## 13.1 Vectors in the Plane

**5.** The initial point of a vector **v** is $(1, 2)$ and its terminal point is $(5, 5)$. (a) Sketch the given directed line segment, (b) write the vector in component form, and (c) sketch the vector with its initial point at the origin.

**Solution:**

(b) We let $P = (1, 2) = (p_1, p_2)$ and $Q = (5, 5) = (q_1, q_2)$. Then the components of $\mathbf{v} = \langle v_1, v_2 \rangle$ are given by

$$v_1 = q_1 - p_1 = 5 - 1 = 4$$
$$v_2 = q_2 - p_2 = 5 - 2 = 3$$

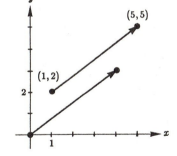

Thus, $\mathbf{v} = \langle 4, 3 \rangle$.

**17.** Find the component form of $\mathbf{v} = \mathbf{u} + 2\mathbf{w}$, where $\mathbf{u} = 2\mathbf{i} - \mathbf{j}$ and $\mathbf{w} = \mathbf{i} + 2\mathbf{j}$. Illustrate the indicated vector operations geometrically.

**Solution:**

Since $\mathbf{u} = 2\mathbf{i} - \mathbf{j}$ and $2\mathbf{w} = 2(\mathbf{i} + 2\mathbf{j}) = 2\mathbf{i} + 4\mathbf{j}$, we have

$$\mathbf{v} = \mathbf{u} + 2\mathbf{w} = (2\mathbf{i} - \mathbf{j}) + (2\mathbf{i} + 4\mathbf{j})$$
$$= (2 + 2)\mathbf{i} + (4 - 1)\mathbf{j} = 4\mathbf{i} + 3\mathbf{j} = \langle 4, 3 \rangle.$$

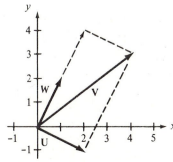

Geometrically, this sum is illustrated in the accompanying figure.

**23.** Find $a$ and $b$ such that $\mathbf{v} = a\mathbf{u} + b\mathbf{w}$ where $\mathbf{v} = \langle 3, 0 \rangle$, $\mathbf{u} = \langle 1, 2 \rangle$, and $\mathbf{w} = \langle 1, -1 \rangle$.

**Solution:**

$$\mathbf{v} = a\mathbf{u} + b\mathbf{w}$$
$$3\mathbf{i} = a(\mathbf{i} + 2\mathbf{j}) + b(\mathbf{i} - \mathbf{j}) = (a+b)\mathbf{i} + (2a-b)\mathbf{j}$$

Equating coefficients we have

$$a + b = 3$$

and

$$2a - b = 0$$

Solving the system of equations simultaneously we obtain $a = 1$ and $b = 2$.

**37.** Demonstrate the triangle inequality using the vectors $\mathbf{u} = \langle 2, 1 \rangle$ and $\mathbf{v} = \langle 5, 4 \rangle$.

**Solution:**

$$\|\mathbf{u}\| = \sqrt{2^2 + 1^2} = \sqrt{5}$$
$$\|\mathbf{v}\| = \sqrt{5^2 + 4^2} = \sqrt{41}$$

Since $\mathbf{u} + \mathbf{v} = \langle 2, 1 \rangle + \langle 5, 4 \rangle = \langle 7, 5 \rangle$ we have

$$\|\mathbf{u} + \mathbf{v}\| = \sqrt{7^2 + 5^2} = \sqrt{74}$$

Therefore,

$$\|\mathbf{u} + \mathbf{v}\| = \sqrt{74} \approx 8.602 \leq 2.236 + 6.403$$
$$\approx \sqrt{5} + \sqrt{41} = \|\mathbf{u}\| + \|\mathbf{v}\|$$

**39.** Find the vector $\mathbf{v}$ with magnitude $\|\mathbf{v}\| = 4$ and in the same direction as $\mathbf{u} = \langle 1, 1 \rangle$.

**Solution:**

$$\mathbf{v} = (\text{magnitude of } \mathbf{v})(\text{unit vector in the direction of } \mathbf{v})$$
$$= 4\left(\frac{\mathbf{u}}{\|\mathbf{u}\|}\right)$$
$$= 4\left(\frac{\mathbf{u}}{\sqrt{2}}\right) = 2\sqrt{2}\mathbf{u} = \langle 2\sqrt{2}, 2\sqrt{2} \rangle$$

**43.** Find a unit vector (a) parallel to and (b) normal to the graph of $f(x) = x^3$ at $(1, 1)$.

**Solution:**

(a)  At $(1, 1)$ the slope of the tangent line is $f'(1) = 3$. Therefore a vector $\mathbf{v} = x\mathbf{i} + y\mathbf{j}$ parallel to the tangent line must have a slope of 3 (see the accompanying figure) and $y = 3x$. Since $\mathbf{v}$ is a unit vector, we have $x^2 + y^2 = 1$. Thus we have

$$x^2 + (3x)^2 = 1$$
$$10x^2 = 1$$
$$x = \pm\frac{1}{\sqrt{10}} \quad \text{and} \quad y = \pm\frac{3}{\sqrt{10}}$$

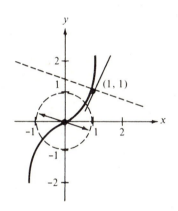

Finally we conclude that

$$\mathbf{v} = \frac{1}{\sqrt{10}}\mathbf{i} + \frac{3}{\sqrt{10}}\mathbf{j} = \left\langle \frac{1}{\sqrt{10}}, \frac{3}{\sqrt{10}} \right\rangle$$

or

$$\mathbf{v} = -\frac{1}{\sqrt{10}}\mathbf{i} - \frac{3}{\sqrt{10}}\mathbf{j} = \left\langle -\frac{1}{\sqrt{10}}, -\frac{3}{\sqrt{10}} \right\rangle$$

(b)  Similarly, a vector $\mathbf{v} = x\mathbf{i} + y\mathbf{j}$ normal to the tangent line must have a slope of $-\frac{1}{3}$ (see accompanying figure), and $x = -3y$. Since $\mathbf{v}$ is a unit vector, we have $x^2 + y^2 = 1$. Thus we have

$$(-3y)^2 + y^2 = 1$$
$$10y^2 = 1$$
$$y = \pm\frac{1}{\sqrt{10}} \quad \text{and} \quad x = \pm 3\sqrt{10}$$

Finally, we conclude that

$$\mathbf{v} = \frac{3}{\sqrt{10}}\mathbf{i} - \frac{1}{\sqrt{10}}\mathbf{j} = \left\langle \frac{3}{\sqrt{10}}, -\frac{1}{\sqrt{10}} \right\rangle$$

or

$$\mathbf{v} = \frac{-3}{\sqrt{10}}\mathbf{i} + \frac{1}{\sqrt{10}}\mathbf{j} = \left\langle -\frac{3}{\sqrt{10}}, \frac{1}{\sqrt{10}} \right\rangle$$

**49.** Find the component form of the vector **v** if it makes an angle of 150° with the positive $x$-axis and has magnitude $\|\mathbf{v}\| = 2$.

**Solution:**

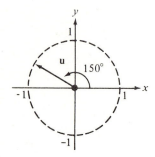

We first find a unit vector **u** making an angle of 150° with the positive $x$-axis. Since **u** is a unit vector, we consider it as the radius of a unit circle as shown in the accompanying figure. Therefore $x = \cos\theta$ and $y = \sin\theta$, and the component form for **u** is

$$\mathbf{u} = x\mathbf{i} + y\mathbf{j} = \cos(150°)\mathbf{i} + \sin(150°)\mathbf{j} = \left\langle -\frac{\sqrt{3}}{2}, \frac{1}{2} \right\rangle$$

Since $\mathbf{v} = 2\,\mathbf{u}$, we have $\mathbf{v} = \langle -\sqrt{3}, 1 \rangle$.

**59.** To carry a 100-lb cylindrical weight, two men lift on the ends of short ropes that are tied to an eyelet on the top center of the cylinder. If one rope makes a 20° angle away from the vertical and the other a 30° angle, find (a) the tension in each rope if the resultant force is vertical, and (b) the vertical component of each man's force.

**Solution:**

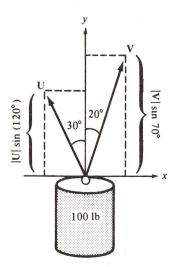

For the weight to be lifted vertically, the sum of the vertical components of **u** and **v** must be 100 and the sum of the horizontal components must be zero. (See the accompanying sketch.) Thus

$$\|\mathbf{u}\| \sin 120° + \|\mathbf{v}\| \sin 70° = 100$$

and

$$\|\mathbf{u}\| \cos 120° + \|\mathbf{v}\| \cos 70° = 0$$

Or, equivalently,

$$\|\mathbf{u}\| \left( \frac{\sqrt{3}}{2} \right) + \|\mathbf{v}\| \sin 70° = 100$$

$$\|\mathbf{u}\| \left( -\frac{1}{2} \right) + \|\mathbf{v}\| \cos 70° = 0$$

If we multiply the second equation by $\sqrt{3}$ and then add the two equations, we have

$$\|\mathbf{v}\|[\sqrt{3}\cos 70° + \sin 70°] = 100$$

$$\|\mathbf{v}\| = \frac{100}{\sqrt{3}\cos 70° + \sin 70°} \approx 65.27 \text{ lb}$$

Substituting 65.27 for $\|\mathbf{v}\|$ into the second equation, we have

$$65.27(\cos 70°) = \frac{\|\mathbf{u}\|}{2}$$

or

$$\|\mathbf{u}\| = 130.54(\cos 70°) \approx 44.65 \text{ lb}$$

(a)   The tension in each rope is

$$\|\mathbf{u}\| = 44.65 \text{ lb}, \qquad \|\mathbf{v}\| = 65.27 \text{ lb}$$

(b)   The vertical component of each man's force is

$$\|\mathbf{u}\| \sin 120° \approx (44.65)(0.8660) = 38.67 \text{ lb}$$
$$\|\mathbf{v}\| \sin 70° \approx (65.27)(0.9397) = 61.33 \text{ lb}$$

(Note that the sum of the two vertical components is 100 lb.)

## 13.2    The Dot Product of Two Vectors

**3.** Given the vectors $\mathbf{u} = \mathbf{i} - 4\mathbf{j}$ and $\mathbf{v} = \frac{1}{2}\mathbf{i} + 3\mathbf{j}$, find (a) $\mathbf{u} \cdot \mathbf{v}$, (b) $\mathbf{u} \cdot \mathbf{u}$, (c) $\|\mathbf{u}\|^2$, (d) $(\mathbf{u} \cdot \mathbf{v})\mathbf{v}$, and (e) $\mathbf{u} \cdot 2\mathbf{v}$.

**Solution:**

(a)   For the vectors $\mathbf{u} = u_1\mathbf{i} + u_2\mathbf{j}$ and $\mathbf{v} = v_1\mathbf{i} + v_2\mathbf{j}$

$$\mathbf{u} \cdot \mathbf{v} = u_1 v_1 + u_2 v_2$$
$$= 1\left(\frac{1}{2}\right) + (-4)(3) = -\frac{23}{2}$$

(b)   $\mathbf{u} \cdot \mathbf{u} = 1(1) + (-4)(-4) = 17$

(c)   $\|\mathbf{u}\|^2 = (1)^2 + (-4)^2 = 17 = \mathbf{u} \cdot \mathbf{u}$

(d)   From part (a) we have $\mathbf{u} \cdot \mathbf{v} = -\frac{23}{2}$. Therefore,

$$(\mathbf{u} \cdot \mathbf{v})\mathbf{v} = -\frac{23}{2}\mathbf{v} = -\frac{23}{4}\mathbf{i} - \frac{69}{2}\mathbf{j}.$$

(e)   From Theorem 13.5 and part (a) we have

$$\mathbf{u} \cdot (2\mathbf{v}) = 2(\mathbf{u} \cdot \mathbf{v}) = 2\left(-\frac{23}{2}\right) = -23$$

**11.** Find the angle $\theta$ between $\mathbf{u} = \cos\dfrac{\pi}{6}\mathbf{i} + \sin\dfrac{\pi}{6}\mathbf{j}$ and

$\mathbf{v} = \cos\dfrac{3\pi}{4}\mathbf{i} + \sin\dfrac{3\pi}{4}\mathbf{j}$.

**Solution:**

Since $\mathbf{u}$ and $\mathbf{v}$ are unit vectors making angles of magnitude $\pi/6$ and $3\pi/4$ with the positive $x$-axis, respectively, the angle $\theta$ between the vectors is given by

$$\theta = \frac{3\pi}{4} - \frac{\pi}{6} = \frac{7\pi}{12}.$$

An alternate method for finding the required angle is to use Theorem 13.6 and the trigonometric identity for the cosine of the difference of two angles.

$$\cos\theta = \frac{\mathbf{u}\cdot\mathbf{v}}{\|\mathbf{u}\|\,\|\mathbf{v}\|}$$
$$= \frac{\cos\frac{\pi}{6}\cos\frac{3\pi}{4} + \sin\frac{\pi}{6}\sin\frac{3\pi}{4}}{1\cdot 1}$$
$$= \cos\left(\frac{3\pi}{4} - \frac{\pi}{6}\right)$$

Therefore, $\theta = \dfrac{3\pi}{4} - \dfrac{\pi}{6} = \dfrac{7\pi}{12}$.

**19.** Determine whether $\mathbf{u} = \langle 2, 18\rangle$ and $\mathbf{v} = \left\langle \dfrac{3}{2}, -\dfrac{1}{6}\right\rangle$ are orthogonal, parallel, or neither.

**Solution:**

Since

$$\mathbf{u}\cdot\mathbf{v} = 2\left(\frac{3}{2}\right) + 18\left(-\frac{1}{6}\right) = 0,$$

the vectors are orthogonal.

**22.** Determine whether $\mathbf{u} = \left\langle -\dfrac{1}{3}, \dfrac{2}{3}\right\rangle$ and $\mathbf{v} = \langle 2, -4\rangle$ are orthogonal, parallel, or neither.

**Solution:**

Since

$$\mathbf{v} = \langle 2, -4\rangle = -6\left\langle -\frac{1}{3}, \frac{2}{3}\right\rangle = -6\,\mathbf{u},$$

the vectors are parallel.

**25.** For the vectors $\mathbf{u} = \langle 2, 3 \rangle$ and $\mathbf{v} = \langle 5, 1 \rangle$, find (a) the vector component of $\mathbf{u}$ along $\mathbf{v}$, and (b) the vector component of $\mathbf{u}$ orthogonal to $\mathbf{v}$.

**Solution:**

(a) The vector component of $\mathbf{u}$ along $\mathbf{v}$ is the projection of $\mathbf{u}$ onto $\mathbf{v}$ and is given by

$$\mathbf{w}_1 = \left( \frac{\mathbf{u} \cdot \mathbf{v}}{\|\mathbf{v}\|^2} \right) \mathbf{v}$$

$$= \left( \frac{13}{26} \right) \langle 5, 1 \rangle = \left\langle \frac{5}{2}, \frac{1}{2} \right\rangle .$$

(b) The vector component of $\mathbf{u}$ orthogonal to $\mathbf{v}$ is given by

$$\mathbf{w}_2 = \mathbf{u} - \mathbf{w}_1$$

$$= \langle 2, 3 \rangle - \left\langle \frac{5}{2}, \frac{1}{2} \right\rangle = \left\langle -\frac{1}{2}, \frac{5}{2} \right\rangle$$

**33.** An object is dragged 10 feet across a floor, using a force of 85 pounds. Find the work done if the direction of the force is $60°$ above the horizontal.

**Solution:**

The work done by a constant force $\mathbf{F}$ as its point of application moves along the vector $\overrightarrow{PQ}$ is given by

$$W = \mathbf{F} \cdot \overrightarrow{PQ}$$

From the accompanying figure we see that

$$\mathbf{F} = 85[(\cos 60°)\mathbf{i} + (\sin 60°)\mathbf{j}]$$

and

$$\overrightarrow{PQ} = 10\mathbf{i}.$$

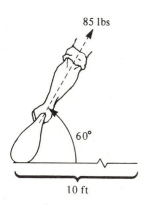

85 lbs

60°

10 ft

Therefore,

$$W = 85[(\cos 60°)(10) + (\sin 60°)(0)] = 425 \text{ ft} \cdot \text{lb}.$$

**35.** What is known about $\theta$, the angle between vectors $\mathbf{u}$ and $\mathbf{v}$, if:
(a) $\mathbf{u} \cdot \mathbf{v} = 0$? (b) $\mathbf{u} \cdot \mathbf{v} > 0$? (c) $\mathbf{u} \cdot \mathbf{v} < 0$?

**Solution:**

Assuming that $\mathbf{u}$ and $\mathbf{v}$ are nonzero vectors and $\mathbf{u} \cdot \mathbf{v} = \|\mathbf{u}\|\,\|\mathbf{v}\| \cos \theta$, we have the following:

(a) $\mathbf{u} \cdot \mathbf{v} = 0$ implies that $\cos \theta = 0$, or $\theta = \pi/2$.

(b) $\mathbf{u} \cdot \mathbf{v} > 0$ implies that $\cos \theta > 0$, or $0 \leq \theta \pi/2$.

(c) $\mathbf{u} \cdot \mathbf{v} < 0$ implies that $\cos \theta < 0$, or $\pi/2 < \theta \leq \pi$.

## 13.3   Vector-Valued Functions

**3.** Find the domain of the vector-valued function $\mathbf{r}(t) = e^t\,\mathbf{i} + \ln t\,\mathbf{j}$.

**Solution:**

The component functions of the vector-valued function

$$\mathbf{r}(t) = f(t)\,\mathbf{i} + g(t)\,\mathbf{j}$$

are the real-valued functions $f$, and $g$. Given the vector-valued function $\mathbf{r}(t) = e^t\,\mathbf{i} + \ln t\,\mathbf{j}$ the component functions and their domains are:

$$f(t) = e^t, \qquad -\infty < t < \infty$$
$$g(t) = \ln t, \qquad 0 < t < \infty$$

The intersection of the domains of $f$, and $g$, is the interval $(0, \infty)$, the domain of $\mathbf{r}$.

**9.** Sketch the curve represented by the vector-valued function $\mathbf{r}(t) = t\,\mathbf{i} + t^2\,\mathbf{j}$ and give the orientation of the curve.

**Solution:**

The parametric equations for the curve are given by

$$x(t) = t \quad \text{and} \quad y(t) = t^2.$$

If we eliminate the parameter from the parametric equations we have

$$y = (t)^2 = x^2.$$

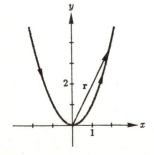

Therefore, the curve is a parabola opening upward with vertex at the origin.

**13.** Evaluate $\lim\limits_{t \to 0} \left( \dfrac{1 - \cos t}{t} \mathbf{i} + t^2 \mathbf{j} \right)$

**Solution:**

$$\lim_{t \to 0} \left( \frac{1 - \cos t}{t} \mathbf{i} + t^2 \mathbf{j} \right) = \left[ \lim_{t \to 0} \frac{1 - \cos t}{t} \right] \mathbf{i} + \left[ \lim_{t \to 0} t^2 \right] \mathbf{j}$$
$$= 0\,\mathbf{i} + 0\,\mathbf{j} = 0$$

**23.** Sketch the graph of the vector-valued function
$\mathbf{r}(t) = \cos t\,\mathbf{i} + \sin t\,\mathbf{j}$ in the $xy$-coordinate plane, and (b) sketch
the vectors $\mathbf{r}(\pi/2)$ and $\mathbf{r}'(\pi/2)$. Position the vectors so that the
initial point of $\mathbf{r}(\pi/2)$ is at the origin and the initial point of
$\mathbf{r}'(\pi/2)$ is at the terminal point of $\mathbf{r}(\pi/2)$.

**Solution:**

(a) Eliminating the parameter from the parametric equations
$x = \cos t$ and $y = \sin t$, we have

$$x^2 + y^2 = \cos^2 t + \sin^2 t = 1.$$

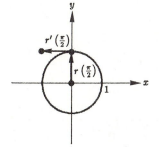

Therefore, the vector-valued function represents a circle of
radius 1 center at the origin. (see the accompanying figure.)

(b)     $\mathbf{r}(t) = \cos t\,\mathbf{i} + \sin t\,\mathbf{j} \quad \Longrightarrow \quad \mathbf{r}(\pi/2) = \mathbf{j}$
          $\mathbf{r}'(t) = -\sin t\,\mathbf{i} + \cos t\,\mathbf{j} \quad \Longrightarrow \quad \mathbf{r}'(\pi/2) = -\mathbf{i}$

**31.** For the vector-valued functions $\mathbf{r}(t) = 3t\,\mathbf{i} + 4t\,\mathbf{j}$ and
$\mathbf{u}(t) = 4t\,\mathbf{i} + t^2\,\mathbf{j}$ find the following:
(a)   $\mathbf{r}'(t)$                             (b)   $D_t[\mathbf{r}(t) \cdot \mathbf{u}(t)]$
(c)   $D_t[3\,\mathbf{r}(t) - \mathbf{u}(t)]$        (d)   $D_t[\|\mathbf{r}(t)\|]$

**Solution:**

(a) Differentiating in a component-by-component basis
produces the following.

$$\mathbf{r}'(t) = 3\,\mathbf{i} + 4\,\mathbf{j}$$

(b)  $\mathbf{r}(t) \cdot \mathbf{u}(t) = 3t(4t) + 4t(t^2) = 12t^2 + 4t^3$
$D_t[\mathbf{r}(t) \cdot \mathbf{u}(t)] = 24t + 12t^2 = 12t(2 + t)$

(c)  $3\,\mathbf{r}(t) - \mathbf{u}(t) = 3(3t\,\mathbf{i} + 4t\,\mathbf{j}) - (4t\,\mathbf{i} + t^2\,\mathbf{j})$
$= 5t\,\mathbf{i} + (12t - t^2)\,\mathbf{j}$
$D_t[3\,\mathbf{r}(t) - \mathbf{u}(t)] = 5\,\mathbf{i} + (12 - 2t)\,\mathbf{j}$

(d)  $\|\mathbf{r}(t)\| = \sqrt{(3t)^2 + (4t)^2} = \sqrt{25t^2} = 5t$
$D_t[\|\mathbf{r}(t)\|] = 5$

**41.** Evaluate the indefinite integral $\displaystyle\int (4 \sin t\,\mathbf{i} + 3 \cos t\,\mathbf{j})\, dt$.

**Solution:**

$$\int (4 \sin t\,\mathbf{i} + 3 \cos t\,\mathbf{j})\, dt = \left[\int 4 \sin t\, dt\right]\mathbf{i} + \left[\int 3 \cos t\, dt\right]\mathbf{j}$$
$$= (-4 \sin t + C_1)\mathbf{i} + (3 \sin t + C_2)\mathbf{j}$$
$$= -4 \sin t\,\mathbf{i} + 3 \sin t\,\mathbf{j} + \mathbf{C}$$

**47.** Find the position vector $\mathbf{r}$ if $\mathbf{r}'(t) = 4e^{2t}\,\mathbf{i} + 3e^t\,\mathbf{j}$ and $\mathbf{r}(0) = 2\,\mathbf{i}$.

**Solution:**

$$\mathbf{r}(t) = \mathbf{i}\int 4e^{2t}\, dt + \mathbf{j}\int 3e^t\, dt = \mathbf{i}[2e^{2t} + C_1] + \mathbf{j}[3e^t + C_2]$$
$$\mathbf{r}(0) = \mathbf{i}[2 + C_1] + \mathbf{j}[3 + C_2] = 2\,\mathbf{i}$$

Therefore, $2 + C_1 = 2$, or $C_1 = 0$. Also, $3 + C_2 = 0$, or $C_2 = -3$. Thus

$$\mathbf{r}(t) = 2e^{2t}\,\mathbf{i} + (3e^t - 3)\mathbf{j} = 2e^{2t}\,\mathbf{i} + 3(e^t - 1)\mathbf{j}$$

## 13.4   Velocity and Acceleration

**3.** The motion of an object in the $xy$-plane is described by the vector-valued function $\mathbf{r}(t) = t^2\mathbf{i} + t\mathbf{j}$. Sketch a graph of the path and sketch the velocity and acceleration vectors at the point $(4, 2)$.

**Solution:**

Eliminating the parameter from the parametric equations $x = t^2$ and $y = t$, we obtain the rectangular equation $x = y^2$. Therefore, the object is moving in a parabolic path. (See the accompanying figure.)
The velocity is given by

$$\mathbf{v}(t) = \mathbf{r}'(t) = 2t\,\mathbf{i} + \mathbf{j}$$

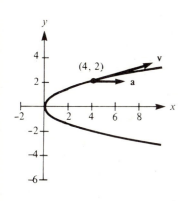

and the acceleration is given by

$$\mathbf{a}(t) = \mathbf{r}''(t) = 2\,\mathbf{i}$$

At the point $(4, 2), t = 2$. Thus

$$\mathbf{v}(2) = 4\,\mathbf{i} + \mathbf{j}$$

and

$$\mathbf{a}(2) = 2\,\mathbf{i}$$

**11.** Use the acceleration function $\mathbf{a}(t) = t\mathbf{i} + t\mathbf{j}$ to find the velocity and position functions if $\mathbf{v}(1) = 5\mathbf{j}$ and $\mathbf{r}(1) = \mathbf{0}$. Find the position at time $t = 2$.

**Solution:**

$$\mathbf{v}(t) = \int \mathbf{a}(t)\, dt + \mathbf{C}$$
$$= \int (t\mathbf{i} + t\mathbf{j})\, dt + \mathbf{C}$$
$$= \frac{1}{2}t^2\mathbf{i} + \frac{1}{2}t^2\mathbf{j} + C_1\mathbf{i} + C_2\mathbf{j}$$
$$\mathbf{v}(1) = (\frac{1}{2} + C_1)\mathbf{i} + (\frac{1}{2} + C_2)\mathbf{j} = 5\mathbf{j}$$

Therefore,

$$\frac{1}{2} + C_1 = 0 \qquad \Longrightarrow \qquad C_1 = -\frac{1}{2}$$

$$\frac{1}{2} + C_2 = 5 \qquad \Longrightarrow \qquad C_2 = \frac{9}{2}$$

Thus, the velocity vector is

$$\mathbf{v}(t) = \left(\frac{t^2}{2} - \frac{1}{2}\right)\mathbf{i} + \left(\frac{t^2}{2} + \frac{9}{2}\right)\mathbf{j}$$

$$\mathbf{r}(t) = \int \mathbf{v}(t)\, dt + \mathbf{C}$$

$$= \int \left(\frac{t^2}{2} - \frac{1}{2}\right) dt\, \mathbf{i} + \int \left(\frac{t^2}{2} + \frac{9}{2}\right) dt\, \mathbf{k} + \mathbf{C}$$

$$= \left(\frac{t^3}{6} - \frac{1}{2}t\right)\mathbf{i} + \left(\frac{t^3}{6} + \frac{9}{2}t\right)\mathbf{j} + C_3\mathbf{i} + C_4\mathbf{j}$$

$$\mathbf{r}(1) = \left(\frac{1}{6} - \frac{1}{2} + C_3\right)\mathbf{i} + \left(\frac{1}{6} + \frac{9}{2} + C_4\right)\mathbf{j}$$

$$= \left(-\frac{1}{3} + C_3\right)\mathbf{i} + \left(\frac{14}{3} + C_4\right)\mathbf{j} = 0$$

This implies that

$$C_3 = \frac{1}{3}, \text{ and } C_4 = \frac{-14}{3}.$$

Thus, the position vector is

$$\mathbf{r}(t) = \left(\frac{t^3}{6} - \frac{1}{2}t + \frac{1}{3}\right)\mathbf{i} + \left(\frac{t^3}{6} + \frac{9}{2}t - \frac{14}{3}\right)\mathbf{j}$$

and

$$\mathbf{r}(2) = \frac{2}{3}\mathbf{i} + \frac{17}{3}\mathbf{j}$$

**15.** A baseball, hit three feet above the ground, leaves the bat at an angle of $45°$ and is caught by an outfielder 300 feet from the plate. What was the initial velocity of the ball and how high did it rise if it was caught 3 feet above the ground?

**Solution:**

Using Theorem 13.10 with $h = 3$, the path of the ball is given by

$$\mathbf{r}(t) = (v_0 \cos 45°)t\, \mathbf{i} + [3 + (v_0 \sin 45°t - 16t^2]\mathbf{j}$$

$$= \left(\frac{tv_0}{\sqrt{2}}\right)\mathbf{i} + \left(3 + \frac{tv_0}{\sqrt{2}} - 16t^2\right)\mathbf{j}$$

We know that the horizontal component is 300 when the vertical component is three. Thus

$$\frac{tv_0}{\sqrt{2}} = 300 \qquad \text{and} \qquad 3 + \frac{tv_0}{\sqrt{2}} - 16t^2 = 3$$

Therefore $t = 300\sqrt{2}/v_0$ and is follows that

$$\frac{300\sqrt{2}}{v_0}\left(\frac{v_0}{\sqrt{2}}\right) - 16\left(\frac{300\sqrt{2}}{v_0}\right)^2 = 0$$

$$300 = \frac{16(300^2)(2)}{v_0^2}$$

$$v_0^2 = 32(300)$$

$$v_0 = \sqrt{9600} = 40\sqrt{6} \approx 97.98 \text{ ft/s}$$

The maximum height is reached when the derivative of the vertical component is zero. Thus we have

$$y(t) = 3 + \frac{tv_0}{\sqrt{2}} - 16t^2 = 3 + \frac{t(40\sqrt{6})}{\sqrt{2}} - 16t^2$$

$$= 3 + 40\sqrt{3}t - 16t^2$$

$$y'(t) = 40\sqrt{3} - 32t = 0$$

$$t = \frac{40\sqrt{3}}{32} = \frac{5\sqrt{3}}{4}$$

Finally, the maximum height is

$$y\left(\frac{5\sqrt{3}}{4}\right) = 3 + 40\sqrt{3}\left(\frac{5\sqrt{3}}{4}\right) - 16\left(\frac{5\sqrt{3}}{4}\right)^2$$

$$= 3 + 150 - 75 = 78 \text{ ft}$$

**23.** The path of a ball is given by the rectangular equation $y = x - 0.005x^2$. Use the result of Exercise 22 to find the vector-valued position function. Then find the speed and direction of the ball at the point when it has traveled a horizontal distance of 60 feet.

**Solution:**

From Exercise 22 we know the equation $y = -0.005x^2 + x$ is of the form

$$y = -\left[\frac{16(\sec^2 \theta)}{v_0^2}\right]x^2 + (\tan \theta)x + h.$$

Thus, $\tan\theta = 1$, and $\theta = 45°$. Since the coefficients of $x^2$ must be equal and $\sec 45° = \sqrt{2}$, we have

$$-0.005 = -\frac{16(\sec 45°)^2}{v_0{}^2}$$

$$v_0{}^2 = \frac{16(2)}{0.005} = 6400$$

$$v_0 = 80 \text{ ft/sec}$$

From Theorem 13.10 the vector-valued function for the path of the ball is given by

$$\mathbf{r}(t) = t(v_0\cos\theta)\mathbf{i} + \left(h + tv_0\sin\theta - \frac{gt^2}{2}\right)\mathbf{j}$$

$$= t(80\cos 45°)\mathbf{i} + [t(80)\sin 45° - 16t^2]$$

$$= 40\sqrt{2}t\,\mathbf{i} + (40\sqrt{2}t - 16t^2)\mathbf{j}.$$

[Since the path of the ball starts at the point $(0,0)$, $h = 0$.] The velocity of the ball is given by

$$\mathbf{v}(t) = \mathbf{r}'(t) = 40\sqrt{2}\,\mathbf{i} + (40\sqrt{2} - 32t)\mathbf{j}.$$

Furthermore, when $x = 60$, we have

$$40\sqrt{2}t = 60 \quad\text{or}\quad t = \frac{60}{40\sqrt{2}} = \frac{3\sqrt{2}}{4}.$$

Thus the velocity when $x = 60$ is given by

$$\mathbf{v}\left(\frac{3\sqrt{2}}{4}\right) = 40\sqrt{2}\,\mathbf{i} + \left[40\sqrt{2} - 32\left(\frac{3\sqrt{2}}{4}\right)\right]\mathbf{j}$$

$$= 40\sqrt{2}\,\mathbf{i} + 16\sqrt{2}\,\mathbf{j} = 8\sqrt{2}(5\,\mathbf{i} + 2\,\mathbf{j})$$

which implies that the speed is

$$\|\mathbf{v}\| = 8\sqrt{2}\sqrt{5^2 + 2^2} = 8\sqrt{2}\sqrt{29} = 8\sqrt{58} \text{ ft/sec.}$$

27. Consider a particle moving on a circular path of radius $b$ described by

$$\mathbf{r}(t) = b\cos\omega t\,\mathbf{i} + b\sin\omega t\,\mathbf{j}$$

where $\omega = d\theta/dt$ is the constant angular velocity. Find the velocity vector and show that it is orthogonal to $\mathbf{r}(t)$.

**Solution:**

The velocity vector is

$$\mathbf{v}(t) = \mathbf{r}'(t) = -b\omega\sin\omega t\,\mathbf{i} + b\omega\cos\omega t\,\mathbf{j}$$

and since

$$\mathbf{r}(t)\cdot\mathbf{v}(t) = -b^2\omega\sin\omega t\cos\omega t + b^2\omega\sin\omega t\cos\omega t = 0$$

we conclude that $\mathbf{r}(t)$ and $\mathbf{v}(t)$ are orthogonal.

**28.** Show that the speed of the particle is $b\omega$ if
$\mathbf{r}(t) = b\cos\omega t\,\mathbf{i} + b\sin\omega t\,\mathbf{j}$.

**Solution:**

$$\text{speed} = ||\mathbf{v}|| = \sqrt{b^2\omega^2\sin^2\omega t + b^2\omega^2\cos^2\omega t}$$
$$= \sqrt{b^2\omega^2[\sin^2\omega t + \cos^2\omega t]} = b\omega$$

**29.** Find the acceleration vector and show that its direction is always toward the center of the circle if
$\mathbf{r}(t) = b\cos\omega t\,\mathbf{i} + b\sin\omega t\,\mathbf{j}$.

**Solution:**

Since

$$\mathbf{a}(t) = \mathbf{r}''(t) = [-b\omega^2\cos\omega t]\,\mathbf{i} - [b\omega^2\sin\omega t]\,\mathbf{j}$$
$$= -b\omega^2[\cos\omega t\,\mathbf{i} + \sin\omega t\,\mathbf{j}]$$
$$= -b\omega^2\,\mathbf{r}(t)$$

we see that $\mathbf{a}(t)$ is a negative multiple of a unit vector from $(0,0)$ to $(\cos\omega t, \sin\omega t)$, and thus $\mathbf{a}(t)$ is directed toward the origin.

**30.** Show that the magnitude of the acceleration vector is $b\omega^2$ if
$\mathbf{r}(t) = b\cos\omega t\,\mathbf{i} + b\sin\omega t\,\mathbf{j}$.

**Solution:**

$$||\mathbf{a}(t)|| = b\omega^2 || \cos\omega t\,\mathbf{i} + \sin\omega t\,\mathbf{j}|| = b\omega^2$$

## 13.5   Tangent Vectors and Normal Vectors

**5.** If $\mathbf{r} = t\,\mathbf{i} + (1/t)\,\mathbf{j}$, find $\mathbf{T}(t)$, $\mathbf{N}(t)$, $\mathbf{a}(t)\cdot\mathbf{T}(t)$, and $\mathbf{a}(t)\cdot\mathbf{N}(t)$ when $t = 1$.

**Solution:**

$$\mathbf{r}(t) = t\,\mathbf{i} + \frac{1}{t}\,\mathbf{j}$$

$$\mathbf{v}(t) = \mathbf{r}'(t) = \mathbf{i} - \frac{1}{t^2}\,\mathbf{j}$$

$$\mathbf{a}(t) = \mathbf{r}''(t) = \frac{2}{t^3}\,\mathbf{j}$$

At $t = 1$, we have

$$\mathbf{v}(1) = \mathbf{i} - \mathbf{j}, \quad \|\mathbf{v}(1)\| = \sqrt{2}, \quad \text{and} \quad \mathbf{a}(1) = 2\mathbf{j}$$

Therefore, when $t = 1$,

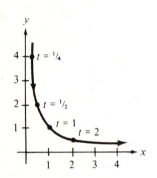

$$\mathbf{T}(1) = \frac{\mathbf{v}(1)}{\|\mathbf{v}(1)\|} = \frac{\mathbf{i}}{\sqrt{2}} - \frac{\mathbf{j}}{\sqrt{2}} = \frac{\sqrt{2}}{2}(\mathbf{i} - \mathbf{j})$$

and since $\mathbf{N}(1)$ points toward the concave side of the curve (see the accompanying figure)

$$\mathbf{N}(1) = \frac{\mathbf{i}}{\sqrt{2}} + \frac{\mathbf{j}}{\sqrt{2}} = \frac{\sqrt{2}}{2}(\mathbf{i} + \mathbf{j}).$$

Furthermore,

$$\mathbf{a}(1) \cdot \mathbf{T}(1) = (2\mathbf{j}) \cdot \left(\frac{\mathbf{i}}{\sqrt{2}} - \frac{\mathbf{j}}{\sqrt{2}}\right) = \frac{-2}{\sqrt{2}} = -\sqrt{2}$$

and

$$\mathbf{a}(1) \cdot \mathbf{N}(1) = (2\mathbf{j}) \cdot \left(\frac{\mathbf{i}}{\sqrt{2}} + \frac{\mathbf{j}}{\sqrt{2}}\right) = \frac{2}{\sqrt{2}} = \sqrt{2}$$

**11.** If $\mathbf{r}(t) = e^t \cos t \, \mathbf{i} + e^t \sin t \, \mathbf{j}$, find $\mathbf{T}(t)$, $\mathbf{N}(t)$, $\mathbf{a}(t) \cdot \mathbf{T}(t)$, and $\mathbf{a}(t) \cdot \mathbf{N}(t)$, when $t = \pi/2$.

**Solution:**

$$\mathbf{r}(t) = e^t \cos t \, \mathbf{i} + e^t \sin t \, \mathbf{j}$$
$$\mathbf{v}(t) = \mathbf{r}'(t) = e^t(\cos t - \sin t)\mathbf{i} + e^t(\cos t + \sin t)\mathbf{j}$$
$$\mathbf{a}(t) = \mathbf{r}''(t)$$
$$= e^t(-\sin t - \cos t + \cos t - \sin t)\mathbf{i}$$
$$+ e^t(-\sin t + \cos t + \cos t + \sin t)\mathbf{j}$$
$$= e^t(-2\sin t)\mathbf{i} + e^t(2\cos t)\mathbf{j}$$

Then at $t = \pi/2$, we have

$$\mathbf{v}(\pi/2) = -e^{\pi/2}\mathbf{i} + e^{\pi/2}\mathbf{j}$$
$$\|\mathbf{v}(\pi/2)\| = e^{\pi/2}\sqrt{2}$$
$$\mathbf{a}(\pi/2) = -2e^{\pi/2}\mathbf{i}$$

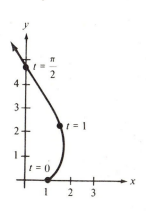

Therefore, at $t = \pi/2$,

$$\mathbf{T}(\pi/2) = \frac{\mathbf{v}(\pi/2)}{\|\mathbf{v}(\pi/2)\|} = \frac{-\mathbf{i}}{\sqrt{2}} + \frac{\mathbf{j}}{\sqrt{2}} = \frac{\sqrt{2}}{2}(-\mathbf{i} + \mathbf{j})$$

and since $\mathbf{N}(\pi/2)$ points toward the concave side of the curve (see the accompanying figure)

$$\mathbf{N}(\pi/2) = -\frac{\sqrt{2}}{2}(\mathbf{i} + \mathbf{j})$$

$$\mathbf{a}(\pi/2) \cdot \mathbf{T}(\pi/2) = (-2e^{\pi/2}\,\mathbf{i}) \cdot \frac{(-\mathbf{i} + \mathbf{j})}{\sqrt{2}} = \sqrt{2}e^{\pi/2}$$

$$\mathbf{a}(\pi/2) \cdot \mathbf{N}(\pi/2) = (-2e^{\pi/2}\,\mathbf{i}) \cdot \frac{(-\mathbf{i} - \mathbf{j})}{\sqrt{2}} = \sqrt{2}e^{\pi/2}$$

**17.** Find the tangential and normal components of acceleration for a projectile fired at an angle $\theta$ with the horizontal and with an initial speed of $v_0$. What are the components when the projectile is at its maximum height?

**Solution:**

From Theorem 13.10, we have

$$\mathbf{r}(t) = (v_0 t \cos \theta)\mathbf{i} + (h + v_0 t \sin \theta - 16t^2)\mathbf{j}$$
$$\mathbf{v}(t) = (v_0 \cos \theta)\mathbf{i} + (v_0 \sin \theta - 32t)\mathbf{j}$$
$$\mathbf{a}(t) = -32\mathbf{j}$$
$$\mathbf{T}(t) = \frac{\mathbf{v}(t)}{\|\mathbf{v}(t)\|} = \frac{(v_0 \cos \theta)\mathbf{i} + (v_0 \sin \theta - 32t)\mathbf{j}}{\sqrt{v_0^2 \cos^2 \theta + (v_0 \sin \theta - 32t)^2}}$$

Since the path of a projectile is concave downward, we choose

$$\mathbf{N}(t) = \frac{(v_0 \sin \theta - 32t)\mathbf{i} + (-v_0 \cos \theta)\mathbf{j}}{\sqrt{v_0^2 \cos^2 \theta + (v_0 \sin \theta - 32t)^2}}$$

Therefore,

$$\mathbf{a}(t) \cdot \mathbf{T}(t) = \frac{-32(v_0 \sin \theta - 32t)}{\sqrt{v_0^2 \cos^2 \theta + (v_0 \sin \theta - 32t)^2}}$$
$$\mathbf{a}(t) \cdot \mathbf{N}(t) = \frac{32v_0 \cos \theta}{\sqrt{v_0^2 \cos^2 \theta + (v_0 \sin \theta - 32t)^2}}$$

The projectile will reach its maximum height when the vertical component of velocity is zero, or

$$v_0 \sin \theta - 32t = 0$$

Therefore, at maximum height $\mathbf{a}(t) \cdot \mathbf{T}(t) = 0$ and $\mathbf{a}(t) \cdot \mathbf{N}(t) = 32$. Thus, all the acceleration is normal to the path at the maximum height.

**20.** An object of mass $m$ moves at a constant speed $v$ in a circular path of radius $r$. The force required to produce the centripetal component of acceleration is called the **centripetal force.** Show that this force is $F = mv^2/r$. Newton's **Law of Universal Gravitaiton** is given by $F = GMm/d^2$, where $d$ is the distance between the centers of the two bodies of mass $M$ and $m$. Use this to show that the speed required for circular motion is $v = \sqrt{GM/r}$.

**Solution:**

Since the motion is circular, the path is described by

$$\mathbf{r}(t) = r\cos\omega t\,\mathbf{i} + r\sin\omega t\,\mathbf{j}$$

Furthermore,

$$\mathbf{v}(t) = -r\omega\sin\omega t\,\mathbf{i} + r\omega\cos\omega t\,\mathbf{j}$$
$$\|\mathbf{v}(t)\| = r\omega\sqrt{1} = r\omega$$

and

$$\mathbf{a}(t) = (-r\omega^2\cos\omega t)\,\mathbf{i} - (r\omega^2\sin\omega t)\,\mathbf{j}$$
$$\|\mathbf{a}(t)\| = r\omega^2\sqrt{1} = r\omega^2$$

Since force equals mass times acceleration, we have

$$F = m[\mathbf{a}(t)] = m(r\omega^2) = \frac{m}{r}(r^2\omega^2) = \frac{mv^2}{r}$$

If we consider the moving object to have a point mass $m$, then in Newton's Law we let $d = r$ and we have

$$\frac{mv^2}{r} = \frac{GMm}{r^2} \quad \text{or} \quad v^2 = \frac{GM}{r} \quad \text{or} \quad v = \sqrt{\frac{GM}{r}}$$

**24.** Use the result of Exercise 20 to find the speed necessary for the circular orbit of a syncom satellite in a geosynchronous orbit $r$ miles above the surface of the earth. Let $GM = 9.56 \times 10^4 \text{mi}^3/\text{s}^2$ and assume the radius of the earth is 4000 miles. [The satellite completes one orbit per sidereal day (23 hours, 56 minutes) and thus appears to remain stationary above a point on the earth.]

**Solution:**

We let $r$ be the radius of the orbit measured from the center of the earth. Then the distance the satellite travels in one day is $d = 2\pi r$ and the speed is $v = 2\pi r/t = 2\pi r/(24)(3600)$. However, from Exercise 34 the speed is also given bby $v = \sqrt{9.56(10^4)}/r$. Solving these two equations simultaneously, we have

$$\frac{2\pi r}{(24)(3600)} = \sqrt{\frac{9.56(10^4)}{r}}$$

$$\frac{4\pi^2 r^2}{(24^2)(3600^2)} = \frac{9.56(10^4)}{r}$$

$$r^3 = \frac{9.56(10^4)(24^2)(3600^2)}{4\pi^2}$$

$$r = \sqrt[3]{\frac{9.56(10^4)(24^2)(3600^2)}{4\pi^2}} \approx 26,245 \text{ mi}$$

Therefore the altitude above the earth is $26,245 - 4,000 = 22,245$ mi, and the speed is

$$v = \frac{\text{distance}}{\text{time}} = \frac{2\pi(26,245)}{(24)(3,600)}$$

$$\approx 1.91 \text{ mi/sec} \approx 6,871 \text{ mi/hr.}$$

## 13.6   Arc Length and Curvature

**3.** Sketch and find the length of the vector-valued function

$$\mathbf{r}(t) = a\cos^3 t\,\mathbf{i} + a\sin^3 t\,\mathbf{j}$$

over the interval $[0, 2\pi]$.

**Solution:**

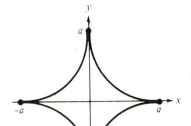

The graph is a hypocycloid of four cusps. (See the accompanying figure.) Since

$$\mathbf{r}\,(t) = 3a\cos^2 t(-\sin t)\mathbf{i} + 3a\sin^2 t(\cos t)\mathbf{j},$$

the arc length of the interval $[0, 2\pi]$ is

$$s = \int_0^{2\pi} \|\mathbf{r}'(t)\|\, dt$$

$$= 4\int_0^{\pi/2} \sqrt{(-3a\cos^2 t\sin t)^2 + (3a\sin^2 t\cos t)^2}\, dt$$

$$= 4\int_0^{\pi/2} \sqrt{9a^2\cos^4 t\sin^2 t + 9a^4\sin^4 t\cos^2 t}\, dt$$

$$= 12a\int_0^{\pi/2} \sin t\cos t\sqrt{\cos^2 t + \sin^2 t}\, dt$$

$$= 12a\int_0^{\pi/2} \sin t\cos t\, dt$$

$$= 6a\big[\sin^2 t\big]_0^{\pi/2} = 6a$$

**11.** Find the curvature and radius of curvature of the plane curve $y = 2x^2 + 3$ at $x = -1$.

**Solution:**

$$y = 2x^2 + 3 \qquad y' = 4x \qquad y'' = 4$$

$$K = \left|\frac{y''}{[1 + (y')^2]^{3/2}}\right| = \frac{4}{[1 + (4x)^2]^{3/2}}$$

When $x = -1$,

$$K = \frac{4}{(1 + 16)^{3/2}} = \frac{4}{17^{3/2}} \approx 0.057$$

The radius of curvature when $x = -1$ is

$$r = \frac{1}{K} = \frac{17^{3/2}}{4} \approx 17.523$$

**23.** Find the curvature $K$ of the curve $\mathbf{r}(t) = t\mathbf{i} + \dfrac{1}{t}\mathbf{j}$ at $t = 1$.

**Solution:**

From Exercise 5, Section 13.5, we have

$$\mathbf{a}(1) \cdot \mathbf{N}(1) = \sqrt{2} \qquad \text{and} \qquad \|\mathbf{v}(1)\|^2 = 2$$

Therefore, the curvature is

$$K = \frac{\mathbf{a}(1) \cdot \mathbf{N}(1)}{\|\mathbf{v}(1)\|} = \frac{\sqrt{2}}{2} \approx 0.707$$

**27.** Find the curvature $K$ of the curve $\mathbf{r}(t) = e^t \cos t\, \mathbf{i} + e^t \sin t\, \mathbf{j}$.

**Solution:**

$$\mathbf{r}(t) = e^t \cos t\, \mathbf{i} + e^t \sin t\, \mathbf{j}$$
$$\mathbf{r}'(t) = e^t(\cos t - \sin t)\mathbf{i} + e^t(\cos t + \sin t)\mathbf{j}$$
$$\|\mathbf{r}'(t)\| = e^t \sqrt{(\cos t - \sin t)^2 + (\cos t + \sin t)^2} = \sqrt{2}\,e^t$$
$$\mathbf{T}(t) = \frac{\mathbf{r}'(t)}{\|\mathbf{r}'(t)\|} = \frac{1}{\sqrt{2}}[(\cos t - \sin t)\mathbf{i} + (\cos t + \sin t)\mathbf{j}]$$
$$\mathbf{T}'(t) = \frac{1}{\sqrt{2}}[(-\sin t - \cos t)\mathbf{i} + (-\sin t + \cos t)\mathbf{j}]$$
$$\|\mathbf{T}'(t)\| = \frac{1}{\sqrt{2}}\sqrt{(-\sin t - \cos t)^2 + (-\sin t + \cos t)^2} = 1$$
$$K = \frac{\|\mathbf{T}'(t)\|}{\|\mathbf{r}'(t)\|} = \frac{1}{\sqrt{2}\,e^t} = \frac{\sqrt{2}}{2}e^{-t}$$

**37.** Find the circle of curvature of the graph of $y = x + (1/x)$ at the point $(1, 2)$.

**Solution:**

$$f(x) = x + \frac{1}{x} \quad f'(x) = 1 - \frac{1}{x^2} = \frac{x^2 - 1}{x^2} \quad f''(x) = \frac{2}{x^3}$$

At the point $(1, 2)$, $f'(1) = 0$ and $f''(1) = 2$. Thus at $(1, 2)$ the curvature is

$$K = \left| \frac{2}{(1 + 0^2)^{3/2}} \right| = 2$$

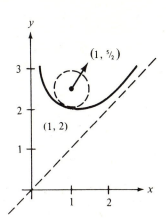

and the radius of curvature is

$$r = \frac{1}{K} = \frac{1}{2}$$

Since the slope at $(1, 2)$ is 0, the normal line at $(1, 2)$ is vertical and the center of the closest circular approximation is $(1, \frac{5}{2})$. (See the accompanying figure.) Finally, the equation of the closest circular approximation is

$$(x - 1)^2 + \left(y - \frac{5}{2}\right)^2 = \left(\frac{1}{2}\right)^2$$

**39.** Show that the curvature is greatest at the endpoints of the major axis and least at the endpoints of the minor axis for the ellipse given by $x^2 + 4y^2 = 4$.

**Solution:**

The endpoints of the major axis are $(\pm 2, 0)$ and the endpoints of the minor axis are $(0, \pm 1)$.

$$x^2 + 4y^2 = 4$$
$$2x + 8yy' = 0$$
$$y' = \frac{-x}{4y}$$
$$y'' = \frac{(4y)(-1) - (-x)(4y')}{16y^2} = \frac{-4y - (x^2/y)}{16y^2}$$
$$= \frac{-(4y^2 + x^2)}{16y^3} = \frac{-4}{16y^3} = \frac{-1}{4y^3}$$

The curvature is given by

$$K = \left| \frac{-1/4y^3}{[1 + (-x/4y)^2]^{3/2}} \right| = \left| \frac{-1}{4y^3[(16y^2 + x^2)/16y^2]^{3/2}} \right|$$
$$= \left| \frac{-16}{(16y^2 + x^2)^{3/2}} \right| = \frac{16}{(12y^2 + 4y^2 + x^2)^{3/2}} = \frac{16}{(12y^2 + 4)^{3/2}}$$

Therefore, since $-1 \le y \le 1$, $K$ is largest when $y = 0$ and smallest when $y = \pm 1$.

## Review Exercises for Chapter 13

**1.** Given the points $P = (1, 2)$, $Q = (4, 1)$, and $R = (5, 4)$, find (a) the component forms of $\mathbf{u} = \overrightarrow{PQ}$ and $\mathbf{v} = \overrightarrow{PR}$, (b) the magnitude of $\mathbf{v}$, (c) $\mathbf{u} \cdot \mathbf{v}$, (d) $2\mathbf{u} + \mathbf{v}$, (e) the vector component of $\mathbf{u}$ in the direction of $\mathbf{v}$, and (f) the vector component of $\mathbf{u}$ orthogonal to $\mathbf{v}$.

**Solution:**

(a)
$$\mathbf{u} = \overrightarrow{PQ} = \langle 4 - 1, 1 - 2 \rangle = \langle 3, -1 \rangle$$
$$\mathbf{v} = \overrightarrow{PR} = \langle 5 - 1, 4 - 2 \rangle = \langle 4, 2 \rangle$$

(b)
$$\|\mathbf{v}\| = \sqrt{4^2 + 2^2} = \sqrt{20} = 2\sqrt{5}$$

(c)
$$\mathbf{u} \cdot \mathbf{v} = 3(4) + (-1)(2) = 12 - 2 = 10$$

(d)
$$2\mathbf{u} + \mathbf{v} = 2\langle 3, -1 \rangle + \langle 4, 2 \rangle$$
$$= \langle 6, -2 \rangle + \langle 4, 2 \rangle = \langle 10, 0 \rangle$$

(e) The vector component of $\mathbf{u}$ in the direction of $\mathbf{v}$ is the projection of $\mathbf{u}$ onto $\mathbf{v}$ and is given by

$$\mathbf{w}_1 = \left( \frac{\mathbf{u} \cdot \mathbf{v}}{\|\mathbf{v}\|^2} \right) \mathbf{v}$$
$$= \left( \frac{10}{20} \right) \langle 4, 2 \rangle = \langle 2, 1 \rangle.$$

(f) The vector component of $\mathbf{u}$ orthogonal to $\mathbf{v}$ is given by

$$\mathbf{w}_2 = \mathbf{u} - \mathbf{w}_1$$
$$= \langle 3, -1 \rangle - \langle 2, 1 \rangle = \langle 1, -2 \rangle$$

**7.** Determine whether $\mathbf{u} = \langle 8, -12 \rangle$ and $\mathbf{v} = \langle -2, 3 \rangle$ are orthogonal, parallel, or neither.

**Solution:**

Since

$$\mathbf{u} = \langle 8, -12 \rangle = -4\langle -2, 3 \rangle = -4\mathbf{v},$$

the vectors are parallel.

**17.** For the vector-valued functions $\mathbf{r}(t) = 3t\,\mathbf{i} + (t-1)\mathbf{j}$ and $\mathbf{u}(t) = \sqrt{t}\,\mathbf{i} + 4t\,\mathbf{j}$ find the following:

(a)   $\mathbf{r}'(t)$                    (b)  $D_t[\mathbf{u}(t) - 2\,\mathbf{r}(t)]$

(c)   $D_t[\mathbf{r}(t) \cdot \mathbf{u}(t)]$          (d)  $D_t[\|\mathbf{r}(t)\|]$

**Solution:**

(a) Differentiating in a component-by-component basis produces the following.

$$\mathbf{r}'(t) = 3\,\mathbf{i} + \mathbf{j}$$

(b)     $\mathbf{u}(t) - 2\,\mathbf{r}(t) = (\sqrt{t}\,\mathbf{i} + 4t\,\mathbf{j}) - 2(3t\,\mathbf{i} + (t-1)\mathbf{j})$

$$= (\sqrt{t} - 6t)\,\mathbf{i} + (2t + 2)\mathbf{j}$$

$$D_t[\mathbf{u}(t) - 2\mathbf{r}(t)] = \left(\frac{1}{2\sqrt{t}} - 6\right)\mathbf{i} + 2\mathbf{j}$$

(c)     $\mathbf{r}(t) \cdot \mathbf{u}(t) = 3t(\sqrt{t}) + (t-1)(4t) = 3t^{3/2} + 4t^2 - 4t$

$$D_t[\mathbf{r}(t) \cdot \mathbf{u}(t)] = \frac{9}{2}\sqrt{t} + 8t - 4$$

(d)     $\|\mathbf{r}(t)\| = \sqrt{(3t)^2 + (t-1)^2}$

$$D_t[\|\mathbf{r}(t)\|] = \frac{1}{2}[9t^2 + (t-1)^2]^{-1/2}[18t + 2(t-1)]$$

$$= \frac{10t - 1}{\sqrt{10t^2 - 2t + 1}}$$

**26.** Given the vector-valued function

$$\mathbf{r}(t) = t\cos t\,\mathbf{i} + t\sin t\,\mathbf{j}$$

find the velocity, speed, and acceleration at any time $t$. Then find $\mathbf{a}\cdot\mathbf{T}$, $\mathbf{a}\cdot\mathbf{N}$, and the curvature at time $t$.

**Solution:**

$$\mathbf{r}(t) = t\cos t\,\mathbf{i} + t\sin t\,\mathbf{j},$$
$$\mathbf{v}(t) = \mathbf{r}'(t) = (-t\sin t + \cos t)\mathbf{i} + (t\cos t + \sin t)\mathbf{j}$$
$$\|\mathbf{v}(t)\| = \text{speed} = \sqrt{(-t\sin t + \cos t)^2 + (t\cos t + \sin t)^2}$$
$$= \sqrt{t^2 + 1}$$
$$\mathbf{a}(t) = \mathbf{r}''(t) = (-t\cos t - 2\sin t)\mathbf{i} + (-t\sin t + 2\cos t)\mathbf{j}$$
$$\mathbf{T}(t) = \frac{\mathbf{v}(t)}{\|\mathbf{v}(t)\|} \frac{(-t\sin t + \cos t)\mathbf{i} + (t\cos t + \sin t)\mathbf{j}}{\sqrt{t^2 + 1}}$$

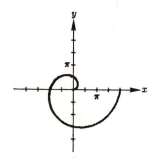

Since the direction of $\mathbf{N}(t)$ is toward the concave side of the curve (see the accompanying figure) and orthogonal to $\mathbf{T}(t)$, we have

$$\mathbf{N}(t) = \frac{-(t\cos t + \sin t)\mathbf{i} + (-t\sin t + \cos t)\mathbf{j}}{\sqrt{t^2 + 1}}.$$

Therefore,

$$\mathbf{a}(t)\cdot\mathbf{T}(t) = \frac{t}{\sqrt{t^2 + 1}}$$
$$\mathbf{a}(t)\cdot\mathbf{N}(t) = \frac{t^2 + 2}{\sqrt{t^2 + 1}}$$

and

$$K = \frac{\mathbf{a}(t)\cdot\mathbf{N}(t)}{\|\mathbf{v}(t)\|} = \frac{t^2 + 2}{(t^2 + 1)^{3/2}}$$

**27.** Find the length of the curve given by the vector-valued function

$$\mathbf{r}(t) = 12t\,\mathbf{i} + 2t^2\,\mathbf{j}$$

over the interval $0 \le t \le 4$.

**Solution:**

$$s = \int_0^4 \|\mathbf{r}'(t)\|\,dt$$

$$= \int_0^4 \sqrt{12^2 + (4t)^2}\,dt$$

$$= \int_0^4 4\sqrt{9 + t^2}\,dt$$

$$= 2\big[t\sqrt{9 + t^2} + 9\ln|t + \sqrt{9 + t^2}|\big]_0^4$$

$$= 2[4(5) + 9\ln(4 + 5) - 9\ln 3]$$

$$= 2(20 + 9\ln 3) \approx 59.775$$

# 14 SOLID ANALYTIC GEOMETRY AND VECTORS IN SPACE

## 14.1 Space Coordinates and Vectors in Space

**7.** Find the lengths of the sides of the triangle with vertices $(1, -3, -2)$, $(5, -1, 2)$, and $(-1, 1, 2)$ and determine whether the triangle is a right triangle, an isosceles triangle, or neither of these.

**Solution:**

Let us denote the three given points by $A, B$, and $C$, respectively. Then

$$|AB| = \sqrt{(5-1)^2 + (-1+3)^2 + (2+2)^2}$$
$$= \sqrt{16 + 4 + 16} = \sqrt{36} = 6$$
$$|AC| = \sqrt{(-1-1)^2 + (1+3)^2 + (2+2)^2}$$
$$= \sqrt{4 + 16 + 16} = \sqrt{36} = 6$$
$$|BC| = \sqrt{(-1-5)^2 + (1+1)^2 + (2-2)^2}$$
$$= \sqrt{36 + 4} = \sqrt{40} = 2\sqrt{10}$$

Since two sides have equal lengths, the triangle is isosceles.

**15.** Find the center and radius of the sphere

$$x^2 + y^2 + z^2 - 2x + 6y + 8z + 1 = 0$$

**Solution:**

Writing the equation in standard form, we have

$$x^2 + y^2 + z^2 - 2x + 6y + 8z + 1 = 0$$
$$(x^2 - 2x) + (y^2 + 6y) + (z^2 + 8z) = -1$$
$$(x^2 - 2x + 1) + (y^2 + 6y + 9) + (z^2 + 8z + 16) = -1 + 1 + 9 + 16$$
$$(x - 1)^2 + (y + 3)^2 + (z + 4)^2 = 25$$

Thus the sphere is centered at $(1, -3, -4)$ with radius 5.

**31.** Find $\mathbf{z}$, where $\mathbf{u} = \langle 1, 2, 3 \rangle$, $\mathbf{w} = \langle 4, 0, -4 \rangle$, and $2\mathbf{z} - 3\mathbf{u} = \mathbf{w}$.

**Solution:**

$$2\mathbf{z} - 3\mathbf{u} = \mathbf{w}$$
$$2\mathbf{z} = 3\mathbf{u} + \mathbf{w}$$
$$= 3(\mathbf{i} + 2\mathbf{j} + 3\mathbf{k}) + (4\mathbf{i} - 4\mathbf{k})$$
$$= 3\mathbf{i} + 6\mathbf{j} + 9\mathbf{k} + 4\mathbf{i} - 4\mathbf{k}$$
$$= 7\mathbf{i} + 6\mathbf{j} + 5\mathbf{k}$$
$$\mathbf{z} = \frac{7}{2}\mathbf{i} + 3\mathbf{j} + \frac{5}{2}\mathbf{k} = \left\langle \frac{7}{2}, 3, \frac{5}{2} \right\rangle$$

**37.** Use vectors to determine whether the points
$(0, -2, -5)$, $(3, 4, 4)$, and $(2, 2, 1)$ are collinear.

**Solution:**

If we denote the three points by $A, B,$ and $C$, respectively, then

$$\overrightarrow{AB} = (3 - 0)\mathbf{i} + (4 + 2)\mathbf{j} + (4 + 5)\mathbf{k} = 3\mathbf{i} + 6\mathbf{j} + 9\mathbf{k}$$
$$\overrightarrow{AC} = (2 - 0)\mathbf{i} + (2 + 2)\mathbf{j} + (1 + 5)\mathbf{k}$$
$$= 2\mathbf{i} + 4\mathbf{j} + 6\mathbf{k}$$
$$= \frac{2}{3}(3\mathbf{i} + 6\mathbf{j} + 9\mathbf{k}) = \frac{2}{3}\overrightarrow{AB}$$

Since $\overrightarrow{AC}$ is a scaler multiple of $\overrightarrow{AB}$, the three points are collinear.

**51.** Find a unit vector (a) in the direction of $\mathbf{u} = \langle 2, -1, 2 \rangle$ and (b) in the opposite direction of $\mathbf{u}$.

**Solution:**

(a) Since the magnitude of $\mathbf{u}$ is

$$\|\mathbf{u}\| = \sqrt{2^2 + (-1)^2 + 2^2} = \sqrt{9} = 3$$

the unit vector in the direction of $\mathbf{u}$ is

$$\frac{\mathbf{u}}{\|\mathbf{u}\|} = \frac{1}{3}\langle 2, -1, 2 \rangle.$$

(b) Since the unit vector in the opposite direction is obtained by multiplying by the scalar -1, we have

$$(-1)\frac{\mathbf{u}}{\|\mathbf{u}\|} = -\frac{1}{3}\langle 2, -1, 2 \rangle.$$

**59.** Use vectors to find the point that lies two-thirds of the way from $P = (4, 3, 0)$ to $Q = (1, -3, 3)$.

**Solution:**

By finding the component form of the vector from $P$ to $Q$, we have

$$\overrightarrow{PQ} = \langle 1 - 4, \ -3 - 3, \ 3 - 0 \rangle = \langle -3, -6, 3 \rangle.$$

Let $R = (x, y, z)$ be a point on the line segment two-thirds of the way from $P$ to $Q$. (See the accompanying figure.) Then

$$\overrightarrow{PR} = \langle x - 4, \ y - 3, \ z - 0 \rangle = \langle x - 4, y - 3, z \rangle$$

and

$$\overrightarrow{PR} = \frac{2}{3}\overrightarrow{PQ}$$

$$= \frac{2}{3}\langle -3, -6, 3 \rangle = \langle -2, -4, 2 \rangle = \langle x - 4, y - 3, z \rangle.$$

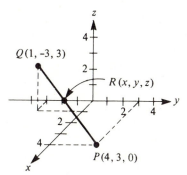

Therefore,

$$\begin{aligned} x - 4 = -2 \quad &\Longrightarrow \quad x = 2 \\ y - 3 = -4 \quad &\Longrightarrow \quad y = -1 \\ z = 2 \end{aligned}$$

and the required point is $(2, -1, 2)$.

**69.** Find the angle $\theta$ between $\mathbf{u} = 3\mathbf{i} + 4\mathbf{j}$ and $\mathbf{v} = -2\mathbf{j} + 3\mathbf{k}$.

**Solution:**

Since

$$\cos \theta = \frac{\mathbf{u} \cdot \mathbf{v}}{\|\mathbf{u}\|\,\|\mathbf{v}\|}$$

we have

$$\cos \theta = \frac{0 - 8 + 0}{\sqrt{24}\sqrt{13}} = \frac{-8\sqrt{13}}{65}$$

and

$$\theta = \arccos\left(\frac{-8\sqrt{13}}{65}\right) \approx 116.3°.$$

**77.** For the vectors $\mathbf{u} = \langle 1, 1, 1\rangle$ and $\mathbf{v} = \langle -2, -1, 1\rangle$ find (a) the vector component of $\mathbf{u}$ along $\mathbf{v}$ and (b) the vector component of $\mathbf{u}$ orthogonal to $\mathbf{v}$.

**Solution:**

(a) The vector component of $\mathbf{u}$ along $\mathbf{v}$ is the projection of $\mathbf{u}$ onto $\mathbf{v}$ and is given by

$$\mathbf{w}_1 = \left(\frac{\mathbf{u} \cdot \mathbf{v}}{\|\mathbf{v}\|^2}\right)\mathbf{v}$$

$$= \left(\frac{-2}{6}\right)\langle -2, -1, 1\rangle = \left\langle \frac{2}{3}, \frac{1}{3}, -\frac{1}{3}\right\rangle$$

(b) The vector component of $\mathbf{u}$ orthogonal to $\mathbf{v}$ is given by

$$\mathbf{w}_2 = \mathbf{u} - \mathbf{w}_1$$

$$= \langle 1, 1, 1\rangle - \left\langle \frac{2}{3}, \frac{1}{3}, -\frac{1}{3}\right\rangle = \left\langle \frac{1}{3}, \frac{2}{3}, \frac{4}{3}\right\rangle$$

**81.** Find the direction cosines of $\mathbf{u} = \mathbf{i} + 2\mathbf{j} + 2\mathbf{k}$ and demonstrate that the sum of the squares of the direction cosines is 1.

**Solution:**

Given a vector $\mathbf{u} = u_1\mathbf{i} + u_2\mathbf{j} + u_3\mathbf{k}$ the direction cosines are

$$\cos\alpha = \frac{u_1}{\|\mathbf{u}\|} \qquad \cos\beta = \frac{u_2}{\|\mathbf{u}\|} \qquad \cos\gamma = \frac{u_3}{\|\mathbf{u}\|}$$

Therefore, for the given vector, we have

$$\cos\alpha = \frac{1}{\sqrt{1^2 + 2^2 + 2^2}} = \frac{1}{\sqrt{9}} = \frac{1}{3}$$
$$\cos\beta = \frac{2}{\sqrt{1^2 + 2^2 + 2^2}} = \frac{2}{\sqrt{9}} = \frac{2}{3}$$
$$\cos\gamma = \frac{2}{\sqrt{1^2 + 2^2 + 2^2}} = \frac{2}{\sqrt{9}} = \frac{2}{3}$$

The sum of the squares of the direction cosines is

$$\cos^2\alpha + \cos^2\beta + \cos^2\gamma = \left(\frac{1}{3}\right)^2 + \left(\frac{2}{3}\right)^2 + \left(\frac{2}{3}\right)^2 = 1$$

**85.** Let $\mathbf{u} = \mathbf{i} + \mathbf{j}$, $\mathbf{v} = \mathbf{j} + \mathbf{k}$, and $\mathbf{w} = a\mathbf{u} + b\mathbf{v}$.

(a) Sketch $\mathbf{u}$ and $\mathbf{v}$.

(b) If $\mathbf{w} = \mathbf{0}$, show that $a$ and $b$ are both zero.

(c) Find $a$ and $b$ such that $\mathbf{w} = \mathbf{i} + 2\mathbf{j} + \mathbf{k}$.

(d) Show that no choice of $a$ and $b$ yields $\mathbf{w} = \mathbf{i} + 2\mathbf{j} + 3\mathbf{k}$

**Solution:**

(a) See accompanying figure.

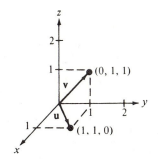

(b) $$\mathbf{w} = a(\mathbf{i} + \mathbf{j}) + b(\mathbf{j} + \mathbf{k}) = \mathbf{0}$$
$$a\mathbf{i} + (a + b)\mathbf{j} + b\mathbf{k} = 0\mathbf{i} + 0\mathbf{j} + 0\mathbf{k}$$

Therefore, $a = b = 0$.

(c) $$\mathbf{w} = a(\mathbf{i} + \mathbf{j}) + b(\mathbf{j} + \mathbf{k}) = \mathbf{i} + 2\mathbf{j} + \mathbf{k}$$
$$a\mathbf{i} + (a + b)\mathbf{j} + b\mathbf{k} = \mathbf{i} + 2\mathbf{j} + \mathbf{k}$$

Therefore, $a = b = 1$

(d)
$$\mathbf{w} = a(\mathbf{i} + \mathbf{j}) + b(\mathbf{j} + \mathbf{k}) = \mathbf{i} + 2\mathbf{j} + 3\mathbf{k}$$
$$a\mathbf{i} + (a + b)\mathbf{j} + b\mathbf{k} = \mathbf{i} + 2\mathbf{j} + 3\mathbf{k}$$

Therefore, $a = 1$, $b = 3$, and $a + b = 2$, which is a contradiction.

## 14.2   The Cross Product of Two Vectors in Space

**11.** If $\mathbf{u} = \langle 1, 1, 1 \rangle$ and $\mathbf{v} = \langle 2, 1, -1 \rangle$, find $\mathbf{u} \times \mathbf{v}$ and show that it is orthogonal to both $\mathbf{u}$ and $\mathbf{v}$.

**Solution:**

First we have

$$\mathbf{u} \times \mathbf{v} = \begin{vmatrix} \mathbf{i} & \mathbf{j} & \mathbf{k} \\ 1 & 1 & 1 \\ 2 & 1 & -1 \end{vmatrix} = \mathbf{i} \begin{vmatrix} 1 & 1 \\ 1 & -1 \end{vmatrix} - \mathbf{j} \begin{vmatrix} 1 & 1 \\ 2 & -1 \end{vmatrix} + \mathbf{k} \begin{vmatrix} 1 & 1 \\ 2 & 1 \end{vmatrix}$$
$$= (-1 - 1)\mathbf{i} - (-1 - 2)\mathbf{j} + (1 - 2)\mathbf{k}$$
$$= -2\mathbf{i} + 3\mathbf{j} - \mathbf{k} = \langle -2, 3, -1 \rangle$$

Now

$$\mathbf{u} \cdot (\mathbf{u} \times \mathbf{v}) = (\mathbf{i} + \mathbf{j} + \mathbf{k}) \cdot (-2\mathbf{i} + 3\mathbf{j} - \mathbf{k}) = -2 + 3 - 1 = 0$$

and

$$\mathbf{v} \cdot (\mathbf{u} \times \mathbf{v}) = (2\mathbf{i} + \mathbf{j} - \mathbf{k}) \cdot (-2\mathbf{i} + 3\mathbf{j} - \mathbf{k}) = -4 + 3 + 1 = 0$$

Therefore $\mathbf{u} \times \mathbf{v}$ is orthogonal to both $\mathbf{u}$ and $\mathbf{v}$.

**19.** Find the area of the parallelogram with vertices $(1,1,1), (2,3,4), (6,5,2)$, and $(7,7,5)$.

**Solution:**

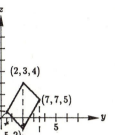

If we denote the four given points by $A, B, C$, and $D$, respectively, we observe from the accompanying figure that the adjacent sides of the parallelograms are given by

$$\overrightarrow{AB} = (2-1)\mathbf{i} + (3-1)\mathbf{j} + (4-1)\mathbf{k} = \mathbf{i} + 2\mathbf{j} + 3\mathbf{k}$$

$$\overrightarrow{AC} = (6-1)\mathbf{i} + (5-1)\mathbf{j} + (2-1)\mathbf{k} = 5\mathbf{i} + 4\mathbf{j} + \mathbf{k}$$

Since the area of the parallelogram is given by $\| \overrightarrow{AB} \times \overrightarrow{AC} \|$, we have

$$\overrightarrow{AB} \times \overrightarrow{AC} = \begin{vmatrix} \mathbf{i} & \mathbf{j} & \mathbf{k} \\ 1 & 2 & 3 \\ 5 & 4 & 1 \end{vmatrix}$$

$$= \begin{vmatrix} 2 & 3 \\ 4 & 1 \end{vmatrix} \mathbf{i} - \begin{vmatrix} 1 & 3 \\ 5 & 1 \end{vmatrix} \mathbf{j} + \begin{vmatrix} 1 & 2 \\ 5 & 4 \end{vmatrix} \mathbf{k}$$

$$= -10\mathbf{i} + 14\mathbf{j} - 6\mathbf{k} = 2(-5\mathbf{i} + 7\mathbf{j} - 3\mathbf{k})$$

$$\| \overrightarrow{AB} \times \overrightarrow{AC} \| = 2\sqrt{25 + 49 + 9} = 2\sqrt{83}.$$

**27.** Find the triple scalar product $\mathbf{u} \cdot (\mathbf{v} \times \mathbf{w})$ if $\mathbf{u} = \langle 2, 0, 1 \rangle, \mathbf{v} = \langle 0, 3, 0 \rangle$, and $\mathbf{w} = \langle 0, 0, 1 \rangle$.

**Solution:**

$$\mathbf{v} \times \mathbf{w} = \begin{vmatrix} \mathbf{i} & \mathbf{j} & \mathbf{k} \\ 0 & 3 & 0 \\ 0 & 0 & 1 \end{vmatrix}$$

$$= \mathbf{i} \begin{vmatrix} 3 & 0 \\ 0 & 1 \end{vmatrix} - \mathbf{j} \begin{vmatrix} 0 & 0 \\ 0 & 1 \end{vmatrix} + \mathbf{k} \begin{vmatrix} 0 & 3 \\ 0 & 0 \end{vmatrix} = 3\mathbf{i}$$

Therefore, the triple scalar product is

$$\mathbf{u} \cdot (\mathbf{v} + \mathbf{w}) = 2(3) + 0(0) + 1(0) = 6$$

**31.** Find the volume of the parallelopiped with vertices $(0,0,0)$, $(3,0,0)$, $(0,5,1)$, $(3,5,1)$, $(2,0,5)$, $(5,0,5)$, $(2,5,6)$, and $(5,5,6)$.

**Solution:**

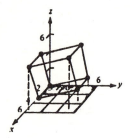

From the accompanying figure we note that the adjacent sides of the parallelopiped are given by $\mathbf{u}, \mathbf{v}$, and $\mathbf{w}$ where

$$\mathbf{u} = 3\mathbf{i}, \quad \mathbf{v} = 5\mathbf{j} + \mathbf{k}, \quad \text{and} \quad \mathbf{w} = 2\mathbf{i} + 5\mathbf{k}.$$

Since the volume of the parallelopiped is the absolute value of the triple scalar product $\mathbf{u} \cdot (\mathbf{v} \times \mathbf{w})$, we have

$$\mathbf{v} \times \mathbf{w} = \begin{vmatrix} \mathbf{i} & \mathbf{j} & \mathbf{k} \\ 0 & 5 & 1 \\ 2 & 0 & 5 \end{vmatrix} = \begin{vmatrix} 5 & 1 \\ 0 & 5 \end{vmatrix} \mathbf{i} - \begin{vmatrix} 0 & 1 \\ 2 & 5 \end{vmatrix} \mathbf{j} + \begin{vmatrix} 0 & 5 \\ 2 & 0 \end{vmatrix} \mathbf{k}$$

$$= 25\mathbf{i} + 2\mathbf{j} - 10\mathbf{k}$$

$$|\mathbf{u} \cdot (\mathbf{v} \times \mathbf{w})| = |3(25) + 0(2) + 0(-10)| = 75$$

**33.** A child applies the brakes on a bicycle by applying a downward force of 20 pounds on the pedal when the crank makes a 40° angle with the horizontal (see figure). Find the torque at $P$ if the crank is 6 inches in length.

**Solution:**

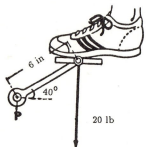

If we represent the 20 pound force as $\mathbf{F} = -20\mathbf{k}$ and the lever as

$$\mathbf{v} = \frac{1}{2}(\cos 40°\mathbf{j} + \sin 40°\mathbf{k}),$$

then the moment of $\mathbf{F}$ about $P$ is given by

$$\mathbf{M} = \mathbf{V} \times \mathbf{F} = \begin{vmatrix} \mathbf{i} & \mathbf{j} & \mathbf{k} \\ 0 & \frac{1}{2}\cos 40° & \frac{1}{2}\sin 40° \\ 0 & 0 & -20 \end{vmatrix} = -10\cos 40°\,\mathbf{i}.$$

The torque is the magnitude of this moment. Thus,

$$\text{torque} = \|\mathbf{M}\| = 10\cos 40° \approx 7.66 \text{ foot-pounds.}$$

## 14.3   Lines and Planes in Space

**5.** Find a set of (a) parametric equation and (b) symmetric equations of the line passing throught the points $(5, -3, -2)$ and $(-\frac{2}{3}, \frac{2}{3}, 1)$.

**Solution:**

(a) If $A = (-\frac{2}{3}, \frac{2}{3}, 1)$ and $B = (5, -3, -2)$, then a direction vector for the line passing throught these points is given by

$$\overrightarrow{AB} = \frac{17}{3}\mathbf{i} - \frac{11}{3}\mathbf{j} - 3\mathbf{k} = \frac{1}{3}(17\mathbf{i} - 11\mathbf{j} - 9\mathbf{k})$$

and a set of direction numbers for the line is $a = 17$, $b = -11$, and $c = -9$. Using the form

$$x = x_1 + at, \qquad y = y_1 + bt, \qquad z = z_1 + ct$$

with $(x_1, y_1, z_1) = (5, -3, -2)$, a set of parametric equation for the line is

$$x = 5 + 17t, \qquad y = -3 - 11t, \qquad z = -2 - 9t.$$

(b) Solving for $t$ in each equation of part (a) gives us

$$t = \frac{x-5}{17} = \frac{y+3}{-11} = \frac{z+2}{-9}$$

Consequently, we have the symmetric form

$$\frac{x-5}{17} = \frac{y+3}{-11} = \frac{z+2}{-9}$$

**13.** Determine if the lines

$$x = 4t + 2 \qquad x = 2s + 2$$
$$y = 3 \qquad y = 2s + 3$$
$$z = -t + 1 \qquad z = s + 1$$

intersect and, if so, find the point of intersection and the cosine of the angle of intersection.

**Solution:**

At the point of intersection, the coordinates for one line equal the corresponding coordinates for the other line. Thus we have the three equations.

(1) $\qquad\qquad 4t + 2 = 2s + 2$

(2) $\qquad\qquad 3 = 2s + 3$

(3) $\qquad\qquad -t + 1 = s + 1$

From equation (2) we find that $s = 0$ and consequently, from equation (3), $t = 0$. Letting $s = t = 0$, we see that equation (1) is satisfied and we conclude that the two lines intersect. Substituting zero for $s$ or for $t$, we obtain the point $(2, 3, 1)$.

To find the cosine of the angle of intersection, we consider the vectors

$$\mathbf{u} = 4\mathbf{i} - \mathbf{k} \quad \text{and} \quad \mathbf{v} = 2\mathbf{i} + 2\mathbf{j} + \mathbf{k}$$

that have the respective directions of the two given lines. Therefore

$$\cos \theta = \frac{|\mathbf{u} \cdot \mathbf{v}|}{\|\mathbf{u}\| \, \|\mathbf{v}\|} = \frac{8 - 1}{\sqrt{17}\sqrt{9}} = \frac{7}{3\sqrt{17}} = \frac{7\sqrt{17}}{51}$$

**21.** Find an equation of the plane throught the points $(0, 0, 0)$, $(1, 2, 3)$, and $(-2, 3, 3)$.

**Solution:**

To use the form

$$a(x - x_1) + b(y - y_1) + c(z - z_1) = 0$$

we need to know a point in the plane and a vector $\mathbf{n}$ that is normal to the plane. To obtain a normal vector, we use the cross product of the vectors $\mathbf{v}_1$ and $\mathbf{v}_2$ from the point $(0,0,0)$ to $(1,2,3)$ and to $(-2,3,3)$, respectively. We have

$$\mathbf{v}_1 = \mathbf{i} + 2\mathbf{j} + 3\mathbf{k} \quad \text{and} \quad \mathbf{v}_2 = -2\mathbf{i} + 3\mathbf{j} + 3\mathbf{k}$$

Thus the vector

$$\mathbf{n} = \mathbf{v}_1 \times \mathbf{v}_2 = \begin{vmatrix} \mathbf{i} & \mathbf{j} & \mathbf{k} \\ 1 & 2 & 3 \\ -2 & 3 & 3 \end{vmatrix} = -3\mathbf{i} - 9\mathbf{j} + 7\mathbf{k}$$

is normal to the given plane. Using the direction numbers from $\mathbf{n}$ and the point $(0,0,0)$, we have

$$-3(x-0) - 9(y-0) + 7(z-0) = 0$$
$$3x + 9y - 7z = 0$$

**27.** Find an equation of the plane determined by the two intersecting lines

$$\frac{x-1}{-2} = y - 4 = z \quad \text{and} \quad \frac{x-2}{-3} = \frac{y-1}{4} = \frac{z-2}{-1}$$

**Solution:**

Writing the equations of the lines in parametric form, we have

$$x = 1 - 2t \qquad x = 2 - 3t$$
$$y = 4 + t \qquad y = 1 + 4t$$
$$z = t \qquad z = 2 - t$$

To find the point of intersection of the lines, we observe that the $z$-coordinate are equal when

$$t = 2 - t \qquad \text{or} \qquad t = 1$$

Therefore the point of intersection is $(-1, 5, 1)$ and occurs when $t = 1$. Since the direction vectors of the given lines are

$$\mathbf{v}_1 = -2\mathbf{i} + \mathbf{j} + \mathbf{k} \qquad \text{and} \qquad \mathbf{v}_2 = -3\mathbf{i} + 4\mathbf{j} - \mathbf{k}$$

the vector $\mathbf{n}$ normal to the plane is

$$\mathbf{n} = \mathbf{v}_1 \times \mathbf{v}_2 = \begin{vmatrix} \mathbf{i} & \mathbf{j} & \mathbf{k} \\ -2 & 1 & 1 \\ -3 & 4 & -1 \end{vmatrix} = -5(\mathbf{i} + \mathbf{j} + \mathbf{k})$$

Therefore the equation of the plane is

$$1(x + 1) + 1(y - 5) + 1(z - 1) = 0$$
$$x + y + z = 5$$

**29.** Find an equation of the plane throught the points $(2, 2, 1)$ and $(-1, 1, -1)$ that is perpendicular to the plane $2x - 3y + z = 3$.

**Solution:**

Let $\mathbf{v}$ be the vector from $(-1, 1, -1)$ to $(2, 2, 1)$, and let $\mathbf{n}$ be a vector normal to the plane $2x - 3y + z = 3$. Then $\mathbf{v}$ and $\mathbf{n}$ both lie in the required plane, where

$$\mathbf{v} = 3\mathbf{i} + \mathbf{j} + 2\mathbf{k} \quad \text{and} \quad \mathbf{n} = 2\mathbf{i} - 3\mathbf{j} + \mathbf{k}$$

The vector

$$\mathbf{v} \times \mathbf{n} = \begin{vmatrix} \mathbf{i} & \mathbf{j} & \mathbf{k} \\ 3 & 1 & 2 \\ 2 & -3 & 1 \end{vmatrix} = 7\mathbf{i} + \mathbf{j} - 11\mathbf{k}$$

is normal to the required plane, and therefore this plane has direction numbers $7, 1$, and $-11$. Finally, since the point $(2, 2, 1)$ lies in the plane, an equation is

$$7(x - 2) + 1(y - 2) - 11(z - 1) = 0$$
$$7x + y - 11z = 5$$

**49.** Find a set of parametric equations for the line of intersection of the planes $3x + 2y - z = 7$ and $x - 4y + 2z = 0$.

**Solution:**

Let $\mathbf{n}_1 = 3\mathbf{i} + 2\mathbf{j} - \mathbf{k}$ and $\mathbf{n}_2 = \mathbf{i} - 4\mathbf{j} + 2\mathbf{k}$ be the normal vectors to the respective planes. The line of intersection of the two planes will have the same direction as the vector $\mathbf{n}_1 \times \mathbf{n}_2$. Since

$$\mathbf{n}_1 \times \mathbf{n}_2 = \begin{vmatrix} \mathbf{i} & \mathbf{j} & \mathbf{k} \\ 3 & 2 & -1 \\ 1 & -4 & 2 \end{vmatrix}$$
$$= 0\mathbf{i} - 7\mathbf{j} - 14\mathbf{k} = -7(0\mathbf{i} + \mathbf{j} + 2\mathbf{k})$$

a set of direction numbers for the line of intersection is $0$, $1$, $2$. By solving the equations for the two planes simultaneously, we can find points on the line of intersection.

$$3x + 2y - z = 7 \Longrightarrow 6x + 4y - 2z = 14$$
$$x - 4y + 2z = 0 \Longrightarrow \quad x - 4y + 2z = 0$$
$$7x = 14 \quad \text{or} \quad x = 2$$

By substituting $2$ for $x$, obtain the equation $2y - z = 1$. If we let $y = 1$, then $z = 1$. Thus $(2, 1, 1)$ lies on the line of intersection, and we conclude that a set of parametric equations for the line of intersection is

$$x = 2, \quad y = 1 + t, \quad z = 1 + 2t$$

**51.** Find the point of intersection (if any) of the line

$$\frac{x - \frac{1}{2}}{1} = \frac{y + \frac{3}{2}}{-1} = \frac{z + 1}{2}$$

and the plane $2x - 2y + z = 12$.

**Solution:**

The parametric equations for the line are

$$x = \frac{1}{2} + t, \qquad y = \frac{-3}{2} - t, \qquad z = -1 + 2t$$

Now is the line intersects the plane, then the values of $x$, $y$, and $z$ must satisfy the equation of the plane. Thus,

$$2x - 2y + z = 12$$

$$2\left(\frac{1}{2} + t\right) - 2\left(\frac{-3}{2} - t\right) + (-1 + 2t) = 12$$

$$6t + 3 = 12$$

$$6t = 9$$

$$t = \frac{3}{2}$$

and we conclude that the point of intersection occurs when $t = \frac{3}{2}$, which yields the point $(2, -3, 2)$.

**61.** Find the distance between the two skew lines

$$\frac{x}{1} = \frac{y}{2} = \frac{z}{3} \qquad \text{and} \qquad \frac{x - 1}{-1} = \frac{y - 4}{1} = \frac{z + 1}{1}$$

**Solution:**

Let

$$\mathbf{n}_1 = \mathbf{i} + 2\mathbf{j} + 3\mathbf{k} \qquad \text{and} \qquad \mathbf{n}_2 = -\mathbf{i} + \mathbf{j} + \mathbf{k}$$

be vectors along the given lines, respectively. Then the vector $\mathbf{n}_1 \times \mathbf{n}_2$ will be orthogonal to both given lines. By choosing an arbitrary vector $\mathbf{v}$ from one line to the other, we can project $\mathbf{v}$ onto $\mathbf{n}_1 \times \mathbf{n}_2$ to find the distance between the two lines. Thus since $(0, 0, 0)$ lies on the first line and $(0, 5, 0)$ lies on the second line, we have $\mathbf{v} = 5\mathbf{j}$. Now the absolute value of the component of **v** in the direction of $\mathbf{n}_1 \times \mathbf{n}_2$ will be the actual distance between the lines. Since

$$\mathbf{n}_1 \times \mathbf{n}_2 = \begin{vmatrix} \mathbf{i} & \mathbf{j} & \mathbf{k} \\ 1 & 2 & 3 \\ -1 & 1 & 1 \end{vmatrix} = -\mathbf{i} - 4\mathbf{j} + 3\mathbf{k}$$

the distance between the lines is

$$|\text{component of } \mathbf{v} \text{ in direction of } \mathbf{n}_1 \times \mathbf{n}_2| = \frac{|\mathbf{v} \cdot (\mathbf{n}_1 \times \mathbf{n}_2)|}{||\mathbf{n}_1 \times \mathbf{n}_2||}$$

$$= \left| \frac{-20}{\sqrt{1 + 16 + 9}} \right|$$

$$= \frac{10\sqrt{26}}{13} \approx 3.92$$

## 14.4   Surfaces in Space

**13.** Describe and sketch the surface defined by the equation $x^2 - y = 0$.

**Solution:**

Since the $z$-coordinate is missing in the equation, the surface is a cylindrical surface with rulings parallel to the $z$-axis. The generating curve is the parabola $y = x^2$ and the surface is called a *parabolic cylinder*. (See accompanying figure.)

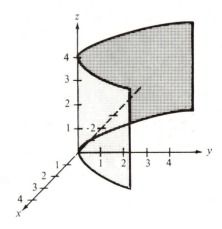

**21.** Identify and sketch the quadric surface given by
$16x^2 - y^2 + 16z^2 = 4$.

**Solution:**

The given equation has the form

$$\frac{x^2}{\frac{1}{4}} - \frac{y^2}{4} + \frac{z^2}{\frac{1}{4}} = 1$$

which is the form for a **hyperboloid of one sheet.** The axis
of the hyperboloid is the $y-$axis. The $xz$-trace $(y = 0)$ is the
circle

$$\frac{x^2}{\frac{1}{4}} + \frac{z^2}{\frac{1}{4}} = 1$$

and the $xy$ and $yz$ traces are the hyperbolas

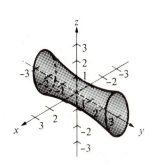

$$\frac{x^2}{\frac{1}{4}} - \frac{y^2}{4} = 1 \quad \text{and} \quad \frac{z^2}{\frac{1}{4}} - \frac{y^2}{4} = 1$$

(See the accompanying figure.)

**31.** Identify and sketch the quadric surface given by $3z = -y^2 + x^2$.

**Solution:**

The given equation can be written as

$$z = \frac{x^2}{3} - \frac{y^2}{3}$$

which has the form for a **hyperbolic paraboloid.**

$xy$-trace $(z = 0)$ :   $y = \pm x$      (intersecting lines)

$xz$-trace $(y = 0)$ :   $z = \frac{1}{3}x^2$      (parabola opening upward)

$yz$-trace $(x = 0)$ :   $z = -\frac{1}{3}y^2$      (parabola opening downward)

Traces parallel to the $xy$-coordinate plane are hyperbolas. For
example, when $z = 3$, we have

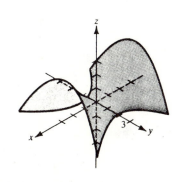

$$3 = \frac{x^2}{3} - \frac{y^2}{3}$$

$$1 = \frac{x^2}{9} - \frac{y^2}{9}$$

**43.** Find an equation for the surface of revolution generated by revolving the graph of $z = 2y$ in the $yz$-plane about the $z$-axis.

**Solution:**

Since we are revolving the curve about the $z-$axis, the equation for the surface of revolution has the form

$$x^2 + y^2 = [a(z)]^2$$

But $y = a(z) = z/2$. Therefore the equation is

$$x^2 + y^2 = \frac{z^2}{4} \qquad \text{or} \qquad 4x^2 + 4y^2 = z^2$$

## 14.5   Curves and Vector-Valued Functions in Space

**3.** Find the domain and the component functions of the vector-valued function $\mathbf{r}(t) = \ln t\,\mathbf{i} - e^t\mathbf{j} - t\,\mathbf{k}$.

**Solution:**

The component functions of the vector-valued function

$$\mathbf{r}(t) = f(t)\,\mathbf{i} + g(t)\,\mathbf{j} + h(t)\,\mathbf{k}$$

are the real-valued functions $f, g$, and $h$. Given the vector-valued function $\mathbf{r}(t) = \ln t\,\mathbf{i} - e^t\mathbf{j} - t\,\mathbf{k}$ the component functions and their domains are:

$$
\begin{aligned}
f(t) &= \ln t, & 0 < t < \infty \\
g(t) &= -e^t, & -\infty < t < \infty \\
h(t) &= -t, & -\infty < t < \infty
\end{aligned}
$$

The intersection of the domains of $f, g$, and $h$, is the interval $(0, \infty)$, the domain of $\mathbf{r}$.

**13.** Sketch the curve represented by the vector-valued function

$$\mathbf{r}(t) = 2\cos t\,\mathbf{i} + 2\sin t\,\mathbf{j} + t\,\mathbf{k}$$

and give the orientation of the curve.

**Solution:**

From the first two parametric equations $x = 2\cos t$ and $y = 2\sin t$, we obtain

$$x^2 + y^2 = 4\cos^2 t + 4\sin^2 t = 4$$

This means that the curve lies on a right circular cylinder of radius 2 centered about the $z$-axis. To locate the curve on this cylinder, we use the third parametric equation $z = t$. Thus the graph of the vector-valued function spirals counterclockwise up the cylinder to produce a circular helix.

**21.** Sketch the curve represented by the intersection of the surfaces $x^2 + y^2 + z^2 = 4$ and $x + z = 2$. Find a vector-valued function for the curve using the parameter $x = 1 + \sin t$.

**Solution:**

The equation $x^2 + y^2 + z^2 = 4$ represents a sphere of radius 2 centered at the origin. The equation $x + z = 2$ represents a plane. We let $x = 1 + \sin t$, then

$$z = 2 - x = 1 - \sin t$$

Substituting into the equation of the sphere, we have

$$x^2 + y^2 + z^2 = 4$$
$$(1 + \sin t)^2 + y^2 + (1 - \sin t)^2 = 4$$
$$2 + 2\sin^2 t + y^2 = 4$$
$$y^2 = 2 - 2\sin^2 t$$
$$y^2 = 2\cos^2 t$$
$$y = \pm\sqrt{2}\cos t$$

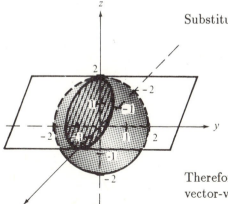

Therefore $x = 1 + \sin t$, $y = \pm\sqrt{2}\cos t$, $z = 1 - \sin t$ and the two vector-valued functions are

$$\mathbf{r}(t) = (1 + \sin t)\mathbf{i} + (\sqrt{2}\cos t)\mathbf{j} + (1 - \sin t)\mathbf{k}$$

and

$$\mathbf{r}(t) = (1 + \sin t)\mathbf{i} - (\sqrt{2}\cos t)\mathbf{j} + (1 - \sin t)\mathbf{k}$$

| $t$ | $-\dfrac{\pi}{2}$ | $-\dfrac{\pi}{6}$ | $0$ | $\dfrac{\pi}{6}$ | $\dfrac{\pi}{2}$ |
|---|---|---|---|---|---|
| $x$ | $0$ | $\dfrac{1}{2}$ | $1$ | $\dfrac{3}{2}$ | $2$ |
| $y$ | $0$ | $\pm\dfrac{\sqrt{6}}{2}$ | $\pm\sqrt{2}$ | $\pm\dfrac{\sqrt{6}}{2}$ | $0$ |
| $z$ | $2$ | $\dfrac{3}{2}$ | $1$ | $\dfrac{1}{2}$ | $0$ |

**23.** Sketch the curve (first octant portion) represented by the intersection of the surfaces $x^2 + z^2 = 4$ and $y^2 + z^2 = 4$. Find a vector-valued function for the curve using the parameter $x = t$.

**Solution:**

The equation $x^2 + z^2 = 4$ represents a cylinder of radius 2 with its axis along the $y$-axis, while the equation $y^2 + z^2 = 4$ represents a cylinder of radius 2 with its axis along the $x$-axis. Thus the curve of intersection lies along both cylinders, as shown in the accompanying sketch. Two points on the curve are $(2, 2, 0)$ and $(0, 0, 2)$. To find a set of parametric equations for the curve of intersection, we subtract the second equation from the first to obtain

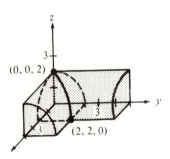

$$x^2 + z^2 = 4$$
$$-(y^2 + z^2 = 4)$$
$$\overline{\phantom{xxxx}}$$
$$x^2 - y^2 = 0 \quad \text{or} \quad y = \pm x$$

Therefore, in the first octant, if we let $x = t$ $(t > 0)$, then parametric equations for the curve are

$$x = t, \qquad y = t, \qquad z = \sqrt{4 - t^2}$$

and the vector-valued function is

$$\mathbf{r}(t) = t\,\mathbf{i} + t\,\mathbf{j} + \sqrt{4 - t^2}\,\mathbf{k}$$

**27.** Evaluate the limit

$$\lim_{t\to 0}\left(t^2\mathbf{i}+3t\mathbf{j}+\frac{1-\cos t}{t}\mathbf{k}\right)$$

**Solution:**

$$\lim_{t\to 0}\left(t^2\mathbf{i}+3t\mathbf{j}+\frac{1-\cos t}{t}\mathbf{k}\right)$$

$$=\left[\lim_{t\to 0}t^2\right]\mathbf{i}+\left[\lim_{t\to 0}3t\right]\mathbf{j}+\left[\lim_{t\to 0}\frac{1-\cos t}{t}\right]\mathbf{k}$$

$$=0\mathbf{i}+0\mathbf{j}+0\mathbf{k}=\mathbf{0}$$

**43.** Find $\mathbf{r}'(t)$ for $\mathbf{r}(t)=\langle t\sin t, t\cos t, t\rangle$.

**Solution:**

$$\mathbf{r}(t)=\langle t\sin t, t\cos t, t\rangle$$

$$\mathbf{r}'(t)=\left\langle\frac{d}{dt}[t\sin t],\frac{d}{dt}[t\cos t],\frac{d}{dt}[t]\right\rangle$$

$$=\langle t\cos t+\sin t, -t\sin t+\cos t, 1\rangle$$

**45.** For the vector-valued functions $\mathbf{r}(t)=t\mathbf{i}+3t\mathbf{j}+t^2\mathbf{k}$ and $\mathbf{u}(t)=4t\mathbf{i}+t^2\mathbf{j}+t^3\mathbf{k}$ find the following:
(a)   $\mathbf{r}'(t)$                              (b)  $\mathbf{r}''(t)$
(c)   $D_t[\mathbf{r}(t)\cdot\mathbf{u}(t)]$                (d)  $D_t[3\mathbf{r}(t)-\mathbf{u}(t)]$
(e)   $D_t[\mathbf{r}(t)\times\mathbf{u}(t)]$              (f)  $D_t[\|\mathbf{r}(t)\|]$

**Solution:**

(a)  Differentiating in a component-by-component basis produces the following.

$$\mathbf{r}'(t)=\mathbf{i}+3\mathbf{j}+2t\mathbf{k}$$

(b)  Differentiating $\mathbf{r}'(t)$ in a component-by‿component basis produces.

$$\mathbf{r}''(t)=2\mathbf{k}$$

(c)
$$\mathbf{r}(t) \cdot \mathbf{u}(t) = t(4t) + 3t(t^2) + t^2(t^3)$$
$$= 4t^2 + 3t^3 + t^5$$
$$D_t[\mathbf{r}(t) \cdot \mathbf{u}(t)] = 8t + 9t^2 + 5t^4$$

(d)
$$3\,\mathbf{r}(t) - \mathbf{u}(t) = 3(t\,\mathbf{i} + 3t\,\mathbf{j} + t^2\,\mathbf{k}) - (4t\,\mathbf{i} + t^2\,\mathbf{j} + t^3\,\mathbf{k})$$
$$= -t\,\mathbf{i} + (9t - t^2)\,\mathbf{j} + (3t^2 - t^3)\,\mathbf{k}$$
$$D_t[3\,\mathbf{r}(t) - \mathbf{u}(t)] = -\mathbf{i} + (9 - 2t)\,\mathbf{j} + (6t - 3t^2)\mathbf{k}$$

(e)
$$\mathbf{r}(t) \times \mathbf{u}(t) = \begin{vmatrix} \mathbf{i} & \mathbf{j} & \mathbf{k} \\ t & 3t & t^2 \\ 4t & t^2 & t^3 \end{vmatrix}$$
$$= \begin{vmatrix} 3t & t^2 \\ t^2 & t^3 \end{vmatrix} \mathbf{i} - \begin{vmatrix} t & t^2 \\ 4t & t^3 \end{vmatrix} \mathbf{j} + \begin{vmatrix} t & 3t \\ 4t & t^2 \end{vmatrix} \mathbf{k}$$
$$= 2t^4\mathbf{i} - (t^4 - 4t^3)\mathbf{j} + (t^3 - 12t^2)\mathbf{k}$$
$$D_t[\mathbf{r}(t) \times \mathbf{u}(t)] = 8t^3\,\mathbf{i} + (12t^2 - 4t^3)\,\mathbf{j} + (3t^2 - 24t)\mathbf{k}$$

(f)
$$\|\mathbf{r}(t)\| = \sqrt{t^2 + (3t)^2 + (t^2)^2}$$
$$= \sqrt{10t^2 + t^4}$$
$$D_t[\|\mathbf{r}(t)\|] = \frac{1}{2}(10t^2 + t^4)^{-1/2}(20t + 4t^3)$$
$$= \frac{10t + 2t^3}{\sqrt{10t^2 + t^4}} = \frac{10 + 2t^2}{\sqrt{10 + t^2}}$$

49. Evaluate the indefinite integral $\displaystyle \int \left( \frac{1}{t}\mathbf{i} + \mathbf{j} - t^{3/2}\,\mathbf{k} \right) dt$.

**Solution:**

$$\int \left( \frac{1}{t}\mathbf{i} + \mathbf{j} - t^{3/2}\,\mathbf{k} \right) dt$$
$$= \left[ \int \frac{1}{t}\,dt \right]\mathbf{i} + \left[ \int dt \right]\mathbf{j} + \left[ \int -t^{3/2}dt \right]\mathbf{k}$$
$$= \ln|t|\,\mathbf{i} + t\,\mathbf{j} - \frac{2}{5}t^{5/2}\,\mathbf{k} + \mathbf{C}$$

**57.** Find the $\mathbf{r}(t)$ if $\mathbf{r}'(t) = te^{-t^2}\mathbf{i} - e^{-t}\mathbf{j} + \mathbf{k}$ and $\mathbf{r}(0) = \frac{1}{2}\mathbf{i} - \mathbf{j} + \mathbf{k}$.

**Solution:**

$$\mathbf{r}(t) = \left[\int te^{-t^2}\,dt\right]\mathbf{i} + \left[\int(-e^{-t})\,dt\right]\mathbf{j} + \left[\int dt\right]\mathbf{k}$$
$$= \left[-\frac{1}{2}e^{-t^2} + C_1\right]\mathbf{i} + [e^{-t} + C_2]\mathbf{j} + [t + C_3]\mathbf{k}$$
$$\mathbf{r}(0) = \left[-\frac{1}{2} + C_1\right]\mathbf{i} + [1 + C_2]\mathbf{j} + C_3\mathbf{k}$$
$$= \frac{1}{2}\mathbf{i} - \mathbf{j} + \mathbf{k}$$

Therefore,

$$-\frac{1}{2} + C_1 = \frac{1}{2} \quad \Longrightarrow \quad C_1 = 1$$
$$1 + C_2 = -1 \quad \Longrightarrow \quad C_2 = -2$$
$$C_3 = 1$$

Thus

$$\mathbf{r}(t) = \left[-\frac{1}{2}e^{-t^2} + 1\right]\mathbf{i} + [e^{-t} - 2]\mathbf{j} + [t + 1]\mathbf{k}$$

**59.** Evaluate the definite integral $\displaystyle\int_0^1 (8t\,\mathbf{i} + t\mathbf{j} - \mathbf{k})\,dt$.

**Solution:**

$$\int_0^1 (8t\,\mathbf{i} + t\mathbf{j} - \mathbf{k})\,dt = \mathbf{i}\int_0^1 8t\,dt + \mathbf{j}\int_0^1 t\,dt - \mathbf{k}\int_0^1 dt$$
$$= \left[4t^2\right]_0^1\mathbf{i} + \left[\frac{1}{2}t^2\right]_0^1\mathbf{j} - \left[t\right]_0^1\mathbf{k}$$
$$= 4\mathbf{i} + \frac{1}{2}\mathbf{j} - \mathbf{k}$$

**65.** Sketch and find the length of the circular helix

$$\mathbf{r}(t) = a\cos t\,\mathbf{i} + a\sin t\,\mathbf{j} + bt\,\mathbf{k}$$

over the interval $[0, 2\pi]$.

**Solution:**

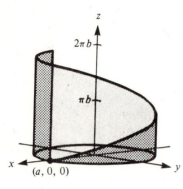

The graph is a circular helix of radius $a$. (See the accompanying figure.)
Since

$$\mathbf{r}'(t) = -a\sin t\,\mathbf{i} + a\cos t\,\mathbf{j} + b\,\mathbf{k}$$

the arc length on the interval $[0, 2\pi]$ is

$$s = \int_0^{2\pi} \|\mathbf{r}'(t)\|\, dt$$

$$= \int_0^{2\pi} \sqrt{(-a\sin t)^2 + (a\cos t)^2 + b^2}\, dt$$

$$= \int_0^{2\pi} \sqrt{a^2(\sin^2 t + \cos^2 t) + b^2}\, dt$$

$$= \int_0^{2\pi} \sqrt{a^2 + b^2}\, dt = [\sqrt{a^2 + b^2}\,t]_0^{2\pi} = 2\pi\sqrt{a^2 + b^2}$$

**72.** Consider the space curve represented by the parametric equations

$$x(t) = \sin t - t\cos t, \quad y(t) = \cos t + t\sin t, \quad z(t) = t^2$$

(a) Express the length of the arc $s$ in the curve as a function of $t$ by evaluating the integral

$$s = \int_0^t \sqrt{[x'(\tau)]^2 + [y'(\tau)]^2 + [z'(\tau)]^2}\, d\tau$$

(b) Solve for $t$ in the relationship derived in part a, and substitute the result into the orginal set of parametric equations. This yields a parametrization of the curve in terms of the arc length parameter $s$.

**Solution:**

(a)
$$s = \int_0^t \sqrt{[x'(\tau)]^2 + [y'(\tau)]^2 + [z'(\tau)]^2} \, d\tau$$

$$= \int_0^t \sqrt{[\tau \sin \tau]^2 + [\tau \cos \tau]^2 + [2\tau]^2} \, d\tau$$

$$= \sqrt{5} \int_0^t \tau \, d\tau$$

$$= \frac{\sqrt{5}}{2} \left[ \tau^2 \right]_0^t = \frac{\sqrt{5}t}{2}$$

(b) Since $s = \sqrt{5}t/2$, we have

$$t = \frac{2s}{\sqrt{5}} = \frac{2\sqrt{5}s}{5}$$

Therefore, the parametrization of the curve in terms of $s$ is

$$x(s) = \sin \left( \frac{2\sqrt{5}s}{5} \right) - \frac{2\sqrt{5}s}{5} \cos \left( \frac{2\sqrt{5}s}{5} \right)$$

$$y(s) = \cos \left( \frac{2\sqrt{5}s}{5} \right) + \frac{2\sqrt{5}s}{5} \sin \left( \frac{2\sqrt{5}s}{5} \right)$$

$$z(s) = \left( \frac{2\sqrt{5}s}{5} \right)^2 = \frac{4}{5}s^2$$

## 14.6   Tangent Vectors, Normal Vectors, and Curvature

**7.** The position function

$$\mathbf{r}(t) = \langle 4t, 3\cos t, 3\sin t \rangle$$

describes the path of an object moving in space. Find the velocity, speed, and acceleration of the object

**Solution:**

Since

$$\mathbf{r}(t) = \langle 4t, 3\cos t, 3\sin t \rangle$$

we have

$$\mathbf{v}(t) = \mathbf{r}'(t) = \langle 4, -3\sin t, 3\cos t \rangle$$

$$\|\mathbf{v}(t)\| = \text{speed} = \sqrt{16 + 9(\sin^2 t + \cos^2 t)} = \sqrt{25} = 5$$

and

$$\mathbf{a}(t) = \mathbf{r}''(t) = \langle 0, -3\cos t, -3\sin t \rangle$$

11. Use the acceleration function $\mathbf{a}(t) = t\mathbf{j} + t\mathbf{k}$ to find the velocity and position functions if $\mathbf{v}(1) = 5\mathbf{j}$ and $\mathbf{r}(1) = \mathbf{0}$. Find the position at time $t = 2$.

**Solution:**

$$\mathbf{v}(t) = \int \mathbf{a}(t)\,dt + \mathbf{C}$$

$$= \int (t\mathbf{j} + t\mathbf{k})\,dt + \mathbf{C}$$

$$= \frac{1}{2}t^2\mathbf{j} + \frac{1}{2}t^2\mathbf{k} + C_1\mathbf{i} + C_2\mathbf{j} + C_3\mathbf{k}$$

$$\mathbf{v}(1) = C_1\mathbf{i} + (\frac{1}{2} + C_2)\mathbf{j} + (\frac{1}{2} + C_3)\mathbf{k} = 5\mathbf{j}$$

Therefore,

$$C_1 = 0$$

$$\frac{1}{2} + C_2 = 5 \quad \Longrightarrow \quad C_2 = \frac{9}{2}$$

$$\frac{1}{2} + C_3 = 0 \quad \Longrightarrow \quad C_3 = -\frac{1}{2}$$

Thus, the velocity vector is

$$\mathbf{v}(t) = \left(\frac{t^2}{2} + \frac{9}{2}\right)\mathbf{j} + \left(\frac{t^2}{2} - \frac{1}{2}\right)\mathbf{k}$$

$$\mathbf{r}(t) = \int \mathbf{v}(t)\,dt + \mathbf{C}$$

$$= \int \left(\frac{t^2}{2} + \frac{9}{2}\right)dt\,\mathbf{j} + \int \left(\frac{t^2}{2} - \frac{1}{2}\right)dt\,\mathbf{k} + \mathbf{C}$$

$$= \left(\frac{t^3}{6} + \frac{9}{2}t\right)\mathbf{j} + \left(\frac{t^3}{6} - \frac{1}{2}t\right)\mathbf{k} + C_4\mathbf{i} + C_5\mathbf{j} + C_6\mathbf{k}$$

$$\mathbf{r}(1) = C_4\mathbf{i} + \left(\frac{1}{6} + \frac{9}{2} + C_5\right)\mathbf{j} + \left(\frac{1}{6} - \frac{1}{2} + C_6\right)\mathbf{k}$$

$$= C_4\mathbf{i} + \left(\frac{14}{3} + C_5\right)\mathbf{j} + \left(-\frac{1}{3} + C_6\right)\mathbf{k} = \mathbf{0}$$

This implies that

$$C_4 = 0, \ C_5 = \frac{-14}{3}, \ \text{and} \ C_6 = \frac{1}{3}.$$

Thus, the position vector is

$$\mathbf{r}(t) = \left(\frac{t^3}{6} + \frac{9}{2}t - \frac{14}{3}\right)\mathbf{j} + \left(\frac{t^3}{6} - \frac{1}{2}t + \frac{1}{3}\right)\mathbf{k}$$

and

$$\mathbf{r}(2) = \frac{17}{3}\mathbf{j} + \frac{2}{3}\mathbf{k}$$

**15.** Find a set of parametric equations for the line tangent to the helix $\mathbf{r}(t) = 2\cos t\,\mathbf{i} + 2\sin t\,\mathbf{j} + t\,\mathbf{k}$ at the point $(2,0,0)$.

**Solution:**

$$\mathbf{r}(t) = (2\cos t)\,\mathbf{i} + (2\sin t)\,\mathbf{j} + t\,\mathbf{k}$$
$$\mathbf{r}'(t) = (-2\sin t)\,\mathbf{i} + (2\cos t)\,\mathbf{j} + \mathbf{k}$$

Since $t = 0$ at the point $(2,0,0)$, the direction vector for the line is given by $\mathbf{r}'(0) = 2\mathbf{j} + \mathbf{k}$, and the parametric representation of the line is

$$x = 2, \qquad y = 2s, \qquad z = s$$

**25.** If $\mathbf{r}(t) = 4t\,\mathbf{i} + 3\cos t\,\mathbf{j} + 3\sin t\,\mathbf{k}$ find $\mathbf{T}(t)$, $\mathbf{N}(t)$, $a_{\mathbf{T}}$, and $a_{\mathbf{N}}$ when $t = \pi/2$.

**Solution:**

$$\mathbf{r}(t) = 4t\,\mathbf{i} + 3\cos t\,\mathbf{j} + 3\sin t\,\mathbf{k}$$
$$\mathbf{v}(t) = \mathbf{r}'(t) = 4\mathbf{i} - 3\sin t\,\mathbf{j} + 3\cos t\,\mathbf{k}$$
$$\|\mathbf{v}(t)\| = \sqrt{16 + 9(\sin^2 t + \cos^2 t)} = \sqrt{25} = 5$$
$$\mathbf{a}(t) = \mathbf{r}''(t) = -3\cos t\,\mathbf{j} - 3\sin t\,\mathbf{k}$$
$$\mathbf{T}(t) = \frac{\mathbf{v}(t)}{\|\mathbf{v}(t)\|} = \frac{1}{5}[4\mathbf{i} - 3\sin t\,\mathbf{j} + 3\cos t\,\mathbf{k}]$$
$$\mathbf{N}(t) = \frac{\mathbf{T}'(t)}{\|\mathbf{T}'(t)\|} = \frac{(\frac{1}{5})[-3\cos t\,\mathbf{j} - 3\sin t\,\mathbf{k}]}{(\frac{1}{5})\sqrt{9(\cos^2 t + \sin^2 t)}}$$
$$= \frac{(-\frac{3}{5})[(\cos t)\,\mathbf{j} + (\sin t)\,\mathbf{k}]}{\frac{3}{5}}$$
$$= -\cos t\,\mathbf{j} - \sin t\,\mathbf{k}$$

Therefore, we have

$$\mathbf{a}(\pi/2) = -3\,\mathbf{k}, \quad \mathbf{T}(\pi/2) = \frac{1}{5}[4\,\mathbf{i} - 3\,\mathbf{j}], \quad \text{and} \quad \mathbf{N}(\pi/2) = \mathbf{k}.$$

Thus,

$$a_{\mathbf{T}} = \mathbf{a}\left(\frac{\pi}{2}\right) \cdot \mathbf{T}\left(\frac{\pi}{2}\right) = 0 \quad \text{and} \quad a_{\mathbf{N}} = \mathbf{a}\left(\frac{\pi}{2}\right) \cdot \mathbf{N}\left(\frac{\pi}{2}\right) = 3$$

**33.** Find the curvature $K$ of the curve $\mathbf{r}(t) = 4t\,\mathbf{i} + 3\cos t\,\mathbf{j} + 3\sin t\,\mathbf{k}$.

**Solution:**

From Exercise 25, we have

$$\|\mathbf{T}'(t)\| = \frac{3}{5}$$

and

$$\|\mathbf{r}'(t)\| = 5$$

Therefore, the curvature is

$$K = \frac{\|\mathbf{T}'(t)\|}{\|\mathbf{r}'(t)\|} = \frac{\frac{3}{5}}{5} = \frac{3}{25}$$

## Review Exercises for Chapter 14

**1.** Given the points $P = (5, 0, 0)$, $Q = (4, 4, 0)$ and $R = (2, 0, 6)$, let $\mathbf{u} = \overrightarrow{PQ}$ and $\mathbf{v} = \overrightarrow{PR}$. Find (a) the component forms of $\mathbf{u}$ and $\mathbf{v}$, (b) $\mathbf{u} \cdot \mathbf{v}$, (c) $\mathbf{u} \times \mathbf{v}$, (d) an equation of the plane containing $P, Q$, and $R$, and (e) a set of parametric equations of the line throught $P$ and $Q$.

**Solution:**

(a) $\qquad \mathbf{u} = \overrightarrow{PQ} = \langle 4 - 5, 4 - 0, 0 - 0 \rangle = \langle -1, 4, 0 \rangle$

$\qquad \mathbf{v} = \overrightarrow{PR} = \langle 2 - 5, 0 - 0, 6 - 0 \rangle = \langle -3, 0, 6 \rangle$

(b)     $\mathbf{u} \cdot \mathbf{v} = (-1)(-3) + (4)(0) + (0)(6) = 3$

(c)     $\mathbf{u} \times \mathbf{v} = \begin{vmatrix} \mathbf{i} & \mathbf{j} & \mathbf{k} \\ -1 & 4 & 0 \\ -3 & 0 & 6 \end{vmatrix} = 24\,\mathbf{i} + 6\,\mathbf{j} + 12\,\mathbf{k}$

$$= 6(4\,\mathbf{i} + \mathbf{j} + 2\,\mathbf{k})$$

(d) A vector normal to the plane is

$$\frac{1}{6}(\mathbf{u} \times \mathbf{v}) = 4\,\mathbf{i} + \mathbf{j} + 2\,\mathbf{k} \quad \text{See part (c)}$$

Therefore, using the point $(5, 0, 0)$, an equation of the plane is

$$4(x - 5) + 1(y - 0) + 2(z - 0) = 0$$
$$4x + y + 2z = 20$$

(e) Since the direction of the line is determined by $\mathbf{u} = \langle -1, 4, 0 \rangle$ [see part (a)], a set of parametric equations of the line passing through the point $(4, 4, 0)$ is

$$x = 4 - t, \qquad y = 4 + 4t, \qquad z = 0$$

Note that when $t = -1$, we have $x = 5$, $y = 0$, and $z = 0$. Thus the line passes through the point $P$.

**15.** Find (a) a set of parametric equations and (b) a set of symmetric equations for the line perpendicular to the $xz$-coordinate plane passing through the point $(1, 2, 3)$.

**Solution:**

(a) Any line perpendicular to the $xz$-plane must have the direction of the vector $\mathbf{v} = 0\,\mathbf{i} + \mathbf{j} + 0\,\mathbf{k}$. Thus direction numbers for the required line are $0, 1, 0$. Since the line passes through the point $(1, 2, 3)$, the parametric equations are

$$x = 1 + (0)t = 1$$
$$y = 2 + t$$
$$z = 3 + (0)t = 3$$

(b) Since two of the direction numbers are zero, there is no symmetric form.

**21.** Find an equation of the plane containing the lines

$$\frac{x-1}{-2} = y = z+1 \quad \text{and} \quad \frac{x+1}{-2} = y-1 = x-2$$

**Solution:**

We first observe that the lines are parallel since they have the same direction numbers, $-2, 1, 1$. Therefore, a vector parallel to the plane is $\mathbf{u} = \langle -2, 1, 1 \rangle$. A point on the first line is $(1, 0, -1)$ and a point on the second line is $(-1, 1, 2)$. The vector $\mathbf{v} = \langle 2, -1, -3 \rangle$ connecting these two points is also parallel to the plane. Thus, a normal to the plane is

$$\mathbf{u} \times \mathbf{v} = \begin{vmatrix} \mathbf{i} & \mathbf{j} & \mathbf{k} \\ -2 & 1 & 1 \\ 2 & -1 & -3 \end{vmatrix} = -2(\mathbf{i} + 2\mathbf{j})$$

Therefore, an equation of the plane is

$$1(x-1) + 2(y-0) + 0(z+1) = 0$$
$$x + 2y = 1$$

**35.** Sketch the graph of the surface $16x^2 + 16y^2 - 9z^2 = 0$.

**Solution:**

The given equation has the form

$$\frac{x^2}{9} + \frac{y^2}{9} - \frac{z^2}{16} = 0$$

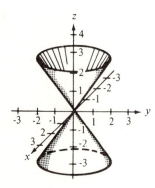

which is the form for a cone. The axis of the cone is the $z$-axis. The $xz$-trace ($y = 0$) is given by

$$\frac{x^2}{9} - \frac{z^2}{16} = 0 \quad \text{or} \quad z = \pm\frac{4}{3}x$$

The $yz$-trace ($x = 0$) is given by

$$\frac{y^2}{9} - \frac{z^2}{16} = 0 \quad \text{or} \quad z = \pm\frac{4}{3}y$$

Traces parallel to the $xy$–plane are circles. For example, when $z = 4$ we have

$$16x^2 + 16y^2 - 9(16) = 0$$
$$x^2 + y^2 = 9$$

59. Find the indefinite integral $\int \| \cos t\,\mathbf{i} + \sin t\,\mathbf{j} + t\,\mathbf{k} \|\, dt$.

**Solution:**

$$\int \| \cos t\,\mathbf{i} + \sin t\,\mathbf{j} + t\,\mathbf{k} \|\, dt$$

$$= \int \sqrt{\cos^2 t + \sin^2 t + t^2}\, dt$$

$$= \int \sqrt{1+t^2}\, dt = \frac{1}{2}(t\sqrt{1+t^2} + \ln |t + \sqrt{1+t^2}|) + C$$

61. Find a set of parametric equations for the line tangent to the space curve given by $\mathbf{r}(t) = 2\cos t\,\mathbf{i} + 2\sin t\,\mathbf{j} + t\,\mathbf{k}$ at the point $t = 3\pi/4$.

**Solution:**

$$\mathbf{r}(t) = 2\cos t\,\mathbf{i} + 2\sin t\,\mathbf{j} + t\,\mathbf{k}$$

$$\mathbf{r}(3\pi/4) = -\sqrt{2}\,\mathbf{i} + \sqrt{2}\,\mathbf{j} + (3\pi/4)\,\mathbf{k}$$

$$\mathbf{r}'(t) = -2\sin t\,\mathbf{i} + 2\cos t\,\mathbf{j} + \mathbf{k}$$

$$\mathbf{r}'(3\pi/4) = -\sqrt{2}\,\mathbf{i} - \sqrt{2}\,\mathbf{j} + \mathbf{k}$$

Therefore, the tangent line must pass through the point $(-\sqrt{2}, \sqrt{2}, 3\pi/4)$ and have direction numbers $a = -\sqrt{2}, b = -\sqrt{2},$ and $c = 1$. Thus, the parametric equations of the line are given by

$$x = -\sqrt{2} - \sqrt{2}t, \quad y = \sqrt{2} - \sqrt{2}t, \quad z = \frac{3\pi}{4} + t.$$

69. If the position function of an object is given by $\mathbf{r}(t) = t\,\mathbf{i} = t^2\mathbf{j} + \frac{1}{2}t^2\,\mathbf{k}$, find $\mathbf{v}(t)$, $\|\mathbf{v}(t)\|$, $\mathbf{a}(t)$, $\mathbf{a} \cdot \mathbf{T}$, $\mathbf{a} \cdot \mathbf{N}$, and the curvature of the path at time $t$.

**Solution:**

$$\mathbf{v}(t) = \mathbf{r}'(t) = \mathbf{i} + 2t\mathbf{j} + t\mathbf{k}$$

$$\|\mathbf{v}(t)\| = \text{speed} = \sqrt{i^2 + (2t)^2 + t^2} = \sqrt{1 + 5t^2}$$

$$\mathbf{a}(t) = \mathbf{r}''(t) = 2\mathbf{j} + \mathbf{k}$$

$$\mathbf{T}(t) = \frac{\mathbf{v}(t)}{\|\mathbf{v}(t)\|} = \frac{\mathbf{i} + 2t\mathbf{j} + t\mathbf{k}}{\sqrt{1 + 5t^2}}$$

$$\mathbf{T}'(t) = \frac{5t}{(1 + 5t^2)^{3/2}}\mathbf{i} + \frac{2}{(1 + 5t^2)^{3/2}}\mathbf{j} + \frac{1}{(1 + 5t^2)^{3/2}}\mathbf{k}$$

$$= \frac{5t\mathbf{i} + 2\mathbf{j} + \mathbf{k}}{(1 + 5t^2)^{3/2}}$$

$$\|\mathbf{T}'(t)\| = \frac{\sqrt{(5t)^2 + 2^2 + 1^2}}{(1 + 5t^2)^{3/2}} = \frac{\sqrt{5}\sqrt{1 + 5t^2}}{(1 + 5t^2)^{3/2}} = \frac{\sqrt{5}}{1 + 5t^2}$$

$$\mathbf{N}(t) = \frac{\mathbf{T}'(t)}{\|\mathbf{T}'(t)\|} = \frac{5t\mathbf{i} + 2\mathbf{j} + \mathbf{k}}{\sqrt{5}\sqrt{1 + 5t^2}}$$

$$\mathbf{a}(t) \cdot \mathbf{T}(t) = \frac{2(2t) + 1(t)}{\sqrt{1 + 5t^2}} = \frac{5t}{\sqrt{1 + 5t^2}}$$

$$\mathbf{a}(t) \cdot \mathbf{N}(t) = \frac{2(2) + 1(1)}{\sqrt{5}\sqrt{1 + 5t^2}} = \frac{5}{\sqrt{5}\sqrt{1 + 5t^2}} = \frac{\sqrt{5}}{\sqrt{1 + 5t^2}}$$

$$K = \frac{\mathbf{a}(t) \cdot \mathbf{N}(t)}{\|\mathbf{v}(t)\|} = \frac{\dfrac{\sqrt{5}}{\sqrt{1 + 5t^2}}}{1 + 5t^2} = \frac{\sqrt{5}}{(1 + 5t^2)^{3/2}}$$

**71.** Find the length of the space curve given by vector-valued function

$$\mathbf{r}(t) = \frac{1}{2}t\mathbf{i} + \sin t\,\mathbf{j} + \cos t\,\mathbf{k}$$

over the interval $0 \le t \le \pi$.

**Solution:**

$$s = \int_0^\pi \|\mathbf{r}'(t)\|\, dt$$

$$= \int_0^\pi \sqrt{(\tfrac{1}{2})^2 + \cos^2 t + (-\sin t)^2}\, dt$$

$$= \frac{\sqrt{5}}{2} \int_0^\pi dt = \frac{\sqrt{5}}{2}[t]_0^\pi = \frac{\sqrt{5}\pi}{2}$$

# 15 FUNCTIONS OF SEVERAL VARIABLES

## 15.1 Introduction to Functions of Several Variables

9. Find the functional values (a) $f(0,4)$ and (b) $f(1,4)$ if

$$f(x,y) = \int_x^y (2t - 3)\, dt$$

**Solution:**

$$f(x,y) = \int_x^y (2t - 3)\, dt = \left[ t^2 - 3t \right]_x^y = (y^2 - 3y) - (x^2 - 3x)$$

(a) $f(0,4) = (16 - 12) - (0 - 0) = 4$

(b) $f(1,4) = (16 - 12) - (1 - 3) = 6$

11. Describe the region $R$, in the $xy$–coordinate plane, that corresponds to the domain of $f(x,y) = \sqrt{4 - x^2 - y^2}$. Find the range of $f(x,y)$.

**Solution:**

Since $f(x,y) = \sqrt{4 - x^2 - y^2}$, we have

$$4 - x^2 - y^2 \geq 0$$
$$4 \geq x^2 + y^2$$

Therefore, the region $R$ is the set of all points inside and on the boundary of the circle $x^2 + y^2 = 4$. The range of $f$ is the set of all real numbers in the interval $[0, 2]$.

**13.** Describe the region $R$, in the $xy$-coordinate plane, that corresponds to the domain of $f(x, y) = \arcsin(x + y)$. Find the range of $f(x, y)$.

**Solution:**

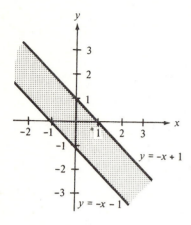

Since $z = \arcsin(x + y)$ implies that $\sin z = x + y$, we conclude that $|x + y| \leq 1$. Therefore, the region $R$ is such that

$$-1 \leq x + y \leq 1$$
$$-1 - x \leq y \leq -x + 1$$

Which means that $R$ lies on and between the parallel lines

$$y = -1 - x \quad \text{and} \quad y = -x + 1$$

as shown in the accompanying figure. The range of the arcsine function is the set of all reals in the interval $[-\pi/2, \pi/2]$.

**37.** (a) Sketch the graph of the surface given by $f(x, y) = x^2 + y^2$.
(b) On the surface of part (a), sketch the graphs of $z = f(1, y)$ and $z = f(x, 1)$.

**Solution:**

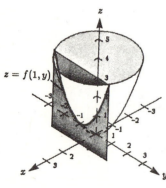

(a) The surface is a paraboloid and its axis is the $z$-axis. Some conventional traces are

$$
\begin{array}{lll}
yz\text{-trace } (x = 0): & z = y^2 & \text{Parabola} \\
xz\text{-trace } (y = 0): & z = x^2 & \text{Parabola} \\
\text{Parallel to } xy\text{-plane } (z = 4): & 4 = x^2 + y^2 & \text{Circle}
\end{array}
$$

The domain of $f$ is the set of all points $(x, y)$ in the $xy$-coordinate plane and the range is the set of all non-negative real numbers. The surface is shown in the accompanying figure.

(b) The trace parallel to the $yz$-coordinate plane when $x = 1$ is the parabola given by

$$z = f(1, y) = 1 + y^2$$

The trace parallel to the $xz$-coordinate plane when $y = 1$ is the parabola given by

$$z = f(x, 1) = x^2 + 1$$

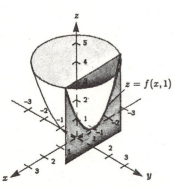

The traces are shown in the accompanying figure.

**43.** Describe the level curves for the function $f(x, y) = x/(x^2 + y^2)$. Sketch the level curves for $c = \pm\frac{1}{2}, \pm 1, \pm\frac{3}{2}, \pm 2$.

**Solution:**

If $f(x, y) = c$, then the level curves are of the form

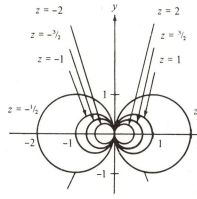

$$c = \frac{x}{x^2 + y^2}$$

$$x^2 + y^2 = \frac{x}{c}$$

$$x^2 - \frac{x}{c} + y^2 = 0$$

$$\left(x^2 - \frac{x}{c} + \frac{1}{4c^2}\right) + y^2 = \frac{1}{4c^2}$$

$$\left(x - \frac{1}{2c}\right)^2 + y^2 = \left(\frac{1}{2x}\right)^2$$

Therefore, each level curve is a circle centered at $(1/2c, 0)$ with radius equal to $1/2c$. For example, if $c = 1$, the level curve has the equation

$$\left(x - \frac{1}{2}\right)^2 + y^2 = \frac{1}{4}$$

The required level curves are shown in the accompanying figure.

## 15.2  Limits and Continuity

**9.** Find

$$\lim_{(x,y)\to(0,1)} \frac{\arcsin(x/y)}{1 + xy}$$

and discuss the continuity of the function.

**Solution:**

Since the limit of a quotient is the quotient of the limits, we have

$$\lim_{(x,y)\to(0,1)} \frac{\arcsin(x/y)}{1 + xy} = \frac{\arcsin 0}{1 + 0} = \frac{0}{1} = 0$$

A rational function is continuous at every point in its domain. Therefore, the given function is continuous for all points $(x, y)$ in the $xy$-plane such that $1 + xy \neq 0$, $y \neq 0$, and $|x/y| \leq 1$.

**15.** Find the limit (if it exists)

$$\lim_{(x,y)\to(0,0)} \frac{x^2 y}{x^4 + y^2}$$

by examining the behavior of the function along the paths (a) $y = ax$, (b) $y = ax^2$, and (c) $y = ax^3$. Assume $a \neq 0$.

**Solution:**

(a)   Along the path $y = ax$ we have

$$\lim_{(x,y)\to(0,0)} \frac{x^2}{x^4 + y^2} = \lim_{(x,ax)\to(0,0)} \frac{x^2(ax)}{x^4 + (ax)^2}$$
$$= \lim_{(x,ax)\to(0,0)} \frac{ax}{x^2 + a^2} = 0$$

(b)   Along the path $y = ax^2$ we have

$$\lim_{(x,y)\to(0,0)} \frac{x^2 y}{x^4 + y^2} = \lim_{(x,ax^2)\to(0,0)} \frac{x^2(ax^2)}{x^4 + (ax^2)^2}$$
$$= \lim_{(x,ax^2)\to(0,0)} \frac{a}{1 + a^2} = \frac{a}{1 + a^2}$$

(c)   Along the path $y = ax^3$ we have

$$\lim_{(x,y)\to(0,0)} \frac{x^2 y}{x^4 + y^2} = \lim_{(x,ax^3)\to(0,0)} \frac{x^2(ax^3)}{x^4 + (ax^3)^2}$$
$$= \lim_{(x,ax^3)\to(0,0)} \frac{ax}{1 + a^2 x^2} = 0$$

Since the limits are not the same along different paths, the limit does not exist.

**25.** Use polar coordinates to find the limit

$$\lim_{(x,y)\to(0,0)} \frac{\sin(x^2 + y^2)}{x^2 + y^2}$$

**Solution:**

We first observe that direct substitution yields the indeterminate from $0/0$. Letting $x = r\cos\theta, y = r\sin\theta$, and $r^2 = x^2 + y^2$, we have

$$\lim_{(x,y)\to(0,0)} \frac{\sin(x^2 + y^2)}{x^2 + y^2} = \lim_{r\to 0} \frac{\sin r^2}{r^2} = 1$$

**35.** Discuss the continuity of the composite function $f \circ g$ if $f(t) = 1/t$ and $g(x, y) = 3x - 2y$.

**Solution:**

$$(f \circ g)(x, y) = f[g(x, y)]$$

$$= \frac{1}{g(x, y)} = \frac{1}{3x - 2y}$$

The composite function is continuous for $y \neq 3x/2$.

**37.** Given the function $f(x, y) = x^2 - 4y$, find

(a) $\displaystyle \lim_{\Delta x \to 0} \frac{f(x + \Delta x, y) - f(x, y)}{\Delta x}$

(b) $\displaystyle \lim_{\Delta y \to 0} \frac{f(x, y + \Delta y) - f(x, y)}{\Delta y}$

**Solution:**

(a) $\displaystyle \lim_{\Delta x \to 0} \frac{f(x + \Delta x, y) - f(x, y)}{\Delta x}$

$$= \lim_{\Delta x \to 0} \frac{[(x + \Delta x)^2 - 4y] - (x^2 - 4y)}{\Delta x}$$

$$= \lim_{\Delta x \to 0} \frac{x^2 + 2x\Delta x + (\Delta x)^2 - 4y - x^2 + 4y}{\Delta x}$$

$$= \lim_{\Delta x \to 0} (2x + \Delta x) = 2x$$

(b) $\displaystyle \lim_{\Delta y \to 0} \frac{f(x, y + \Delta y) - f(x, y)}{\Delta y}$

$$= \lim_{\Delta y \to 0} \frac{x^2 - 4(y + \Delta y) - (x^2 - 4y)}{\Delta y}$$

$$= \lim_{\Delta y \to 0} \frac{x^2 - 4y - 4\Delta y - x^2 + 4y}{\Delta y}$$

$$= \lim_{\Delta y \to 0} (-4) = -4$$

## 15.3   Partial Derivatives

**11.** Find the first partial derivatives with respect to $x$ and with respect to $y$ for

$$z = \ln \frac{x+y}{x-y}$$

**Solution:**

Using the properties of the logarithm function we rewrite the function and obtain

$$z = \ln(x+y) - \ln(x-y)$$

Considering $y$ to be a constant and differentiating with respect to $x$ we have

$$\frac{\partial z}{\partial x} = \frac{1}{x+y}(1) - \frac{1}{x-y}(1) = \frac{-2y}{x^2 - y^2}$$

Now considering $x$ to be a constant and differentiating with respect to $y$ we have

$$\frac{\partial z}{\partial y} = \frac{1}{x+y}(1) - \frac{1}{x-y}(-1) = \frac{2x}{x^2 - y^2}$$

**19.** Find the first partial derivatives with respect to $x$ and with respect to $y$ for

$$z = e^y \sin xy$$

**Solution:**

First, considering $y$ to be constant, we have

$$\frac{\partial z}{\partial x} = e^y (\cos xy)(y) = ye^y \cos xy$$

Now considering $x$ to be constant and using the Product Rule, we have

$$\frac{\partial z}{\partial y} = e^y (\cos xy)(x) + (\sin xy)(e^y)(1) = e^y (x \cos xy + \sin xy)$$

**23.** Given the function $f(x,y) = \arctan(y/x)$, find $f_x(2,-2)$ and $f_y(2,-2)$.

**Solution:**

First, considering $y$ to be constant, we have

$$\frac{\partial z}{\partial x} = \frac{1}{1+(y^2/x^2)}\left(\frac{-y}{x^2}\right) = \frac{-y}{x^2+y^2}$$

Now considering $x$ to be constant, we have

$$\frac{\partial z}{\partial y} = \frac{1}{1+(y^2/x^2)}\left(\frac{1}{x}\right) = \frac{x}{x^2+y^2}$$

Therefore,

$$f_x(2,-2) = \frac{1}{4} \quad \text{and} \quad f_y(2,-2) = \frac{1}{4}$$

**33.** Find all the second partial derivatives for $z = x^2 - 2xy + 3y^2$.

**Solution:**

The first partials are

$$\frac{\partial z}{\partial x} = 2x - 2y \quad \text{and} \quad \frac{\partial z}{\partial y} = -2x + 6y \ .$$

The second partials are

$$\frac{\partial^2 z}{\partial x^2} = \frac{\partial}{\partial x}\left[\frac{\partial z}{\partial x}\right] = 2 \qquad \frac{\partial^2 z}{\partial y\,\partial x} = \frac{\partial}{\partial y}\left[\frac{\partial z}{\partial x}\right] = -2$$

$$\frac{\partial^2 z}{\partial y^2} = \frac{\partial}{\partial y}\left[\frac{\partial z}{\partial y}\right] = 6 \qquad \frac{\partial^2 z}{\partial x\,\partial y} = \frac{\partial}{\partial x}\left[\frac{\partial z}{\partial y}\right] = -2$$

[Note that the mixed partials are equals.]

**37.** Find all the second derivatives for

$$z = \arctan \frac{y}{x}$$

**Solution:**

From Exercise 23, we know the first partial derivatives are

$$\frac{\partial z}{\partial 2x} = \frac{-y}{x^2 + y^2} \qquad \text{and} \qquad \frac{\partial z}{\partial y} = \frac{x}{x^2 + y^2}$$

The second partials are

$$\frac{\partial^2 z}{\partial x^2} = \frac{\partial}{\partial x}\left[\frac{\partial z}{\partial x}\right] = \frac{(x^2 + y^2)(0) - (-y)(2x)}{(x^2 + y^2)^2} = \frac{2xy}{(x^2 + y^2)^2}$$

$$\frac{\partial^2 z}{\partial y\,\partial x} = \frac{\partial}{\partial y}\left[\frac{\partial z}{\partial x}\right] = \frac{(x^2 + y^2)(-1) - (-y)(2y)}{(x^2 + y^2)^2} = \frac{y^2 - x^2}{(x^2 + y^2)^2}$$

$$\frac{\partial^2 z}{\partial y^2} = \frac{\partial}{\partial y}\left[\frac{\partial z}{\partial y}\right] = \frac{(x^2 + y^2)(0) - x(2y)}{(x^2 + y^2)^2} = \frac{-2xy}{(x^2 + y^2)^2}$$

$$\frac{\partial^2 z}{\partial x\,\partial y} = \frac{\partial}{\partial x}\left[\frac{\partial z}{\partial y}\right] = \frac{(x^2 + y^2)(1) - x(2x)}{(x^2 + y^2)^2} = \frac{y^2 - x^2}{(x^2 + y^2)^2}$$

[Note that the mixed partials are equal.]

**49.** Show that the mixed partials $f_{xyy}$, $f_{yxy}$, and $f_{yyx}$ are equal if $f(x, y, z) = e^{-x} \sin yz$.

**Solution:**

$$f_x(x, y, z) = -e^{-x} \sin yz$$
$$f_{xy}(x, y, z) = -e^{-x}(\cos yz)(z) = -ze^{-x} \cos yz$$
$$f_{xyy}(x, y, z) = -ze^{-x}(-\sin yz)(z) = z^2 e^{-x} \sin yz$$
$$f_y(x, y, z) = e^{-x}(\cos yz)(z) = ze^{-x} \cos yz$$
$$f_{yx}(x, y, z) = -ze^{-x} \cos yz$$
$$f_{yxy}(x, y, z) = -ze^{-x}(-\sin yz)(z) = z^2 e^{-x} \sin yz$$
$$f_y(x, y, z) = e^{-x}(\cos yz)(z) = ze^{-x} \cos yz$$
$$f_{yy}(x, y, z) = ze^{-x}(-\sin yz)(z) = -z^2 e^{-x} \sin yz$$
$$f_{yyx}(x, y, z) = -z^2 e^{-x}(-1) \sin yz = z^2 e^{-x} \sin yx$$

Therefore,

$$f_{xyy}(x, y, z) = f_{yxy}(x, y, z) = f_{yyx}(x, y, z) = z^2 e^{-x} \sin yx.$$

**55.** Show that $z = \sin(x - ct)$ is a solution of the equation $\partial^2 z / \partial t^2 = c^2 (\partial^2 z / \partial x^2)$.

**Solution:**

$$\frac{\partial z}{\partial x} = \cos(x - ct) \qquad \text{and} \qquad \frac{\partial^2 z}{\partial x^2} = -\sin(x - ct)$$

$$\frac{\partial z}{\partial t} = -c \cos(x - ct) \qquad \text{and} \qquad \frac{\partial^2 z}{\partial t^2} = -c^2 \sin(x - ct)$$

Therefore,

$$\frac{\partial^2 z}{\partial t^2} = -c^2 \sin(x - ct) = c^2 \frac{\partial^2 z}{\partial x^2}.$$

**65.** Sketch the curve formed by the intersection of the paraboloid $z = 9x^2 - y^2$ and the plane $y = 3$. Find the slope of the curve at the point $(1, 3, 0)$.

**Solution:**

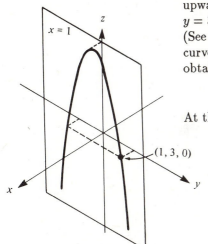

The graph of the equation $z = 9x^2 - y^2$ is a hyperbolic paraboloid. The $xy$-trace ($z = 0$) consists of the intersecting lines $y = \pm 3x$. The $yz$-trace, $z = -y^2$, is a parabola opening downward and the $xz$-trace, $z = 9x^2$, is a parabola opening upward. The curve of intersection of the paraboloid and plane $y = 3$ is given by $z = 9x^2 - 9$, is a parabola opening downward. (See the accompanying figure.) Since $y$ is a constant on the curve of intersection, we differentiate with respect to $x$ to obtain

$$\frac{\partial z}{\partial x} = 18x$$

At the point $(1, 3, 0)$ the slope is

$$\frac{\partial z}{\partial x} = 18(1) = 18$$

## 15.4    Differentials

---

5. Find the total differential for $z = x \cos y - y \cos x$.

**Solution:**

$$dz = \frac{\partial z}{\partial x} \, dx + \frac{\partial z}{\partial y} \, dy$$
$$= (\cos y + y \sin x) \, dx + (-x \sin y - \cos x) \, dy$$
$$= (\cos y + y \sin x) \, dx - (x \sin y + \cos x) \, dy$$

9. Find the total differential for $u = (x + y)/(z - 2y)$.

**Solution:**

Since $u = (x + y)/(z - 2y)$, we have

$$du = \frac{\partial u}{\partial x} \, dx + \frac{\partial u}{\partial y} \, dy + \frac{\partial u}{\partial z} \, dz$$
$$= \frac{1}{z - 2y} \, dx + \frac{(z - 2y)(1) - (x + y)(-2)}{(z - 2y)^2} \, dy + \frac{0 - (x + y)(1)}{(z - 2y)^2} \, dz$$
$$= \frac{1}{z - 2y} \, dx + \frac{2x + z}{(z - 2y)^2} \, dy - \frac{x + y}{(z - 2y)^2} \, dz$$

11. Given the function $f(x, y) = 9 - x^2 - y^2$, (a) evaluate $f(1, 2)$ and $f(1.05, 2.1)$ and calculate $\Delta z$, and (b) use the total differential $dz$ to approximate $\Delta z$.

**Solution:**

(a)
$$f(1, 2) = 9 - 1^1 - 2^2 = 4$$
$$f(1.05, 2.1) = 9 - (1.05)^2 - (2.1)^2 = 3.4875$$
$$\Delta z = f(1.05, 2.1) - f(1, 2) = -0.5125$$

(b)
$$dz = f_x(x, y) \, dx + f_y(x, y) \, dy$$
$$= -2x \, dx - 2y \, dy$$

Letting $x = 1$, $y = 2$, $dx = 0.05$, and $dy = 0.1$, we have

$$dz = -2(1)(0.05) - 2(2)(0.1) = -0.5$$

**17.** Find the values of $\epsilon_1$ and $\epsilon_2$ (see the definition of differentiability) for $f(x,y) = x^2 - 2x + y$ and verify that $\epsilon_1 \to 0$ and $\epsilon_2 \to 0$ as $(\Delta x, \Delta y) \to (0,0)$.

**Solution:**

$$
\begin{aligned}
\Delta z &= f(x + \Delta x, y + \Delta y) - f(x,y) \\
&= (x + \Delta x)^2 - 2(x + \Delta x) + (y + \Delta y) - (x^2 - 2x + y) \\
&= x^2 + 2x(\Delta x) + (\Delta x)^2 - 2x - 2\Delta x + y + \Delta y - x^2 + 2x - y \\
&= (2x - 2)\Delta x + (1)\Delta y + \Delta x(\Delta x) + 0(\Delta y) \\
&= f_x(x,y)\Delta x + f_y(x,y)\Delta y + \epsilon_1(\Delta x) + \epsilon_2(\Delta y)
\end{aligned}
$$

Therefore, $\epsilon_1 = \Delta x$ and $\epsilon_2 = 0$. As $(\Delta x, \Delta y) \to (0,0)$, $\epsilon_1 \to 0$ and $\epsilon_2 \to 0$.

**21.** The radius $r$ and height $h$ of a right circular cylinder are measured with a possible error of 4% and 2%, respectively. Approximate the maximum possible percentage error in measuring the volume.

**Solution:**

First we consider the percentage errors in $r$ and $h$ as

$$
\frac{dr}{r} = \pm 4\% = \pm 0.04 \quad \text{and} \quad \frac{dh}{h} = \pm 2\% = \pm 0.02
$$

Now since $V = \pi r^2 h$, we have

$$
dV = 2\pi rh\, dr + \pi r^2\, dh
$$

or the percentage error in $V$ is

$$
\begin{aligned}
\frac{dV}{V} &= \frac{2\pi rh\, dr}{\pi r^2 h} + \frac{\pi r^2\, dh}{\pi r^2 h} \\
&= 2\frac{dr}{r} + \frac{dh}{h} = 2(\pm 0.04) \pm 0.02 = \pm 0.10 = \pm 10\%
\end{aligned}
$$

## 15.5    Chain Rules for Functions of Several Variables

5. If $w = x^2 + y^2 + z^2$, $x = e^t \cos t$, $y = e^t \sin t$, and $z = e^t$, find $dw/dt$ by using the Chain Rule.

**Solution:**

$$\frac{dw}{dt} = \frac{\partial w}{\partial x}\frac{dx}{dt} + \frac{\partial w}{\partial y}\frac{dy}{dt} + \frac{\partial w}{\partial z}\frac{dz}{dt}$$
$$= 2x(-e^t \sin t + e^t \cos t) + 2y(e^t \cos t + e^t \sin t) + 2ze^t$$
$$= 2e^t[x(\cos t - \sin t) + y(\cos t + \sin t) + z]$$
$$= 2e^t[e^t \cos t(\cos t - \sin t) + e^t \sin t(\cos t + \sin t) + e^t]$$
$$= 4e^{2t}$$

9. If $w = x^2 - y^2$, $x = s \cos t$, and $y = s \sin t$, find $\partial w/\partial s$ and $\partial w/\partial t$ by using the Chain Rule. Evaluate each partial derivative when $s = 3$ and $t = \pi/4$.

**Solution:**

By the Chain Rule,

$$\frac{\partial w}{\partial s} = \frac{\partial w}{\partial x}\frac{\partial x}{\partial s} + \frac{\partial w}{\partial y}\frac{\partial y}{\partial s} \quad \text{and} \quad \frac{\partial w}{\partial t} = \frac{\partial w}{\partial x}\frac{\partial x}{\partial t} + \frac{\partial w}{\partial y}\frac{\partial y}{\partial t}$$

Therefore,

$$\frac{\partial w}{\partial s} = 2x(\cos t) + (-2y)(\sin t)$$
$$= (2s \cos t)(\cos t) - (2s \sin t)(\sin t)$$
$$= 2s(\cos^2 t - \sin^2 t) = 2s \cos 2t$$

when $s = 3$ and $t = \pi/4$, $\partial w/\partial s = 2(3)\cos(\pi/2) = 0$. Similarly,

$$\frac{\partial w}{\partial t} = 2x(-s \sin t) - 2y(s \cos t)$$
$$= 2s \cos t(-s \sin t) - 2s \sin t(s \cos t)$$
$$= -4s^2 \sin t \cos t = -2s^2 \sin 2t$$

when $s = 3$ and $t = \pi/4$, $\partial w/\partial t = -2(9)\sin(\pi/2) = -18$.

**13.** If $w = xy + xz + yz$, $x = t - 1$, $y = t^2 - 1$, and $z = t$, find $dw/dt$ (a) by the Chain Rule and (b) by converting $w$ to a function of $t$ before differentiating.

**Solution:**

(a) By the Chain Rule we have

$$\frac{dw}{dt} = \frac{\partial w}{\partial x}\frac{dx}{dt} + \frac{\partial w}{\partial y}\frac{dy}{dt} + \frac{\partial w}{\partial z}\frac{dz}{dt}$$
$$= (y + z)(1) + (x + z)(2t) + (x + y)(1)$$
$$= (t^2 - 1 + t)(1) + (t - 1 + t)(2t) + (t - 1 + t^2 - 1)(1)$$
$$= 6t^2 - 3 = 3(2t^2 - 1)$$

(b) By writing $w$ as a function of $t$ before differentiating, we have

$$w = (t - 1)(t^2 - 1) + (t - 1)t + (t^2 - 1)t$$
$$= 2t^3 - 3t + 1$$
$$\frac{dw}{dt} = 6t^2 - 3 = 3(2t^2 - 1)$$

**17.** If $w = \arctan(y/x)$, $x = r\cos\theta$, and $y = r\sin\theta$, find $\partial w/\partial r$ and $\partial w/\partial\theta$ (a) by the Chain Rule and (b) by converting $w$ to a function of $r$ and $\theta$ before differentiating.

**Solution:**

(a) First we calculate $\partial w/\partial x$ and $\partial w/\partial y$.

$$w = \arctan\frac{y}{x}$$
$$\frac{\partial w}{\partial x} = \frac{-y/x^2}{1 + (y^2/x^2)} = \frac{-y/x^2}{(x^2 + y^2)/x^2} = \frac{-y}{x^2 + y^2}$$
$$\frac{\partial w}{\partial y} = \frac{1/x}{1 + (y^2/x^2)} = \frac{1/x}{(x^2 + y^2)/x^2} = \frac{x}{x^2 + y^2}$$

Now by the Chain Rule,

$$\frac{\partial w}{\partial r} = \frac{\partial w}{\partial x}\frac{\partial x}{\partial r} + \frac{\partial w}{\partial y}\frac{\partial y}{\partial r}$$
$$= \frac{-y}{x^2 + y^2}(\cos\theta) + \frac{x}{x^2 + y^2}(\sin\theta)$$
$$= \frac{x\sin\theta - y\cos\theta}{x^2 + y^2} = \frac{r\cos\theta\sin\theta - r\sin\theta\cos\theta}{r^2} = 0$$

Furthermore,

$$\frac{\partial w}{\partial \theta} = \frac{\partial w}{\partial x}\frac{\partial x}{\partial \theta} + \frac{\partial w}{\partial y}\frac{\partial y}{\partial \theta}$$

$$= \frac{-y}{x^2 + y^2}(-r\sin\theta) + \frac{x}{x^2 + y^2}(r\cos\theta)$$

$$= \frac{-r\sin\theta(-r\sin\theta) + r\cos\theta(r\cos\theta)}{r^2} = \frac{r^2}{r^2} = 1$$

(b) Since

$$w = \arctan\frac{y}{x} = \arctan\frac{r\sin\theta}{r\cos\theta} = \arctan(\tan\theta) = \theta + n\pi$$

we have

$$\frac{\partial w}{\partial r} = 0 \quad \text{and} \quad \frac{\partial w}{\partial \theta} = 1$$

**19.** Differntiate implicitly to find the first partial derivatives of $z$ if $x^2 + y^2 + z^2 = 25$.

**Solution:**

Let

$$F(x, y, z) = x^2 + y^2 + z^2 - 25.$$

Then,

$$F_x(x, y, z) = 2x$$
$$F_y(x, y, z) = 2y$$
$$F_z(x, y, z) = 2z$$

Now by Theorem 15.8 we have

$$\frac{\partial z}{\partial x} = -\frac{F_x(x, y, z)}{F_z(x, y, z)} = -\frac{2x}{2z} = -\frac{x}{z}$$

$$\frac{\partial z}{\partial y} = -\frac{F_y(x, y, z)}{F_z(x, y, z)} = -\frac{2y}{2z} = -\frac{y}{z}$$

**27.** The length, width, and depth of a rectangular chamber are increasing at the rate of 3 ft/min, 2 ft/min, and $\frac{1}{2}$ ft/min, respectively. Find the rate at which the volume and surface area are changing the instant the length, width, and depth are 10 ft, 6 ft, and 4 ft, respectively.

**Solution:**

Let the volume of the chamber be given by $V = xyz$. Then

$$\frac{dV}{dt} = V_x\frac{dx}{dt} + V_y\frac{dy}{dt} + V_z\frac{dz}{dt} = yz\frac{dx}{dt} + xz\frac{dy}{dt} + xy\frac{dz}{dt}$$

$$= (6)(4)(3) + (10)(4)(2) + (10)(6)\left(\frac{1}{2}\right) = 182 \ \text{ft}^3/\text{min}$$

Let the surface area of the chamber be given by $S = 2(xy + yz + xz)$. Then

$$\frac{dS}{dt} = S_x\frac{dx}{dt} + S_y\frac{dy}{dt} + S_z\frac{dz}{dt}$$

$$= 2\left[(y+z)\frac{dx}{dt} + (x+z)\frac{dy}{dt} + (x+y)\frac{dz}{dt}\right]$$

$$= 2\left[(6+4)(3) + (10+4)(2) + (10+6)\left(\frac{1}{2}\right)\right]$$

$$= 132 \ \text{ft}^2/\text{min}$$

**33.** A function $f(x,y)$ is **homogeneous of degree $n$** if

$$f(tx, ty) = t^n f(x, y)$$

Find the degree of the homogeneous function

$$f(x, y) = x^3 - 3xy^2 + y^3$$

and show that

$$x\,f_x(x, y) + y\,f_y(x, y) = nf(x, y)$$

**Solution:**

$$f(tx, ty) = (tx)^3 - 3(tx)(ty)^2 + (ty)^3$$
$$= t^3x^3 - 3t^3xy^2 + t^3y^3$$
$$= t^3(x^2 - 3xy^2 + y^3) = t^3 f(x, y)$$

Therefore, the function is homogeneous of degree 3.

$$x\, f_x(x,y) + y\, f_y(x,y) = x(3x^2 - 3y^2) + y(-6xy + 3y^2)$$
$$= 3x^3 - 3xy^2 - 6xy^2 + 3y^3$$
$$= 3x^3 - 9xy^2 + 3y^3$$
$$= 3(x^3 - 3xy^2 + y^3) = 3f(x,y)$$

## 15.6   Directional Derivatives and Gradients

5. Find the directional derivative of the function
$g(x,y) = \sqrt{x^2 + y^2}$ at the point $P = (3,4)$ in the direction of
$\mathbf{v} = 3\mathbf{i} - 4\mathbf{j}$.

**Solution:**

We begin by finding a unit vector, $\mathbf{u}$ in the direction of $\mathbf{v}$

$$\mathbf{u} = \frac{\mathbf{v}}{||\mathbf{v}||} = \frac{3}{5}\mathbf{i} - \frac{4}{5}\mathbf{j} = \cos\theta\,\mathbf{i} + \sin\theta\,\mathbf{j}$$

Thus

$$\cos\theta = \frac{3}{5} \quad\text{and}\quad \sin\theta = -\frac{4}{5}$$

By Theorem 15.9 we have

$$D_{\mathbf{u}}f(x,y) = f_x(x,y)\cos\theta + f_y(x,y)\sin\theta$$
$$= \frac{x}{\sqrt{x^2+y^2}}\left(\frac{3}{5}\right) + \frac{y}{\sqrt{x^2+y^2}}\left(-\frac{4}{5}\right)$$
$$= \frac{1}{5\sqrt{x^2+y^2}}(3x - 4y)$$

Therefore,

$$D_{\mathbf{u}}f(3,4) = \frac{-7}{25}$$

(Note that $D_{\mathbf{u}}f(x,y) = \nabla f(x,y) \cdot \mathbf{u}$ where

$$\nabla f(x,y) = \frac{x}{\sqrt{x^2+y^2}}\mathbf{i} + \frac{y}{\sqrt{x^2+y^2}}\mathbf{j}.)$$

9. Find the directional derivative of $f(x, y, z) = xy + yz + xz$ at the point $P = (1, 1, 1)$ in the direction of $\mathbf{v} = 2\mathbf{i} + \mathbf{j} - \mathbf{k}$.

**Solution:**

We begin by finding $\nabla f(x, y, z)$ and a unit vector, $\mathbf{u}$, in the direction of $\mathbf{v}$.

$$\nabla f(x, y, z) = f_x(x, y, z)\mathbf{i} + f_y(x, y, z)\mathbf{j} + f_z(x, y, z)\mathbf{k}$$
$$= (y + z)\mathbf{i} + (x + z)\mathbf{j} + (x + y)\mathbf{k}$$

and

$$\mathbf{u} = \frac{\mathbf{v}}{\|\mathbf{v}\|} = \frac{\sqrt{6}}{6}(2\mathbf{i} + \mathbf{j} - \mathbf{k})$$

Therefore,

$$D_{\mathbf{u}}f(x, y, z) = \nabla f(x, y, z) \cdot \mathbf{u}$$

$$= \frac{\sqrt{6}}{6}[2(y + z) + (x + z) - (x + y)]$$

$$= \frac{\sqrt{6}}{6}(y + 3z)$$

and

$$D_{\mathbf{u}}f(1, 1, 1) = \frac{4\sqrt{6}}{6} = \frac{2\sqrt{6}}{3}$$

17. Find the directional derivative of the function $f(x, y) = x^2 + 4y^2$ at the point $P = (3, 1)$ in the direction of $Q = (1, -1)$.

**Solution:**

A vector in the specified direction is

$$\overrightarrow{PQ} = \mathbf{v} = (1 - 3)\mathbf{i} + (-1 - 1)\mathbf{j} = -2\mathbf{i} - 2\mathbf{j}$$

and a unit vector in this direction is

$$\mathbf{u} = \frac{\mathbf{v}}{\|\mathbf{v}\|} = \frac{-2}{\sqrt{8}}\mathbf{i} - \frac{2}{\sqrt{8}}\mathbf{j} = -\frac{1}{\sqrt{2}}\mathbf{i} - \frac{1}{\sqrt{2}}\mathbf{j}$$

Since $\nabla f(x, y) = f_x(x, y)\mathbf{i} + f_y(x, y)\mathbf{j} = 2x\mathbf{i} + 8y\mathbf{j}$, the gradient at $(3, 1)$ is

$$\nabla f(3, 1) = 6\mathbf{i} + 8\mathbf{j}$$

Consequently, at $(3, 1)$ the directional derivative is

$$D_{\mathbf{u}}f(3, 1) = \nabla f(3, 1) \cdot \mathbf{u}$$

$$= (6\mathbf{i} + 8\mathbf{j}) \cdot \left(-\frac{\sqrt{2}}{2}\mathbf{i} - \frac{\sqrt{2}}{2}\mathbf{j}\right)$$

$$= -3\sqrt{2} - 4\sqrt{2} = -7\sqrt{2}$$

**23.** Find the gradient of the function $h(x, y) = x \tan y$ and the maximum value of the directional derivative at the point $P = (2, \pi/4)$.

**Solution:**

The gradient vector is given by

$$\nabla f(x, y) = f_x(x, y)\,\mathbf{i} + f_y(x, y)\,\mathbf{j} = \tan y\,\mathbf{i} + x \sec^2 y\,\mathbf{j}$$

and at the point $P = (2, \pi/4)$ we have

$$\nabla f(2, \pi/4) = \tan \frac{\pi}{4}\,\mathbf{i} + 2\sec^2 \frac{\pi}{4}\,\mathbf{j} = \mathbf{i} + 4\mathbf{j}$$

Hence, it follows that the maximum value of the directional derivative at the point $P = (2, \pi/4)$ is

$$\|\nabla f(2, \pi/4)\| = \sqrt{17}$$

**27.** Find the gradient of the function $f(x, y, z) = \sqrt{x^2 + y^2 + z^2}$ and the maximum value of the directional derivative at the point $(1, 4, 2)$.

**Solution:**

The gradient vector is given by

$$
\begin{aligned}
&\nabla f(x, y, z) \\
&= f_x(x, y, z)\,\mathbf{i} + f_y(x, y, z)\,\mathbf{j} + f_z(x, y, z)\,\mathbf{k} \\
&= \frac{x}{\sqrt{x^2 + y^2 + z^2}}\,\mathbf{i} + \frac{y}{\sqrt{x^2 + y^2 + z^2}}\,\mathbf{j} + \frac{z}{\sqrt{x^2 + y^2 + z^2}}\,\mathbf{k} \\
&= \frac{x\,\mathbf{i} + y\,\mathbf{j} + z\,\mathbf{k}}{\sqrt{x^2 + y^2 + z^2}}
\end{aligned}
$$

and at the point $P = (1, 4, 2)$ we have

$$\nabla f(1, 4, 2,) = \frac{1}{\sqrt{21}}(\mathbf{i} + 4\mathbf{j} + 2\mathbf{k}).$$

Hence, the maximum value of the directional derivative at the the point $P = (1, 4, 2)$ is

$$\|\nabla f(1, 4, 2)\| = \frac{1}{\sqrt{21}}\sqrt{1 + 16 + 4} = 1$$

**33.** If $f(x, y) = 3 - (x/3) - (y/2)$, find $D_{\mathbf{u}}f(3, 2)$ if (a) $\theta = 4\pi/3$ and (b) $\theta = -\pi/6$

**Solution:**

(a) By Theorem 15.9 the directional derivative is

$$D_{\mathbf{u}}f(x, y) = f_x(x, y)\cos\theta + f_y(x, y)\sin\theta$$
$$= -\frac{1}{3}\cos\theta - \frac{1}{2}\sin\theta$$

For $\theta = 4\pi/3$, $x = 3$, and $y = 2$, we have

$$D_{\mathbf{u}}f(3, 2) = -\frac{1}{3}\cos\frac{4\pi}{3} - \frac{1}{2}\sin\frac{4\pi}{3}$$
$$= -\frac{1}{3}\left(-\frac{1}{2}\right) - \frac{1}{2}\left(-\frac{\sqrt{3}}{2}\right) = \frac{2 + 3\sqrt{3}}{12}$$

(b) For $\theta = -\frac{\pi}{6}$, $x = 3$, and $y = 2$, we have

$$D_{\mathbf{u}}f(3, 2) = -\frac{1}{3}\cos\left(-\frac{\pi}{6}\right) - \frac{1}{2}\sin\left(-\frac{\pi}{6}\right)$$
$$= -\frac{1}{3}\left(\frac{\sqrt{3}}{2}\right) - \frac{1}{2}\left(-\frac{1}{2}\right) = \frac{3 - 2\sqrt{3}}{12}$$

**35.** If $f(x, y) = 3 - (x/3) - (y/2)$, find $D_{\mathbf{v}}f(3, 2)$ if (a) $\mathbf{v}$ is the vector from $(1, 2)$ to $(-2, 6)$ and (b) $\mathbf{v}$ is the vector from $(3, 2)$ to $(4, 5)$.

**Solution:**

(a) Let $\mathbf{u}$ be a unit vector in the direction of $\mathbf{v}$. Then

$$\mathbf{v} = (-2 - 1)\mathbf{i} + (6 - 2)\mathbf{j} = -3\mathbf{i} + 4\mathbf{j}$$

and

$$\mathbf{u} = \frac{\mathbf{v}}{\|\mathbf{v}\|} = \frac{-3\mathbf{i} + 4\mathbf{j}}{\sqrt{25}} = -\frac{3}{5}\mathbf{i} + \frac{4}{5}\mathbf{j}$$

At $(3, 2)$

$$\nabla f(3, 2) = f_x(3, 2)\mathbf{i} + f_y(3, 2)\mathbf{j} = -\frac{1}{3}\mathbf{i} - \frac{1}{2}\mathbf{j}$$

Therefore, the directional derivative in the direction of **v** is

$$\nabla f(3,2) \cdot \mathbf{u} = \left(-\frac{1}{3}\right)\left(-\frac{3}{5}\right) + \left(-\frac{1}{2}\right)\left(\frac{4}{5}\right) = \frac{1}{5} - \frac{2}{5} = -\frac{1}{5}$$

(b) Let **u** be a unit vector in the direction of **v**. Then

$$\mathbf{v} = (4-3)\mathbf{i} + (5-2)\mathbf{j} = \mathbf{i} + 3\mathbf{j}$$

and

$$\mathbf{u} = \frac{\mathbf{v}}{\|\mathbf{v}\|} = \frac{\mathbf{i} + 3\mathbf{j}}{\sqrt{10}} = \frac{\sqrt{10}}{10}\mathbf{i} + \frac{3\sqrt{10}}{10}\mathbf{j}$$

Therefore, the directional derivative in the direction of **v** is

$$\nabla f(3,2) \cdot \mathbf{u} = \left(-\frac{1}{3}\right)\left(\frac{\sqrt{10}}{10}\right) + \left(-\frac{1}{2}\right)\left(\frac{3\sqrt{10}}{10}\right) = -\frac{11\sqrt{10}}{60}$$

**37.** If $f(x,y) = 3 - (x/3) - (y/2)$, find the maximum value of the directional derivative at $(3,2)$.

**Solution:**

By Theorem 15.11 the maximum value of the directional derivative is $\|\nabla f(3,2)\|$.

$$f(x,y) = 3 - \frac{x}{3} - \frac{y}{2}$$

$$\nabla f(x,y) = f_x(x,y)\mathbf{i} + f_y(x,y)\mathbf{j} = -\frac{1}{3}\mathbf{i} - \frac{1}{2}\mathbf{j}$$

Therefore, the maximum value of the directional derivative at $(3,2)$ is

$$\|\nabla f(3,2)\| = \sqrt{\frac{1}{9} + \frac{1}{4}} = \frac{\sqrt{13}}{6}$$

**49.** Use the gradient to find a unit normal vector to the graph of $9x^2 + 4y^2 = 40$ at $P = (2, -1)$. Sketch the results.

**Solution:**

The ellipse given by the equation $9x^2 + 4y^2 = 40$ corresponds to the level curve with $c = 0$ to the function

$$f(x, y) = 9x^2 + 4y^2 - 40$$

By Theorem 15.12, $\nabla f(x_0, y_0)$ yields a normal vector to the level curve at the point $(x_0, y_0)$. Therefore,

$$\nabla f(x, y) = 18x\,\mathbf{i} + 8y\,\mathbf{j}$$

and at $(2, -1)$, a normal vector is

$$\nabla f(2, -1) = 36\,\mathbf{i} - 8\,\mathbf{j} = 4(9\,\mathbf{i} - 2\,\mathbf{j})$$

Since, $\|\nabla f(2, -1)\| = 4\sqrt{9^2 + (-2)^2} = 4\sqrt{85}$, a unit normal vector is

$$\frac{\sqrt{85}}{85}(9\,\mathbf{i} - 2\,\mathbf{j}).$$

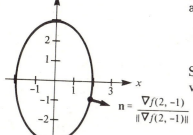

$$\mathbf{n} = \frac{\nabla f(2, -1)}{\|\nabla f(2, -1)\|}$$

**51.** The temperature field at any point in a plate is given by $T = x/(x^2 + y^2)$. Find the direction of greatest increase in heat at the point $(3, 4)$.

**Solution:**

The direction of greatest increase in temperature at $(3, 4)$ will be the direction of the gradient $\nabla T(x, y)$ at that point. Since

$$T_x(x, y) = \frac{(x^2 + y^2)(1) - x(2x)}{(x^2 + y^2)^2} = \frac{y^2 - x^2}{(x^2 + y^2)^2}$$

$$T_y(x, y) = \frac{(x^2 + y^2)(0) - x(2y)}{(x^2 + y^2)^2} = \frac{-2xy}{(x^2 + y^2)^2}$$

the gradient at $(3, 4)$ is

$$\nabla T(3, 4) = T_x(3, 4)\,\mathbf{i} + T_y(3, 4)\,\mathbf{j}$$

$$= \frac{7}{(25)^2}\,\mathbf{i} - \frac{24}{(25)^2}\,\mathbf{j} = \frac{1}{625}(7\,\mathbf{i} - 24\,\mathbf{j})$$

## 15.7   Tangent Planes and Normal Lines

**7.** Find a unit normal vector to the surface $z - x \sin y = 4$ at the point $P = (6, \pi/6, 7)$.

**Solution:**

We begin by writing the equation for the surface as a function of three variables obtaining

$$F(x, y, z) = z - x \sin y - 4$$

By Theorem 15.13, a normal vector to the surface, $F(x, y, z) = 0$, at $(x_0, y_0, z_0)$ is given by $\nabla F(x_0, y_0, z_0)$.

$$\nabla F(x, y, z) = F_x(x, y, z)\mathbf{i} + F_y(x, y, z)\mathbf{j} + F_z(x, y, z)\mathbf{k}$$
$$= -\sin y\,\mathbf{i} - x \cos y\,\mathbf{j} + \mathbf{k}$$

$$\nabla F(6, \pi/6, 7) = -\frac{1}{2}\mathbf{i} - 3\sqrt{3}\mathbf{j} + \mathbf{k}$$

Now, the unit normal vector to the surface is

$$\frac{\nabla F(6, \pi/6, 7)}{\|\nabla F(6, \pi/6, 7)\|} = \frac{\sqrt{113}}{113}(-\mathbf{i} - 6\sqrt{3}\mathbf{j} + 2\mathbf{k})$$

**13.** Find the equation of the tangent plane for the function $f(x, y) = y/x$ at the point $(1, 2, 2)$.

**Solution:**

We begin by writing the level surface equation

$$F(x, y, z) = f(x, y) - z = \frac{y}{x} - z = 0$$

Since

$$F_x(x, y, z) = \frac{-y}{x^2}, \quad F_y(x, y, z) = \frac{1}{x}, \quad \text{and} \quad F_z(x, y, z) = -1$$

we have

$$F_x(1, 2, 2) = -2, \quad F_y(1, 2, 2) = 1, \quad \text{and} \quad F_z(1, 2, 2) = -1$$

Therefore, by Theorem 15.13, the equation of the tangent plane at $(1, 2, 2)$ is

$$[F_x(1, 2, 2)](x - 1) + [F_y(1, 2, 2)](y - 2) + [F_z(1, 2, 2)](z - 2) = 0$$
$$-2(x - 1) + 1(y - 2) - (z - 2) = 0$$
$$-2x + y - z + 2 = 0$$
$$2x - y + z = 2$$

**23.** Find the equation of the tangent plane to $xy^2 + 3x - z^2 = 4$ at the point $(2, 1, -2)$.

**Solution:**

Let $F(x, y, z) = xy^2 + 3x - z^2 - 4$. Then $\nabla F(2, 1, -2)$ is normal to the tangent plane at $(2, 1, -2)$.

$$\nabla F(x, y, z) = F_x(x, y, z)\mathbf{i} + F_y(x, y, z)\mathbf{j} + F_z(x, y, z)\mathbf{k}$$
$$= (y^2 + 3)\mathbf{i} + 2xy\mathbf{j} - 2z\,\mathbf{k}$$
$$\nabla F(2, 1, -2) = 4\mathbf{i} + 4\mathbf{j} + 4\mathbf{k}$$

Therefore, the equation of the tangent plane is

$$4(x - 2) + 4(y - 1) + 4(z + 2) = 0$$
$$x + y + z = 1$$

**25.** Find an equation for the tangent plane and find symmetric equations of the normal line to the surface $x^2 + y^2 + z = 9$ at the point $(1, 2, 4)$.

**Solution:**

Let $F(x, y, z) = x^2 + y^2 + z - 9$. Then $\nabla F(1, 2, 4)$ is normal to the surface at $(1, 2, 4)$.

$$\nabla F(x, y, z) = F_x(x, y, z)\mathbf{i} + F_y(x, y, z)\mathbf{j} + F_z(x, y, z)\mathbf{k}$$
$$= 2x\,\mathbf{i} + 2y\mathbf{j} + \mathbf{k}$$
$$\nabla F(1, 2, 4) = 2\mathbf{i} + 4\mathbf{j} + \mathbf{k}$$

Therefore, the equation of the tangent plane is

$$2(x - 1) + 4(y - 2) + 1(z - 4) = 0$$
$$2x + 4y + z = 14$$

Now since a normal line at $(1, 2, 4)$ has the same direction as $\nabla F(1, 2, 4)$, the direction numbers for this line are $2, 4$, and $1$. Therefore, symmetric equations for a normal line at $(1, 2, 4)$ are

$$\frac{x - 1}{2} = \frac{y - 2}{4} = \frac{z - 4}{1}$$

**29.** Find an equation for the tangent plane and find symmetric equations of the normal line to the surface $z = \arctan(y/x)$ at the point $(1, 1, \pi/4)$.

**Solution:**

We let $F(x, y, z) = \arctan(y/x) - z$. Then

$$\nabla F(x, y, z) = F_x(x, y, z)\mathbf{i} + F_y(x, y, z)\mathbf{j} + F_z(x, y, z)\mathbf{k}$$

$$= \frac{-y}{x^2 + y^2}\mathbf{i} + \frac{x}{x^2 + y^2}\mathbf{j} - \mathbf{k}$$

$$\nabla F(1, 1, \pi/4) = -\frac{1}{2}\mathbf{i} + \frac{1}{2}\mathbf{j} - \mathbf{k} = -\frac{1}{2}(\mathbf{i} - \mathbf{j} + 2\mathbf{k})$$

Since $\nabla F(1, 1, \pi/4)$ is normal to the surface at the point $(1, 1, \pi/4)$, an equation of the tangent plane is

$$(x - 1) - (y - 1) + 2\left(z - \frac{\pi}{4}\right) = 0$$

$$x - y + 2z = \frac{\pi}{2}$$

and symmetric equations for the normal line are

$$\frac{x - 1}{1} = \frac{y - 1}{-1} = \frac{z - (\pi/4)}{2}$$

**35.** For the surfaces described by $x^2 + y^2 = 5$ and $z = x$, (a) find symmetric equations of the tangent line to their curve of intersection at the point $(2, 1, 2)$, and (b) find the cosine of the angle between the gradients of the surfaces at the same point.

**Solution:**

Let $f(x, y, z) = x^2 + y^2 - 5$ and $g(x, y, z) = x - z$. Then at $(2, 1, 2)$

$$\nabla f(x, y, z) = 2x\mathbf{i} + 2y\mathbf{j} = 4\mathbf{i} + 2\mathbf{j} \quad \text{and} \quad \nabla g(x, y, z) = \mathbf{i} - \mathbf{k}$$

(a) Since $\nabla f$ and $\nabla g$ are each normal to their respective surfaces, then the vector $\nabla f \times \nabla g$ will be tangent to both surfaces at the point $(2, 1, 2)$ on the curve of intersection. Therefore, from

$$\nabla f \times \nabla g = \begin{vmatrix} \mathbf{i} & \mathbf{j} & \mathbf{k} \\ 4 & 2 & 0 \\ 1 & 0 & -1 \end{vmatrix} = -2\mathbf{i} + 4\mathbf{j} - 2\mathbf{k}$$

we obtain the direction numbers $-2, 4, -2$ or $1, -2, 1$.
Hence symmetric equations for the tangent line at $(2, 1, 2)$
are

$$\frac{x-2}{1} = \frac{y-1}{-2} = \frac{z-2}{1}$$

(b) The angle between $\nabla f$ and $\nabla g$ at $(2, 1, 2)$ is such that

$$\cos\theta = \frac{\nabla f \cdot \nabla g}{\|\nabla f\| \|\nabla g\|} = \frac{4+0-0}{\sqrt{20}\sqrt{2}} = \frac{4}{\sqrt{40}} = \frac{4}{2\sqrt{10}} = \frac{\sqrt{10}}{5}$$

Therefore, the surfaces are **not** orthogonal at the point of
intersection.

45. Show that the tangent plane to the hyperboloid

$$\frac{x^2}{a^2} + \frac{y^2}{b^2} + \frac{z^2}{c^2} = 1$$

at the point $(x_0, y_0, z_0)$ can be written in the form

$$\frac{x_0 x}{a^2} + \frac{y_0 y}{b^2} + \frac{z_0 z}{c^2} = 1$$

**Solution:**

We let $F(x, y, z) = \dfrac{x^2}{a^2} + \dfrac{y^2}{b^2} + \dfrac{z^2}{c^2} - 1$. Then

$$\nabla F(x, y, z) = F_x(x, y, z)\,\mathbf{i} + F_y(x, y, z)\,\mathbf{j} + f_z(x, y, z)\,\mathbf{k}$$

$$= \frac{2x}{a^2}\mathbf{i} + \frac{2y}{b^2}\mathbf{j} + \frac{2z}{c^2}\mathbf{k}$$

$$\nabla F(x_0, y_0, z_0) = 2\left[\frac{x_0}{a^2}\mathbf{i} + \frac{y_0}{b^2}\mathbf{j} + \frac{z_0}{c^2}\mathbf{k}\right]$$

Now since $\nabla F(x_0, y_0, z_0)$ is normal to the surface at the point
$(x_0, y_0, z_0)$, an equation of the tangent plane is

$$\frac{x_0}{a^2}(x - x_0) + \frac{y_0}{b^2}(y - y_0) + \frac{z_0}{c^2}(z - z_0) = 0$$

$$\left[\frac{x_0 x}{a^2} + \frac{y_0 y}{b^2} + \frac{z_0 z}{c^2}\right] - \left[\frac{x_0^2}{a^2} + \frac{y_0^2}{b^2} + \frac{z_0^2}{c^2}\right] = 0$$

$$\frac{x_0 x}{a^2} + \frac{y_0 y}{b^2} + \frac{z_0 z}{c^2} = 1$$

## 15.8 Extrema of Functions of Two Variables

**7.** Examine the function $h(x, y) = x^2 - y^2 - 2x - 4y - 4$ for relative extrema and saddle points.

**Solution:**

Since

$$h_x(x, y) = 2x - 2 = 2(x - 1) = 0 \quad \text{when} \quad x = 1$$

and

$$h_y(x, y) = -2y - 4 = -2(y + 2) = 0 \quad \text{when} \quad y = -2$$

we have one critical point, $(1, -2)$. Since

$$h_{xx}(x, y) = 2, \quad h_{yy}(x, y) = -2, \quad h_{xy}(x, y) = 0$$

we have

$$d = h_{xx}(1, -2)h_{yy}(1, -2) - [h_{xy}(1, -2)]^2 = -4 - 0 = -4 < 0$$

Therefore, by part 3 of Theorem 15.17, the critical point $(1, -2)$ yields the saddle point $(1, -2, -1)$.

**11.** Examine the function $f(x, y) = x^3 - 3xy + y^3$ for relative extrema and saddle points.

**Solution:**

Since

$$f_x(x, y) = 3x^2 - 3y \quad \text{and} \quad f_y(x, y) = -3x + 3y^2$$

we can solve the system of equation

$$3x^2 - 3y = 0 \quad \text{and} \quad 3y^2 - 3x = 0$$

by substitution. From the first equation we see that $y = x^2$. Making this substitution for $y$ in the second equation yields

$$3(x^2)^2 - 3x = 0$$
$$x^4 - x = 0$$
$$x(x^3 - 1) = 0$$

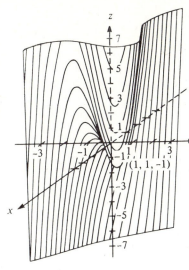

Therefore, the critical points are $(0,0)$ and $(1,1)$. Since

$$f_{xx}(x,y) = 6x, \quad f_{yy}(x,y) = 6y, \quad f_{xy}(x,y) = -3$$

we have $f_{xx}(0,0) = 0$ and

$$d = f_{xx}(0,0) - [f_{xy}(0,0)]^2 = 0 - 9 < 0$$

Therefore, by Theorem 15.17, part 3, we conclude that $(0,0,0)$ is a saddle point of $f$. At $(1,1)$ we have $f(1,1) = -1$, $f_{xx}(1,1) = 6 > 0$ and

$$d = f_{xx}(1,1)f_{yy}(1,1) - [f_{xy}(1,1)]^2 = (6)(6) - 9 > 0.$$

By part 1 of Theorem 15.17, the point $(1,1,-1)$ is a relative minimum. (See the accompanying figure.)

21. Find the absolute extrema of $f(x,y) = x^2 + 2xy + y^2$ over the region $R = \{(x,y) : x^2 + y^2 \le 8\}$.

**Solution:**

We first observe that $f$ is a perfect square trinomial and can be written as

$$f(x,y) = (x+y)^2$$

Therefore, the minimum value of $f$ is zero and occurs at all points in $R$ such that $y = -x$. From the partial derivatives, we have

$$f_x(x,y) = 2(x+y) = f_y(x,y)$$

Each point lying on the line $y = -x$ in the $xy$–coordinate plane is a critical point. However, from the discussion above, we know these points always yield the minimum value of zero. Thus, by Theorem 15.15, the maximum must occur at point(s) on the boundary of $R$. There the function $f$ will be maximum when the absolute value of $x + y$ is maximum. Thus, we have

$$f(-2,-2) = f(2,2) = 16$$

as the absolute maxima of $f$ over the region $R$.

**27.** For the function $f(x, y) = (x^2 + y^2)^{2/3}$, find the critical points and text for relative extrema. List the critical points for which the Second-Partials Test fails.

**Solution:**

Since $x^2 + y^2 \geq 0$ for all $x$ and $y$, we observe that the absolute minimum of $f$ is $f(0, 0) = 0$. From the first partial derivatives we have

$$f_x(x, y) = \frac{2}{3}(x^2 + y^2)^{-1/3}(2x) = \frac{4x}{3\sqrt[3]{x^2 + y^2}}$$

and

$$f_y(x, y) = \frac{2}{3}(x^2 + y^2)^{-1/3}(2y) = \frac{4y}{3\sqrt[3]{x^2 + y^2}}$$

Since neither first partial derivative exists at $(0, 0)$, it is a critical point. Moreover,

$$f_{xx}(x, y) = \frac{4(x^2 + 3y^2)}{9(x^2 + y^2)^{4/3}}$$

also does not exist at $(0, 0)$ and thus the Second-Partials Test fails.

**31.** Find the relative extrema of the function

$$f(x, y, z) = x^2 + y^2 + 2xz - 4yz + 10z$$

**Solution:**

Setting each of the first partial derivatives of $f$ equal to zero we have the following system of equations

(1)    $f_x(x, y, z) = 2x + 2z = 0 \quad \Longrightarrow \quad x = -z$

(2)    $f_y(x, y, z) = 2y - 4z = 0 \quad \Longrightarrow \quad y = 2z$

(3)    $f_z(x, y, z) = 2x - 4y + 10 = 0$

Substituting the expression for $x$ and $y$ from Equations 1 and 2 into Equation 3 we have

$$2(-z) - 4(2z) + 10 = 0$$
$$-10z = -10$$
$$z = 1$$

Therefore, by substituting $z = 1$ back into Equations 1 and 2 we obtain the critical point $(-1, 2, 1)$. By examining the behaviour of the function in the neighborhood of the critical number we conclude that $f(-1, 2, 1) = 5$ is a relative minimum.

## 15.9    Applications of Extrema of Functions of Two Variables

**7.** Find three positive numbers whose sum is 30 and the sum of whose squares is minimum.

**Solution:**

Let $x, y$, and $z$ be the numbers and let $s = x^2 + y^2 + z^2$. Since $x + y + z = 30$, we wish to minimize

$$s = x^2 + y^2 + (30 - x - y)^2$$

Setting the first partial derivatives equal to zero we have

$$s_x = 2x - 2(30 - x - y) = 0 \implies 2x + y = 30$$
$$s_y = 2y - 2(30 - x - y) = 0 \implies 2x + 4y = 60$$

Subtracting the first equation from the second we obtain the equation $-3y = -30$ or $y = 10$. Thus, we have the critical values $x = y = z = 10$. These values give us the desired minimum, since $s_{xx}(10, 10) = 4 > 0$ and

$$s_{xx}(10, 10)s_{yy}(10, 10) - [s_{xy}(10, 10)]^2 = (4)(4) - [2]^2 > 0$$

**11.** A waterline is built from point $P$ to point $S$ and must pass through regions where construction costs differ (see figure). Find $x$ and $y$ so the total cost $C$ will be minimum if the cost per mile in dollars is $3K$ from $P$ to $Q$, $2K$ from $Q$ to $R$, and $K$ from $R$ to $S$.

**Solution:**

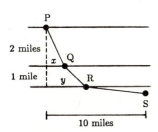

$$C = \text{(cost per mile)(distance from } P \text{ to } Q)$$
$$+ \text{(cost per mile)(distance from } Q \text{ to } R)$$
$$+ \text{(cost per mile)(distance from } R \text{ to } S)$$
$$= 3K\sqrt{x^2 + 4} + 2K\sqrt{(y - x)^2 + 1} + K(10 - y)$$

Setting the first partials equal to zero we obtain the system

$$\frac{\partial C}{\partial x} = K\left[\frac{3x}{\sqrt{x^2+4}} + \frac{-2(y-x)}{\sqrt{(y-x)^2+1}}\right] = 0$$

$$\frac{\partial C}{\partial y} = K\left[\frac{2(y-x)}{\sqrt{(y-x)^2+1}} - 1\right] = 0$$

From the equation $\partial C/\partial y = 0$ we have,

$$\frac{2(y-x)}{\sqrt{(y-x)^2+1}} = 1$$

$$2(y-x) = \sqrt{(y-x)^2+1}$$

$$4(y-x)^2 = (y-x)^2 + 1$$

$$3(y-x)^2 = 1$$

$$y - x = \pm\frac{1}{\sqrt{3}} \implies y = x \pm \frac{1}{\sqrt{3}}$$

Substituting this result with the positive root into the equation $\partial C/\partial x$ we obtain

$$\frac{3x}{\sqrt{x^2+4}} = \frac{2(y-x)}{\sqrt{(y-x)^2+1}}$$

$$\frac{3x}{\sqrt{x^2+4}} = \frac{2/\sqrt{3}}{\sqrt{4/3}}$$

$$3x = \sqrt{x^2+4}$$

$$9x^2 = x^2 + 4$$

$$8x^2 = 4$$

Therefore, we have

$$x = \frac{1}{\sqrt{2}} \approx 0.707 \ \text{mi}$$

$$y = x + \frac{\sqrt{3}}{3} = \frac{3\sqrt{2}+2\sqrt{3}}{6} \approx 1.284 \ \text{miles}$$

**17.** An eaves trough with trapezoidal cross sections is formed by turning up the edges of a sheet of aluminum which is $w$ inches wide. Find the cross section of maximum area.

**Solution:**

From the accompanying figure we observe that the area of a trapezoidal cross section is given by

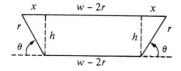

$$A = h\left[\frac{(w - 2r) + [(w - 2r) + 2x]}{2}\right]$$

$$= (w - 2r + x)h$$

where $x = r\cos\theta$ and $h = r\sin\theta$. Substituting these expressions for $x$ and $h$, we have

$$A(r, \theta) = (w - 2r + r\cos\theta)(r\sin\theta)$$
$$= wr\sin\theta - 2r^2\sin\theta + r^2\sin\theta\cos\theta$$

Now

$$A_r(r, \theta) = w\sin\theta - 4r\sin\theta + 2r\sin\theta\cos\theta$$
$$= \sin\theta(w - 4r + 2r\cos\theta) = 0 \quad\Longrightarrow\quad w = r(4 - 2\cos\theta)$$

and

$$A_\theta(r, \theta) = wr\cos\theta - 2r^2\cos\theta + r^2\cos 2\theta = 0.$$

Substituting the expression for $w$ from $A_r(r, \theta) = 0$ into the equation $A_\theta(r, \theta) = 0$, we have

$$r^2(4 - 2\cos\theta)\cos\theta - 2r^2\cos\theta + r^2(2\cos^2\theta - 1) = 0$$

$$r^2(2\cos\theta - 1) = 0 \text{ or } \cos\theta = \frac{1}{2}$$

The first partial derivatives are zero when $\theta = \pi/3$ and $r = w/3$. (Ignore the solution $r = \theta = 0$.) Thus, the trapezoid of maximum area occurs when each edge of width $w/3$ is turned up 60° from the horizontal.

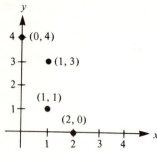

**25.** For the points in the accompanying figure, (a) find the least squares regression line, and (b) calculate $S$, the sum of squared errors.

**Solution:**

(a)

| $x$ | $y$ | $xy$ | $x^2$ |
|---|---|---|---|
| 0 | 4 | 0 | 0 |
| 1 | 3 | 3 | 1 |
| 1 | 1 | 1 | 1 |
| 2 | 0 | 0 | 4 |
| $\sum x_i = 4$ | $\sum y_i = 8$ | $\sum x_i y_i = 4$ | $\sum x_i^2 = 6$ |

By Theorem 15.18, we have

$$a = \frac{n\sum x_i y_i - \sum x_i \sum y_i}{n\sum x_i^2 - \left(\sum x_i\right)^2} = \frac{4(4) - 4(8)}{4(6) - 4^2} = -2$$

$$b = \frac{1}{n}\left(\sum y_i - a\sum x_i\right) = \frac{1}{4}[8 + 2(4)] = 4$$

Therefore, the least squares regression line is

$$f(x) = -2x + 4.$$

(b)     $S = \sum [f(x_i) - y_i]^2$
$$= (4-4)^2 + (2-3)^2 + (2-1)^2 + (0-0)^2 = 2$$

**33.** Use the result of Exercise 21 to find the least squares regression quadratic for the points $(0,0), (2,2), (3,6)$, and $(4,12)$.

**Solution:**

From Exercise 21, we have that the least squares regression quadratic for the points $(x_1, y_1), (x_2, y_2), \ldots, (x_n, y_n)$ is

$$y = ax^2 + bx + c$$

where $a$, $b$, and $c$ are the solutions to the system

$$a \sum_{i=1}^{n} x_i{}^4 + b \sum_{i=1}^{n} x_i{}^3 + c \sum_{i=1}^{n} x_i{}^2 = \sum_{i=1}^{n} x_i{}^2 y_i$$

$$a \sum_{i=1}^{n} x_i{}^3 + b \sum_{i=1}^{n} x_i{}^2 + c \sum_{i=1}^{n} x_i = \sum_{i=1}^{n} x_i y_i$$

$$a \sum_{i=1}^{n} x_i{}^2 + b \sum_{i=1}^{n} x_i + cn = \sum_{i=1}^{n} y_i$$

For the given points, we have

$$\sum x_i = 9, \quad \sum x_i{}^2 = 29, \quad \sum x_i{}^3 = 99, \quad \sum x_i{}^4 = 353$$

$$\sum y_i = 20, \quad \sum x_i y_i = 70, \quad \sum x_i{}^2 y_i = 192$$

The resulting system of equations is

$$353a + 99b + 29c = 192$$
$$99a + 29b + 9c = 70$$
$$29a + 9b + 4c = 20$$

Solving this system we have $a = 1$, $b = -1$, and $c = 0$. Therefore, the least squares regression quadratic is

$$y = x^2 - x$$

## 15.10   Lagrange Multipliers

**5.** Use Lagrange multipliers to find the maximum of $f(x, y) = x^2 - y^2$ subject to the constraint $y - x^2 = 0$. Assume $x$ and $y$ are positive.

**Solution:**

First, we let
$$g(x, y) = y - x^2.$$

Then, since $\nabla f(x, y) = 2x\,\mathbf{i} - 2y\,\mathbf{j}$ and $\lambda \nabla g(x, y) = \lambda(-2x\,\mathbf{i} + \mathbf{j})$, we have the following system of equations.

$$2x = -2\lambda x \qquad f_x(x, y) = \lambda g_x(x, y)$$
$$-2y = \lambda \qquad f_y(x, y) = \lambda g_y(x, y)$$
$$y - x^2 = 0 \qquad \text{Constraint}$$

From the first equation we have $\lambda = -1$. Using this result in the second and third equations we have $x = \sqrt{2}/2$ and $y = 1/2$. Therefore, the maximum value of $f$, subject to the given constraint is

$$f\left(\frac{\sqrt{2}}{2}, \frac{1}{2}\right) = \frac{1}{4}.$$

**17.** Use Lagrange multipliers to find the maximum of $f(x, y, z) = xyz$, subject to the constraints $x + y + z = 32$ and $x - y + z = 0$. Assume $x$, $y$, and $z$ are positive.

**Solution:**

In this case we have two constraints which we will denote by $g$ and $h$.

$$g(x, y, z) = x + y + z - 32$$
$$h(x, y, z) = x - y + z$$

Since $\nabla f(x, y, z) = yz\,\mathbf{i} + xz\,\mathbf{j} + xy\,\mathbf{j}$, $\lambda\nabla g(x, y, z) = \lambda\mathbf{i} + \lambda\mathbf{j} + \lambda\mathbf{k}$, and $\mu\nabla h(x, y, z) = \mu\mathbf{i} - \mu\mathbf{j} + \mu\mathbf{k}$, we have the following system of equations.

$$yz = \lambda + \mu \qquad f_x(x, y, z) = \lambda g_x(x, y, z) + \mu h_x(x, y, z)$$
$$xz = \lambda - \mu \qquad f_y(x, y, z) = \lambda g_y(x, y, z) + \mu h_y(x, y, z)$$
$$xy = \lambda + \mu \qquad f_z(x, y, z) = \lambda g_z(x, y, z) + \mu h_z(x, y, z)$$
$$x + y + z = 32 \qquad \text{Constraint 1}$$
$$x - y + z = 0 \qquad \text{Constraint 2}$$

From the first and third equations we have

$$yz = xy \quad \text{or} \quad z = x.$$

Substituting this result into the second constraint, we have

$$y = 2x.$$

Finally, substituting these two results into the first constraint we obtain

$$4x = 32 \quad \Longrightarrow x = 8$$
$$y = 2x = 16$$
$$z = x = 8$$

Therefore, the maximum value of $f$, subject to the given constraints is

$$f(8, 16, 8) = 8(16)(8) = 1024.$$

**20.** Use Lagrange multipliers to find any extrema of $f(x, y, z) = xyz$ subject to the constraints $x^2 + z^2 = 5$ and $x - 2y = 0$. Assume $x$, $y$, and $z$ are positive.

**Solution:**

In this case we have two constraints given by $f$ and $g$.

$$g(x, y, z) = x^2 + z^2 - 5$$
$$h(x, y, z) = x - 2y$$

Since $\nabla f(x, y, z) = yz\,\mathbf{i} + xz\,\mathbf{j} + xy\,\mathbf{k}$, $\lambda \nabla g(x, y, z) = 2\lambda x\,\mathbf{i} + 2\lambda z\,\mathbf{k}$, and $\mu \nabla h(x, y, z) = \mu\,\mathbf{i} - 2\mu\,\mathbf{j}$, we have the following system of equations.

$$yz = 2\lambda x + \mu \qquad f_x(x, y, z) = \lambda g_x(x, y, z) + \mu h_x(x, y, z)$$
$$xz = -2\mu \qquad f_y(x, y, z) = \lambda g_y(x, y, z) + \mu h_y(x, y, z)$$
$$xy = 2\lambda z \qquad f_z(x, y, z) = \lambda g_z(x, y, z) + \mu h_z(x, y, z)$$
$$x^2 + z^2 = 5 \qquad \text{Constraint 1}$$
$$x - 2y = 0 \qquad \text{Constraint 2}$$

From the second constraint we observe that $x = 2y$. Substituting this into second, third and fourth equations above, we obtain

$$2yz = -2\mu \quad \Longrightarrow \quad \mu = -yz$$
$$2y^2 = 2\lambda z \quad \Longrightarrow \quad \lambda = \frac{y^2}{z}$$
$$4y^2 + z^2 = 5 \quad \Longrightarrow \quad z^2 = 5 - 4y^2$$

Now we substitute these results into the first equation above to obtain

$$yz = 2\left(\frac{y^2}{z}\right)(2y) + (-yz)$$

$$2yz^2 - 4y^3 = 0$$
$$2y(z^2 - 2y^2) = 0$$
$$2y(5 - 4y^2 - 2y^2) = 0$$
$$2y(5 - 6y^2) = 0$$

$$y = 0 \quad \text{or} \quad y = \pm\sqrt{\frac{5}{6}}$$

For $y = 0$, we have $x = 0$ and $z = \pm\sqrt{5}$. For $y = \pm\sqrt{\frac{5}{6}}$, we have $x = \pm2\sqrt{\frac{5}{6}}$ and $z = \pm\sqrt{\frac{10}{6}} = \pm\sqrt{\frac{5}{3}}$. Finally, since we require that $x$, $y$, and $z$ be positive we have

$$f\left(2\sqrt{\frac{5}{6}}, \sqrt{\frac{5}{6}}, \sqrt{\frac{5}{3}}\right) = \left(2\sqrt{\frac{5}{6}}\right)\left(\sqrt{\frac{5}{6}}\right)\sqrt{\frac{5}{3}} = \frac{5}{3}\sqrt{\frac{5}{3}}$$

as the maximum value of $f$. Note that if we did not require $x, y$, and $z$ to be positive, and if we wanted all the extrema, then we would have

$$f\left(-2\sqrt{\frac{5}{6}}, -\sqrt{\frac{5}{6}}, \sqrt{\frac{5}{3}}\right) = \left(-2\sqrt{\frac{5}{6}}\right)\left(-\sqrt{\frac{5}{6}}\right)\sqrt{\frac{5}{3}} = \frac{5}{3}\sqrt{\frac{5}{3}}$$

and

$$f\left(\pm2\sqrt{\frac{5}{6}}, \pm\sqrt{\frac{5}{6}}, -\sqrt{\frac{5}{3}}\right) = -\frac{5}{3}\sqrt{\frac{5}{3}}$$

as the maximum and minimum values, respectively, of $f$.

23. A cargo container (in the shape of a rectangular solid) must have a volume of 480 cubic feet. Use Lagrange multipliers to find the dimensions of the container of this size which has minimum cost if the bottom will cost $5 per square foot to construct and the sides and top will cost $3 per square foot to construct.

**Solution:**

Letting $x, y$, and $z$ be the length, width and height of the solid, respectively, we are required to minimize the cost function

$$C(x, y, z) = 5xy + 3(2yz + xy + 2xz) = 8xy + 6yz + 6xz$$

subject to the constraint

$$xyz = 480.$$

First, we write the constraint as

$$g(x, y, z) = xyz - 480.$$

Then, since $\nabla C(x, y, z) = (8y + 6z)\mathbf{i} + (8x + 6z)\mathbf{j} + (6y + 6x)\mathbf{k}$ and $\lambda \nabla g(x, y, z) = \lambda(yz\,\mathbf{i} + xz\,\mathbf{j} + xy\,\mathbf{k})$, we obtain the following system of equations.

$$8y + 6z = \lambda yz \qquad C_x(x, y, z) = \lambda g_x(x, y, z)$$
$$8x + 6z = \lambda xz \qquad C_y(x, y, z) = \lambda g_y(x, y, z)$$
$$6y + 6x = \lambda xy \qquad C_z(x, y, z) = \lambda g_z(x, y, z)$$
$$xyz - 480 = 0 \qquad \text{Constraint}$$

We now multiply the first equation by $x$, the second by $-y$ and add to obtain

$$6xy - 6yz = 0 \quad \Longrightarrow \quad y = x.$$

Next we multiply the first equation by $x$, the third by $-z$ and add to obtain

$$8xy - 6yz = 0 \quad \Longrightarrow \quad z = \frac{4}{3}x.$$

Finally, we substitute these results into constraint to obtain

$$x(x)\left(\frac{4}{3}x\right) = 480$$
$$x^3 = 360$$
$$x = y = \sqrt[3]{360} \quad \text{and} \quad z = \frac{4}{3}\sqrt[3]{360}.$$

**31.** Use Lagrange multipliers to find the minimum distance from the point $(2, 1, 1)$ to the plane $x + y + z = 1$.

**Solution:**

Let $(x, y, z)$ be an arbitrary point in the given plane. Then

$$s = \sqrt{(x - 2)^2 + (y - 1)^2 + (z - 1)^2}$$

represents the distance between $(2, 1, 1)$ and a point in the plane. To simplify our calculations, we will minimize $s^2$ rather than $s$. With $g(x, y, z) = x + y + z - 1$ as the constraint, we have $\nabla s^2 = 2(x - 2)\mathbf{i} + 2(y - 1)\mathbf{j} + 2(z - 1)\mathbf{k}$ and $\lambda \nabla g(x, y, z) = \lambda \mathbf{i} + \lambda \mathbf{j} + \lambda \mathbf{k}$. Therefore,

$$2(x - 2) = \lambda \qquad s_x^{\,2}(x, y, z) = \lambda g_x(x, y, z)$$
$$2(y - 1) = \lambda \qquad s_y^{\,2}(x, y, z) = \lambda g_y(x, y, z)$$
$$2(z - 1) = \lambda \qquad s_z^{\,2}(x, y, z) = \lambda g_z(x, y, z)$$
$$x + y + z - 1 = 0 \qquad \text{Constraint}$$

From the first three equations, we conclude that

$$\lambda = 2(x - 2) = 2(y - 1) = 2(z - 1)$$

or that $x = y + 1$ and $z = y$. Therefore, from the constraint, we have

$$(y + 1) + y + y - 1 = 3y = 0 \quad \text{or} \quad y = 0.$$

Thus $x = 1$ and $z = 0$ and the point $(1, 0, 0)$ in the plane $x + y + z = 1$, is closest to the given point $(2, 1, 1)$. The minimum distance is

$$s = \sqrt{(1 - 2)^2 + (0 - 1)^2 + (0 - 1)^2} = \sqrt{3}.$$

## Review Exercises for Chapter 15

**13.** Find the first partial derivatives of

$$g(x, y) = \frac{xy}{x^2 + y^2}$$

**Solution:**

Using the Quotient Rule we have

$$g_x(x, y) = \frac{(x^2 + y^2)y - xy(2x)}{(x^2 + y^2)^2} = \frac{y(y^2 - x^2)}{(x^2 + y^2)^2}$$

$$g_y(x, y) = \frac{(x^2 + y^2)x - xy(2y)}{(x^2 + y^2)^2} = \frac{x(x^2 - y^2)}{(x^2 + y^2)^2}$$

**23.** Find all second partial derivatives for $h(x, y) = x \sin y + y \cos x$ and verify that the second mixed partials are equal.

**Solution:**

Since $h(x, y) = x \sin y + y \cos x$ we have

$$h_x(x, y) = \sin y - y \sin x \qquad h_{xx}(x, y) = -y \cos x$$

$$h_y(x, y) = x \cos y + \cos x \qquad h_{yy}(x, y) = -x \sin y$$

Furthermore,

$$h_{xy}(x, y) = \cos y - \sin x \quad \text{and} \quad h_{yx}(x, y) = \cos y - \sin x$$

**27.** Show that the function $z = y/(x^2 + y^2)$ satisfies the **Laplace Equation**

$$\frac{\partial^2 z}{\partial x^2} + \frac{\partial^2 z}{\partial y^2} = 0$$

**Solution:**

$$\frac{\partial z}{\partial x} = \frac{-2xy}{(x^2 + y^2)^2} = -2y\left[\frac{x}{(x^2 + y^2)^2}\right]$$

$$\frac{\partial^2 z}{\partial x^2} = -2y\left[\frac{(x^2 + y^2)^2 - x(2)(x^2 + y^2)(2x)}{(x^2 + y^2)^4}\right]$$

$$= \frac{2y(3x^2 - y^2)}{(x^2 + y^2)^3}$$

$$\frac{\partial z}{\partial y} = \frac{(x^2 + y^2) - y(2y)}{(x^2 + y^2)^2} = \frac{x^2 - y^2}{(x^2 + y^2)^2}$$

$$\frac{\partial^2 z}{\partial y^2} = \frac{(x^2 + y^2)^2(-2y) - (x^2 - y^2)(2)(x^2 + y^2)(2y)}{(x^2 + y^2)^4}$$

$$= \frac{-2y(3x^2 - y^2)}{(x^2 + y^2)^3}$$

Therefore,

$$\frac{\partial^2 z}{\partial x^2} + \frac{\partial^2 z}{\partial y^2} = \frac{2y(3x^2 - y^2)}{(x^2 + y^2)^3} + \frac{(-2y)(3x^2 - y^2)}{(x^2 + y^2)^3} = 0$$

**31.** Find $\partial u/\partial r$ and $\partial u/\partial t$ if $u = x^2 + y^2 + z^2$, $x = r\cos t$, $y = r\sin t$, and $z = t$. (a) Use the Chain Rule and (b) check the result by substituting for $x$, $y$, and $z$ before differentiating.

**Solution:**

(a) By the Chain Rule,

$$\frac{\partial u}{\partial r} = \frac{\partial u}{\partial x}\frac{\partial x}{\partial r} + \frac{\partial u}{\partial y}\frac{\partial y}{\partial r} + \frac{\partial u}{\partial z}\frac{\partial z}{\partial r}$$

$$= 2x(\cos t) + 2y(\sin t) + 2z(0)$$

$$= 2r\cos t(\cos t) + 2r\sin t(\sin t)$$

$$= 2r(\cos^2 t + \sin^2 t) = 2r$$

and

$$\frac{\partial u}{\partial t} = \frac{\partial u}{\partial x}\frac{\partial x}{\partial t} + \frac{\partial u}{\partial y}\frac{\partial y}{\partial t} + \frac{\partial u}{\partial z}\frac{\partial z}{\partial t}$$
$$= 2x(-r\sin t) + 2y(r\cos t) + 2z(1)$$
$$= 2r\cos t(-r\sin t) + 2r\sin t(r\cos t) + 2t$$
$$= -2r^2\sin t\cos t + 2r^2\sin t\cos t + 2t = 2t$$

(b) By first substituting for $x$, $y$, and $z$, we have

$$u = r^2\cos^2 t + r^2\sin^2 t + t^2$$
$$= r^2(\cos^2 t + \sin^2 t) + t^2 = r^2 + t^2$$

Therefore,

$$\frac{\partial u}{\partial r} = 2r \quad \text{and} \quad \frac{\partial u}{\partial t} = 2t$$

**37.** Find the gradient and the maximum value of the directional derivative of the function

$$f(x,y) = \frac{y}{x^2 + y^2}$$

at the point $(1, 1)$.

**Solution:**

The gradient is given by

$$\nabla f(x,y) = f_x(x,y)\mathbf{i} + f_y(x,y)\mathbf{j} = \frac{-2xy}{(x^2 + y^2)^2}\mathbf{i} + +\frac{x^2 - y^2}{(x^2 + y^2)^2}\mathbf{j}$$

At the point $(1, 1)$ the gradient is

$$\nabla f(1,1) = \frac{-2(1)(1)}{(1^2 + 1^2)^2}\mathbf{i} + \frac{1^2 - 1^2}{(1^2 + 1^2)^2}\mathbf{j} = -\frac{1}{2}\mathbf{i} = \langle -1/2, 0\rangle.$$

The maximum value of the directional derivative at the point $(1, 1)$ is given by

$$\|\nabla f(1,1)\| = \frac{1}{2}$$

**43.** For the surface defined by $f(x,y) = 9 + 4x - 6y - x^2 - y^2$, find an equation of the tangent plane and parametric equations for the normal line at the point $(2, -3, 4)$.

**Solution:**

We start by defining a function $F$ and finding its first partial derivatives.

$$F(x, y, z) = z - f(x, y) = z - 9 - 4x + 6y + x^2 + y^2$$
$$F_x(x, y, z) = -4 + 2x$$
$$F_y(x, y, z) = 6 + 2y$$
$$F_z(x, y, z) = 1$$

Then at $(2, -3, 4)$ we have $F_x(2, -3, 4) = 0$, $F_y(2, -3, 4) = 0$, and $F_z(2, -3, 4) = 1$. Therefore, by Theorem 15.13 an equation of the tangent plane at $(2, -3, 4)$ is

$$0(x - 2) + 0(y + 3) + 1(z - 4) = 0 \quad \text{or} \quad z = 4$$

Furthermore, a normal line at $(2, -3, 4)$ has direction numbers $0, 0, 1$ and its parametric equation are

$$x = 2, \quad y = -3, \quad z = 4 + t$$

**49.** Locate and classify any extrema of the function

$$f(x, y) = xy + \frac{1}{x} + \frac{1}{y}.$$

**Solution:**

We begin by setting the first partials of $f$ equal to zero.

$$f(x, y) = xy + \frac{1}{x} + \frac{1}{y}$$
$$f_x(x, y) = y - \frac{1}{x^2} = 0 \quad \Longrightarrow \quad x^2 y = 1$$
$$f_y(x, y) = x - \frac{1}{y^2} = 0 \quad \Longrightarrow \quad xy^2 = 1$$

Thus, $x^2y = xy^2$ or $x = y$ and substitution into $f_x(x,y) = 0$ yields

$$f_x(x,y) = y - \frac{1}{x^2}$$

$$= x - \frac{1}{x^2}$$

$$= \frac{x^3 - 1}{x^2} = 0 \implies x = 1..$$

Therefore, the critical point is $(1,1)$. We now use the Second Derivative Test and obtain

$$f_{xx} = \frac{2}{x^3}, \quad f_{xy} = 1, \quad f_{yy} = \frac{2}{y^3}.$$

At the critical point $(1,1)$, we have
$f(1,1) = 3$, $f_{xx}(1,1) = 2 > 0$ and
$f_{xx}(1,2)f_{yy}(1,1) - (f_{xy}(1,1))^2 = 3 > 0$. Thus, $(1,1,3)$ is a relative minimum.

**51.** Use Lagrange multipliers to locate and identify any extrema of the function $f(x,y) = x^2y$ subject to the constraint $x + 2y = 2$.

**Solution:**

First, we let
$$g(x,y) = x + 2y - 2.$$

Then, since $\nabla f(x,y) = 2xy\,\mathbf{i} + x^2\,\mathbf{j}$ and $\lambda\nabla g(x,y) = \lambda\mathbf{i} + 2\lambda\mathbf{j}$ we obtain the following system of equations

$$\begin{aligned}
2xy &= \lambda & f_x(x,y) &= \lambda g_x(x,y) \\
x^2 &= 2\lambda & f_y(x,y) &= \lambda f_y(x,y) \\
x + 2y - 2 &= 0 & &\text{Constraint}
\end{aligned}$$

Substitution of the expression for $\lambda$ from the first equation into the second equation we have

$$x^2 = 2(2xy)$$
$$x^2 - 4xy = 0$$
$$x(x - 4y) = 0 \implies x = 0 \text{ or } x = 4y.$$

Now, we substitute these values for $x$ into the constraint. When $x = 0$ we have
$$2y - 2 = 0 \implies y = 1.$$

Therefore, the relative minimum of $f$ is

$$f(0, 1) = 0.$$

When $x = 4y$ we have

$$4y + 2y - 2 = 0$$

$$6y = 2 \quad \Longrightarrow \quad y = \frac{1}{3} \text{ and } x = \frac{4}{3}.$$

Thus, the relative maximum of $f$ is

$$f\left(\frac{4}{3}, \frac{1}{3}\right) = \frac{16}{27}$$

**55.** The volume of a right circular cone is $V = \frac{1}{3}\pi r^2 h$. Find the maximum approximate error in the volume due to possible errors of $\frac{1}{8}$ inch in the measured values of $r$ and $h$, if these values are found to be 2 and 5 inches, respectively.

**Solution:**

Using the total differential we have

$$V = \frac{1}{3}\pi r^2 h$$

$$dV = V_r \, dr + V_h \, dh$$

$$= \frac{2}{3}\pi r h \, dr + \frac{1}{3}\pi r^2 \, dh.$$

Now, letting $r = 2, h = 5$, and $dr = dh = \pm\frac{1}{8}$ we obtain the maximum approximate error.

$$dV = \frac{2}{3}\pi(2)(5)\left(\pm\frac{1}{8}\right) + \frac{1}{3}\pi(2)^2\left(\pm\frac{1}{8}\right)$$

$$= \pm\frac{5}{6}\pi \pm \frac{1}{6}\pi = \pm\pi \text{ in}^3.$$

# 16 MULTIPLE INTEGRATION

## 16.1 Iterated Integrals and Area in the Plane

**7.** Evaluate

$$\int_{e^y}^{y} \frac{y \ln x}{x} \, dx$$

**Solution:**

Consider the integral in the form

$$\int_{e^y}^{y} y(\ln x)\left(\frac{1}{x}\right) dx$$

Then integration with respect to $x$ yields

$$\left[\frac{y(\ln x)^2}{2}\right]_{e^y}^{y} = \frac{y}{2}[(\ln y)^2 - (\ln e^y)^2] = \frac{y}{2}[(\ln y)^2 - y^2]$$

**13.** Evaluate $\int_{1}^{2} \int_{0}^{4} (x^2 - 2y^2 + 1) \, dx \, dy$.

**Solution:**

$$\int_{1}^{2} \int_{0}^{4} (x^2 - 2y^2 + 1) \, dx \, dy = \int_{1}^{2} \left[\frac{x^3}{3} - 2xy^2 + x\right]_{0}^{4} dy$$

$$= \int_{1}^{2} \left(\frac{64}{3} - 8y^2 + 4\right) dy$$

$$= \int_{1}^{2} \left(\frac{76}{3} - 8y^2\right) dy$$

$$= \frac{1}{3}\left[76y - 8y^3\right]_{1}^{2}$$

$$= \frac{1}{3}[152 - 64 - 76 + 8] = \frac{20}{3}$$

**19.** Evaluate $\displaystyle\int_0^{\pi/2}\int_0^{\sin\theta}\theta r\,dr\,d\theta$.

**Solution:**

$$\int_0^{\pi/2}\int_0^{\sin\theta}\theta r\,dr\,d\theta = \int_0^{\pi/2}\left[\frac{\theta r^2}{2}\right]_0^{\sin\theta}d\theta$$

$$= \frac{1}{2}\int_0^{\pi/2}\theta\sin^2\theta\,d\theta$$

$$= \frac{1}{2}\int_0^{\pi/2}\theta\left(\frac{1-\cos 2\theta}{2}\right)d\theta$$

$$= \frac{1}{4}\int_0^{\pi/2}(\theta-\theta\cos 2\theta)\,d\theta$$

Using integration by parts or Integration Formula 53 at the back of the text, we have

$$\int_0^{\pi/2}\int_0^{\sin\theta}\theta r\,dr\,d\theta = \frac{1}{4}\left[\frac{\theta^2}{2}-\left(\frac{1}{4}\cos 2\theta+\frac{\theta}{2}\sin 2\theta\right)\right]_0^{\pi/2}$$

$$= \frac{1}{4}\left(\frac{\pi^2}{8}+\frac{1}{4}-0-0+\frac{1}{4}+0\right)$$

$$= \frac{1}{4}\left(\frac{\pi^2}{8}+\frac{1}{2}\right)=\frac{\pi^2}{32}+\frac{1}{8}$$

**23.** Evaluate the improper iterated integral
$$\int_0^\infty\int_0^\infty xye^{-(x^2+y^2)}dx\,dy.$$

**Solution:**

$$\int_0^\infty\int_0^\infty xye^{-(x^2+y^2)}\,dx\,dy = -\frac{1}{2}\int_0^\infty\int_0^\infty ye^{-(x^2+y^2)}(-2x)dx\,dy$$

$$= -\frac{1}{2}\int_0^\infty ye^{-(x^2+y^2)}\bigg]_0^\infty dy$$

$$= -\frac{1}{2}\int_0^\infty -ye^{-y^2}\,dy$$

$$= -\frac{1}{4}\int_0^\infty e^{-y^2}(-2y)\,dy$$

$$= -\frac{1}{4}\left[e^{-y^2}\right]_0^\infty = \frac{1}{4}$$

**35.** Sketch the region $R$ whose area is given by the iterated integral

$$\int_0^1 \int_{y^2}^{\sqrt[3]{y}} dx\, dy$$

Switch the order of integration, and show that both orders yield the same area.

**Solution:**

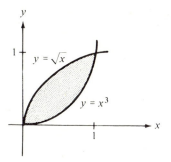

From the limits of integration, we know that when $0 \le y \le 1$, then $y^2 \le x \le \sqrt[3]{y}$, which means that the region $R$ is bounded by the curves $x = y^2$ and $x = \sqrt[3]{y}$. Therefore, the region $R$ can be sketched as in the accompanying figure.

If we interchange the order of integration so that $x$ is the outer variable, we see that $0 \le x \le 1$. Solving for $y$ in the equations $x = y^2$ and $x = \sqrt[3]{y}$, we have $y = \sqrt{x}$ and $y = x^3$. Thus $x^3 \le y \le \sqrt{x}$, and the area of $R$ is given by the iterated integral

$$\int_0^1 \int_{x^3}^{\sqrt{x}} dy\, dx$$

Evaluating each iterated integral, we have

$$\int_0^1 \int_{y^2}^{\sqrt[3]{y}} dx\, dy = \int_0^1 (\sqrt[3]{y} - y^2)\, dy = \left[\frac{3}{4}y^{4/3} - \frac{1}{3}y^3\right]_0^1 = \frac{5}{12}$$

$$\int_0^1 \int_{x^3}^{\sqrt{x}} dy\, dx = \int_0^1 (\sqrt{x} - x^3)\, dx = \left[\frac{2}{3}x^{3/2} - \frac{1}{4}x^4\right]_0^1 = \frac{5}{12}$$

**51.** Evaluate the iterated integral $\displaystyle\int_0^1 \int_y^1 \sin x^2\, dx\, dy$.

**Solution:**

Since it is not possible to perform the inner integration, it is necessary to switch the order of integration. From the given limits of integration, we know that

$$y \le x \le 1 \qquad \text{(Inner limits of integration)}$$

which means that the region $R$ is bounded on the left by the line $x = y$ and on the right by $x = 1$. Furthermore, since

$$0 \le y \le 1 \qquad \text{(Outer limits of integration)}$$

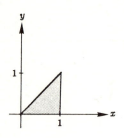

we know that $R$ is bounded below by the $x$-axis as shown in the accompanying figure. Now to charge the order of integration to $dy\,dx$ we observe that the outer limits have the constant bounds $0 \le x \le 1$ and the inner limits have bounds $0 \le y \le x$. Therefore,

$$\int_0^1 \int_y^1 \sin x^2 dx\,dy = \int_0^1 \int_0^x \sin x^2 dy\,dx$$

$$= \int_0^1 \left[y \sin x^2\right]_0^x dx$$

$$= \int_0^1 x \sin x^2\,dx$$

$$= \frac{1}{2}\left[-\cos x^2\right]_0^1 = \frac{1}{2}(1 - \cos 1)$$

## 16.2   Double Integrals and Volume

**3.** Sketch the region $R$ and evaluate the integral

$$\int_0^6 \int_{y/2}^3 (x + y)dx\,dy$$

**Solution:**

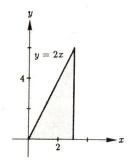

From the given limits of integration, we know that

$$\frac{y}{2} \le x \le 3 \qquad \text{(Inner limits of integration)}$$

which means that the region $R$ is bounded on the left by $x = y/2$ and on the right by $x = 3$. Furthermore, since

$$0 \le y \le 6 \qquad \text{(Outer limits of integration)}$$

we know that $R$ is bounded by the $x$-axis as shown in the accompanying figure.

$$\int_0^6 \int_{y/2}^3 (x+y)dx\,dy = \int_0^6 \left(\frac{1}{2}x^2 + xy\right)\Big]_{y/2}^3 dy$$

$$= \int_0^6 \left[\left(\frac{9}{2} + 3y\right) - \left(\frac{y^2}{8} + \frac{y^2}{2}\right)\right] dy$$

$$= \int_0^6 \left(\frac{9}{2} + 3y - \frac{5y^2}{8}\right) dy$$

$$= \left[\frac{9}{2}y + \frac{3}{2}y^2 - \frac{5}{24}y^3\right]_0^6 = 36$$

**9.** Given

$$\int_R \int \frac{y}{x^2+y^2}\,dA$$

where R is the triangle bounded by $y = x$, $y = 2x$, and $x = 2$, set up the integrals for $dx\,dy$ and for $dy\,dx$ and use the most convenient order to evaluate the integral over the region $R$.

**Solution:**

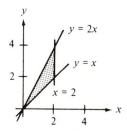

$$\int_R \int \frac{y}{x^2+y^2}\,dx\,dy$$

$$= \int_0^2 \int_{y/2}^y \frac{y}{x^2+y^2}\,dx\,dy + \int_2^4 \int_{y/2}^2 \frac{y}{x^2+y^2}\,dx\,dy$$

and

$$\int_R \int \frac{y}{x^2+y^2}\,dy\,dx = \int_0^2 \int_x^{2x} \frac{y}{x^2+y^2}\,dy\,dx$$

$$= \frac{1}{2}\int_0^2 \left[\ln\left(x^2+y^2\right)\right]_x^{2x} dx$$

$$= \frac{1}{2}\int_0^2 \left[\ln\left(5x^2\right) - \ln\left(2x^2\right)\right] dx$$

$$= \frac{1}{2}\int_0^2 \ln\left(\frac{5}{2}\right) dx$$

$$= \frac{1}{2}\left[x\ln\left(\frac{5}{2}\right)\right]_0^2 = \ln\left(\frac{5}{2}\right)$$

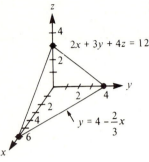

2x + 3y + 4z = 12

$y = 4 - \frac{2}{3}x$

**17.** Use a double integral to find the volume of the solid in the accompanying figure.

**Solution:**

By letting $z = 0$, we see that the base of the solid is the triangle in the $xy$-plane bounded by the graphs of $2x + 3y = 12$, $x = 0$, and $y = 0$. This plane region is both vertically and horizontally simple and we choose the order $dy\,dx$.

$$\text{variable bounds for } y : 0 \le y \le 4 - \frac{2}{3}x$$
$$\text{constant bounds for } x : 0 \le x \le 6$$

Therefore, the volume is given by

$$V = \int_0^6 \int_0^{4-(2/3)x} \left( 3 - \frac{1}{2}x - \frac{3}{4}y \right) dy\,dx$$

$$= \int_0^6 \left( 3y - \frac{1}{2}xy - \frac{3}{8}y^2 \right) \Bigg]_0^{4-(2/3)x} dx$$

$$= \int_0^6 \left( 6 - 2x + \frac{1}{6}x^2 \right) dx$$

$$= \left[ 6x - x^2 + \frac{1}{18}x^3 \right]_0^6 = 12$$

**29.** Use a double integral to find the volume of the first octant portion of the solid of intersection of the cylinders $x^2 + z^2 = 1$ and $y^2 + z^2 = 1$.

**Solution:**

In the accompanying figure we show the solid in the first octant. We divide this solid in two equal parts by the plane $y = x$ and thus find $\frac{1}{2}$ of the total volume. Therefore, we integrate the function $z = \sqrt{1 - x^2}$ over the triangle bounded by $y = 0$, $y = x$, and $x = 1$.

$$\text{Constant bounds for } x : 0 \le x \le 1$$
$$\text{Variable bounds for } y : 0 \le y \le x$$

$z = \sqrt{1 - x^2}$

$y = x$

$$V = 2 \int_0^1 \int_0^x \sqrt{1 - x^2}\, dy\, dx$$

$$= 2 \int_0^1 x\sqrt{1 - x^2}\, dx$$

$$= \left[ -\frac{2}{3}(1 - x^2)^{3/2} \right]_0^1 = \frac{2}{3}.$$

Thus the volume is twice that of Exercise 22.

**33.** Use Wallis's Formula as an aid in finding the volume of the solid bounded by the paraboloid $z = 4 - x^2 - y^2$ and the $xy$-plane.

**Solution:**

Because of the symmetry of the paraboloid, we find the volume of the solid in the first octant (one-fourth the total volume). Thus we integrate over the first-quadrant portion of the circle in the accompanying figure.

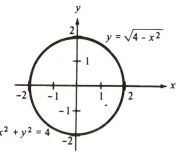

$$\text{Constant bounds for } x : 0 \le x \le 2$$
$$\text{Variable bounds for } y : 0 \le y \le \sqrt{4 - x^2}$$

Therefore,

$$V = 4 \int_0^2 \int_0^{\sqrt{4-x^2}} (4 - x^2 - y^2)\, dy\, dx$$

$$= 4 \int_0^2 \left[ 4y - x^2 y - \frac{1}{3}y^3 \right]_0^{\sqrt{4-x^2}} dx$$

$$= 4 \int_0^2 \left[ 4\sqrt{4 - x^2} - x^2\sqrt{4 - x^2} - \frac{1}{3}(4 - x^2)^{3/2} \right] dx$$

$$= \frac{8}{3} \int_0^2 (4 - x^2)^{3/2}\, dx$$

Now we let $x = 2\sin\theta$. Then, $\sqrt{4 - x^2} = 2\cos\theta$, $dx = 2\cos\theta\, d\theta$, and

$$V = \frac{8}{3} \int_0^{\pi/2} (2\cos\theta)^3(2\cos\theta)\, d\theta = \frac{128}{3} \int_0^{\pi/2} \cos^4\theta\, d\theta$$

$$= \frac{128}{3}\left( \frac{1\cdot 3}{2\cdot 4}\cdot\frac{\pi}{2} \right) \qquad \text{by Wallis's Formula}$$

$$= 8\pi$$

**45.** Evaluate the iterated integral $\int_0^{\ln 10} \int_{e^x}^{10} \frac{1}{\ln y}\, dy\, dx$.

**Solution:**

Since it is not possible to perform the inner integration, it is necessary to switch the order of integration. From the given limits of integration, we know that

$$e^x \le y \le 10 \qquad \text{(Inner limits of integration)}$$

and

$$0 \le x \le \ln 10 \qquad \text{(Outer limits of integration)}$$

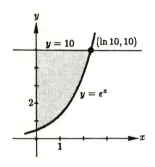

The region $R$ is shown in the accompanying figure. We observe from the figure that if the order of integration is reversed, the outer limits of integration are $1 \le y \le 10$ and the inner limits of integration are $0 \le x \le \ln y$. Therefore,

$$\int_0^{\ln 10} \int_{e^x}^{10} \frac{1}{\ln y}\, dy\, dx = \int_1^{10} \int_0^{\ln y} \frac{1}{\ln y}\, dx\, dy$$

$$= \int_1^{10} \left[ \frac{x}{\ln y} \right]_0^{\ln y} dy$$

$$= \int_1^{10} 1\, dy = \left[ y \right]_1^{10} = 9$$

## 16.3   Change of Variables: Polar Coordinates

**5.** Evaluate the double integral

$$\int_0^{\pi/2} \int_0^{1+\sin\theta} \theta\, dr\, d\theta$$

and sketch the region $R$.

**Solution:**

From the given limits of integration, we know that the inner limits are $0 \le r \le 1 + \sin\theta$ and the outer limits are $0 \le \theta \le \pi/2$. Hence, the region $R$ is the first quadrant portion of the cardiod $r = 1 + \sin\theta$ as shown in the accompanying figure.

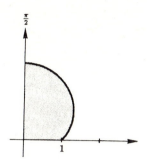

$$\int_0^{\pi/2} \int_0^{1+\sin\theta} \theta \, dr \, d\theta = \int_0^{\pi/2} [r\theta]_0^{1+\sin\theta} \, d\theta$$

$$= \int_0^{\pi/2} (\theta + \theta\sin\theta) \, d\theta$$

$$= \left[ \frac{\theta^2}{2} + \sin\theta - \theta\cos\theta \right]_0^{\pi/2}$$

$$= \frac{\pi^2}{8} + 1$$

[Note: To evaluate $\int \theta\sin\theta \, d\theta$ we use integration by parts with $u = \theta$ and $dv = \sin\theta \, d\theta$.]

**17.** Evaluate the double integral

$$\int_0^2 \int_0^{\sqrt{2x-x^2}} xy \, dy \, dx$$

by changing to polar coordinates.

**Solution:**

From the limits of integration we have

$$0 \le x \le 2$$
$$0 \le y \le \sqrt{2x - x^2} = \sqrt{1 - (x-1)^2}$$

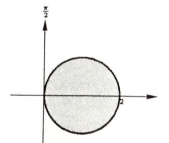

Therefore, the region of integration is bounded by the semicircle in the first quadrant with radius 1 and center $(1,0)$ as shown in the accompanying figure. In polar coordinates the bounds are

$$0 \le \theta \le \frac{\pi}{2}$$
$$0 \le r \le \cos\theta$$

Consequently, the double integral in polar coordinates is

$$\int_0^{\pi/2} \int_0^{2\cos\theta} \overbrace{(r\cos\theta}^{x}\,\overbrace{(r\sin\theta)}^{y}\,\overbrace{r\,dr\,d\theta}^{dA}$$

$$= \int_0^{\pi/2} \int_0^{2\cos\theta} r^3 \sin\theta \cos\theta \, dr \, d\theta$$

$$= \frac{1}{4} \int_0^{\pi/2} \left[ r^4 \sin\theta\cos\theta \right]_0^{2\cos\theta} d\theta$$

$$= 4 \int_0^{\pi/2} \cos^5\theta \sin\theta \, d\theta$$

$$= -\frac{4}{6} \left[ \cos^6\theta \right]_0^{\pi/2} = \frac{2}{3}$$

**19.** Combine the sum of the double integrals

$$\int_0^2 \int_0^x \sqrt{x^2+y^2}\,dy\,dx + \int_0^{2\sqrt{2}} \int_0^{\sqrt{8-x^2}} \sqrt{x^2+y^2}\,dy\,dx$$

into a single double integral by using polar coordinates.
Evaluate the resulting double integral.

**Solution:**

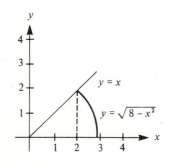

From the accompanying figure we can see that $R$ has the bounds

$$0 \le y \le x \qquad\qquad 0 \le x \le 2$$
$$0 \le y \le \sqrt{8-x^2} \qquad 2 \le x \le 2\sqrt{2}$$

and these bounds form a sector of a circle. In polar coordinates the bounds are

$$0 \le r \le 2\sqrt{2} \quad \text{and} \quad 0 \le \theta \le \frac{\pi}{4}$$

where the integrand is $\sqrt{x^2+y^2} = r$. Consequently, the double integral in polar coordinates is

$$\int_0^{\pi/4} \int_0^{2\sqrt{2}} (r)\overbrace{r\,dr\,d\theta}^{dA} = \int_0^{\pi/4} \left[ \frac{1}{3}r^3 \right]_0^{2\sqrt{2}} d\theta$$

$$= \frac{16\sqrt{2}}{3} \int_0^{\pi/4} d\theta = \frac{4\sqrt{2}\pi}{3}$$

**23.** Use polar coordinates to evaluate $\int_R \int f(x,y)\,dA$, where $f(x,y) = \arctan(y/x)$ and $R$ is such that $x^2 + y^2 \le 1$, $0 \le x$, and $0 \le y$.

**Solution:**

Since $x^2 + y^2 = r^2 \le 1$, $\tan\theta = y/x$, and $dA = r\,dr\,d\theta$, we have

$$\int_0^1 \int_0^{\sqrt{1-x^2}} \arctan\left(\frac{y}{x}\right) dy\,dx = \int_0^{\pi/2} \int_0^1 \theta r\,dr\,d\theta$$

$$= \frac{1}{2}\int_0^{\pi/2} [\theta r^2]_0^1\,d\theta$$

$$= \frac{1}{2}\int_0^{\pi/2} \theta\,d\theta = \frac{1}{4}[\theta^2]_0^{\pi/2} = \frac{\pi^2}{16}$$

**29.** Use a double integral in polar coordinates to find the volume of the solid inside the hemisphere $z = \sqrt{16 - x^2 - y^2}$ and the cylinder $x^2 + y^2 - 4x = 0$.

**Solution:**

Writing the equation for the cylinder in polar form we have $r = 4\cos\theta$. We can see from the accompanying figure that in polar coordinates $R$ has the bounds

$$0 \le r \le 4\cos\theta \quad \text{and} \quad -\frac{\pi}{2} \le \theta \le \frac{\pi}{2}$$

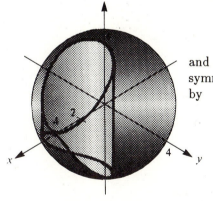

and $z = \sqrt{16 - x^2 - y^2} = \sqrt{16 - r^2}$. Since the solid is symmetric with respect to the $xz$-plane, the volume $V$ is given by

$$V = 2\int_0^{\pi/2} \int_0^{4\cos\theta} \sqrt{16 - r^2}\,r\,dr\,d\theta$$

$$= -\frac{2}{3}\int_0^{\pi/2} [(16 - 16\cos^2\theta)^{3/2} - 64]\,d\theta$$

$$= \frac{128}{3}\int_0^{\pi/2} (1 - \sin^3\theta)\,d\theta = \frac{64}{9}(3\pi - 4)$$

## 16.4   Center of Mass and Moments of Inertia

1. Find the mass and center of mass for the rectangular lamina with vertices $(0,0), (a,0), (0,b),$ and $(a,b)$ if (a) $\rho = k$ and (b) $\rho = ky$.

**Solution:**

(a)  $m = \int_0^a \int_0^b k \, dy \, dx = kab$

$M_x = \int_0^a \int_0^b ky \, dy \, dx = \dfrac{kab^2}{2}$

$M_y = \int_0^a \int_0^b kx \, dy \, dx = \dfrac{ka^2b}{2}$

$\overline{x} = \dfrac{M_y}{m} = \dfrac{ka^2b/2}{kab} = \dfrac{a}{2}, \qquad \overline{y} = \dfrac{M_x}{m} = \dfrac{kab^2/2}{kab} = \dfrac{b}{2}$

(b)  $m = \int_0^a \int_0^b ky \, dy \, dx = \dfrac{kab^2}{2}$

$M_x = \int_0^a \int_0^b ky^2 \, dy \, dx = \dfrac{kab^3}{3}$

$M_y = \int_0^a \int_0^b kxy \, dy \, dx = \dfrac{ka^2b^2}{4}$

$\overline{x} = \dfrac{M_y}{m} = \dfrac{ka^2b^2/4}{kab^2/2} = \dfrac{a}{2}, \qquad \overline{y} = \dfrac{M_y}{m} = \dfrac{kab^3/3}{kab^2/2} = \dfrac{2}{3}b$

15. Find the mass and center of mass of the lamina of density $\rho = ky$ and bounded by $y = \sin(\pi x/L)$, $y = 0$, $x = 0$, and $x = L$.

**Solution:**

$$m = \int_0^L \int_0^{\sin(\pi x/L)} ky \, dy \, dx$$

$$= \frac{k}{2} \int_0^L \sin^2\left(\frac{\pi x}{L}\right) dx$$

$$= \frac{k}{4} \int_0^L \left[1 - \cos\left(\frac{2\pi x}{L}\right)\right] dx$$

$$= \frac{k}{4} \left[ x - \frac{L}{2\pi} \sin \left( \frac{2\pi x}{L} \right) \right]_0^L = \frac{kL}{4}$$

$$M_x = \int_0^L \int_0^{\sin(\pi x/L)} ky^2 \, dy \, dx$$

$$= \frac{k}{3} \int_0^L \sin^3 \left( \frac{\pi x}{L} \right) dx$$

$$= \frac{k}{3} \int_0^L \left[ \sin \left( \frac{\pi x}{L} \right) - \cos^2 \left( \frac{\pi x}{L} \right) \sin \left( \frac{\pi x}{L} \right) \right] dx$$

$$= \frac{kL}{3\pi} \left[ -\cos \left( \frac{\pi x}{L} \right) + \frac{1}{3} \cos^3 \left( \frac{\pi x}{L} \right) \right]_0^L = \frac{4kL}{9\pi}$$

Now by symmetry we have $\bar{x} = \dfrac{L}{2}$ and

$$\bar{y} = \frac{M_x}{m} = \frac{4kL/9\pi}{kL/4} = \frac{16}{9\pi}.$$

19. Find the mass and center of mass of one petal $\left( -\frac{\pi}{6} \leq \theta \leq \frac{\pi}{6} \right)$ of the rose curve $r = 2\cos\theta$. (Hint: Use Simpson's Rule with $n = 6$ to approximate $M_y$.)

**Solution:**

Since the lamina is of uniform density and symmetric to the polar axis (see accompanying figure) we have $\bar{y} = 0$.

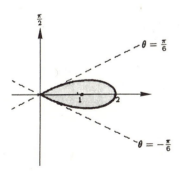

$$m = \int_R \int k \, dA$$

$$= \int_{-\pi/6}^{\pi/6} \int_0^{2\cos 3\theta} kr \, dr \, d\theta \qquad \text{(polar coordinates)}$$

$$= \frac{k}{2} \int_{-\pi/6}^{\pi/6} \left[ r^2 \right]_0^{2\cos 3\theta} d\theta$$

$$= 2k \int_{-\pi/6}^{\pi/6} \cos^2 3\theta \, d\theta$$

$$= k \int_{-\pi/6}^{\pi/6} (1 + \cos 6\theta) \, d\theta = \frac{k\pi}{3}$$

$$M_y = \int_R \int kx \, dA$$

$$= k \int_{-\pi/6}^{\pi/6} \int_0^{2\cos 3\theta} (r\cos\theta)\, r\, dr\, d\theta \qquad \text{(polar coordinates)}$$

$$= \frac{k}{3} \int_{-\pi/6}^{\pi/6} \left[ r^3 \cos\theta \right]_0^{2\cos 3\theta} d\theta$$

$$= \frac{8k}{3} \int_{-\pi/6}^{\pi/6} \cos^3 3\theta \, \cos\theta \, d\theta$$

$$\approx 1.17k \qquad\qquad\qquad \text{(By Simpson's Rule)}$$

Therefore, $\bar{x} = \dfrac{M_y}{m} \approx \dfrac{1.17k}{k\pi/3} \approx 1.12$

**29.** Find $I_x$, $I_y$, $I_0$, $\bar{\bar{x}}$, and $\bar{\bar{y}}$ for the lamina with density $\rho = kx$ and bounded by the graphs of $y = 4 - x^2$, $y = 0$, and $x > 0$.

**Solution:**

Since $\rho = kx$, we have

$$m = \int_0^2 \int_0^{4-x^2} kx \, dy \, dx$$

$$= k \int_0^2 xy \Big]_0^{4-x^2} dx$$

$$= k \int_0^2 x(4 - x^2) \, dx = \left[ -\frac{k}{4}(4 - x^2)^2 \right]_0^2 = 4k$$

Furthermore,

$$I_x = \int_R \int y^2 \rho \, dA = \int_0^2 \int_0^{4-x^2} kxy^2 \, dy \, dx$$

$$= \frac{k}{3} \int_0^2 xy^3 \Big]_0^{4-x^2} dx$$

$$= \frac{k}{3} \int_0^2 x(4 - x^2)^3 \, dx$$

$$= \left[ -\frac{k}{24}(4 - x^2)^4 \right]_0^2 = \frac{32k}{3}$$

and

$$I_y = \int_R \int x^2 \rho \, dA = \int_0^2 \int_0^{4-x^2} kx^3 \, dy \, dx$$

$$= k \int_0^2 \left[ x^3 y \right]_0^{4-x^2} dx$$

$$= k \int_0^2 (4x^3 - x^5) \, dx$$

$$= k \left[ x^4 - \frac{1}{6} x^6 \right]_0^2 = \frac{16k}{3}$$

Therefore, $I_0 = I_x + I_y = 16k$. Finally,

$$\bar{\bar{x}} = \sqrt{\frac{I_y}{m}} = \sqrt{\frac{16k/3}{4k}} = \frac{2}{\sqrt{3}} = \frac{2\sqrt{3}}{3}$$

$$\bar{\bar{y}} = \sqrt{\frac{I_x}{m}} = \sqrt{\frac{32k/3}{4k}} = \frac{4}{\sqrt{6}} = \frac{2\sqrt{6}}{3}$$

**35.** Find the moment of inertia of a uniform circular disk $x^2 + y^2 = b^2$ about the line $x = a \, (a > b)$.

**Solution:**

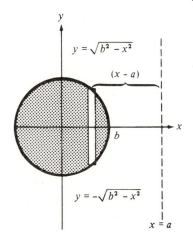

$$I = \int_R \int (\text{distance})^2 \text{mass}$$

$$= \int_{-r}^{r} \int_{-\sqrt{b^2 - x^2}}^{\sqrt{b^2 - x^2}} (x - a)^2 (k) \, dy \, dx$$

$$= 2k \int_0^{\pi} \int_0^{b} (r \cos \theta - a)^2 \, r \, dr \, d\theta \quad \text{(Polar coordinates)}$$

$$= 2k \int_0^{\pi} \int_0^{b} (r^3 \cos^2 \theta - 2r^2 \cos \theta + ra^2) \, dr \, d\theta$$

$$= 2k \int_0^{\pi} \left[ \frac{b^4}{4} \cos^2 \theta - \frac{2}{3} b^3 \cos \theta + \frac{a^2 b^2}{2} \right] d\theta$$

$$= 2k \left[ \frac{b^4}{8} \left( \theta + \frac{1}{2} \sin 2\theta \right) - \frac{2b^3}{3} \sin \theta + \frac{a^2 b^2}{2} \theta \right]_0^{\pi}$$

$$= 2k \left[ \frac{b^4 \pi}{8} + \frac{a^2 b^2 \pi}{2} \right] = \frac{kb^2 \pi}{4} (b^2 + 4a^2)$$

## 16.5 Surface Area

**3.** Find the area of the surface $f(x, y) = 8 + 2x + 2y$ over the region $R = \{(x, y) : x^2 + y^2 \leq 4\}$.

**Solution:**

The first partial derivatives of $f$ are

$$f_x(x, y) = 2 \quad \text{and} \quad f_y(x, y) = 2$$

and from the formula for surface area we have

$$\sqrt{1 + [f_x(x, y)]^2 + [f_y(x, y)]^2} = 3$$

Therefore, the surface area is given by

$$S = \int_R \int 3 \, dA = 3 \int_R \int dA = 3(\text{area of } R) = 3(4\pi) = 12\pi$$

**15.** Find the area of the surface $f(x, y) = \sqrt{x^2 + y^2}$ over the region $R = \{(x, y) : 0 \leq f(x, y) \leq 1\}$.

**Solution:**

We first observe that we are to find the surface area of that part of the cone $f(x, y) = \sqrt{x^2 + y^2}$ inside the cylinder $x^2 + y^2 = 1$. Therefore, the region $R$ in the $xy$-plane is a circle of radius 1 centered at the origin. Furthermore, the first partial derivatives of $f$ are

$$f_x(x, y) = \frac{x}{\sqrt{x^2 + y^2}} \quad \text{and} \quad f_y(x, y) = \frac{y}{\sqrt{x^2 + y^2}}$$

and from the formula for surface area, we have

$$\sqrt{1 + [f_x(x, y)]^2 + [f_y(x, y)]^2} = \sqrt{1 + \frac{x^2 + y^2}{x^2 + y^2}} = \sqrt{2}$$

Therefore, the surface area is given by

$$S = \int_R \int \sqrt{2} \, dA = \sqrt{2} \int_R \int dA = \sqrt{2}(\text{area of } R) = \sqrt{2}\pi$$

**19.** Find the surface area of the solid of intersection of the cylinders $x^2 + z^2 = 1$ and $y^2 + z^2 = 1$.

**Solution:**

In the accompanying figure we show the surface in the first octant. We divide this surface into two equal parts by the plane $y = x$ and thus find $\frac{1}{16}$ of the total surface area. Therefore, we find the area of the surface $z = \sqrt{1 - x^2}$ over the triangle bounded by $y = 0$, $y = x$, and $x = a$.

$$\frac{\partial z}{\partial x} = \frac{-x}{\sqrt{1 - x^2}} \quad \text{and} \quad \frac{\partial z}{\partial y} = 0$$

Therefore,

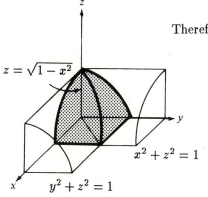

$$S = 16 \int_0^1 \int_0^x \sqrt{1 + \left(\frac{\partial z}{\partial x}\right)^2 + \left(\frac{\partial z}{\partial y}\right)^2}\, dy\, dx$$

$$= 16 \int_0^1 \int_0^x \sqrt{1 + \frac{x^2}{1 - x^2}}\, dy\, dx$$

$$= 16 \int_0^1 \int_0^x \frac{1}{\sqrt{1^2 - x^2}}\, dy\, dx$$

$$= 16 \int_0^1 \frac{x}{\sqrt{1 - x^2}}\, dx = \left[-16\sqrt{1 - x^2}\right]_0^1 = 16$$

**23.** Set up the double integral which gives the area of the surface on the graph of $f(x, y) = e^{-x} \sin y$ over the region $R = \{(x, y) : x^2 + y^2 \le 4\}$.

**Solution:**

The first partial derivatives of $f$ are

$$f_x(x, y) = -e^{-x} \sin y \quad \text{and} \quad f_y(x, y) = e^{-x} \cos y$$

and from the formula for surface area, we have

$$\sqrt{1 + [f_x(x, y)]^2 + [f_y(x, y)]^2} = \sqrt{1 + e^{-2x} \sin^2 y + e^{-2x} \cos^2 y}$$

$$= \sqrt{1 + e^{-2x}}$$

We integrate over the circle $x^2 + y^2 = 4$.

Constant bounds for $x$ :  $\qquad -2 \le x \le 2$

Variable bounds for $y$ :  $\qquad -\sqrt{4 - x^2} \le y \le \sqrt{4 - x^2}$

Therefore,

$$S = \int_{-2}^2 \int_{-\sqrt{4-x^2}}^{\sqrt{4-x^2}} \sqrt{1 + e^{-2x}}\, dy\, dx$$

## 16.6    Triple Integrals and Applications

9. Evaluate $\displaystyle\int_0^2 \int_{-\sqrt{4-x^2}}^{\sqrt{4-x^2}} \int_0^{x^2} x\, dz\, dy\, dx$.

**Solution:**

$$\int_0^2 \int_{-\sqrt{4-x^2}}^{\sqrt{4-x^2}} \int_0^{x^2} x\, dz\, dy\, dx$$

$$= \int_0^2 \int_{-\sqrt{4-x^2}}^{\sqrt{4-x^2}} \left[ xz \right]_0^{x^2} dy\, dx$$

$$= \int_0^2 \int_{-\sqrt{4-x^2}}^{\sqrt{4-x^2}} x^3\, dy\, dx$$

$$= \int_{-\pi/2}^{\pi/2} \int_0^2 (r\cos\theta)^3\, r\, dr\, d\theta \quad \text{(polar coordinates)}$$

$$= \frac{32}{5} \int_{-\pi/2}^{\pi/2} \cos^3\theta\, d\theta$$

$$= \frac{32}{5} \left[ \sin\theta - \frac{1}{3}\sin^3\theta \right]_{-\pi/2}^{\pi/2} = \frac{128}{15}$$

13. Sketch the solid region whose volume is given by the triple integral

$$\int_0^1 \int_y^1 \int_0^{\sqrt{1-y^2}} dz\, dx\, dy$$

and rewrite the integral using $dz\, dy\, dx$ as the order of integration.

**Solution:**

We have

$$\text{Constant bounds on } y: \quad 0 \le y \le 1$$
$$\text{Variable bounds on } x: \quad y \le x \le 1$$
$$\text{Variable bounds on } z: \quad 0 \le z \le \sqrt{1-y^2}$$

From the upper bound on $z$, we have

$$z = \sqrt{1-y^2} \quad \text{or} \quad y^2 + z^2 = 1$$

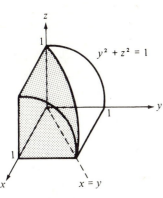

a cylinder of radius 1 with the $x$-axis as its axis. Therefore, the triple integral gives the volume of the solid in the first octant bounded by the graphs of $z = \sqrt{1 - y^2}$, $z = 0$, $x = y$, and $x = 1$. From the accompanying sketch of the solid we observe the following bounds when we change the order of integration to $dz\,dy\,dx$.

Constant bounds on $x$ :   $0 \le x \le 1$

Variable bounds on $y$ :   $0 \le y \le x$

Variable bounds on $z$ :   $0 \le z \le \sqrt{1 - y^2}$

Therefore, the integral is $\displaystyle\int_0^1 \int_0^x \int_0^{\sqrt{1-y^2}} dz\,dy\,dx$.

**21.** Use a triple integral to find the volume of the solid bounded by the cylinders $z = 4 - x^2$ and $y = 4 - x^2$ in the first octant.

**Solution:**

In the first octant we have $0 \le z \le 4 - x^2$, $0 \le y \le 4 - x^2$, and $0 \le x \le 2$. Therefore, the volume is

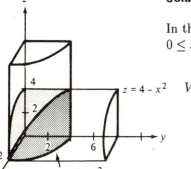

$$V = \int_0^2 \int_0^{4-x^2} \int_0^{4-x^2} dz\,dy\,dx = \int_0^2 \int_0^{4-x^2} (4 - x^2)\,dy\,dx$$

$$= \int_0^2 (4 - x^2)^2\,dx = \int_0^2 (16 - 8x^2 + x^4)\,dx$$

$$= \left[ 16x - \frac{8x^3}{3} + \frac{x^5}{5} \right]_0^2 = 32 - \frac{64}{3} + \frac{32}{5} = \frac{256}{15}$$

**27.** Find the mass and center of mass of the solid with density $\rho(x, y, z) = kxy$ and bounded by the graphs of $x = 0$, $x = b$, $y = 0$, $y = b$, $z = 0$, and $z = b$.

**Solution:**

The mass of the cube is

$$m = \int_0^b \int_0^b \int_0^b kxy\,dz\,dy\,dx = k \int_0^b \int_0^b bxy\,dy\,dx$$

$$= kb \int_0^b \left[ \frac{xy^2}{2} \right]_0^b dx = \frac{kb^3}{2} \int_0^b x\,dx = \frac{kb^3}{4} \left[ x^2 \right]_0^b = \frac{kb^5}{4}$$

Furthermore,

$$M_{yz} = \int_0^b \int_0^b \int_0^b x(kxy)\, dz\, dy\, dx = k \int_0^b \int_0^b x^2 y(b)\, dy\, dx$$

$$= kb \int_0^b \left[\frac{x^2 y^2}{2}\right]_0^b dx = \frac{kb^3}{2} \int_0^b x^2\, dx = \frac{kb^3}{6}\left[x^3\right]_0^b = \frac{kb^6}{6}$$

By the symmetry of the cube and of $\rho = kxy$, we have $M_{xz} = M_{yz} = kb^6/6$. Moreover,

$$M_{xy} = \int_0^b \int_0^b \int_0^b z(kxy)\, dz\, dy\, dx = k \int_0^b \int_0^b \left[\frac{xyz^2}{2}\right]_0^b dy\, dx$$

$$= \frac{kb^2}{2} \int_0^b \int_0^b xy\, dy\, dx = \frac{kb^2}{2} \int_0^b \left[\frac{xy^2}{2}\right]_0^b dx$$

$$= \frac{kb^4}{4} \int_0^b x\, dx = \frac{kb^4}{4}\left[\frac{x^2}{2}\right]_0^b = \frac{kb^6}{8}$$

Finally

$$\bar{x} = \frac{M_{yz}}{m} = \frac{kb^6/6}{kb^5/4} = \frac{2b}{3}$$

$$\bar{y} = \bar{x} = \frac{2b}{3}$$

$$\bar{z} = \frac{M_{xy}}{m} = \frac{kb^6/8}{kb^5/4} = \frac{b}{2}$$

37. Verify the moments of inertia

$$I_x = I_z = \frac{1}{12}m(3a^2 + L^2) \quad \text{and} \quad I_y = \frac{1}{2}ma^2$$

for the cylinder of uniform density in the accompanying figure.

**Solution:**

The solid is a right circular of radius $a$, length $L$, and uniform density $\rho(x, y, z) = k$. Thus the mass of the cylinder is

$$m = k(\text{volume}) = k\pi a^2 L$$

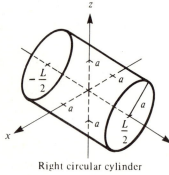

Right circular cylinder

Now

$$I_x = \int_{-a}^{a} \int_{-\sqrt{a^2-x^2}}^{\sqrt{a^2-x^2}} \int_{-L/2}^{L/2} (y^2 + z^2)k\, dy\, dz\, dx$$

$$= 8k \int_{0}^{a} \int_{0}^{\sqrt{a^2-x^2}} \int_{0}^{L/2} (y^2 + z^2)\, dy\, dz\, dx \qquad \text{(By symmetry)}$$

$$= \frac{kL}{3} \int_{0}^{a} \int_{0}^{\sqrt{a^2-x^2}} (L^2 + 12z^2)\, dz\, dx$$

$$= \frac{kL}{3} \int_{0}^{a} [L^2\sqrt{a^2 - x^2} + 4(a^2 - x^2)^{3/2}]\, dx$$

Now let $x = a\sin\theta$. Then $\sqrt{a^2 - x^2} = a\cos\theta$, $dx = a\cos\theta\, d\theta$, and

$$I_x = \frac{ka^2 L^3}{3} \int_{0}^{\pi/2} \cos^2\theta\, d\theta + \frac{4ka^4 L}{3} \int_{0}^{\pi/2} \cos^4\theta\, d\theta$$

By Wallis's Formula we have

$$I_x = \frac{ka^2 L^3}{3} \left(\frac{1}{2}\right)\left(\frac{\pi}{2}\right) + \frac{4ka^4 L}{3} \left(\frac{1}{2}\right)\left(\frac{3}{4}\right)\left(\frac{\pi}{2}\right)$$

$$= \frac{k\pi a^2 L}{12}(L^2 + 3a^2) = \frac{1}{12}m(3a^2 + L^2)$$

By symmetry $I_x = I_z$. To find $I_y$, we change only the integrand in the triple integral given above and obtain

$$I_y = 8k \int_{0}^{a} \int_{0}^{\sqrt{a^2-x^2}} \int_{0}^{L/2} (x^2 + z^2)\, dy\, dz\, dx$$

Proceeding with integrations similar to those given above yields

$$I_y = \frac{k\pi a^4 L}{2} = k\pi a^2 L \left(\frac{a^2}{2}\right) = \frac{1}{2}ma^2$$

## 16.7   Cylindrical and Spherical Coordinates

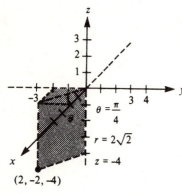

**5.** Convert the point $(2, -2, -4)$ from rectangular to cylindrical coordinates.

**Solution:**

Since $x = 2, y = -2$, and $z = -4$, we have

$$r = \pm\sqrt{x^2 + y^2} = \pm\sqrt{4 + 4} = \pm 2\sqrt{2}$$

$$\theta = \arctan\frac{y}{x} = \arctan\frac{-2}{2} = \arctan(-1) = -\frac{\pi}{4} \quad \text{or} \quad \frac{3\pi}{4}$$

$$z = -4$$

Therefore, one set of corresponding cylindrical coordinates for the given point is $(r, \theta, z) = (2\sqrt{2}, -\pi/4, -4)$. (See the accompanying sketch.) Another set of cylindrical coordinates for the given point would be $(-2\sqrt{2}, 3\pi/4, -4)$.

**11.** Convert the point $(4, 7\pi/6, 3)$ from cylindrical to rectangular coordinates.

**Solution:**

Since $(4, 7\pi/6, 3) = (r, \theta, z)$, we have

$$x = r\cos\theta = 4\cos\frac{7\pi}{6} = -2\sqrt{3}$$

$$y = r\sin\theta = 4\sin\frac{7\pi}{6} = -2$$

$$z = 3$$

Therefore, in rectangular coordinates the point is $(-2\sqrt{3}, -2, 3)$.

**15.** Convert the point $(-2, 2\sqrt{3}, 4)$ from rectangular to spherical coordinates.

**Solution:**

Since $(-2, 2\sqrt{3}, 4) = (x, y, z)$, we have

$$\rho = \sqrt{x^2 + y^2 + z^2} = \sqrt{4 + 12 + 16} = 4\sqrt{2}$$

$$\theta = \arctan \frac{y}{x} = \arctan \frac{2\sqrt{3}}{-2} = \frac{2\pi}{3}$$

$$\phi = \arccos \frac{z}{\rho} = \arccos \frac{4}{4\sqrt{2}} = \frac{\pi}{4}$$

Therefore in spherical coordinates the point is $(4\sqrt{2}, 2\pi/3, \pi/4)$.

**19.** Convert the point $(4, \pi/6, \pi/4)$ from spherical to rectangular coordinates.

**Solution:**

Since $(4, \pi/6, \pi/4) = (\rho, \theta, \phi)$, we have

$$x = \rho \sin \phi \cos \theta$$

$$= 4 \sin \frac{\pi}{4} \cos \frac{\pi}{6} = 4 \left( \frac{\sqrt{2}}{2} \right) \left( \frac{\sqrt{3}}{2} \right) = \sqrt{6}$$

$$y = \rho \sin \phi \sin \theta$$

$$= 4 \sin \frac{\pi}{4} \sin \frac{\pi}{6} = 4 \left( \frac{\sqrt{2}}{2} \right) \left( \frac{1}{2} \right) = \sqrt{2}$$

$$z = \rho \cos \phi = 4 \cos \frac{\pi}{4} = 4 \left( \frac{\sqrt{2}}{2} \right) = 2\sqrt{2}$$

Therefore, in rectangular coordinates, the given point is $(\sqrt{6}, \sqrt{2}, 2\sqrt{2})$.

**27.** Convert the point $(4, -\pi/6, 6)$ from cylindrical to spherical coordinates.

**Solution:**

Since $(4, -\pi/6, 6) = (r, \theta, z)$, we have

$$\rho^2 = x^2 + y^2 + z^2 = r^2 + z^2 = 16 + 36 = 52$$

or

$$\rho = 2\sqrt{13} > 0$$

Furthermore,

$$\cos \phi = \frac{z}{\sqrt{x^2 + y^2 + z^2}} = \frac{z}{\sqrt{r^2 + z^2}} = \frac{6}{\sqrt{52}} = \frac{3}{\sqrt{13}}$$

or

$$\phi = \arccos \frac{3}{\sqrt{13}}$$

Therefore, in spherical coordinates, the given point is $(2\sqrt{13}, -\pi/6, \arccos(3/\sqrt{13}))$.

**47.** Find an equation in rectangular coordinates for the equation $r = 2\sin\theta$ in cylindrical coordinates, and sketch its graph.

**Solution:**

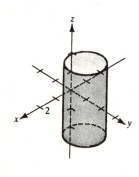

Since $x^2 + y^2 = r^2$ and $y = r\sin\theta$, we have

$$r = 2\sin\theta$$
$$r^2 = 2r\sin\theta$$
$$x^2 + y^2 = 2y$$
$$x^2 + y^2 - 2y + 1 = 1$$
$$x^2 + (y-1)^2 = 1$$

Therefore, the graph of the equation is a circular cylinder with rulings parallel to the $z$-axis.

**55.** Find an equation in rectangular coordinates for $\rho = 4\cos\phi$.

**Solution:**

Since

$$\cos \phi = \frac{z}{\sqrt{x^2 + y^2 + z^2}}$$

we write

$$\frac{\rho}{4} = \cos\phi = \frac{z}{\sqrt{x^2 + y^2 + z^2}}$$

Furthermore, since

$$\rho = \sqrt{x^2 + y^2 + z^2}$$

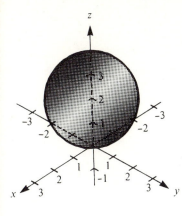

we have

$$\frac{\sqrt{x^2 + y^2 + z^2}}{4} = \frac{z}{\sqrt{x^2 + y^2 + z^2}}$$

$$x^2 + y^2 + z^2 = 4z$$

$$x^2 + y^2 + z^2 - 4z = 0$$

By completing the square on variable $z$, we have the equation

$$x^2 + y^2 + (z - 2)^2 = 4$$

Which represents a sphere of radius 2, centered at $(0, 0, 2)$.

**63.** Find an equation in (a) cylindrical coordinates and (b) spherical coordinates for the equation $x^2 + y^2 = 4y$.

**Solution:**

(a) Since $x^2 + y^2 = r^2$ and $y = 4 \sin \theta$, the equation in cylindrical coordinates is

$$r^2 = 4r \sin \theta \qquad \text{or} \qquad r = 4 \sin \theta$$

(b) In spherical coordinates, since $x^2 + y^2 = r^2 = \rho^2 \sin^2 \phi$ and $y = \rho \sin \phi \sin \theta$, we have

$$x^2 + y^2 = 4y$$

$$\rho^2 \sin^2 \phi = 4\rho \sin \phi \sin \theta$$

$$\rho \sin \phi = 4 \sin \theta$$

$$\rho = 4 \sin \theta \csc \phi$$

**69.** Sketch the solid that has the following description in spherical coordinates: $0 \le \theta \le 2\pi$, $0 \le \phi \le \pi/6$, and $0 \le \rho \le a \sec \phi$.

**Solution:**

From the constraint $0 \le \rho \le a \sec \phi$, we have

$$0 \le \rho \le a \left( \frac{1}{\cos \phi} \right)$$

$$0 \le \rho \cos \phi \le a$$

$$0 \le z \le a$$

## 16.8   Triple Integrals in Cylindrical and Spherical Coordinates

---

**3.** Evaluate the triple integral $\displaystyle\int_0^{\pi/2}\int_0^{2\cos^2\theta}\int_0^{4-r^2} r\sin\theta\,dz\,dr\,d\theta$.

**Solution:**

$$\int_0^{\pi/2}\int_0^{2\cos^2\theta}\int_0^{4-r^2} r\sin\theta\,dz\,dr\,d\theta$$

$$=\int_0^{\pi/2}\int_0^{2\cos^2\theta}\Big[rz\sin\theta\Big]_0^{4-r^2}\,dr\,d\theta$$

$$=\int_0^{\pi/2}\int_0^{2\cos^2\theta} r(4-r^2)\sin\theta\,dr\,d\theta$$

$$=\int_0^{\pi/2}\left[-\frac{1}{4}(4-r^2)^2\sin\theta\right]_0^{2\cos^2\theta}\,d\theta$$

$$=\int_0^{\pi/2}(8\cos^4\theta-4\cos^8\theta)\sin\theta\,d\theta$$

$$=\left[-\frac{8}{5}\cos^5\theta+\frac{4}{9}\cos^9\theta\right]_0^{\pi/2}=\frac{52}{45}$$

**9.** Sketch the solid region whose volume is given by the integral

$$\int_0^{\pi/2}\int_0^3\int_0^{e^{-r^2}} r\,dz\,dr\,d\theta$$

and evaluate the integral.

**Solution:**

We first observe that the triple integral is written in terms of cylindrical coordinates. From the limits of integration we have

Constant bounds on $\theta$ :   $0\le\theta\le\pi/2$

Constant bounds on $r$ :   $0\le r\le 3$

Variable bounds on $z$ :   $0\le z\le e^{-r^2}$

The limits on $r$ determine a circular cylinder of radius 3 having the $z$-axis as its axis. The limits on $\theta$ and $z$ restrict us to the first-octant portion of the cylinder bounded by the surface

$$z = e^{-r^2} = e^{-(x^2+y^2)}$$

The solid is shown in the accompanying figure. The value of the integral is given by

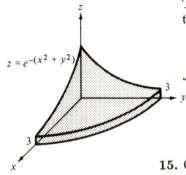

$$\int_0^{\pi/2} \int_0^3 \int_0^{e^{-r^2}} r \, dz \, dr \, d\theta = \int_0^{\pi/2} \int_0^3 re^{-r^2} \, dr \, d\theta$$

$$= -\frac{1}{2}(e^{-9} - 1) \int_0^{\pi/2} d\theta = \frac{\pi}{4}(1 - e^{-9})$$

**15.** Convert the integral

$$\int_{-a}^a \int_{-\sqrt{a^2-x^2}}^{\sqrt{a^2-x^2}} \int_a^{a+\sqrt{a^2-x^2-y^2}} x \, dz \, dy \, dx$$

from rectangular coordinates to both cylindrical and spherical coordinates and evaluate the simplest integral.

**Solution:**

We first observe that the solid $S$ over which we are integrating is a hemisphere, the top half of the sphere of radius $a$ and center $(0, 0, a)$. Projecting the solid onto the $xy$-plane forms a circle of radius $a$. Thus we have

Constant bounds on $\theta$ :   $0 \leq \theta \leq 2\pi$

Constant bounds on $r$ :   $0 \leq r \leq a$

Variable bounds on $z$ :   $a \leq z \leq a + \sqrt{a^2 - x^2 - y^2}$

$$= a + \sqrt{a^2 - r^2}$$

Finally, since $x = r\cos\theta$, we obtain the integral in cylindrical coordinates:

$$\int_0^{2\pi} \int_0^a \int_a^{\sqrt{a^2-r^2}} (r\cos\theta) r \, dz \, dr \, d\theta$$

$$= \int_0^{2\pi} \int_0^a \int_a^{a+\sqrt{a^2-r^2}} r^2 \cos\theta \, dz \, dr \, d\theta$$

To write the integral in spherical coordinates, we first write the equation of the sphere in spherical coordinates:

$$z = a + \sqrt{a^2 - x^2 - y^2}$$
$$x^2 + y^2 + (z - a)^2 = a^2$$
$$x^2 + y^2 + z^2 - 2az + a^2 = a^2$$
$$\rho^2 - 2a(\rho \cos \phi) = 0 \qquad \text{(Since } z = \rho \cos \phi)$$
$$\rho = 2a \cos \phi$$

We next write the equation of the plane $z = a$ in spherical coordinates and obtain

$$z = a$$
$$\rho \cos \phi = a$$
$$\rho = a \sec \phi$$

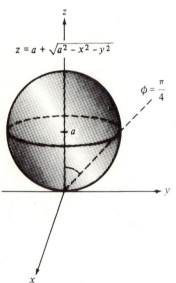

From the accompanying figure we see that the bounds on $\phi$ are $0 \le \phi \le \pi/4$. Now since $x = \rho \sin \phi \cos \theta$, we write the integral in spherical coordinates:

$$\int_0^{\pi/4} \int_0^{2\pi} \int_{a \sec \phi}^{2a \cos \phi} (\rho \sin \phi \cos \theta)(\rho^2 \sin \phi) \, d\rho \, d\theta \, d\phi$$

$$= \frac{1}{4} \int_0^{\pi/4} \int_0^{2\pi} \sin^2 \phi \cos \theta [(2a \cos \phi)^4 - (a \sec \phi)^4] \, d\theta \, d\phi$$

$$= \frac{a^4}{4} \int_0^{\pi/4} \left[ (16 \sin^2 \phi \cos^4 \phi - \sin^2 \phi \sec^4 \phi) \sin \theta \right]_0^{2\pi} d\phi$$

$$= \frac{a^4}{4} \int_0^{\pi/4} 0 \, d\phi = 0$$

**25.** Use a triple integral in cylindrical coordinates to find the volume of the solid inside both the sphere $x^2 + y^2 + z^2 = a^2$ and the cylinder $[x - (a/2)]^2 + y^2 = (a/2)^2$.

**Solution:**

The accompanying figure shows the sphere and cylinder. We will find the volume of the solid by using the cylindrical coordinate system. Thus the equation of the cylinder is given by

$$r = a \cos \theta, \quad (-\pi/2 \le \theta \le \pi/2)$$

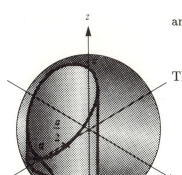

and the equation of the sphere is given by

$$r^2 + z^2 = a^2 \quad \text{or} \quad z = \pm\sqrt{a^2 - r^2}$$

Therefore, the volume is

$$V = \int_{-\pi/2}^{\pi/2} \int_0^{a\cos\theta} \int_{-\sqrt{a^2-r^2}}^{\sqrt{a^2-r^2}} r\, dz\, dr\, d\theta$$

$$= 4\int_0^{\pi/2} \int_0^{a\cos\theta} r\sqrt{a^2 - r^2}\, dr\, d\theta$$

$$= -\frac{4}{3}\int_0^{\pi/2} \left[(a^2 - a^2\cos^2\theta)^{3/2} - a^3\right] d\theta$$

$$= \frac{4a^3}{3}\int_0^{\pi/2} (1 - \sin^3\theta)\, d\theta = \frac{4a^3}{3}\left(\frac{\pi}{2} - \frac{2}{3}\right)$$

**32.** Use spherical coordinates to find the center of mass of the solid
of uniform density lying between two concentric hemispheres of
radii $r$ and $R$, where $r < R$.

**Solution:**

First, without loss of generality, we can position the
hemispheres with their centers at the origin and bases on the
$xy$-coordinate plane. Then, by symmetry, $\bar{x} = \bar{y} = 0$. Since the
solid has uniform density $k$, the mass is given by

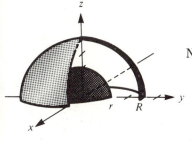

$$m = k(\text{volume}) = k\left(\frac{2}{3}\pi R^3 - \frac{2}{3}\pi r^3\right) = \frac{2}{3}k\pi(R^3 - r^3)$$

Now, by symmetry, we have

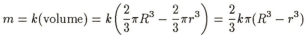

$$M_{xy} = \int\int_Q\int z(\text{density})\, dV$$

$$= 4k\int_0^{\pi/2} \int_0^{\pi/2} \int_r^R \rho^3\cos\phi\sin\phi\, d\rho\, d\theta\, d\phi$$

$$= \frac{1}{2}k(R^4 - r^4)\int_0^{\pi/2} \int_0^{\pi/2} \sin 2\phi\, d\theta\, d\phi$$

$$= \frac{1}{4}k\pi(R^4 - r^4)\int_0^{\pi/2} \sin 2\phi\, d\phi$$

$$= -\frac{1}{8}k\pi(R^4 - r^4)\left[\cos 2\phi\right]_0^{\pi/2} = \frac{1}{4}k\pi(R^4 - r^4)$$

Therefore, $\bar{z} = \dfrac{M_{xy}}{m} = \dfrac{k\pi(R^4 - r^4)/4}{2k\pi(R^3 - r^3)/3} = \dfrac{3(R^4 - r^4)}{8(R^3 - r^3)}.$

## 16.9   Change of Variables: Jacobians

**7.** Find the Jacobian for the change of variables

$$x = e^u \sin v \qquad y = e^u \cos v$$

**Solution:**

From the definition we evaluate the determinant

$$\frac{\partial(x, y)}{\partial(u, v)} = \begin{vmatrix} \dfrac{\partial x}{\partial u} & \dfrac{\partial y}{\partial u} \\[2mm] \dfrac{\partial x}{\partial v} & \dfrac{\partial y}{\partial v} \end{vmatrix} = \begin{vmatrix} e^u \sin v & e^u \cos v \\[1mm] e^u \cos v & -e^u \sin v \end{vmatrix}$$

$$= -e^{2u} \sin^2 v - e^{2u} \cos^2 v = -e^{2u}$$

**13.** Evaluate

$$\int_R \int 4(x + y)e^{x-y} \, dy \, dx$$

using the change of variables $x = \frac{1}{2}(u + v)$ and $y = \frac{1}{2}(u - v)$ and letting $R$ be the region in the accompanying figure.

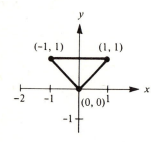

**Solution:**

We begin by solving for $u$ and $v$ in terms of $x$ and $y$ and obtain

$$u = x + y \qquad \text{and} \qquad v = x - y$$

Since the region $R$ is bounded by the lines $x + y = 0$, $x - y = 0$ and $y = 1$, the bounds for the region $S$ in the $uv$-plane are $u = 0, v = 0$, and $v = u - 2$. The Jacobian for the change of variables is

$$\frac{\partial(x, y)}{\partial(u, v)} = \begin{vmatrix} \dfrac{\partial x}{\partial u} & \dfrac{\partial y}{\partial u} \\[2mm] \dfrac{\partial x}{\partial v} & \dfrac{\partial y}{\partial v} \end{vmatrix} = \begin{vmatrix} \dfrac{1}{2} & \dfrac{1}{2} \\[2mm] \dfrac{1}{2} & -\dfrac{1}{2} \end{vmatrix} = -\frac{1}{2}$$

Therefore, by Theorem 16.6 we obtain

$$\int_R \int 4(x+y)e^{x-y}\, dy\, dx = \int_S \int 4ue^v \left|\frac{\partial(x,y)}{\partial(u,v)}\right| dv\, du$$

$$= 2\int_0^2 \int_{u-2}^0 ue^v\, dv\, du$$

$$= 2\int_0^2 (u - ue^{u-2})\, du$$

$$= 2\left[\frac{1}{2}u^2 - e^{u-2}(u-1)\right]_0^2 = 2(1 - e^{-2})$$

**19.** Use a change of variables to evaluate

$$\int_R \int \sqrt{(x-y)(x+4y)}\, dy\, dx$$

where $R$ is the region bounded by the parallelogram with vertices $(0,0), (1,1), (5,0),$ and $(4,-1)$.

**Solution:**

The region $R$ is bounded by the graphs of $x - y = 0$, $x - y = 5$, $x + 4y = 0$ and $x + 4y = 5$. (See the accompanying figure.) By letting $u = x - y$ and $v = x + 4y$, we have

$$x = \frac{1}{5}(4u + v) \qquad \text{and} \qquad y = -\frac{1}{5}(u - v)$$

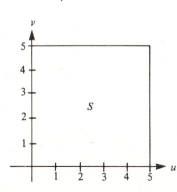

Thus, the Jacobian is

$$\frac{\partial(x,y)}{\partial(u,v)} = \begin{vmatrix} \dfrac{\partial x}{\partial u} & \dfrac{\partial y}{\partial u} \\[2ex] \dfrac{\partial x}{\partial v} & \dfrac{\partial y}{\partial v} \end{vmatrix} = \begin{vmatrix} \dfrac{4}{5} & -\dfrac{1}{5} \\[2ex] \dfrac{1}{5} & \dfrac{1}{5} \end{vmatrix} = \frac{1}{5}$$

Therefore, by Theorem 16.6 we obtain

$$\int_R \int \sqrt{(x-y)(x+4y)}\, dy\, dx = \int_S \int \sqrt{uv} \left|\frac{\partial(x,y)}{\partial(u,v)}\right| dv\, du$$

$$= \frac{1}{5}\int_0^5 \int_0^5 \sqrt{uv}\, dv\, du$$

$$= \frac{2\sqrt{5}}{3}\int_0^5 \sqrt{u}\, du = \frac{100}{9}$$

## Review Exercises for Chapter 16

**7.** Evaluate

$$\int_0^h \int_0^x \sqrt{x^2 + y^2}\, dy\, dx$$

by using the coordinate system that makes the integration easiest.

**Solution:**

We choose to use polar coordinates since $r = \sqrt{x^2 + y^2}$. To rewrite the limits, we first find the equation of the line $x = h$ in polar coordinates.

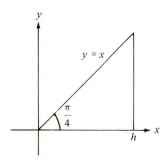

$$x = h$$
$$r \cos \theta = h$$
$$r = h\left(\frac{1}{\cos \theta}\right) = h \sec \theta$$

Therefore,

$$\int_0^h \int_0^x \sqrt{x^2 + y^2}\, dy\, dx = \int_0^{\pi/4} \int_0^{h \sec \theta} (r) r\, dr\, d\theta$$

$$= \frac{h^3}{3} \int_0^{\pi/4} \sec^3 \theta\, d\theta$$

$$= \frac{h^3}{3} \left[ \frac{\sec \theta \tan \theta}{2} + \frac{1}{2} \ln | \sec \theta + \tan \theta | \right]_0^{\pi/4}$$

$$= \frac{h^3}{6} [\sqrt{2} + \ln (\sqrt{2} + 1)]$$

**17.** If $R$ is the larger region between the circle $x^2 + y^2 = 25$ and the line $x = 3$, write the limits for the double integral $\int_R \int f(x, y)\, dA$ for both orders of integration. Compute the area by letting $f(x, y) = 1$.

**Solution:**

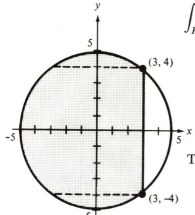

$$\int_R \int f(x,y)\,dA = \int_{-5}^{3} \int_{-\sqrt{25-x^2}}^{\sqrt{25-x^2}} f(x,y)\,dy\,dx$$

$$= \int_{-5}^{-4} \int_{-\sqrt{25-y^2}}^{\sqrt{25-y^2}} f(x,y)\,dx\,dy + \int_{-4}^{4} \int_{-\sqrt{25-y^2}}^{3} f(x,y)\,dx\,dy$$

$$+ \int_{4}^{5} \int_{-\sqrt{25-y^2}}^{\sqrt{25-y^2}} f(x,y)\,dx\,dy$$

The area of $R$ is

$$A = 2 \int_{-5}^{3} \int_{0}^{\sqrt{25-x^2}} dy\,dx = 2 \int_{-5}^{3} \sqrt{25-x^2}\,dx$$

$$= 2 \left(\frac{1}{2}\right) \left[ x\sqrt{25-x^2} + 25 \arcsin\left(\frac{x}{5}\right) \right]_{-5}^{3}$$

$$= 3(4) + 25 \arcsin\left(\frac{3}{5}\right) - 0 - 25\left(-\frac{\pi}{2}\right)$$

$$= 12 + \frac{25\pi}{2} + 25 \arcsin\left(\frac{3}{5}\right) \approx 67.36$$

(Note: The area of the entire circle is $25\pi \approx 78.54$.)

27. Write the equation $r^2(\cos^2 \theta - \sin^2 \theta) + z^2 = 1$ in rectangular coordinates.

**Solution:**

We rewrite the equation as

$$r^2 \cos^2 \theta - r^2 \sin^2 \theta + z^2 = 1$$

Now since $x = r\cos\theta$ and $y = r\sin\theta$, we have

$$x^2 - y^2 + z^2 = 1$$

as the rectangular equation.

**41.** Find the center of mass of the portion of the solid of constant density in the first octant bounded by $x^2 + y^2 + z^2 = a^2$.

**Solution:**

The solution $S$ is the first-octant portion of a sphere of radius $a$ and therefore, because of its symmetry, the coordinates of the center of mass are equal. Let $k$ be the constant density; then the mass is

$$m = k(\text{volume}) = k\left(\frac{1}{8}\right)\left(\frac{4}{3}\pi a^3\right) = \frac{k}{6}\pi a^3$$

Now,

$$M_{xy}$$

$$= \int \int_S \int z(\text{density})\, dS \qquad \text{(rectangular coordinates)}$$

$$= k \int_0^{\pi/2} \int_0^{\pi/2} \int_0^a (\rho \cos \phi)(\rho^2 \sin \phi)\, d\rho\, d\theta\, d\phi \quad \text{(spherical coordinates)}$$

$$= \frac{ka^4}{4} \int_0^{\pi/2} \int_0^{\pi/2} \cos \phi \sin \phi\, d\theta\, d\phi$$

$$= \frac{k\pi a^4}{8} \int_0^{\pi/2} \cos \phi \sin \phi\, d\phi$$

$$= \frac{k\pi a^4}{8} \left[\frac{1}{2} \sin^2 \phi\right]_0^{\pi/2} = \frac{k\pi a^4}{16}$$

Therefore,

$$\bar{x} = \bar{y} = \bar{z} = \frac{M_{xy}}{m} = \frac{k\pi a^4/16}{k\pi a^3/6} = \frac{3a}{8}$$

**43.** Find the area of the surface on the function $f(x, y) = 16 - x^2 - y^2$ and over the region $R = \{(x, y) : x^2 + y^2 \le 16\}$.

**Solution:**

Since $z = 16 - x^2 - y^2$, we have

$$\frac{\partial z}{\partial x} = -2x \quad \text{and} \quad \frac{\partial z}{\partial y} = -2y$$

$$S = \int_R \int \sqrt{1 + \left(\frac{\partial z}{\partial x}\right)^2 + \left(\frac{\partial z}{\partial y}\right)^2}\, dy\, dx$$

$$= \int_{-4}^{4} \int_{-\sqrt{16-x^2}}^{\sqrt{16-x^2}} \sqrt{1 + 4x^2 + 4y^2}\, dy\, dx$$

$$= 4 \int_{0}^{4} \int_{0}^{\sqrt{16-x^2}} \sqrt{1 + 4(x^2 + y^2)}\, dy\, dx$$

$$= \frac{1}{2} \int_{0}^{\pi/2} \int_{0}^{4} \sqrt{1 + 4r^2}\,(8r)\, dr\, d\theta$$

$$= \frac{1}{3} \int_{0}^{\pi/2} (65^{3/2} - 1)\, d\theta = \frac{\pi}{6}(65\sqrt{65} - 1)$$

# 17 VECTOR ANALYSIS

## 17.1  Vector Fields

**5.** Sketch several representative vectors in the vector field

$$\mathbf{F}(x, y) = x\mathbf{i} + y\mathbf{j}$$

### Solution:

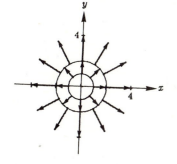

We will plot vectors of equal magnitude and, in this case, they lie along circles given by

$$\|\mathbf{F}(x, y)\| = \sqrt{x^2 + y^2} = c \quad \Longrightarrow \quad x^2 + y^2 = c^2$$

For $c = 1$, we sketch several vectors $x\mathbf{i} + y\mathbf{j}$ of magnitude 1 on the circle given $x^2 + y^2 = 1$. For $c = 4$, we sketch several vectors $x\mathbf{i} + y\mathbf{j}$ of magnitude 2 on the circle given by $x^2 + y^2 = 4$. (See the accompanying figure.)

**23.** Determine if $\mathbf{F}(x, y) = xe^{x^2y}(2y\mathbf{i} + x\mathbf{j})$ is conservative and, if so, find the potential function $f(x, y)$.

### Solution:

Since $\mathbf{F}(x, y) = 2xye^{x^2y}\mathbf{i} + x^2e^{x^2y}\mathbf{j}$, it follows from Theorem 17.1, that $\mathbf{F}$ is conservative since

$$\frac{\partial}{\partial y}[2xye^{x^2y}] = 2x^3ye^{x^2y} + 2xe^{x^2y} = \frac{\partial}{\partial x}[x^2e^{x^2y}]$$

Now, if $f$ is a function such that
$\nabla f(x, y) = f_x(x, y)\mathbf{i} + f_y(x, y)\mathbf{j}$, then we have

$$f_x(x, y) = 2xye^{x^2y} \quad \text{and} \quad f_y(x, y) = x^2e^{x^2y}$$

To reconstruct the function $f$ from these two partial derivatives, we integrate $f_x(x, y)$ with respect to $x$ and $f_y(x, y)$ with respect to $y$ as follows

$$f(x, y) = \int f_x(x, y)\, dx = \int 2xye^{x^2y}\, dx = e^{x^2y} + g(y) + K$$

$$f(x, y) = \int f_y(x, y)\, dy = \int x^2 e^{x^2y}\, dy = e^{x^2y} + h(x) + K$$

Now, we reconcile the differences between these two expressions for $f(x, y)$ by determining that $g(y) = h(x) = 0$. Therefore, we have

$$f(x, y) = e^{x^2y} + K$$

29. Find the curl of the vector field $\mathbf{F}(x, y, z) = xyz\,\mathbf{i} + y\,\mathbf{j} + z\,\mathbf{k}$ at the point $(1, 2, 1)$.

**Solution:**

By definition of curl, we have

$$\mathbf{curl}\ \mathbf{F}(x, y, z)$$

$$= \begin{vmatrix} \mathbf{i} & \mathbf{j} & \mathbf{k} \\ \dfrac{\partial}{\partial x} & \dfrac{\partial}{\partial y} & \dfrac{\partial}{\partial z} \\ xyz & y & z \end{vmatrix}$$

$$= \begin{vmatrix} \dfrac{\partial}{\partial y} & \dfrac{\partial}{\partial z} \\ y & z \end{vmatrix} \mathbf{i} - \begin{vmatrix} \dfrac{\partial}{\partial x} & \dfrac{\partial}{\partial z} \\ xyz & z \end{vmatrix} \mathbf{j} + \begin{vmatrix} \dfrac{\partial}{\partial x} & \dfrac{\partial}{\partial y} \\ xyz & y \end{vmatrix} \mathbf{k}$$

$$= (0 - 0)\,\mathbf{i} - (0 - xy)\,\mathbf{j} + (0 - xz)\,\mathbf{k} = xy\,\mathbf{j} - xz\,\mathbf{k}$$

Therefore, $\mathbf{curl}\ \mathbf{F}(1, 2, 1) = 2\mathbf{j} - \mathbf{k}$.

33. Find the curl of the vector field

$$\mathbf{F}(x, y, z) = \left(\arctan \frac{x}{y}\right)\mathbf{i} + (\ln \sqrt{x^2 + y^2})\,\mathbf{j} + \mathbf{k}.$$

**Solution:**

$$\text{curl } \mathbf{F}(x, y, z) = \begin{vmatrix} \mathbf{i} & \mathbf{j} & \mathbf{k} \\ \dfrac{\partial}{\partial x} & \dfrac{\partial}{\partial y} & \dfrac{\partial}{\partial z} \\ \arctan \dfrac{x}{y} & \ln \sqrt{x^2 + y^2} & 1 \end{vmatrix}$$

$$= (0 - 0)\mathbf{i} - (0 - 0)\mathbf{j}$$

$$+ \left[ \frac{x}{x^2 + y^2} - \frac{-x/y^2}{1 + (x^2/y^2)} \right] \mathbf{k} = \frac{2x}{x^2 + y^2} \mathbf{k}$$

**37.** Determine if $\mathbf{F}(x, y, z) = \sin y \, \mathbf{i} - x \cos y \, \mathbf{j} + \mathbf{k}$ is conservative and, if so, find the potential function $f(x, y, z)$.

**Solution:**

Using Theorem 17.2, we have

$$\mathbf{F}(x, y, z) = \sin y \, \mathbf{i} - x \cos y \, \mathbf{j} + \mathbf{k} = M \mathbf{i} + N \mathbf{j} + P \mathbf{k}$$

Since

$$\frac{\partial}{\partial y} = 0 = \frac{\partial N}{\partial z}, \quad \frac{\partial P}{\partial x} = 0 = \frac{\partial M}{\partial z},$$

$$\frac{\partial N}{\partial x} = -\cos y \neq \cos y = \frac{\partial M}{\partial y},$$

it follows that $\mathbf{F}$ is not conservative.

**41.** Determine if $\mathbf{F}(x, y, z) = (1/y)\mathbf{i} - (x/y^2)\mathbf{j} + (2z - 1)\mathbf{k}$ is conservative and if so find the potential function $f(x, y, z)$.

**Solution:**

Using Theorem 17.2, we have

$$\mathbf{F}(x, y, z) = \frac{1}{y}\mathbf{i} - \frac{x}{y^2}\mathbf{j} + (2x - 1)\mathbf{k} = M \mathbf{i} + N \mathbf{j} + P \mathbf{k}$$

Since

$$\frac{\partial P}{\partial y} = 0 = \frac{\partial N}{\partial z}, \quad \frac{\partial P}{\partial x} = 0 = \frac{\partial M}{\partial z}, \quad \frac{\partial N}{\partial x} = -\frac{1}{y^2} = \frac{\partial M}{\partial y},$$

it follows that $\mathbf{F}$ is conservative. Now, if $f$ is a function such that $\mathbf{F}(x, y, z) = \nabla f(x, y, z)$, then

$$f_x(x, y, z) = \frac{1}{y}, \quad f_y(x, y, z) = -\frac{x}{y^2}, \quad f_z(x, y, z) = 2z - 1$$

and by integrating with respect to $x$, $y$, and $z$ respectively, we obtain

$$f(x, y, z) = \int M \, dx = \int \frac{1}{y} \, dx = \frac{x}{y} + g(y, z) + K$$

$$f(x, y, z) = \int N \, dy = \int -\frac{x}{y^2} \, dy = \frac{x}{y} + h(x, z) + K$$

$$f(x, y, z) = \int P \, dz = \int (2z - 1) \, dz = z^2 - z + k(x, y) + K$$

By comparing these three versions of $f$, we conclude that

$$f(x, y, z) = \frac{x}{y} + z^2 - z + K$$

**45.** Find $\mathbf{curl}(\mathbf{curl}\ \mathbf{F})$ if $\mathbf{F}(x, y, z) = xyz\,\mathbf{i} + y\,\mathbf{j} + z\,\mathbf{k}$.

**Solution:**

From Exercise 29 we have $\mathbf{curl}\ \mathbf{F}(x, y, z) = xy\,\mathbf{j} - xz\,\mathbf{k}$. Thus,

$$\mathbf{curl}[\mathbf{curl}\ \mathbf{F}(x, y, z)] = \begin{vmatrix} \mathbf{i} & \mathbf{j} & \mathbf{k} \\ \dfrac{\partial}{\partial x} & \dfrac{\partial}{\partial y} & \dfrac{\partial}{\partial z} \\ 0 & xy & -xz \end{vmatrix}$$

$$= (0 - 0)\mathbf{i} - (-z - 0)\mathbf{j} + (y - 0)\mathbf{k}$$
$$= z\,\mathbf{j} + y\,\mathbf{k}$$

**49.** Find the divergence of the vector field

$$\mathbf{F}(x, y, z) = \sin x\,\mathbf{i} + \cos y\,\mathbf{j} + z^2\,\mathbf{k}.$$

**Solution:**

By definition, we have

$$\operatorname{div} \mathbf{F}(x, y, z) = \frac{\partial}{\partial x}[\sin x] + \frac{\partial}{\partial y}[\cos y] + \frac{\partial}{\partial z}[z^2]$$

$$= \cos x - \sin y + 2z$$

**57.** Find the divergence of the curl of the vector field

$$\mathbf{F}(x, y, z) = xyz\,\mathbf{i} + y\mathbf{j} + z\,\mathbf{k}.$$

**Solution:**

From Exercise 29 we have **curl** $\mathbf{F}(x, y, z) = xy\mathbf{j} - xz\,\mathbf{k}$. By definition, we have

$$\text{div}(\textbf{curl F}) = \frac{\partial}{\partial x}[0] + \frac{\partial}{\partial y}[xy] + \frac{\partial}{\partial z}[-xz] = x - x = 0$$

## 17.2   Line Integrals

**9.** Evaluate $\displaystyle\int_C (x^2 + y^2 + z^2)\,ds$ along the path

$C : \mathbf{r}(t) = \sin t\,\mathbf{i} + \cos t\,\mathbf{j} + 8t\,\mathbf{k}$ for $0 \le t = \pi/2$.

**Solution:**

Since $r'(t) = \cos t\,\mathbf{i} - \sin t\,\mathbf{j} + 8\,\mathbf{k}$, we have

$$\begin{aligned}
ds = \|\mathbf{r}'(t)\|\,dt &= \sqrt{[x'(t)]^2 + [y'(t)]^2 + [z'(t)]^2}\,dt \\
&= \sqrt{\cos^2 t + (-\sin t)^2 + 8^2}\,dt = \sqrt{65}\,dt.
\end{aligned}$$

It follows that

$$\begin{aligned}
\int_C (x^2 + y^2 + z^2)\,ds &= \int_0^{\pi/2} (\sin^2 t + \cos^2 t + 64t^2)\sqrt{65}\,dt \\
&= \sqrt{65} \int_0^{\pi/2} (1 + 64t^2)\,dt \\
&= \sqrt{65}\left[t + \frac{64}{3}t^3\right]_0^{\pi/2} = \frac{\sqrt{65}\pi}{6}(3 + 16\pi^2)
\end{aligned}$$

**13.** Evaluate $\int_C (x^2 + y^2)\, ds$ along the path $C : x^2 + y^2 = 1$ from $(1,0)$ counterclockwise to $(0,1)$.

**Solution:**

Since the path is one-fourth the unit circle, it can be represented by $\mathbf{r}(t) = \cos t\, \mathbf{i} + \sin t\, \mathbf{j}$ for $0 \le t \le \pi/2$. Therefore,

$$\mathbf{r}'(t) = -\sin t\, \mathbf{i} + \cos t\, \mathbf{j}$$

$$ds = \|\mathbf{r}'(t)\|\, dt = \sqrt{(-\sin t)^2 + (\cos t)^2}\, dt = dt.$$

It follows that

$$\int_C (x^2 + y^2)\, ds = \int_0^{\pi/2} (\cos^2 t + \sin^2 t)\, dt$$

$$= \int_0^{\pi/2} dt = \frac{\pi}{2}.$$

**17.** Evaluate $\int_C (x + 4\sqrt{y})\, ds$ counterclockwise around the triangle with vertices $(0,0), (1,0)$, and $(0,1)$.

**Solution:**

Path $C$ has parts as shown in the accompanying figure.

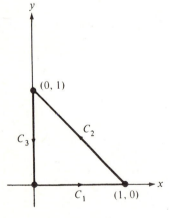

$$C_1 : \quad x = t, \qquad\qquad\qquad y = 0, \qquad 0 \le t \le 1$$
$$ds = \sqrt{1 + 0}\, dt = dt$$
$$C_2 : \quad x = 1 - t, \qquad\qquad y = t, \qquad 0 \le t \le 1$$
$$ds = \sqrt{1 + 1}\, dt = \sqrt{2}\, dt$$
$$C_3 : \quad x = 0, \qquad\qquad\qquad y = 1 - t, \quad 0 \le t \le 1$$
$$ds = \sqrt{0 + 1}\, dt = dt$$

Therefore,

$$\int_C (x + 4\sqrt{y})\, dx$$

$$= \int_0^1 t\, dt + \int_0^1 [(1 - t) + 4\sqrt{t}]\sqrt{2}\, dt + \int_0^1 4\sqrt{1 - t}\, dt$$

$$= \left[\frac{t^2}{2}\right]_0^1 + \sqrt{2}\left[t - \frac{t^2}{2} + \frac{8}{3}t^{3/2}\right]_0^1 - \left[\frac{8}{3}(1 - t)^{3/2}\right]_0^1$$

$$= \frac{1}{2} + \sqrt{2}\left(\frac{19}{6}\right) + \frac{8}{3} = \frac{19}{6}(1 + \sqrt{2})$$

**21.** Evaluate $\displaystyle\int_C \mathbf{F} \cdot d\mathbf{r}$ for $\mathbf{F}(x, y) = 3x\,\mathbf{i} + 4y\,\mathbf{j}$ and
$\mathbf{r}(t) = (2\cos t)\mathbf{i} + (2\sin t)\mathbf{j}$ for $0 \le t \le \pi/2$.

**Solution:**

Since $x = 2\cos t$ and $y = 2\sin t$, we have
$d\mathbf{r} = (-2\sin t\,\mathbf{i} + 2\cos t\,\mathbf{j})\,dt$ and $\mathbf{F}(x, y) = 6\cos t\,\mathbf{i} + 8\sin t\,\mathbf{j}$.
Therefore, with $0 \le t \le \dfrac{\pi}{2}$, we have

$$\int_C \mathbf{F} \cdot d\mathbf{r} = \int_0^{\pi/2} (-12\sin t \cos t + 16 \sin t \cos t)\,dt$$

$$= 4 \int_0^{\pi/2} \sin t \cos t\,dt = 4\left[\frac{\sin^2 t}{2}\right]_0^{\pi/2} = 2$$

**33.** Evaluate $\displaystyle\int_C (2x - y)\,dx + (x + 3y)\,dy$ along the parabolic path
$x = t$ and $y = 2t^2$ from $(0, 0)$ to $(2, 8)$.

**Solution:**

Path $C$ is given by $x = t$, $y = 2t^2$, with $0 \le t \le 2$, $dx = dt$, and
$dy = 4t\,dt$. Therefore,

$$\int_C (2x - y)\,dx + (x + 3y)\,dy = \int_0^2 (2t - 2t^2)\,dt + (t + 6t^2)4t\,dt$$

$$= \int_0^2 (2t + 2t^2 + 24t^3)\,dt$$

$$= \left[t^2 + \frac{2t^3}{3} + 6t^4\right]_0^2$$

$$= 4 + \frac{16}{3} + 96 = \frac{316}{3}$$

**37.** Find the work done by the force field $\mathbf{F}(x, y) = -x\,\mathbf{i} - 2y\,\mathbf{j}$ on an
object moving along the path given by $y = x^3$ from $(0, 0)$ to
$(2, 8)$.

**Solution:**

Since the path $C$ is given by $x = t$ and $y = t^3$, with $0 \le t \le 2$,
we have $\mathbf{F}(t) = -t\,\mathbf{i} - 2t^3\,\mathbf{j}$ and $\mathbf{r}'(t) = \mathbf{i} + 3t^2\,\mathbf{j}$. Therefore,

$$W = \int_C \mathbf{F}(t) \cdot \mathbf{r}'(t)\,dt = \int_0^2 (-t - 6t^5)\,dt = \left[-\frac{t^2}{2} - t^6\right]_0^2 = -66$$

**45.** If the tangent vector $\mathbf{r}'(t)$ is orthogonal to the force vector $\mathbf{F}$ then

$$\int_C \mathbf{F} \cdot d\mathbf{r} = 0$$

regardless of the initial and terminal points of $C$. Demonstrate this for

$$\mathbf{F}(x, y) = (x^3 - 2x^2)\mathbf{i} + \left(x - \frac{y}{2}\right)\mathbf{j} \quad \text{and} \quad \mathbf{r}(t) = t\mathbf{i} + t^2\mathbf{j}$$

**Solution:**

Since the path $C$ is given by $x = t$ and $y = t^2$ we have

$$\mathbf{F}(x, y) = (t^3 - 2t^2)\mathbf{i} + \left(t - \frac{t^2}{2}\right)\mathbf{j}$$

Also,

$$d\mathbf{r} = (\mathbf{i} + 2t\,\mathbf{j})\,dt$$

$$\int_C \mathbf{F} \cdot d\mathbf{r} = \int_C (t^3 - 2t^2 + 2t^2 - t^3)\,dt = \int_C 0\,dt = 0$$

**49.** Find the area of the lateral surface under the function $f(x, y) = xy$ and over the circle $x^2 + y^2 = 1$ from $(1, 0)$ to $(0, 1)$.

**Solution:**

We represent $C$ parametrically as $x = \cos t$ and $y = \sin t$ for $0 \le t \le \pi/2$. Then we have

$$f(x, y) = xy = \cos t \sin t$$

and

$$ds = \sqrt{[x'(t)]^2 + [y'(t)]^2}\,dt = \sqrt{(-\sin t)^2 + (\cos t)^2}\,dt = dt$$

Therefore,

$$\text{area} = \int_C f(x, y)\,ds$$

$$= \int_0^{\pi/2} \cos t \sin t\,dt = \left[\frac{1}{2}\sin^2 t\right]_0^{\pi/2} = \frac{1}{2}$$

## 17.3  Conservative Vector Fields and Independence of Path

3. Show that the value of $\displaystyle\int_C \mathbf{F} \cdot d\mathbf{r}$ is the same for each parametric representation of $C$ if $\mathbf{F}(x, y) = y\mathbf{i} - x\mathbf{j}$.

   (a) $\mathbf{r}_1(\theta) = \sec\theta\,\mathbf{i} + \tan\theta\,\mathbf{j}, \qquad 0 \le \theta \le \pi/3$

   (b) $\mathbf{r}_2(t) = \sqrt{t+1}\,\mathbf{i} + \sqrt{t}\,\mathbf{j}, \qquad 0 \le t \le 3$

**Solution:**

(a) For $\mathbf{r}_1(\theta) = \sec\theta\,\mathbf{i} + \tan\theta\,\mathbf{j}$, we have $\mathbf{F}(x, y) = \tan\theta\,\mathbf{i} - \sec\theta\,\mathbf{j}$ and $\mathbf{r}_1'(\theta) = \sec\theta\tan\theta\,\mathbf{i} + \sec^2\theta\,\mathbf{j}$. Therefore,

$$
\begin{aligned}
\int_C \mathbf{F} \cdot d\mathbf{r}_1 &= \int_0^{\pi/3} \mathbf{F} \cdot \mathbf{r}_1'\, d\theta \\
&= \int_0^{\pi/3} (\sec\theta\tan^2\theta - \sec^3\theta)\, d\theta \\
&= -\int_0^{\pi/3} \sec\theta\, d\theta \\
&= \left[-\ln|\sec\theta + \tan\theta|\right]_0^{\pi/3} = -\ln(2 + \sqrt{3})
\end{aligned}
$$

(b ) For $\mathbf{r}_2(t) = \sqrt{t+1}\,\mathbf{i} + \sqrt{t}\,\mathbf{j}$, we have $\mathbf{F}(t) = \sqrt{t}\,\mathbf{i} - \sqrt{t+1}\,\mathbf{j}$ and $\mathbf{r}_2'(t) = \dfrac{1}{2\sqrt{t+1}}\,\mathbf{i} + \dfrac{1}{2\sqrt{t}}\,\mathbf{j}$. Therefore,

$$
\begin{aligned}
\int_C \mathbf{F} \cdot d\mathbf{r}_2 &= \int_0^3 \mathbf{F} \cdot \mathbf{r}_2'\, dt \\
&= \int_0^3 \left(\frac{\sqrt{t}}{2\sqrt{t+1}} - \frac{\sqrt{t+1}}{2\sqrt{t}}\right) dt = \frac{1}{2}\int_0^3 \frac{-1}{\sqrt{t+1}\sqrt{t}}\, dt
\end{aligned}
$$

By letting $u = \sqrt{t+1}$, we have $u^2 = t+1$, $u^2 - 1 = t$, $2u\,du = dt$, and $\sqrt{u^2 - 1} = \sqrt{t}$. Thus,

$$
\begin{aligned}
\frac{1}{2}\int_0^3 \frac{-1}{\sqrt{t+1}\sqrt{t}}\, dt &= -\int_1^2 \frac{du}{\sqrt{u^2 - 1}} \\
&= \left[-\ln|u + \sqrt{u^2 + 1}|\right]_1^2 = -\ln(2 + \sqrt{3})
\end{aligned}
$$

**11.** Find the value of the line integral

$$\int_C 2xy\,dx + (x^2 + y^2)\,dy$$

along the paths

(a)    $C$ : ellipse    $(x^2/25) + (y^2/16) = 1$ from $(5,0)$ to $(0,4)$.

(b)    $C$ : parabola    $y = 4 - x^2$ from $(2,0)$ to $(0,4)$.

**Solution:**

(a) We first observe that the vector field
$\mathbf{F}(x,y) = 2xy\,\mathbf{i} + (x^2 + y^2)\,\mathbf{j}$ is conservative, since

$$\frac{\partial}{\partial y}[2xy] = 2x = \frac{\partial}{\partial x}[x^2 + y^2]$$

Therefore, the line integral is independent of path and we can replace the path along the ellipse from $(5,0)$ to $(0,4)$ with a path which will simplify the integration. We select the path along the coordinate axes from $(5,0)$ to $(0,0)$ and then from $(0,0)$ to $(0,4)$. Along the path from $(5,0)$ to $(0,0)$ we have $y = 0$ and $dy = 0$. On the path from $(0,0)$ to $(0,4)$ we have $x = 0$ and $dx = 0$. Hence,

$$\int_C 2xy\,dx + (x^2 + y^2)\,dy$$

$$= \int_5^0 0\,dx + (x^2)0 + \int_0^4 (0)(0) + (0 + y^2)\,dy$$

$$= \left[\frac{y^3}{3}\right]_0^4 = \frac{64}{3}$$

(b) Using the same method as in part (a) we replace the path along the parabola by the path along the axes from $(2,0)$ to $(0,0)$ and then from $(0,0)$ to $(0,4)$. Thus we have

$$\int_C 2xy\,dx + (x^2 + y^2)\,dy$$

$$= \int_2^0 0\,dx + (x^2 + 0)(0) + \int_0^4 (0)(0) + (0 + y^2)\,dy$$

$$= \left[\frac{y^3}{3}\right]_0^4 = \frac{64}{3}$$

[Since the line integral is path-independent, we could have used the Fundamental Theorem.]

15. Find the value of the line integral $\int_C \mathbf{F} \cdot d\mathbf{r}$ where

$$\mathbf{F}(x, y, z) = (2y + x)\mathbf{i} + (x^2 - z)\mathbf{j} + (2y - 4z)\mathbf{k}$$

and
(a)   $\mathbf{r}_1(t) = t\mathbf{i} + t^2\mathbf{j} + \mathbf{k}$ \qquad\qquad $0 \le t \le 1$
(b)   $\mathbf{r}_2(t) = t\mathbf{i} + t\mathbf{j} + (2t - 1)^2\mathbf{k}$ \qquad $0 \le t \le 1$

**Solution:**

(a) Along the path $\mathbf{r}_1(t)$, we have

$$\mathbf{F}(x, y, z) = (2t^2 + t)\mathbf{i} + (t^2 - 1)\mathbf{j} + (2t^2 - 4)\mathbf{k}$$

and

$$d\mathbf{r}_1 = \mathbf{r}_1'(t)\, dt = \mathbf{i} + 2t\mathbf{j}$$

Therefore,

$$\int_C \mathbf{F} \cdot d\mathbf{r}_1 = \int_0^1 (2t^3 + 2t^2 - t)\, dt = \frac{2}{3}$$

(b) Along the path $\mathbf{r}_2(t)$, we have

$$\mathbf{F}(x, y, z) = (2t + t)\mathbf{i} + [t^2 - (2t - 1)^2]\mathbf{j} + [2t - 4(2t - 1)^2]\mathbf{k}$$
$$= 3t\mathbf{i} + (-3t^2 + 4t - 1)\mathbf{j} + (-16t^2 + 18t - 4)\mathbf{k}$$

and

$$d\mathbf{r}_2(t) = \mathbf{i} + \mathbf{j} + 4(2t - 1)\mathbf{k}$$

Therefore,

$$\int_C \mathbf{F} \cdot d\mathbf{r}_2 = \int_0^1 (-128t^3 + 205t^2 - 97t + 15)\, dt = \frac{17}{6}$$

**23.** Use the Fundamental Theorem to evaluate

$$\int_C e^x \sin y \, dx + e^x \cos y \, dy$$

where $C$ is one arch of the cycloid $x = \theta - \sin \theta$, $y = 1 - \cos \theta$ from $(0,0)$ to $(2\pi, 0)$.

**Solution:**

Since

$$\frac{\partial}{\partial y}[e^x \sin y] = e^x \cos y = \frac{\partial}{\partial x}[e^x \cos y],$$

the integral is path-independent. Therefore, we can evaluate the line integral by using the Fundamental Theorem. We begin by finding the potential function for the vector field $\mathbf{F}(x, y) = e^x \sin y \, \mathbf{i} + e^x \cos y \, \mathbf{j}$. If $f$ is a potential function of $\mathbf{F}$, then

$$f_x(x, y) = e^x \sin y \qquad \text{and} \qquad f_y(x, y) = e^x \cos y$$

and we have

$$f(x, y) = \int f_x(x, y) \, dx = \int e^x \sin y \, dx = e^x \sin y + g(y) + K$$

$$f(x, y) = \int f_y(x, y) \, dy = \int e^x \cos y \, dy = e^x \sin y + h(x) + K$$

Now, we reconcile the differences between these two expressions for $f(x, y)$ by determining that $g(y) = h(x) = 0$. Therefore, we have

$$f(x, y) = e^x \sin y + K$$

and

$$\int_C e^x \sin y \, dx + e^x \cos y \, dy = f(2\pi, 0) - f(0, 0)$$

$$= e^{2\pi}(0) - e^0(0) = 0$$

[Since this line integral is path-independent we could have integrated along the $x$-axis from $(0,0)$ to $(2\pi, 0)$ with $y = 0$ and $dy = 0$. This would have given the result of zero immediately.]

## 17.4   Green's Theorem

**3.** Verify Green's Theorem by evaluating both integrals

$$\int_C y^2\, dx + x^2\, dy = \int_R \int \left( \frac{\partial N}{\partial x} - \frac{\partial M}{\partial y} \right) dA$$

where $C$ is the boundary of the region lying between $y = x$ and $y = x^2/4$.

**Solution:**

(a) As a line integral we define $C_1$ and $C_2$ as shown in the accompanying figure.

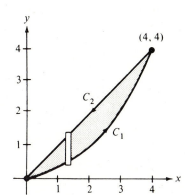

$$C_1 : \ x = t, \qquad y = \frac{t^2}{4}, \qquad dx = dt,\ dy = \frac{t}{2}\, dt, \quad 0 \le t \le 4$$
$$C_2 : \ x = 4 - t, \ \ y = 4 - t, \qquad dx = dy = -dt, \qquad \qquad 0 \le t \le 4$$

Thus,

$$\int_C y^2\, dx + x^2\, dy$$

$$= \int_{C_1} \left( \frac{t^4}{16} + \frac{t^3}{2} \right) dt + \int_{C_2} 2(4 - t)^2(-dt)$$

$$= \int_0^4 \left( \frac{t^4}{16} + \frac{t^3}{2} \right) dt + 2 \int_0^4 (4 - t)^2(-1)\, dt$$

$$= \left[ \frac{t^5}{5(16)} + \frac{t^4}{2(4)} + \frac{2(4 - t)^3}{3} \right]_0^4$$

$$= \frac{64}{5} + 32 - \frac{128}{3} = \frac{32}{15}$$

(b) By Green's Theorem we have

$$\int_R \int \left( \frac{\partial N}{\partial x} - \frac{\partial M}{\partial y} \right) dA$$

$$= \int_R \int (2x - 2y)\, dy\, dx$$

$$= \int_0^4 \int_{x^2/4}^x (2x - 2y)\, dy\, dx$$

$$= \int_0^4 \left[ 2xy - y^2 \right]_{x^2/4}^x dx$$

$$= \int_0^4 \left( x^2 - \frac{x^3}{2} + \frac{x^4}{16} \right) dx$$

$$= \left[ \frac{x^3}{3} - \frac{x^4}{8} + \frac{x^5}{80} \right]_0^4 = \frac{63}{3} - 32 + \frac{64}{5} = \frac{32}{15}$$

**7.** Use Green's theorem to evaluate the integral

$$\int_C (y - x)\, dx + (2x - y)\, dy$$

where $C$ is the boundary of the region lying inside the rectangle with vertices $(5,3), (-5,3), (-5,-3)$, and $(5,-3)$ and outside the square with vertices $(1,1), (-1,1), (-1,-1)$, and $(1,-1)$.

**Solution:**

Using $M(x,y) = y - x$, $N(x,y) = 2x - y$, and the accompanying figure, we have

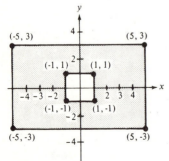

$$\int_C (y - x)\, dx + (2x - y)\, dy = \int_R \int \left( \frac{\partial N}{\partial x} - \frac{\partial M}{\partial y} \right) dA$$

$$= \int_R \int (2 - 1)\, dA$$

$$= \text{area of region}$$

$$= (\text{area of rectangle}) - (\text{area of square})$$

$$= 6(10) - 2(2) = 56$$

**13.** Use Green's Theorem to evaluate

$$\int_C 2\arctan \frac{y}{x}\, dx + \ln\left(x^2 + y^2\right) dy$$

where $C$ is the boundary of the ellipse
$x = 4 + 2\cos\theta,\ y = 4 + \sin\theta$.

**Solution:**

By Green's Theorem, we have

$$\int_C M\, dx + N\, dy = \int_C 2\arctan \frac{y}{x}\, dx + \ln\left(x^2 + y^2\right) dy$$

$$= \int_R\int \left(\frac{\partial N}{\partial x} - \frac{\partial M}{\partial y}\right) dA$$

$$= \int_R\int \left[\frac{2x}{x^2 + y^2} - \frac{2(1/x)}{1 + (y/x)^2}\right] dA$$

$$= \int_R\int \left[\frac{2x}{x^2 + y^2} - \frac{2x}{x^2 + x^2}\right] dA = 0$$

**25.** Use a line integral to find the area of the region $R$ bounded by $y = 2x + 1$ and $y = 4 - x^2$. (See accompanying figure.)

**Solution:**

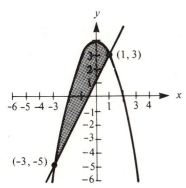

The region $R$ (see the accompanying figure) is enclosed by the path $C$ given by

$$C_1:\ x = t,\ y = 2t + 1, \qquad\qquad -3 \le t \le 1$$
$$dx = dt,\ dy = 2\, dt$$
$$C_2:\ x = 1 - t,\ y = 4 - (1 - t)^2 = 3 + 2t - t^2, \quad 0 \le t \le 4$$
$$dx = -\, dt,\ dy = (2 - 2t)\, dt$$

Therefore, by Theorem 17.9 the area of $R$ is

$$A = \frac{1}{2} \int_C x\,dy - y\,dx$$

$$= \frac{1}{2} \int_{C_1} t(2\,dt) - (2t+1)\,dt$$

$$+ \frac{1}{2} \int_{C_2} (1-t)(2-2t)\,dt - (3+2t-t^2)(-dt)$$

$$= \frac{1}{2} \int_{-3}^{1} (-1)\,dt + \frac{1}{2} \int_0^4 (5 - 2t + t^2)\,dt$$

$$= \left[ -\frac{t}{2} \right]_{-3}^{1} + \frac{1}{2} \left[ 5t - t^2 + \frac{t^3}{3} \right]_0^4$$

$$= -2 + \frac{1}{2} \left( 20 - 16 + \frac{64}{3} \right) = \frac{32}{3}$$

**31.** The centroid of the region having area $A$ and bounded by the simple closed path $C$ is

$$\bar{x} = \frac{1}{2A} \int_C x^2\,dy \quad \text{and} \quad \bar{y} = \frac{-1}{2A} \int_C y^2\,dx$$

Find the centroid of the region $R$ bounded by the graphs of $y = x^3$, $y = x$, $0 \le x \le 1$.

**Solution:**

The region $R$ (see the accompanying figure) enclosed by the path $C$ given by

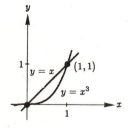

$$C_1 : \ x = t, \ y = t^3, \qquad\qquad dx = dt, \ dy = 3t^2\,dt, \ \ 0 \le t \le 1$$
$$C_2 : \ x = 1-t, \ y = 1-t, \ \ dx = -dt, \ dy = -dt, \ 0 \le t \le 1$$

Therefore, by Theorem 17.9 the area of $R$ is

$$A = \frac{1}{2} \int_C x\,dy - y\,dx$$

$$= \frac{1}{2} \int_{C_1} t(3t^2)\,dt - t^3\,dt + \frac{1}{2} \int_{C_2} (1-t)(-dt) - (1-t)(-dt)$$

$$= \frac{1}{2} \int_0^1 2t^3\,dt = \left[ \frac{1}{4} t^4 \right]_0^1 = \frac{1}{4}$$

Thus,

$$\bar{x} = \frac{1}{2A} \int_C x^2 \, dy = 2 \left[ \int_{C_1} t^2 (3t^2) \, dt + \int_{C_2} (1-t)^2 (-dt) \right]$$

$$= 2 \int_0^1 [3t^4 - (1-t)^2] \, dt$$

$$= 2 \left[ \frac{3}{5} t^5 + \frac{1}{3} (1-t)^3 \right]_0^1$$

$$= 2 \left[ \frac{3}{5} - \frac{1}{3} \right] = \frac{8}{15}$$

and

$$\bar{y} = -\frac{1}{2A} \int_C y^2 \, dx = -2 \left[ \int_{C_1} t^6 \, dt + \int_{C_2} (1-t)^2 (-dt) \right]$$

$$= -2 \int_0^1 [t^6 - (1-t)^2] \, dt$$

$$= -2 \left[ \frac{1}{7} t^7 + \frac{1}{3} (1-t)^3 \right]_0^1$$

$$= -2 \left[ \frac{1}{7} - \frac{1}{3} \right] = \frac{8}{21}$$

**35.** The area of a plane region in polar coordinates is

$$A = \frac{1}{2} \int_C r^2 \, d\theta$$

Find the area of the region $R$ bounded by the inner loop of the limaçon $r = 1 + 2 \cos \theta$. (See accompanying figure.)

**Solution:**

The inner loop of $r = 1 + 2 \cos \theta$ starts at $\theta = 2\pi/3$ and ends at $\theta = 4\pi/3$. Hence the area enclosed by this inner loop is

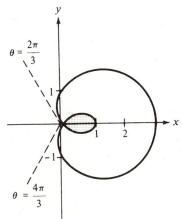

$$A = \frac{1}{2} \int_{2\pi/3}^{4\pi/3} (1 + 2 \cos \theta)^2 \, d\theta$$

$$= \frac{1}{2} \int_{2\pi/3}^{4\pi/3} \left[ 1 + 4 \cos \theta + 4 \left( \frac{1 + \cos 2\theta}{2} \right) \right] d\theta$$

$$= \frac{1}{2} \int_{2\pi/3}^{4\pi/3} (3 + 4 \cos \theta + 2 \cos 2\theta) \, d\theta$$

$$= \frac{1}{2} \left[ 3\theta + 4 \sin \theta + \sin 2\theta \right]_{2\pi/3}^{4\pi/3}$$

$$= \pi - \frac{3\sqrt{3}}{2}$$

## 17.5   Surface Integrals

**7.** Evaluate the surface integral

$$\int_S \int xy \, dS$$

over the surface given by $z = 9 - x^2$ for $0 \le x \le 2$, and $0 \le y \le x$.

**Solution:**

We write the equation for the surface $S$ as $z = 9 - x^2 = g(x, y)$ so that $g_x(x, y) = -2x$ and $g_y(x, y) = 0$, and obtain

$$\sqrt{1 + [g_x(x, y)]^2 + [g_y(x, y)]^2} = \sqrt{1 + 4x^2}$$

Using Theorem 17.10 and the accompanying figure, we have

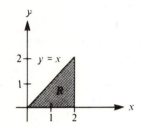

$$\int_S \int xy \, dS = \int_R \int f(x, y, g(x, y))\sqrt{1 + [g_x(x, y)]^2 + [g_y(x, y)]^2} \, dA$$

$$= \int_0^2 \int_0^x xy\sqrt{1 + 4x^2} dy \, dx$$

$$= \frac{1}{2} \int_0^2 x^3 \sqrt{1 + 4x^2} \, dx$$

Now, we use trigonometric substitution letting $2x = \tan\theta$ and $2 \, dx = \sec^2\theta \, d\theta$ to obtain

$$\int_S \int xy \, dS = \frac{1}{32} \int_0^{\arctan 4} \tan^3\theta \sec^3\theta \, d\theta$$

$$= \frac{1}{32} \left[ \frac{1}{5} \sec^5\theta - \frac{1}{3}\sec^3\theta \right]_0^{\arctan 4}$$

$$= \frac{391\sqrt{17} + 1}{240}$$

**11.** Evaluate the surface integral

$$\int_S \int \sqrt{x^2 + y^2 + z^2}\, dS$$

over the surface $z = \sqrt{x^2 + y^2}$ for $x^2 + y^2 \le 4$.

**Solution:**

We can write $S$ as $z = g(x, y) = \sqrt{x^2 + y^2}$, so that

$$g_x(x, y) = \frac{x}{\sqrt{x^2 + y^2}} \quad \text{and} \quad g_y(x, y) = \frac{y}{\sqrt{x^2 + y^2}} \quad \text{and obtain}$$

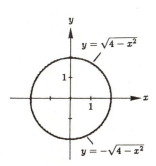

$$\sqrt{1 + [g_x(x, y)]^2 + [g_y(x, y)]^2} = \sqrt{1 + \frac{x^2}{x^2 + y^2} + \frac{y^2}{x^2 + y^2}} = \sqrt{2}$$

Using Theorem 17.10 and the accompanying figure, we obtain

$$\int_S \int \sqrt{x^2 + y^2 + z^2}\, dS$$

$$= \int_R \int f(x, y, g(x, y)) \sqrt{1 + [g_x(x, y)]^2 + [g_y(x, y)]^2}\, dA$$

$$= \int_{-2}^{2} \int_{-\sqrt{4-x^2}}^{\sqrt{4-x^2}} \sqrt{x^2 + y^2 + (\sqrt{x^2 + y^2})^2}\, \sqrt{2}\, dy\, dx$$

$$= 2 \int_{-2}^{2} \int_{-\sqrt{4-x^2}}^{\sqrt{4-x^2}} \sqrt{x^2 + y^2}\, dy\, dx$$

$$= 2 \int_0^{2\pi} \int_0^2 r(r\, dr\, d\theta) \quad \text{(polar coordinates)}$$

$$= 2 \int_0^{2\pi} \left[\frac{r^3}{3}\right]_0^2 d\theta = \frac{16}{3} \int_0^{2\pi} d\theta = \frac{32\pi}{3}$$

**17.** Find $\int_S \int \mathbf{F} \cdot \mathbf{N}\, dS$ (the flux of $\mathbf{F}$ through $S$ where $\mathbf{N}$ is the unit upper normal to $S$) if $\mathbf{F}(x, y, z) = x\,\mathbf{i} + y\,\mathbf{j} + z\,\mathbf{k}$ and $S$ is the surface given by $z = 9 - x^2 - y^2$ for $0 \le z$.

**Solution:**

The vector field $\mathbf{F}$, over the surface $S$, is given by

$$\mathbf{F}(x, y, z) = x\,\mathbf{i} + y\,\mathbf{j} + z\,\mathbf{k} = x\,\mathbf{i} + y\,\mathbf{j} + (9 - x^2 - y^2)\,\mathbf{k}$$

We write the equation for the surface $S$ as
$z = 9 - x^2 - y^2 = g(x, y)$ so that $g_x(x, y) = -2x$ and
$g_y(x, y) = -2y$. Then by Theorem 17.11 we have

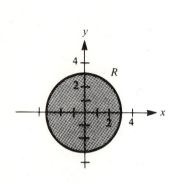

$$\int_S \int \mathbf{F} \cdot \mathbf{N} \, dS$$

$$= \int_R \int \mathbf{F} \cdot [-g_x(x, y) \mathbf{i} - g_y(x, y) \mathbf{j} + \mathbf{k}] \, dA$$

$$= \int_R \int [x\mathbf{i} + y\mathbf{j} + (9 - x^2 - y^2) \mathbf{k}] \cdot (2x\mathbf{i} + 2y\mathbf{j} + \mathbf{k}) \, dA$$

$$= \int_R \int (9 + x^2 + y^2) \, dA$$

$$= 4 \int_0^{\pi/2} \int_0^3 (9 + r^2) r \, dr \, d\theta \qquad \text{(polar coordinates)}$$

$$= \frac{243\pi}{2}$$

27. Find $I_z$ for the lamina $x^2 + y^2 = a^2$ $(0 \le z \le h)$ of uniform density 1.

**Solution:**

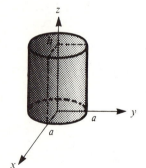

Note that $S$ does not define $z$ as a function of $x$ and $y$. Hence, we project onto the $xz$-plane, so that $y = \sqrt{a^2 - x^2} = g(x, z)$ and obtain

$$\sqrt{1 + [g_x(x, z)]^2 + [g_z(x, z)]^2} = \sqrt{1 + \frac{x^2}{a^2 - x^2}} = \frac{a}{\sqrt{a^2 - x^2}}$$

Therefore,

$$I_z = \int_S \int (x^2 + y^2)(1) \, dS$$

$$= \int_R \int a^2 \sqrt{1 + [g_x(x, z)]^2 + [g_z(x, z)]^2} \, dA$$

$$= 4a^2 \int_0^a \int_0^h \frac{a}{\sqrt{a^2 - x^2}} \, dz \, dx$$

$$= 4a^3 h \int_0^a \frac{1}{\sqrt{a^2 - x^2}} \, dx$$

$$= 4a^3 h \left[ \arcsin \frac{x}{a} \right]_0^a = 2\pi a^3 h$$

## 17.6   Divergence Theorem

**2.** Verify the Divergence Theorem by evaluating $\int_S\int \mathbf{F} \cdot \mathbf{N}\, dS$
as a surface integral and as a triple integral where
$\mathbf{F}(x, y, z) = 2x\,\mathbf{i} - 2y\,\mathbf{j} + z^2\,\mathbf{k}$ and $S$ is the cylinder given by
$x^2 + y^2 = 1$ for $0 \le z \le h$.

**Solution:**

As a *surface integral*, we have

$$S_1 : \ x^2 + y^2 \le 1, \qquad z = h$$
$$S_2 : \ x^2 + y^2 = 1, \qquad 0 \le z \le h$$
$$S_3 : \ x^2 + y^2 \le 1, \qquad z = 0$$

For the surface $S_1$ we have $\mathbf{N}_1(x, y, z) =$
$\mathbf{k}$, $\mathbf{F}(x, y, z) \cdot \mathbf{N}_1(x, y, z) = z^2$, and $dS_1 = dA$. Therefore,

$$\int_{S_1}\int \mathbf{F} \cdot \mathbf{N}\, dS = \int_{S_1}\int z^2\, dS = \int_{S_1}\int h^2\, dS = \pi h^2$$

For the surface $S_2$ we have

$$\mathbf{N}_2(x, y, z) = \frac{2x\,\mathbf{i} + 2y\,\mathbf{j}}{\sqrt{4x^2 + 4y^2}} = x\,\mathbf{i} + y\,\mathbf{j}$$

and

$$\mathbf{F}(x, y, z) \cdot \mathbf{N}_2(x, y, z) = 2x^2 - 2y^2 = 2x^2 - 2(1 - x^2) = 4x^2 - 2$$

Note that $S_2$ does **not** define $z$ as a function of $x$ and $y$. Hence,
we project onto the $xz$-plane so that $y = \sqrt{1 - x^2} = g(x, z)$ and
obtain

$$dS_2 = \sqrt{1 + [g_x(x, z)]^2 + [g_z(x, z)]^2}\, dx\, dz = \frac{1}{\sqrt{1 - x^2}}\, dx\, dz$$

Therefore,

$$\int_{S_2}\int \mathbf{F} \cdot \mathbf{N}\, dS = \int_{S_2}\int \frac{4x^2 - 2}{\sqrt{1 - x^2}} dx\, dz$$

$$= \int_0^h \int_{-1}^1 \left[ \frac{4x^2}{\sqrt{1 - x^2}} - \frac{2}{\sqrt{1 - x^2}} \right] dx\, dz$$

$$= \int_0^h \left[ 2 \arcsin x - 2x\sqrt{1 - x^2} - 2 \arcsin x \right]_{-1}^1 dz$$

$$= \int_0^h 0\, dz = 0$$

For the surface $S_3$ we have $\mathbf{N}_3(x, y, z) = -\mathbf{k}$ and $\mathbf{F}(x, y, z) \cdot \mathbf{N}(x, y, z) = -z^2$. Since $z = 0$ on $S_3$, we have

$$\int_{S_3}\!\!\int \mathbf{F} \cdot \mathbf{N}\, dS = 0$$

Thus, the surface integral for the cylinder $S$ is

$$\int_S\!\!\int \mathbf{F} \cdot \mathbf{N}\, dS = \pi h^2 + 0 + 0 = \pi h^2$$

Using the *Divergence Theorem*, we have

$$\operatorname{div} \mathbf{F}(x, y, z) = 2 - 2 + 2z = 2z$$

Thus,

$$
\begin{aligned}
\int\!\!\int_Q\!\!\int 2z\, dV &= 4 \int_0^1 \int_0^{\sqrt{1-x^2}} \int_0^h 2z\, dz\, dy\, dx \\
&= 4 \int_0^1 \int_0^{\sqrt{1-x^2}} \left[z^2\right]_0^h dy\, dx \\
&= 4 \int_0^1 \int_0^{\sqrt{1-x^2}} h^2\, dy\, dx \\
&= 4h^2 \int_0^1 [y]_0^{\sqrt{1-x^2}} dx = 4h^2 \int_0^1 \sqrt{1 - x^2}\, dx \\
&= 4h^2 (\text{area of quarter circle of radius } 1) \\
&= 4h^2 \left(\frac{\pi}{4}\right) = \pi h^2
\end{aligned}
$$

**9.** Use the Divergence Theorem to evaluate $\displaystyle\int\!\!\int \mathbf{F} \cdot \mathbf{N}\, dS$, where $\mathbf{F}(x, y, z) = x\mathbf{i} + y\mathbf{j} + z\mathbf{k}$ and $S$ is the sphere given by $x^2 + y^2 + z^2 = 4$.

**Solution:**

Since $\operatorname{div} \mathbf{F}(x, y, z) = 1 + 1 + 1 = 3$, we have

$$
\begin{aligned}
\int\!\!\int_Q\!\!\int \operatorname{div} \mathbf{F}\, dV &= 3 \int\!\!\int_Q\!\!\int dV \\
&= 3(\text{volume of sphere of radius } 2) \\
&= 3\left[\frac{4\pi 2^3}{3}\right] = 32\pi
\end{aligned}
$$

**17.** Evaluate $\displaystyle\int_S\int \mathbf{curl\,F}\cdot\mathbf{N}\,dS$ where
$\mathbf{F}(x,y,z)=(4xy+z^2)\mathbf{i}+(2x^2+6yz)\mathbf{j}+2xz\,\mathbf{k}$ and $S$ the closed surface of the solid bounded by the graphs of $x=4$, $z=9-y^2$, and the coordinate planes.

**Solution:**

Using the Divergence Theorem, we have

$$\int_S\int \mathbf{curl\,F}\cdot\mathbf{N}\,dS = \int\int_Q\int \text{div}(\mathbf{curl\,F})\,dV$$

$$\mathbf{curl\,F}(x,y,z)=\begin{vmatrix} \mathbf{i} & \mathbf{j} & \mathbf{k} \\ \dfrac{\partial}{\partial x} & \dfrac{\partial}{\partial y} & \dfrac{\partial}{\partial z} \\ 4xy+z^2 & 2x^2+6yz & 2xz \end{vmatrix}$$

$$= -6y\,\mathbf{i}-(2z-2z)\mathbf{j}+(4x-4x)\mathbf{k}=-6y\,\mathbf{i}$$

Therefore, div $\mathbf{curl\,F}(x,y,z)=0$  and

$$\int_S\int \mathbf{curl\,F}\cdot\mathbf{N}\,dS = \int\int_Q\int \text{div}(\mathbf{curl\,F})\,dV = 0.$$

## 17.7   Stokes' Theorem

**9.** Verify Stokes' Theorem by evaluating $\displaystyle\int_C \mathbf{F}\cdot\mathbf{T}\,ds$
as a line integral and as a double integral, where
$\mathbf{F}(x,y,z)=xyz\mathbf{i}+y\mathbf{j}+z\,\mathbf{k}$ and $S$ is the portion of the plane
$3x+4y+2z=12$ lying in the first octant.

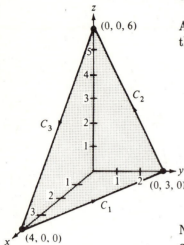

**Solution:**

As a *line integral*, we integrate along the three paths shown in the accompanying figure and obtain

$$\int_C \mathbf{F} \cdot \mathbf{T} \, ds = \int_C xyz \, dx + y \, dy + z \, dz$$

$$= \int_{C_1} 0 \, dx + y \, dy + 0 \, dz + \int_{C_2} 0 \, dx + y \, dy + z \, dz$$

$$+ \int_{C_3} 0 \, dx + 0 \, dy + z \, dz$$

$$= \int_0^3 y \, dy + \int_3^0 y \, dy + \int_0^6 z \, dz + \int_6^0 z \, dz = 0$$

Now, to evaluate the *double integral*, we begin by finding **curl F**.

$$\mathbf{curl\, F}(x, y, z) = \begin{vmatrix} \mathbf{i} & \mathbf{j} & \mathbf{k} \\ \dfrac{\partial}{\partial x} & \dfrac{\partial}{\partial y} & \dfrac{\partial}{\partial z} \\ xyz & y & z \end{vmatrix} = xy\mathbf{j} - xz\mathbf{k}$$

By Theorem 17.11 for an upward normal **N**, we obtain

$$\int_S \int (\mathbf{curl\, F}) \cdot \mathbf{N} \, dS$$

$$= \int_R \int (xy\mathbf{j} - xz\mathbf{k}) \cdot (3\mathbf{i} + 4\mathbf{j} + 2\mathbf{k}) \, dA$$

$$= \int_R \int (4xy - 2xz) \, dA$$

$$= \int_0^4 \int_0^{3(4-x)/4} \left[ 4xy - 2x\left(6 - 2y - \frac{3x}{2}\right) \right] dy \, dx$$

$$= \int_0^4 \int_0^{3(4-x)/4} (8xy + 3x^2 - 12x) \, dy \, dx$$

$$= \int_0^4 \left( 36x - 18x^2 + \frac{9x^3}{4} + 9x^2 - \frac{9x^3}{4} - 36x + 9x^2 \right) dx$$

$$= \int_0^4 (0) \, dx = 0$$

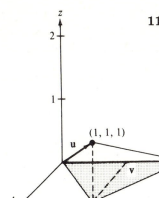

**11.** Use Stokes' Theorem to evaluate $\int_C \mathbf{F} \cdot d\mathbf{r}$ for $\mathbf{F}(x, y, z) = 2y\,\mathbf{i} + 3z\,\mathbf{j} - x\,\mathbf{k}$, where $C$ is the triangle whose vertices are $(0,0,0), (0,2,0)$, and $(1,1,1)$. (See the accompanying figure.)

**Solution:**

$$\operatorname{curl} \mathbf{F} = \begin{vmatrix} \mathbf{i} & \mathbf{j} & \mathbf{k} \\ \dfrac{\partial}{\partial x} & \dfrac{\partial}{\partial y} & \dfrac{\partial}{\partial z} \\ 2y & 3z & -x \end{vmatrix}$$

$$= (0 - 3)\,\mathbf{i} - (-1 - 0)\,\mathbf{j} + (0 - 2)\,\mathbf{k} = -3\,\mathbf{i} + \mathbf{j} - 2\,\mathbf{k}$$

Using the coordinates of the vertices of the triangle we obtain the vectors $\mathbf{u}$ and $\mathbf{v}$ forming two of its edges. They are $\mathbf{u} = \mathbf{i} + \mathbf{j} + \mathbf{k}$ and $\mathbf{v} = 0\mathbf{i} + 2\mathbf{j} + \mathbf{k} = 2\mathbf{j}$. Therefore, a vector normal to the surface is given by

$$\mathbf{u} \times \mathbf{v} = \begin{vmatrix} \mathbf{i} & \mathbf{j} & \mathbf{k} \\ 1 & 1 & 1 \\ 0 & 2 & 0 \end{vmatrix}$$

$$= -2\,\mathbf{i} + 2\,\mathbf{k}$$

and a unit vector normal to the surface is

$$\mathbf{N} = \frac{\mathbf{u} \times \mathbf{v}}{\|\mathbf{u} \times \mathbf{v}\|} = \frac{-2\,\mathbf{i} + 2\,\mathbf{k}}{2\sqrt{2}} = \frac{-\mathbf{i} + \mathbf{k}}{\sqrt{2}}.$$

Thus, the surface (plane) is given by $f(x, y, z) = -x + z$, and we have $dS = \sqrt{1 + 1}\,dA = \sqrt{2}\,dA$. We conclude that

$$\int_C \mathbf{F} \cdot d\mathbf{r} = \int_S \int (\operatorname{curl} \mathbf{F}) \cdot \mathbf{N}\,dS = \int_R \int \frac{(3 - 2)}{\sqrt{2}} \sqrt{2}\,dA$$

$$= \int_R \int dA = \text{area of triangle} = \left(\frac{1}{2}\right)(2)(1) = 1$$

**17.** Use Stokes' Theorem to evaluate $\displaystyle\int_C \mathbf{F} \cdot d\mathbf{r}$ for

$\mathbf{F}(x,y,z) = -\ln\sqrt{x^2 + y^2}\,\mathbf{i} + \arctan\dfrac{x}{y}\,\mathbf{j} + \mathbf{k}$, where $S$ is the first octant portion of the plane $z = 9 - 2x - 3y$ over one petal of the rose curve $r = 2\sin 2\theta$.

**Solution:**

$$\operatorname{curl}\mathbf{F} = \begin{vmatrix} \mathbf{i} & \mathbf{j} & \mathbf{k} \\ \dfrac{\partial}{\partial x} & \dfrac{\partial}{\partial y} & \dfrac{\partial}{\partial z} \\ -\ln\sqrt{x^2+y^2} & \arctan\dfrac{x}{y} & 1 \end{vmatrix}$$

$$= \left[\frac{1/y}{1+(x^2/y^2)} + \frac{y}{x^2+y^2}\right]\mathbf{k} = \left[\frac{2y}{x^2+y^2}\right]\mathbf{k}$$

Since $S$ is the first octant portion of the plane $z = 9 - 2x - 3y$ over one petal of $r = 2\sin 2\theta$, we have

$$\mathbf{N} = \frac{2\mathbf{i}+3\mathbf{j}+\mathbf{k}}{\sqrt{14}}$$

and

$$dS = \sqrt{1+(-2)^2+(-3)^2}\,dA = \sqrt{14}\,dA$$

Therefore,

$$\iint_S \operatorname{curl}\mathbf{F}\cdot\mathbf{N}\,dS$$

$$= \iint_R \frac{2y}{x^2+y^2}\,\frac{1}{\sqrt{14}}\sqrt{14}\,dA$$

$$= \iint_R \frac{2y}{x^2+y^2}\,dA$$

$$= \int_0^{\pi/2}\int_0^{2\sin 2\theta} \frac{2r\sin\theta}{r^2}\,r\,dr\,d\theta \qquad \text{(polar coordinates)}$$

$$= \int_0^{\pi/2}\int_0^{4\sin\theta\cos\theta} 2\sin\theta\,dr\,d\theta$$

$$= \int_0^{\pi/2} 8\sin^2\theta\cos\theta\,d\theta = \left[\frac{8\sin^3\theta}{3}\right]_0^{\pi/2} = \frac{8}{3}$$

## 17.8   Review Exercises for Chapter 17

9. Determine if $\mathbf{F}(x, y, z) = (yz\,\mathbf{i} - xz\,\mathbf{j} - xy\,\mathbf{k})/y^2 z^2$ is conservative and if it is, find the potential function.

**Solution:**

$\operatorname{curl}\mathbf{F}(x, y, z)$

$$= \begin{vmatrix} \mathbf{i} & \mathbf{j} & \mathbf{k} \\ \dfrac{\partial}{\partial x} & \dfrac{\partial}{\partial y} & \dfrac{\partial}{\partial z} \\ \dfrac{1}{yz} & \dfrac{-x}{y^2 z} & \dfrac{-x}{yz^2} \end{vmatrix}$$

$$= \left( \frac{x}{y^2 z^2} - \frac{x}{y^2 z^2} \right)\mathbf{i} - \left( \frac{-1}{yz^2} - \frac{-1}{yz^2} \right)\mathbf{j} + \left( \frac{-1}{y^2 z} - \frac{-1}{y^2 z} \right)\mathbf{k} = 0$$

Therefore, $\mathbf{F}$ is conservative. Now, if $f$ is a function such that $\mathbf{F}(x, y, z) = \nabla f(x, y, z)$, then

$$f_x(x, y, z) = \frac{1}{yz}, \quad f_y(x, y, z) = -\frac{x}{y^2 z}, \quad f_z(x, y, z) = -\frac{x}{yz^2}$$

and by integrating with respect to $x, y$, and $z$ separately, we obtain

$$f(x, y, z) = \int \frac{1}{yz}\, dx = \frac{x}{yz} + g(y, z) + K$$

$$f(x, y, z) = \int -\frac{x}{y^2 z}\, dy = \frac{x}{yz} + h(x, z) + K$$

$$f(x, y, z) = \int -\frac{x}{yz^2}\, dz = \frac{x}{yz} + k(x, y) + K$$

By comparing these three versions of $f(x, y, z)$, we conclude that $g(y, z) = h(x, z) = k(x, y) = 0$, and

$$f(x, y, z) = \frac{x}{yz} + K$$

**13.** Find the divergence and the curl of the vector field

$$\mathbf{F}(x, y, z) = (\cos y + y \cos x)\mathbf{i} + (\sin x - x \sin y)\mathbf{j} + xyz\,\mathbf{k}.$$

**Solution:**

$$\text{div}\,\mathbf{F}(x, y, z) = \frac{\partial}{\partial x}[\cos y + y \cos x] + \frac{\partial}{\partial y}[\sin x - x \sin y] + \frac{\partial}{\partial z}[xyz]$$

$$= -y \sin x - x \cos y + xy$$

$$\text{curl}\,\mathbf{F}(x, y, z)$$

$$= \begin{vmatrix} \mathbf{i} & \mathbf{j} & \mathbf{k} \\ \dfrac{\partial}{\partial x} & \dfrac{\partial}{\partial y} & \dfrac{\partial}{\partial z} \\ \cos y + y \cos x & \sin x - x \sin y & xyz \end{vmatrix}$$

$$= (xz - 0)\mathbf{i} - (yz - 0)\mathbf{j} + (\cos x - \sin y + \sin y - \cos x)\mathbf{k}$$

$$= xz\,\mathbf{i} - yz\,\mathbf{j}$$

**23.** Evaluate $\displaystyle\int_C (2x - y)\,dx + (x + 3y)\,dy$:

(a) $C$ is the line segment from $(0, 0)$ to $(2, -3)$.

(b) $C$ is one counterclockwise revolution on the circle $x = 3\cos t$ and $y = 3\sin t$.

**Solution:**

(a) $C: x = t,\ y = -3t/2,\ 0 \le t \le 2.$ Then

$$\int_C (2x - y)\,dx + (x + 3y)\,dy = \int_0^2 \left[ \frac{7t}{2}\,dt + \left(-\frac{7t}{2}\right)\left(-\frac{3}{2}\,dt\right) \right]$$

$$= \int_0^2 \frac{35}{4} t\,dt = \left[ \frac{35}{8} t^2 \right]_0^2 = \frac{35}{2}$$

(b) $C: x = 3\cos t,\ y = 3\sin t,\ 0 \le t \le 2\pi.$ Then

$$\int_C (2x - y)\,dx + (x + 3y)\,dy$$

$$= \int_0^{2\pi} [(6\cos t - 3\sin t)(-3\sin t) + (3\cos t + 9\sin t)(3\cos t)]\,dt$$

$$= \int_0^{2\pi} (9\sin t \cos t + 9)\,dt = \left[ \frac{9\sin^2 t}{2} + 9t \right]_0^{2\pi} = 18\pi$$

**29.** Evaluate $\displaystyle\int_C \mathbf{F} \cdot d\mathbf{r}$, where $\mathbf{F}(x, y, z) = (y-z)\mathbf{i} + (z-x)\mathbf{j} + (x-y)\mathbf{k}$ and $C$ is the curve of the intersection of the paraboloid $z = x^2 + y^2$ and the plane $x + y = 0$ from the point $(-2, 2, 8)$ to $(2, -2, 8)$.

**Solution:**

On the curve of intersection $z = x^2 + (-x)^2 = 2x^2$. Hence $C$ is given by $x = t$, $y = -t$, $z = 2t^2$, $-2 \le t \le 2$, and we have

$$\mathbf{r}(t) = t\,\mathbf{i} - t\,\mathbf{j} + 2t^2\,\mathbf{k} \quad \text{and} \quad d\mathbf{r} = (\mathbf{i} - \mathbf{j} + 4t\,\mathbf{k})\,dt$$

Therefore, with

$$\mathbf{F}(x, y, z) = (y - z)\mathbf{i} + (z - x)\mathbf{j} + (x - y)\mathbf{k}$$
$$= (-t - 2t^2)\mathbf{i} + (2t^2 - t)\mathbf{j} + 2t\,\mathbf{k}$$

we have

$$\int_C \mathbf{F} \cdot d\mathbf{r} = \int_{-2}^{2} (-2t^2 - t - 2t^2 + t + 8t^2)\,dt$$
$$= \int_{-2}^{2} 4t^2\,dt = \left[\frac{4t^3}{3}\right]_{-2}^{2} = \frac{64}{3}$$

**32.** Use the Fundamental Theorem to evaluate

$$\int_{(0,0,1)}^{(4,4,4)} y\,dx + x\,dy + \frac{1}{z}\,dz.$$

**Solution:**

Since

$$\frac{\partial}{\partial y}\left[\frac{1}{z}\right] = 0 = \frac{\partial}{\partial z}[x], \quad \frac{\partial}{\partial x}\left[\frac{1}{z}\right] = 0 = \frac{\partial}{\partial z}[y], \quad \frac{\partial}{\partial x}[x] = 1 = \frac{\partial}{\partial y}[y],$$

the vector field $\mathbf{F}(x, y, z) = y\mathbf{i} + x\mathbf{j} + (1/z)\mathbf{k}$ is conservative. By integrating $y$, $x$, and $1/z$ with respect to $x$, $y$, and $z$, respectively, we obtain the potential function $f(x, y, z) = xy + \ln z + C$.

Therefore, by the Fundamental Theorem (Theorem 17.5), we have

$$\int_{(0,0,1)}^{(4,4,4)} y\,dx + x\,dy + \frac{1}{z}\,dz = f(4, 4, 4) - f(0, 0, 1) = 16 + \ln 4$$

**37.** Use Green's Theorem to evaluate $\displaystyle\int_C xy\,dx + x^2\,dy$ where $C$ is the boundary of the region between the graphs of $y = x^2$ and $y = x$.

**Solution:**

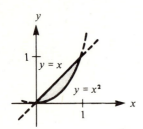

By Green's Theorem and the accompanying figure, we have

$$\int_C M(x,y)\,dx + N(x,y)\,dy = \int_C xy\,dx + x^2\,dy$$

$$= \int_R \int \left[ \frac{\partial N}{\partial x} - \frac{\partial M}{\partial y} \right] dA$$

$$= \int_0^1 \int_{x^2}^x x\,dy\,dx$$

$$= \int_0^1 (x^2 - x^3)\,dx = \frac{1}{12}$$

# 18 DIFFERENTIAL EQUATIONS

## 18.1 Definitions and basic concepts

**7.** Classify the differential equation $(y'')^2 + 3y' - 4y = 0$ according to type, order and degree (when applicable).

**Solution:**

This is an ordinary differential equation and the term $(y'')^2$ indicates that the order is 2 and the degree is 2.

**17.** Verify that $u = e^{-t} \sin bx$ is a solution to the differential equation

$$b^2 \frac{\partial u}{\partial t} = \frac{\partial^2 u}{\partial x^2}.$$

**Solution:**

Since $y = e^{-t} \sin bx$, we have

$$\frac{\partial u}{\partial t} = -e^{-t} \sin bx$$

$$\frac{\partial u}{\partial x} = e^{-t}(\cos bx)b = be^{-t} \cos bx$$

$$\frac{\partial^2 u}{\partial x^2} = be^{-t}(-\sin bx)b = -b^2 e^{-t} \sin bx$$

Therefore,

$$b^2 \frac{\partial u}{\partial t} = b^2(-e^{-t} \sin bx) = -b^2 e^{-t} \sin bx = \frac{\partial^2 u}{\partial x^2}$$

**21.** Determine if $y = e^{-2x}$ is a solution to the differential equation $y^{(4)} - 16y = 0$.

**Solution:**

Since $y = e^{-2x}$, we have

$$y' = -2e^{-2x}$$
$$y'' = 4e^{-2x}$$
$$y''' = -8e^{-2x}$$
$$y^{(4)} = 16e^{-2x}$$

Therefore,

$$y^{(4)} - 16y = 16e^{-2x} - 16(e^{-2x}) = 0$$

and the function is a solution to the differential equation.

**35.** Verify that $y = C_1 \sin 3x + C_2 \cos 3x$ is a solution to the differential equation $y'' + 9y = 0$. Then find the particular solution satisfying the initial conditions $y = 2$ and $y' = 1$ when $x = \pi/6$.

**Solution:**

Since   $y = C_1 \sin 3x + C_2 \cos 3x,$   we have

$$y' = 3C_1 \cos 3x - 3C_2 \sin 3x$$
$$y'' = -9C_1 \sin 3x - 9C_2 \cos 3x$$

Therefore,

$$y'' + 9y = -9C_1 \sin 3x - 9C_2 \cos 3x + 9(C_1 \sin 3x + C_2 \cos 3x) = 0$$

and the function is a solution to the given differential equation. Furthermore, since $y = 2$ and $y' = 1$ when $x = \pi/6$, we have

$$2 = C_1 \sin 3\left(\frac{\pi}{6}\right) + C_2 \cos 3\left(\frac{\pi}{6}\right) \quad \Longrightarrow \quad 2 = C_1(1) + C_2(0)$$

$$1 = 3C_1 \cos 3\left(\frac{\pi}{6}\right) - 3C_2 \sin 3\left(\frac{\pi}{6}\right) \quad \Longrightarrow \quad 1 = 3C_1(0) - 3C_2(1)$$

Therefore, $C_1 = 2$, $C_2 = -\dfrac{1}{3}$  and the particular solution is

$$y = 2 \sin 3x - \frac{1}{3} \cos 3x.$$

## 18.2   Separation of Variables in First-Order Equations

**3.** Solve the differential equation $(2 + x)y' = 3y$ by separation of variables.

**Solution:**

We begin by separating the variables as follows.

$$(2 + x)y' = 3y$$
$$(2 + x)\frac{dy}{dx} = 3y$$
$$\frac{1}{y}\,dy = \frac{3}{2 + x}\,dx$$

Then by integration we obtain

$$\int \frac{1}{y}\,dy = 3\int \frac{1}{2 + x}\,dx$$
$$\ln|y| = 3\ln|2 + x| + \ln C$$
$$\ln y = \ln|C(2 + x)^3|$$
$$y = C(2 + x)^3$$

**9.** Solve the differential equation $y(x + 1) + y' = 0$, $y(-2) = 1$, by separation of variables.

**Solution:**

First, we have

$$y(x + 1) + y' = 0$$
$$\frac{dy}{dx} = -y(x + 1)$$
$$\frac{dy}{y} = -(x + 1)\,dx$$

Integration yields

$$\int \frac{dy}{y} = -\int (x + 1)\,dx$$
$$\ln|y| = -\frac{(x + 1)^2}{2} + C_1$$
$$y = e^{C_1 - (x+1)^2/2} = Ce^{-(x+1)^2/2}$$

Since $y = 1$ when $x = -2$, it follows that

$$1 = Ce^{-(-2+1)^2/2} = Ce^{-1/2} \qquad \text{or} \qquad C = e^{1/2}$$

Therefore, the particular solution is

$$y = e^{1/2}e^{-(x+1)^2/2} = e^{-(x^2+2x)/2}$$

**25.** Solve the homogeneous differential equation $y' = xy/(x^2 - y^2)$.

**Solution:**

Letting $y = vx$, we have

$$y' = \frac{xy}{x^2 - y^2}$$

$$v + x\frac{dv}{dx} = \frac{x(vx)}{x^2 - v^2x^2} = \frac{v}{1 - v^2}$$

$$x\frac{dv}{dx} = \frac{v}{1 - v^2} - v = \frac{v^3}{1 - v^2}$$

$$x\,dv = \frac{v^3}{1 - v^2}\,dx$$

Separating variables, we obtain

$$\frac{1 - v^2}{v^3}\,dv = \frac{dx}{x}$$

$$\int\left(v^{-3} - \frac{1}{v}\right)dv = \int \frac{dx}{x}$$

$$\frac{v^{-2}}{-2} - \ln|v| = \ln|x| + \ln|C_1|$$

$$\frac{1}{-2v^2} = \ln|v| + \ln|x| + \ln|C_1| = \ln|C_1vx|$$

$$\frac{x^2}{-2y^2} = \ln|C_1y|$$

$$e^{-x^2/2y^2} = C_1y$$

Finally, the general solution can be written as

$$y = Ce^{-x^2/2y^2}$$

**29.** Solve the homogeneous differential equation

$$\left[ x \sec \left( \frac{y}{x} \right) + y \right] dx - x \, dy = 0, \qquad y(1) = 0$$

**Solution:**

Letting $y = vx$, we have

$$\left[ x \sec \left( \frac{vx}{x} \right) + vx \right] dx - x(v \, dx + x \, dv) = 0$$

$$x \sec v \, dx + vx \, dx - vx \, dx - x^2 dx = 0$$

$$x \sec v \, dx = x^2 \, dv$$

$$\frac{x}{x^2} dx = \frac{dv}{\sec v}$$

Integration yields

$$\int \frac{dx}{x} = \int \cos v \, dv$$

$$\ln |x| = \sin v + C_1 = \sin \frac{y}{x} + C_1$$

Since $y = 0$ when $x = 1$, we have

$$0 = \sin (0) + C_1 = C_1$$

and it follows that

$$\ln |x| = \sin \frac{y}{x} \qquad \text{or} \qquad x = e^{\sin (y/x)}$$

**35.** Find the orthogonal trajectories of the family of curves $y^2 = Cx^3$, and sketch several members of each family.

**Solution:**

First, we solve for $C$ in the given equation and obtain

$$C = \frac{y^2}{x^3}$$

Then, differentiating $y^2 = Cx^3$ implicitly with respect to $x$ and substituting the expression for $C$ given above, we have

$$2yy' = 3Cx^2 = 3 \left( \frac{y^2}{x^3} \right) x^2 = \frac{3y^2}{x}$$

Therefore,

$$\frac{dy}{dx} = \frac{3y}{2x} \qquad \text{Slope of given family}$$

Since $dy/dx$ represents the slope of the given family of curves at $(x, y)$, it follows that the orthogonal family has the negative reciprocal slope, and we write

$$\frac{dy}{dx} = -\frac{2x}{3y} \qquad \text{Slope of orthogonal family}$$

Now, we find the orthogonal family by separating variables and integrating

$$3 \int y \, dy = -2 \int x \, dx$$

$$\frac{3}{2} y^2 = -x^2 + K$$

$$2x^2 + 3y^2 = K$$

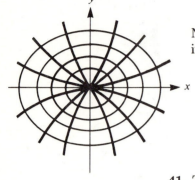

**41.** The rate of decomposition of radioactive radium is proportional to the amount present at a given instant. Find the percentage of a present amount that remains after 25 years if the half-life of radioactive radium is 1600 years.

**Solution:**

Letting $y$ represent the percentage amount of radioactive radium present at time $t$, we have the differential equation

$$y' = ky$$

and the general solution has the form

$$y = Ce^{kt} = C(e^k)^t$$

Since $y = 100\% = 1$ when $t = 0$, we have

$$1 = Ce^0 = C$$

Furthermore, $y = \frac{1}{2}$ when $t = 1600$. Hence

$$\frac{1}{2} = e^{k(1600)} = (e^k)^{1600} \quad \text{or} \quad e^k = \left(\frac{1}{2}\right)^{1/1600}$$

and we have

$$y = \left(\frac{1}{2}\right)^{t/1600}$$

Finally, when $t = 25$, we have

$$y = \left(\frac{1}{2}\right)^{25/1600} = \left(\frac{1}{2}\right)^{1/64}$$

$$= 0.989 = 98.9\% \text{ of the original amount}$$

## 18.3   Exact First-Order Equations

**3.** Test the differential equation

$$(3y^2 + 10xy^2)\, dx + (6xy - 2 + 10x^2y)\, dy = 0$$

for exactness and solve the equation if it is exact.

**Solution:**

Since

$$M(x,y)\, dx + N(x,y)\, dy = (3y^2 + 10xy^2)\, dx$$
$$+ (6xy - 2 + 10x^2y)\, dy = 0,$$

we have

$$\frac{\partial M}{\partial y} = 6y + 20xy = \frac{\partial N}{\partial x}$$

and the equation is exact. We next find a function $f$ such that $f_x(x,y) = M(x,y)$ and $f_y(x,y) = N(x,y)$. Partial integration yields

$$f(x,y) = \int (3y^2 + 10xy^2)\, dx = 3xy^2 + 5x^2y^2 + g(y)$$

Then

$$f_y(x,y) = 6xy + 10x^2y + g'(y) = \overbrace{6xy - 2 + 10x^2y}^{N(x,y)}$$

Therefore,

$$g'(y) = -2$$

and

$$g(y) = \int -2\, dy = -2y + C_1$$

Thus the general solution is

$$f(x,y) = C \qquad \text{or} \qquad 3xy^2 + 5x^2y^2 - 2y = C$$

**9.** Test the differential equation

$$\frac{1}{(x-y)^2}(y^2\,dx + x^2\,dy) = 0$$

for exactness and solve the equation if it is exact.

**Solution:**

Since

$$M(x,y)\,dx + N(x,y)\,dy = \frac{y^2}{(x-y)^2}\,dx + \frac{x^2}{(x-y)^2}\,dy = 0,$$

we have

$$\frac{\partial M}{\partial y} = \frac{(x-y)^2(2y) - y^2(2)(x-y)(-1)}{(x-y)^4}$$

$$= \frac{2xy}{(x-y)^3}$$

$$\frac{\partial N}{\partial x} = \frac{(x-y)^2(2x) - x^2(2)(x-y)(1)}{(x-y)^4}$$

$$= \frac{-2xy}{(x-y)^3}$$

Since $\dfrac{\partial M}{\partial y} \neq \dfrac{\partial N}{\partial x}$, the differential equation is not exact.

**13.** Find the particular solution of the differential equation

$$\frac{1}{x^2 + y^2}(x\,dx + y\,dy) = 0$$

that satisfies the boundary condition $y(0) = 4$.

**Solution:**

Since $M(x,y) = \dfrac{x}{x^2 + y^2}$ and $N(x,y) = \dfrac{y}{x^2 + y^2}$, we have

$$\frac{\partial M}{\partial y} = \frac{-2xy}{(x^2 + y^2)^2} = \frac{\partial N}{\partial x}$$

and the equation is exact. We next find a function $f$ such that $f_x(x,y) = M(x,y)$ and $f_y(x,y) = N(x,y)$.

$$f(x,y) = \int \frac{x}{x^2 + y^2}\,dx = \ln\sqrt{x^2 + y^2} + g(y)$$

$$f(x,y) = \int \frac{y}{x^2 + y^2}\,dy = \ln\sqrt{x^2 + y^2} + h(x)$$

Therefore, $g(y) = h(x) = C_1$ and the general solution to the differential equation is

$$f(x, y) = C \quad \text{or} \quad \ln \sqrt{x^2 + y^2} = C$$

Finally, since $y = 4$ when $x = 0$, we have $\ln 4 = C$, and the particular solutin is $\ln \sqrt{x^2 + y^2} = \ln 4$ or $x^2 + y^2 = 16$.

**21.** Solve the differential equation

$$(x + y)\, dx + \tan x\, dy = 0$$

by using an integrating factor that is a function of $x$ or $y$ alone.

**Solution:**

Since

$$\frac{(\partial M/\partial y) - (\partial N/\partial x)}{N} = \frac{1 - \sec^2 x}{\tan x} = h(x)$$

and

$$\int \frac{1 - \sec^2 x}{\tan x}\, dx = -\int \frac{\tan^2 x}{\tan x}\, dx = -\int \tan x\, dx = \ln |\cos x|,$$

it follows that

$$e^{\int h(x)dx} = e^{\ln |\cos x|} = \cos x$$

is an integrating factor. Thus

$$(x + y) \cos x\, dx + \cos x \tan x\, dy = 0$$
$$(x + y) \cos x\, dx + \sin x\, dy = 0$$

is exact and we have

$$f(x, y) = \int \sin x\, dy = y \sin x + g(x)$$

and

$$f_x(x, y) = y \cos x + g'(x) = \overbrace{(x + y) \cos x}^{M(x,y)}$$

Therefore,

$$g'(x) = x \cos x$$

$$g(x) = \int x \cos x\, dx = x \sin x + \cos x + C_1 \qquad \text{Integration by parts}$$

and

$$f(x, y) = y \sin x + x \sin x + \cos x + C_1$$
$$= (x + y) \sin x + \cos x + C_1$$

Finally the general solution is

$$f(x, y) = C \quad \text{or} \quad (x + y) \sin x + \cos x = C$$

**29.** Use the integrating factor $x^{-2}y^{-3}$ to find the general solution of the differential equation

$$(-y^5 + x^2 y)\, dx + (2xy^4 - 2x^3)\, dy = 0.$$

**Solution:**

Multiplying the given differential equation by the integrating factor $x^{-2}y^{-3}$ yields

$$\left(-\frac{y^2}{x^2} + \frac{1}{y^2}\right) dx + \left(\frac{2y}{x} - \frac{2x}{y^3}\right) dy = 0.$$

Since

$$\frac{\partial M}{\partial y} = -\frac{2y}{x^2} - \frac{2}{y^3} = \frac{\partial N}{\partial x},$$

the differential equation is exact. We next find a function $f$ such that $f_x(x,y) = M(x,y)$ and $f_y(x,y) = N(x,y)$.

$$f(x,y) = \int \left(-\frac{y^2}{x^2} + \frac{1}{y^2}\right) dx = \frac{y^2}{x} + \frac{x}{y^2} + g(y)$$

$$f(x,y) = \int \left(\frac{2y}{x} - \frac{2x}{y^3}\right) dy = \frac{y^2}{x} + \frac{x}{y^2} + h(x)$$

Therefore, $g(y) = h(x) = C_1$ and the general solution to the differential equation is

$$f(x,y) = C \quad \text{or} \quad \frac{y^2}{x} + \frac{x}{y^2} = C.$$

## 18.4   First-Order Linear Differential Equations

**5.** Solve the first-order linear differential equation $y' - y = \cos x$.

**Solution:**

The equation $y' - y = \cos x$ is linear with $P(x) = -1$. Thus

$$\int P(x)\, dx = -x \quad \text{and} \quad e^{\int P(x)dx} = e^{-x}$$

Therefore, by Theorem 18.4 the general solution has the form

$$ye^{-x} = \int e^{-x} \cos x\, dx + C = \frac{1}{2}(e^{-x} \sin x - e^{-x} \cos x) + C$$

$$y = \frac{1}{2}(\sin x - \cos x) + Ce^x$$

9. Solve the first-order linear differential equation
$$(x-1)y' + y = x^2 - 1.$$

**Solution:**

In standard form the equation is
$$y' + \left(\frac{1}{x-1}\right)y = x+1$$

Thus,
$$\int P(x)\,dx = \int \frac{1}{x-1}\,dx = \ln|x-1|$$
and the integrating factor is
$$e^{\int P(x)\,dx} = e^{\ln|x-1|} = x - 1$$

Therefore, by Theorem 18.4 the general solution has the form
$$y(x-1) = \int (x-1)(x+1)\,dx$$
$$= \int (x^2 - 1)\,dx = \frac{x^3}{3} - x + C_1$$
$$y = \frac{x^3 - 3x + C}{3(x-1)}$$

13. Solve the differential equation $y' + y\tan x = \sec x + \cos x$. Find the particular solution if $y = 1$ when $x = 0$.

**Solution:**

The integrating factor is given by
$$e^{\int P\,dx} = e^{\int \tan x\,dx} = e^{\ln|\sec x|} = \sec x$$

Therefore, by Theorem 18.4 the general solution has the form
$$y\sec x = \int \sec x(\sec x + \cos x)\,dx$$
$$= \int (\sec^2 x + 1)\,dx$$
$$= \tan x + x + C$$
$$y = \cos x(\tan x + x + C) = \sin x + x\cos x + C\cos x$$

Since $y = 1$ when $x = 0$, we have
$$1 = \sin 0 + 0\cos 0 + C\cos 0$$
$$1 = C$$

Therefore, the particular solution is $y = \sin x + (x+1)\cos x$.

**19.** Solve the Bernoulli differential equation

$$y' + \left(\frac{1}{x}\right)y = xy^2$$

**Solution:**

For the given Bernoulli equation we have $n = 2, 1 - n = -1$, and $P(x) = 1/x$. Thus we have

$$\int (1-n)P(x)\,dx = -\int \frac{1}{x}\,dx = -\ln|x|$$

$$e^{-\ln|x|} = \frac{1}{x}$$

Since $Q(x) = x$, the general solution is

$$y^{-1}e^{-\ln|x|} = \int (-1)xe^{-\ln|x|}\,dx$$

$$\frac{1}{xy} = \int (-1)\,dx = -x + C$$

$$\frac{1}{y} = -x^2 + Cx$$

$$y = \frac{1}{Cx - x^2}$$

**29.** Glucose is added intraveneously to the bloodstream at the rate $q$ units per minute, and the body removes glucose from the bloodstream at a rate proportional to the amount present. Assume $Q(t)$ is the amount of glucose in the bloodstream at time $t$.

(a) Determine the differential equation describing the rate of change with respect to time of glucose in the bloodstream.

(b) Solve the differential equation, letting $Q = Q_0$ when $t = 0$.

(c) Find the limit of $Q(t)$ as $t \to \infty$.

**Solution:**

(a) Let $Q(t)$ be the amount of glucose in the blood stream at any time. The rate of change of $Q$ is given by

$$\frac{dQ}{dt} = (\text{rate adminsitered}) - (\text{rate removed})$$

$$\frac{dQ}{dt} = q - kQ$$

$$\frac{dQ}{dt} + kQ = q$$

Where $t$ is time in minutes and $k$ is a constant of proportionality.

(b) To solve this linear equation we first find the integrating factor

$$e^{\int P(t)dt} = e^{\int kdt} = e^{kt}$$

Therefore, by Theorem 18.4 the general solution has the form

$$Qe^{kt} = \int qe^{kt} = \frac{q}{k}e^{kt} + C$$

$$Q(t) = \frac{q}{k} + Ce^{-kt}$$

Since $Q(0) = Q_0$, we have

$$Q_0 = \frac{q}{k} + C$$

$$Q_0 - \frac{q}{k} = C$$

$$Q(t) = \frac{q}{k} + (Q_0 - \frac{q}{k})e^{-kt}$$

(c)    $$\lim_{t \to \infty} Q(t) = \lim_{t \to \infty} \left[ \frac{q}{k} + \left( Q_0 - \frac{q}{k} \right)e^{-kt} \right] = \frac{q}{k}$$

**32.** A 200-gal tank is full of a solution containing 25 lb of concentrate. Starting at time $t = 0$, distilled water is admitted to the tank at the rate of 10 gal/min, and the well-stirred solution is withdrawn at the same rate.

(a) Find the amount of the concentrate in the solution as a function of $t$.

(b) Find the time at which concentrate must be added if it is done when the amount of concentrate in the tank reaches 15 lb.

**Solution:**

(a) Using Exercise 31, we have $r_1 = 10$, $r_2 = 10$, $q_1 = 0$, and $v_0 = 200$. Thus we have

$$\frac{dQ}{dt} + \frac{10Q}{200 + (0)t} = 0$$

$$\frac{dQ}{dt} = \frac{-10Q}{200} = -\frac{Q}{20}$$

Separating variables, we have

$$\frac{dQ}{Q} = -\frac{dt}{20}$$

$$\ln|Q| = -\frac{t}{20} + C_1$$

$$Q = e^{C_1 - (t/20)} = Ce^{-t/20}$$

(b) When $Q = 15$, we have

$$15 = 25e^{-t/20}$$

$$\frac{3}{5} = e^{-t/20}$$

$$\ln 3 - \ln 5 = -\frac{t}{20}$$

$$t = 20(\ln 5 - \ln 3) \approx 10.2 \text{ min}$$

**35.** Solve the differential equation $\dfrac{dy}{dx} = \dfrac{e^{2x+y}}{e^{x-y}}$ by any appropriate method.

**Solution:**

Separating variables yields

$$e^{2x+y} \, dx = e^{x-y} \, dy$$

$$e^{2x} e^y \, dx = e^x e^{-y} \, dy$$

$$e^x \, dx = e^{-2y} \, dy$$

$$e^x = -\frac{1}{2} e^{-2y} + C_1$$

$$2e^x + e^{-2y} = C$$

**41.** Solve the differential equation $2xy \, dx + (x^2 + \cos y) \, dy = 0$ by any appropriate method.

**Solution:**

Since

$$\frac{\partial M}{\partial y} = 2x = \frac{\partial N}{\partial x}$$

the equation is exact. Thus

$$f(x,y) = \int 2xy \, dx + g(y) = x^2 y + g(y)$$

$$f_y(x,y) = x^2 + g'(y) = \overbrace{x^2 + \cos y}^{N(x,y)}$$

Therefore,

$$g(y) = \int \cos y \, dy = \sin y + C_1$$

and

$$f(x,y) = x^2 y + \sin y + C_1$$

Hence, the general solution is

$$x^2 y + \sin y = C$$

47. Solve the differential equation $(x^2 y^4 - 1) \, dx + x^3 y^3 \, dy = 0$ by any appropriate method.

**Solution:**

In linear form we have

$$x^3 y^3 \frac{dy}{dx} + x^2 y^4 = 1$$

$$\frac{dy}{dx} + \left(\frac{1}{x}\right) y = x^{-3} y^{-3}$$

which is a Bernoulli equation with $P(x) = 1/x$, $Q(x) = x^{-3}$, and $n = -3$. Thus

$$\int (1-n) P(x) \, dx = \int 4 \left(\frac{1}{x}\right) dx = 4 \ln |x|$$

and

$$e^{4 \ln |x|} = x^4$$

Thus, the general solution is

$$y^4 x^4 = \int 4(x^{-3})(x^4) \, dx = 4 \int x \, dx = 2x^2 + C$$

$$x^4 y^4 - 2x^2 = C$$

## 18.5    Second-Order Homogeneous Linear Equations

3. Find the general solution of the linear differentia equation
$y'' - y' - 6y = 0$.

**Solution:**

The characteristic equation

$$m^2 - m - 6 = 0 \quad \text{or} \quad (m-3)(m+2) = 0$$

has two distinct real roots, $m_1 = 3$ and $m_2 = -2$ (Case I). Thus
the general solution is

$$y = C_1 e^{m_1 x} + C_2 e^{m_2 x} = C_1 e^{3x} + C_2 e^{-2x}$$

7. Find the general solution of the linear differential equation
$y'' + 6y' + 9y = 0$.

**Solution:**

The characteristic equation

$$m^2 + 6m + 9 = 0 \quad \text{or} \quad (m+3)^2 = 0$$

has equal roots, $m_1 = m_2 = -3$ (Case II). Thus the general
solution is

$$y = C_1 e^{m_1 x} + C_2 x e^{m_1 x} = C_1 e^{-3x} + C_2 x e^{-3x} = (C_1 + C_2 x)e^{-3x}$$

17. Find the general solution of the linear differential equation
$y'' - 3y' + y = 0$.

**Solution:**

The characteristic equation

$$m^2 - 3m + 1 = 0$$

has the distinct real roots (using the quadratic formula)

$$m = \frac{3 \pm \sqrt{9 - 4}}{2} = \frac{3 \pm \sqrt{5}}{2}$$

Thus, the general solution is

$$y = C_1 e^{(3+\sqrt{5})x/2} + C_2 e^{(3-\sqrt{5})x/2}$$

**19.** Find the general solution of the linear differential equation $9y'' - 12y' + 11y = 0$.

**Solution:**

The characteristic equation

$$9m^2 - 12m + 11 = 0$$

has complex roots (Case III).

$$m = \frac{12 \pm \sqrt{144 - 4(9)(11)}}{2(9)} = \frac{12 \pm \sqrt{-252}}{18}$$

$$= \frac{12 \pm 6i\sqrt{7}}{18} = \frac{2}{3} \pm \frac{\sqrt{7}}{3} i$$

Thus, $\alpha = 2/3$ and $\beta = \sqrt{7}/3$ and the general solution is

$$y = C_1 e^{2x/3} \cos \frac{\sqrt{7}x}{3} + C_2 e^{2x/3} \sin \frac{\sqrt{7}x}{3}$$

$$= e^{2x/3} \left[ C_1 \cos \frac{\sqrt{7}x}{3} + C_2 \sin \frac{\sqrt{7}x}{3} \right]$$

**33.** Describe the motion of a 32-pound weight suspended on a spring. Assume that the weight stretches the spring $\frac{2}{3}$ foot from its natural position, it is pulled $\frac{1}{2}$ foot below the equilibrium and released, and the motion takes place in a medium which furnishes a damping force of magnitude $\frac{1}{8}$ its speed at all times.

**Solution:**

By Hooke's Law, $32 = k(2/3)$ so that $k = 48$. Moreover, since the weight $w$ is given by $mg$, it follows that $m = w/g = \frac{32}{32} = 1$. Also the damping force is given by $-\frac{1}{8}(dy/dt)$. Thus the differential equation for the oscillations of the weight is

$$\frac{d^2y}{dt^2} = -\frac{1}{8}\left(\frac{dy}{dt}\right) - 48y$$

$$\frac{d^2y}{dt^2} + \frac{1}{8}\left(\frac{dy}{dt}\right) + 48y = 0$$

The characteristic equation is

$$8m^2 + m + 384 = 0$$

with complex roots

$$m = -\frac{1}{16} \pm \frac{\sqrt{12,287}i}{16}$$

Therefore, the general solution is

$$y(t) = e^{-t/16}\left(C_1 \cos \frac{\sqrt{12,287}t}{16} + C_2 \sin \frac{\sqrt{12,287}t}{16}\right)$$

Using the initial conditions, we have

$$y(0) = C_1 = \frac{1}{2}$$

$$y'(0) = e^{-t/16}\left[\left(-\frac{\sqrt{12,287}}{16}C_1 - \frac{C_2}{16}\right)\sin \frac{\sqrt{12,287}t}{16}\right.$$

$$\left. + \left(\frac{\sqrt{12,287}}{16}C_2 - \frac{C_1}{16}\right)\cos \frac{\sqrt{12,287}t}{16}\right]$$

$$y'(0) = \frac{\sqrt{12,287}}{16}C_2 - \frac{C_1}{16} = 0 \implies C_2 = \frac{\sqrt{12,287}}{24,574}$$

and the particular solution

$$y(t) = \frac{e^{-t/16}}{2}\left[\cos \frac{\sqrt{12,287}t}{16} + \frac{\sqrt{12,287}}{12,287}\sin \frac{\sqrt{12,287}t}{16}\right]$$

**39.** Use the Wronskian to verify the linear independence of the functions $y_1 = e^{ax}\sin bx$ and $y_2 = e^{ax}\cos bx$.

**Solution:**

$$W(y_1, y_2) = \begin{vmatrix} y_1 & y_2 \\ y_1' & y_2' \end{vmatrix}$$

$$= \begin{vmatrix} e^{ax}\sin bx & e^{ax}\cos bx \\ e^{ax}(b\cos bx + a\sin bx) & e^{ax}(a\cos bx - b\sin bx) \end{vmatrix}$$

$$= e^{2ax}\begin{vmatrix} \sin bx & \cos bx \\ b\cos bx + a\sin bx & a\cos bx - b\sin bx \end{vmatrix}$$

$$= e^{2ax}(a\sin bx \cos bx - b\sin^2 bx - b\cos^2 bx - a\sin bx \cos bx)$$

$$= -be^{2ax}$$

Since $W(y_1, y_2) \neq 0$, $y_1$ and $y_2$ are linearly independent.

## 18.6   Second-Order Nonhomogeneous Linear Equations

**3.** Solve the differential equation $y'' + y = x^3$, $y(0) = 1$, $y'(0) = 0$ by the method of undetermined coefficients.

**Solution:**

The characteristic equation $m^2 + 1 = 0$ has roots $m = \pm i$. Thus we have

$$y_h = C_1 \cos x + C_2 \sin x$$

Since $F(x) = x^3$ we choose $y_p$ to be

$$y_p = A + Bx + Cx^2 + Dx^3$$

Thus $y_p' = B + 2Cx + 3Dx^2$ and $y_p'' = 2C + 6Dx$. Substitution into the given differential equation yields

$$2C + 6Dx + A + Bx + Cx^2 + Dx^3 = x^3$$

Therefore,

$$2C + A = 0, \qquad 6D + B = 0, \qquad C = 0, \qquad D = 1$$

from which it follows that $A = 0$ and $B = -6$. Thus the general solution is

$$y = y_h + y_p = C_1 \cos x + C_2 \sin x - 6x + x^3$$

Now since $y(0) = 1$, $y'(0) = 0$, and $y' = -C_1 \sin x + C_2 \cos x - 6 + 3x^2$, we have

$$1 = C_1(1) + C_2(0) = C_1$$
$$0 = -C_1(0) + C_2(1) - 6 \qquad \text{or} \qquad C_2 = 6$$

Finally, the particular solution is

$$y = \cos x + 6 \sin x - 6x + x^3$$

**11.** Solve the differential equation $y'' + 9y = \sin 3x$ by the method of undetermined coefficients.

**Solution:**

The characteristic equation $m^2 + 9 = 0$ has roots $m = \pm 3i$ and we have
$$y_h = C_1 \cos 3x + C_2 \sin 3x$$

Since $F(x) = \sin 3x$, we consider

$$y_p = A \cos 3x + B \sin 3x$$

However, the terms of $y_p$ are *not* independent of those of $y_h$. Thus we multiply both terms by $x$ and write

$$y_p = Ax \cos 3x + Bx \sin 3x$$
$$y_p{'} = -3Ax \sin 3x + A \cos 3x + 3Bx \cos 3x + B \sin 3x$$
$$y_p{''} = -6A \sin 3x + 6B \cos 3x - 9Ax \cos 3x - 9Bx \sin 3x$$

Therefore, substitution into the given differential equation yields
$$y_p{''} + 9y_p = -6A \sin 3x + 6B \cos 3x = \sin 3x$$

which means that $A = -\frac{1}{6}$ and $B = 0$. Therefore, the general solution is

$$y = y_h + y_p = C_1 \cos 3x + C_2 \sin 3x - \frac{x}{6} \cos 3x$$
$$= \left( C_1 - \frac{x}{6} \right) \cos 3x + C_2 \sin 3x$$

**19.** Solve the differential equation $y'' + 4y = \csc 2x$ by the method of variation of parameters.

**Solution:**

The characteristic equation, $m^2 + 4 = 0$, has solutions $m = \pm 2i$. Hence,
$$y_h = C_1 \cos 2x + C_2 \sin 2x$$

Replacing $C_1$ and $C_2$ by $u_1$ and $u_2$, respectively, we write

$$y_p = u_1 \cos 2x + u_2 \sin 2x.$$

By the method of variation of parameters we obtain the system

$$u_1' \cos 2x + u_2' \sin 2x = 0$$
$$u_1'(-2\sin 2x) + u_2'(2\cos 2x) = \csc 2x$$

Multiplying the first equation by $2\sin 2x$ and the second by $\cos 2x$, and then adding the equations yields $u_2' = \frac{1}{2}\cot 2x$. Then, by substitution in the first equation, we have $u_1' = -\frac{1}{2}$. Integration yields

$$u_1 = \int -\frac{1}{2}dx = -\frac{x}{2}$$

$$u_2 = \int \frac{1}{2}\cot 2x\,dx = \frac{1}{4}\ln|\sin 2x|$$

and it follows that

$$y = y_h + y_p$$
$$= C_1 \cos 2x + C_2 \sin 2x - \frac{x}{2}\cos 2x + \frac{1}{4}\sin 2x \ln|\sin 2x|$$
$$= \left(C_1 - \frac{x}{2}\right)\cos 2x + \left(C_2 + \frac{1}{4}\ln|\sin 2x|\right)\sin 2x$$

**25.** The oscillating motion of a 24 pound weight suspended by a spring is given by

$$\frac{24}{32}y'' + 48y = \frac{24}{32}(48\sin 4t)$$

where $t$ is time in seconds and $y$ is the displacement in feet. Solve the differential equation if the initial displacement is $y(0) = \frac{1}{4}$ and the initial velocity is $y'(0) = 0$.

**Solution:**

$$\frac{24}{32}y'' + 48y = \frac{24}{32}(48\sin 4t)$$
$$y'' + 64y = 48\sin 4t$$

The characteristic equation, $m^2 + 64 = 0$, has roots $m = \pm 8i$ and we have

$$y_h = C_1\cos 8t + C_2\sin 8t.$$

Since the derivatives of even order of the sine function is a sine function, we let $y_p = A \sin 4t$. Therefore, $y_p' = 4A \cos 4t$ and $y_p'' = -16A \sin 4t$ and we have

$$y_p'' + 64y_p = -16A \sin 4t + 64A \sin 4t = 48A \sin 4t = 48 \sin 4t.$$

Thus, $A = 1$ and

$$y = y_h + y_p = C_1 \cos 8t + C_2 \sin 8t + \sin 4t$$
$$y' = y_h' + y_p' = -8C_1 \sin 8t + 8C_2 \cos 8t + 4 \cos 4t.$$

Since $y = \dfrac{1}{4}$ and $y' = 0$ when $t = 0$, we have

$$\frac{1}{4} = C_1$$

$$0 = 8C_2 + 4 \qquad \text{or} \qquad C_2 = -\frac{1}{2}.$$

Therefore,

$$y = \frac{1}{4} \cos 8t - \frac{1}{2} \sin 8t + \sin 4t$$

## 18.7   Series Solutions of Differential Equations

**3.** Use power series to solve the differential equation $y'' - 9y = 0$.

**Solution:**

Assume $y = \displaystyle\sum_{n=0}^{\infty} a_n x^n$ is a solution. Then,

$$y' = \sum_{n=0}^{\infty} na_n x^{n-1} \quad \text{and} \quad y'' = \sum_{n=0}^{\infty} n(n-1)a_n x^{n-2}$$

Thus the equation $y'' - 9y = 0$ is written as

$$\sum_{n=0}^{\infty} n(n-1)a_n x^{n-2} - 9 \sum_{n=0}^{\infty} a_n x^n = 0$$

$$\sum_{n=0}^{\infty} n(n-1)a_n x^{n-2} = \sum_{n=0}^{\infty} 9a_n x^n$$

Adjusting the indices by replacing $n$ by $n + 2$ in the left sum, we obtain

$$\sum_{n=-2}^{\infty} (n + 2)(n + 1)a_{n+2}x^n = \sum_{n=0}^{\infty} 9a_n x^n$$

Now by equating coefficients, we have

$$(n + 2)(n + 1)a_{n+2} = 9a_n$$

from which we obtain the recursion formula

$$a_{n+2} = \frac{9a_n}{(n + 2)(n + 1)} \qquad (n \geq 0)$$

Thus the coefficients of the series solution are

$$a_2 = \frac{9}{2}a_0 = \frac{3^2}{2}a_0 \qquad\qquad a_3 = \frac{9}{6}a_1 = \frac{3^2}{3 \cdot 2}a_1$$

$$a_4 = \frac{9}{12}a_2 = \frac{3^4}{4 \cdot 3 \cdot 2}a_0 \qquad a_5 = \frac{9}{20}a_3 = \frac{3^4}{5 \cdot 4 \cdot 3 \cdot 2}a_1$$

$$a_{2k} = \frac{(3)^{2k}}{(2k)!}a_0 \qquad\qquad a_{2k+1} = \frac{(3)^{2k}}{(2k + 1)!}a_1$$

In this case we can write the general solution as the sum of two power series: one for the even-powered terms and one for the odd-powered terms. Thus we have

$$y = \left( a_0 x^0 + \frac{3^2}{2!}a_0 x^2 + \frac{3^4}{4!}a_0 x^4 + \cdots \right)$$

$$+ \left( a_1 x + \frac{3^2}{3!}a_1 x^3 + \frac{3^4}{5!}a_1 x^5 + \cdots \right)$$

$$= a_0 \left[ 1 + \frac{(3x)^2}{2!} + \frac{(3x)^4}{4!} + \cdots \right] + \frac{a_1}{3} \left[ 3x + \frac{(3x)^3}{3!} + \frac{(3x)^5}{5!} + \cdots \right]$$

$$= a_0 \sum_{k=0}^{\infty} \frac{(3x)^{2k}}{(2k)!} + \frac{a_1}{3} \sum_{k=0}^{\infty} \frac{(3x)^{2k+1}}{(2k + 1)!}$$

We observe that $y'' - 9y = 0$ is a second-order homogeneous differential equation with characteristic equation $m^2 - 9 = 0$. Thus, the general solution is

$$y = C_1 e^{3x} + C_2 e^{-3x}$$

We reconcile this form of the solution with the series solution given above.

$$y = C_1 e^{3x} + C_2 e^{-3x}$$

$$= C_1 \sum_{n=0}^{\infty} \frac{(3x)^n}{n!} + C_2 \sum_{n=0}^{\infty} \frac{(-3x)^n}{n!}$$

$$= C_1 \left[ 1 + (3x) + \frac{(3x)^2}{2!} + \cdots \right]$$

$$+ C_2 \left[ 1 + (-3x) + \frac{(-3x)^2}{2!} + \cdots \right]$$

$$= C_1 \left[ 1 + \frac{(3x)^2}{2!} + \frac{(3x)^4}{4!} + \cdots \right]$$

$$+ C_1 \left[ (3x) + \frac{(3x)^3}{3!} + \frac{(3x)^5}{5!} + \cdots \right]$$

$$+ C_2 \left[ 1 + \frac{(-3x)^2}{2!} + \frac{(-3x)^4}{4!} + \cdots \right]$$

$$+ C_2 \left[ (-3x) + \frac{(-3x)^3}{3!} + \frac{(-3x)^5}{5!} + \cdots \right]$$

$$= (C_1 + C_2) \left[ 1 + \frac{(3x)^2}{2!} + \frac{(3x)^4}{4!} + \cdots \right]$$

$$+ (C_1 - C_2) \left[ (3x) + \frac{(3x)^3}{3!} + \cdots \right]$$

$$= a_0 \sum_{k=0}^{\infty} \frac{(3x)^{2k}}{(2k)!} + \frac{a_1}{3} \sum_{k=0}^{\infty} \frac{(3x)^{2k+1}}{(2k+1)!}$$

where

$$C_1 + C_2 = a_0 \qquad \text{and} \qquad C_1 - C_2 = \frac{a_1}{3}$$

**9.** Use power series to solve the differential equation $y'' - xy' = 0$.

**Solution:**

Assume $y = \sum_{n=0}^{\infty} a_n x^n$ is a solution. Then

$$y' = \sum_{n=0}^{\infty} n a_n x^{n-1} \qquad \text{and} \qquad y'' = \sum_{n=0}^{\infty} n(n-1) a_n x^{n-2}$$

Thus the equation $y'' - xy' = 0$ is written as

$$\sum_{n=0}^{\infty} n(n-1)a_n x^{n-2} - \sum_{n=0}^{\infty} na_n x^n = 0$$

$$\sum_{n=0}^{\infty} n(n-1)a_n x^{n-2} = \sum_{n=0}^{\infty} na_n x^n$$

Adjusting the indices by replacing $n$ by $n+2$ in the left sum, we obtain

$$\sum_{n=0}^{\infty} (n+2)(n+1)a_{n+2} x^n = \sum_{n=0}^{\infty} na_n x^n$$

Now by equating coefficients, we have

$$(n+2)(n+1)a_{n+2} = na_n$$

$$a_{n+2} = \frac{na_n}{(n+2)(n+1)} \qquad \text{(Recursion Formula)}$$

Thus the coefficients of the series soltuion are·

$$a_2 = \frac{0}{2}a_0 = 0 \qquad\qquad a_3 = \frac{1}{3\cdot 2}a_1$$

$$a_4 = \frac{2}{12}a_2 = 0 \qquad a_5 = \frac{3}{5\cdot 4}a_3 = \frac{1\cdot 3}{2\cdot 3\cdot 4\cdot 5\cdot}a_1$$

$$a_7 = \frac{5}{7\cdot 6}a_5 = \frac{1\cdot 3\cdot 5}{2\cdot 3\cdot 4\cdot 5\cdot 6\cdot 7}a_1$$

$$a_{2k} = 0 \qquad\qquad a_{2k+1} = \frac{1\cdot 3\cdot 5\cdots (2k-1)}{1\cdot 2\cdot 3\cdot 4\cdot 5\cdots (2k+1)}a_1$$

$$= \frac{a_1}{2\cdot 4\cdot 6\cdots (2k)(2k+1)}$$

$$= \frac{a_1}{2^k(1\cdot 2\cdot 3\cdots k)(2k+1)}$$

$$= \frac{a_1}{2^k k!(2k+1)}$$

Since the coefficients of the terms with even powers of $x$ are all zero, the solution is

$$y = a_1 \sum_{k=0}^{\infty} \frac{x^{2k+1}}{2^k k!(2k+1)}$$

**15.** Use Taylor's Theorem to find the series solution of $y'' - 2xy = 0$ given that $y(0) = 1$ and $y'(0) = -3$. Use the first six terms of the series to approximate $y$ when $x = \frac{1}{4}$.

**Solution:**

Taylor's Theorem for $c = 0$ is

$$y = y(0) + y'(0)x + \frac{y''(0)}{2!}x^2 + \frac{y'''(0)}{3!}x^3 + \cdots$$

Since $y'' = 2xy$, $y(0) = 1$ and $y'(0) = -3$, we have

$$y(0) = 1$$
$$y'(0) = -3$$

$$y'' = 2xy \qquad\qquad y''(0) = 0$$
$$y''' = 2xy' + 2y \qquad\qquad y'''(0) = 2$$
$$y^{(4)} = 2xy'' + 4y' \qquad\qquad y^{(4)}(0) = -12$$
$$y^{(5)} = 2xy''' + 6y'' \qquad\qquad y^{(5)}(0) = 0$$
$$y^{(6)} = 2xy^{(4)} + 8y''' \qquad\qquad y^{(6)}(0) = 16$$
$$y^{(7)} = 2xy^{(5)} + 10y^{(4)} \qquad\qquad y^{(7)}(0) = -120$$

Therefore, the first six terms of the Taylor series are

$$y = 1 - 3x + 0x^2 + \frac{2}{3!}x^3 - \frac{12}{4!}x^4 + 0x^5 + \frac{16}{6!}x^6 - \frac{120}{7!}x^7 + \cdots$$

$$= 1 - 3x + \frac{1}{3}x^3 - \frac{1}{2}x^4 + \frac{1}{45}x^6 - \frac{1}{42}x^7 + \cdots$$

Finally, at $x = \frac{1}{4}$ we have

$$y = 1 - \frac{3}{4} + \frac{1}{3 \cdot 4^3} - \frac{1}{2 \cdot 4^4} + \frac{1}{45 \cdot 4^6} + \frac{1}{42 \cdot 4^7} \approx 0.253$$

**17.** Use the differential equation $y' - y = 0$ to verify that the series

$$y = \sum_{n=0}^{\infty} \frac{x^n}{n!}$$

converges to the function $y = e^x$ on the interval $(-\infty, \infty)$.

**Solution:**

Since the solution to a differential equation is unique, we only need to show that both the series and the function are solutions to the given differential equation. Beginning with the series we have

$$y = \sum_{n=0}^{\infty} \frac{x^n}{n!}$$

$$y' = \sum_{n=0}^{\infty} \frac{nx^{n-1}}{n!}$$

$$= \sum_{n=1}^{\infty} \frac{x^{n-1}}{(n-1)!}$$

$$= \sum_{n=0}^{\infty} \frac{x^n}{n!} = y$$

Therefore, for the function represented by the series we have $y' - y = 0$. Using the exponential function we have

$$y = e^x = y'.$$

Therefore, it also is a solution to the differential equation $y' - y = 0$.

# Review Exercises for Chapter 18

**5.** Find the general solution of the first-order differential equation

$$\frac{dy}{dx} - \frac{y}{x} = \frac{x}{y}$$

**Solution:**

The given equation

$$\frac{dy}{dx} - \left(\frac{1}{x}\right) y = xy^{-1}$$

is a Bernoulli equation with $P(x) = -1/x$, $n = -1$, and $1 - n = 2$. Thus we have

$$\int (1 - n)P(x)\, dx = \int -\frac{2}{x}\, dx = -2\ln|x|$$

$$e^{\int (1-n)P(x)\,dx} = e^{-2\ln|x|} = \frac{1}{x^2}$$

Since $Q(x) = x$, the general solution is

$$y^2 \left(\frac{1}{x^2}\right) = \int (2)(x)\left(\frac{1}{x^2}\right) dx = \int \frac{2}{x}\, dx = 2\ln|x| + C$$

or

$$y^2 = 2x^2 \ln|x| + Cx^2 = x^2 \ln x^2 + Cx^2$$

**11.** Find the general solution of the first-order differential equation

$$(2x - 2y^3 + y)\, dx + (x - 6xy^2)\, dy = 0$$

**Solution:**

Since

$$\frac{\partial M}{\partial y} = 1 - 6y^2 = \frac{\partial N}{\partial x}$$

the equation is exact. Therefore, there exists a function $f$ such that

$$f_x(x, y) = 2x - 2y^3 + y \qquad \text{and} \qquad f_y(x, y) = x - 6xy^2$$

Integration yields

$$f(x, y) = \int (2x - 2y^3 + y)\, dx = x^2 - 2xy^3 + xy + g(y)$$

$$f(x, y) = \int (x - 6xy^2)\, dy = xy - 2xy^3 + h(x)$$

Reconciling these two version of $f$, we have $h(x) = x^2$, $g(y) = C_1$, and

$$f(x, y) = x^2 - 2xy^3 + xy + C_1$$

Therefore, the general solution is

$$x^2 - 2xy^3 + xy = C$$

**27.** Find the general solution of the first-order differential equation $(1 + x^2)\, dy = (1 + y^2)\, dx$.

**Solution:**

We can separate the variables and obtain

$$\frac{dy}{1 + y^2} = \frac{dx}{1 + x^2}$$

$$\int \frac{dy}{1 + y^2} = \int \frac{dx}{1 + x^2}$$

$$\arctan y = \arctan x + C_1$$

Now taking the tangent of this equation and using the identity for $\tan(A - B)$, we have

$$\tan(\arctan y - \arctan x) = \tan C_1$$

$$\frac{\tan \arctan y - \tan \arctan x}{1 + \tan(\arctan y)\tan(\arctan x)} = C$$

$$\frac{y - x}{1 + xy} = C$$

as the general solution.

**35.** Find the general solution of the second-order differential equation $y'' - 2y' + y = 2xe^x$.

**Solution:**

Using the method of undetermined coefficients, we have

$$m^2 - 2m + 1 = 0, \qquad m_1 = m_2 = 1$$

and

$$y_h = C_1 e^x + C_2 x e^x.$$

Since $F(x) = 2xe^x$, we consider

$$y_p = (A + Bx)e^x = Ae^x + Bxe^x.$$

However, since $y_h$ already contains both of these terms, we multiply by $x^2$ to obtain

$$y_p = Ax^2 e^x + Bx^3 e^x = e^x(Ax^2 + Bx^3)$$

$$y_p' = 2Axe^x + (3B + A)x^2 e^x + Bx^3 e^x$$

$$y_p'' = 2Ae^x + (4A + 6B)xe^x + (A + 6B)x^2 e^x + Bx^3 e^x.$$

Substituting into the given differential equation, we have

$$2Ae^x + (4A + 6B)xe^x + (A + 6B)x^2e^x + Bx^3e^x - 4Axe^x$$
$$-2(3B + A)x^2e^x - 2Bx^3e^x + Ax^2e^x + Bx^3e^x = 2xe^x$$

or

$$2Ae^x + 6Bxe^x = 2xe^x.$$

Equating coefficients, we obtain $A = 0$ and $6B = 2$, or $B = \frac{1}{3}$. Therefore, $y_p = \frac{1}{3}x^3e^x$, and the general solution is

$$y = y_h + y_p$$
$$= C_1e^x + C_2xe^x + \frac{1}{3}x^3e^x$$
$$= \left(C_1 + C_2x + \frac{1}{3}x^3\right)e^x.$$

# 19  TRUE OR FALSE
## QUESTIONS FOR REVIEW

There is a set of true-false questions for each chapter of the
text. These questions are designed to identify common errors,
emphasize the hypotheses and conclusions of theorems, and
reinforce calculus concepts. Read each question and determine
whether the statement is true or false. If it is false, explain why
or give an example which shows that the statement is false. The
answers to all the true-false questions begin on page 599.

## Chapter 1: The Cartesian Plane and Functions

---

**1.** ____ If $a$ and $b$ are any two real numbers, then $a < b$ or $a > b$.

**2.** ____ $\pi = 355/113$.

**3.** ____ The absolute value of every real number is positive.

**4.** ____ The absolute value of every real number is nonnegative.

**5.** ____ If $x < 0$, $\sqrt{x^2} = -x$.

**6.** ____ If $ab < 0$, then the point $(a, b)$ lies in either quadrant II or
quadrant IV.

**7.** ____ If $\sqrt{(x_2 - x_1)^2 + (y_2 - y_1)^2} = |y_2 - y_1|$, then the points $(x_1, y_1)$
and $(x_2, y_2)$ both lie on the same vertical line.

**8.** ____ If $(x_0, y_0)$ is equidistant from $(x_1, y_1)$ and $(x_2, y_2)$, then $(x_0, y_0)$
must be the midpoint of the line segment joining $(x_1, y_1)$ and
$(x_2, y_2)$.

**9.** ____ The distance between the points $(a + b, a)$ and $(a - b, a)$ is $2b$.

**10.** ___ If the distance between two points is zero, then the two points must coincide.

**11.** ___ If $ab = 0$, then the point $(a, b)$ lies on the $x$-axis or the $y$-axis.

**12.** ___ The graph of an equation must contain an infinite number of solution points.

**13.** ___ If the graph of $y = x^2 + a$ has only one $x$-intercept, then $a = 0$.

**14.** ___ If a graph is symmetric with respect to the origin, then it is also symmetric with respect to either the $x$-axis or the $y$-axis.

**15.** ___ If $(1, -2)$ is a point on a graph that is symmetric with respect to the $x$-axis, then $(-1, -2)$ is also a point on the graph.

**16.** ___ It $(1, -2)$ is a point on a graph that is symmetric with respect to the $y$-axis, then $(-1, -2)$ is also a point on the graph.

**17.** ___ If $b^2 - 4ac > 0$ and $a \neq 0$, then the graph of $y = ax^2 + bx + c$ has two $x$-intercepts.

**18.** ___ If $b^2 - 4ac = 0$ and $a \neq 0$, then the graph of $y = ax^2 + bx + c$ has only one $x$-intercept.

**19.** ___ A horizontal line has a slope of zero.

**20.** ___ If a line contains points in both the first and third quadrants, then its slope must be positive.

**21.** ___ The equation of any line can be written in point-slope form.

**22.** ___ The lines represented by $ax + by = c_1$ and $bx - ay = c_2$ are perpendicular.

**23.** ___ The equation of any line can be written in general form.

**24.** ___ It is possible for two mutually perpendicular lines to both have positive slope.

**25.** ___ If two distinct lines have the same slope, they must be parallel.

**26.** ___ It is possible to construct a circle whose radius and circumference are both rational.

**27.** ____ If $ax^2 + by^2 + cx + dy + e = 0$ represents the equation of a circle, then $a$ and $b$ must be equal.

**28.** ____ Any three distinct points determine a circle.

**29.** ____ If the domain of a function consists of a single number, then its range must also consist of only one number.

**30.** ____ If the range of a function consists of a single number, then its domain must also consist of only one number.

**31.** ____ If $f(a) = f(b)$, then $a = b$.

**32.** ____ A vertical line can intersect the graph of a function at most once.

**33.** ____ If $f(x) = f(-x)$ for all $x$ in the domain of $f$, then the graph of $f$ is symmetric with respect to the $y$-axis.

**34.** ____ If $f$ is a function, then $f(ax) = af(x)$.


## Chapter 2: Limits and Their Properties


**1.** ____ If $\lim_{x \to c} f(x) = L$, then $f(c) = L$.

**2.** ____ If $f(x) = g(x)$ for all real numbers other than $x = 0$, and $\lim_{x \to 0} f(x) = L$, then $\lim_{x \to 0} g(x) = L$.

**3.** ____ If $\lim_{x \to 0} f(x) = 0$, then there must exist a number $c$ such that $f(c) < 0.001$.

**4.** ____ For polynomial functions the limits from the right and left must exist and are equal to each other.

**5.** ____ If $f$ is undefined for $x = c$, then the limit of $f(x)$ as $x \to c$ does not exist.

**6.** ____ If $\lim_{x \to c} f(x) = L$ and $f(c) = L$, then $f$ is continuous at $c$.

**7.** ____ If $f(x) = g(x)$ for $x \neq c$ and $f(c) \neq g(c)$, then either $f$ or $g$ must be discontinuous at $c$.

**8.** ___ A rational function can have infinitely many discontinuities.

**9.** ___ The function $f(x) = |x - 1|/(x - 1)$ is continuous.

**10.** ___ If $p(x)$ is a polynomial, then the function given by $f(x) = p(x)/(x - 1)$ has a vertical asymptote at $x = 1$.

**11.** ___ If $\lim_{x \to c} f(x) = \infty$ and $\lim_{x \to c} g(x) = \infty$, then $\lim_{x \to c}[f(x) - g(x)] = 0$.

**12.** ___ The graph of $f$ may cross a vertical asymptote of $f$.

**13.** ___ Polynomial functions have no vertical asymptotes.

**14.** ___ If $f$ has a vertical asymptote at $x = 0$, then $f$ is undefined at $x = 0$.

**15.** ___ If $f$ is continuous on $(-\infty, \infty)$, then it has no vertical asymptotes.

## Chapter 3: Differentiation

**1.** ___ The slope of the graph of $y = x^2$ is different at every point on the curve.

**2.** ___ The slope of the graph of $y = x^3$ is different at every point on the curve.

**3.** ___ The tangent line to a curve at a point can touch the curve at only one point.

**4.** ___ The equation of the line that is tangent to the graph of $y = x^2$ at the point $(-1, 1)$ is $y - 1 = 2x(x + 1)$.

**5.** ___ If the derivative of a function is zero at a point, then the tangent line at that point is horizontal.

**6.** ___ The functions given by $f(x) = x^2$ and $g(x) = x^2 + 2$ have the same derivative.

**7.** ___ The average rate of change of $y$ with respect to $x$ is given by

$$\frac{\Delta y}{\Delta x} = \frac{f(x + \Delta x) - f(x)}{\Delta x}$$

**8.** ___ The average rate of change approaches the instantaneous rate of change as $\Delta x$ approaches zero.

**9.** ___ The average rate of change is always larger than the instantaneous rate of change.

**10.** ___ The average rate of change can be equal to the instantaneous rate of change.

**11.** ___ If $f'(x) = g'(x)$, then $f(x) = g(x)$.

**12.** ___ If $f(x) = g(x) + c$, then $f'(x) = g'(x)$.

**13.** ___ $\dfrac{d}{dx}[\sqrt{cx}] = c\dfrac{d}{dx}[\sqrt{x}]$

**14.** ___ If $y = \pi^2$, then $dy/dx = 2\pi$.

**15.** ___ If $y = x/\pi$, then $dy/dx = 1/\pi$.

**16.** ___ If $f(x)$ is an $n$th-degree polynomial, the $f^{(n)}(x) = 0$.

**17.** ___ If $f(x)$ is an $n$th-degree polynomial, then $f^{(n+1)}(x) = 0$.

**18.** ___ If $y = 1/f(x)$ then $y' = 1/f'(x)$.

**19.** ___ The second derivative represents the rate of change of the first derivative.

**20.** ___ If the velocity of an object is constant, then its acceleration is zero.

**21.** ___ If a function is continuous, then it is differentiable.

**22.** ___ If a function is differentiable, then it is continuous.

**23.** ___ If $f$ is differentiable at $x = c$, then $f'$ is differentiable at $x = c$.

**24.** ___ If $f$ and $g$ are differentiable, then the quotient $f/g$ is differentiable.

**25.** ——  If the graph of a function possesses a tangent line at a point, then it is differentiable at that point.

**26.** ——  If $y = f(x)g(x)$, then $dy/dx = f'(x)g'(x)$.

**27.** ——  If $y = f(x)g(x)$, then $y'' = f(x)g''(x) + g(x)f''(x)$.

**28.** ——  If $(x+1)^2$ is a factor of $f(x)$, then $(x+1)$ is a factor of $f'(x)$.

**29.** ——  If $y = (x+1)(x+2)(x+3)(x+4)$, then $d^5y/dx^5 = 0$.

**30.** ——  If $f'(c)$ and $g'(c)$ are zero and $h(x) = f(x)g(x)$, then $h'(c) = 0$.

**31.** ——  If $y = (1-x)^{1/2}$, then $y' = \frac{1}{2}(1-x)^{-1/2}$.

**32.** ——  If $y$ is a differentiable function of $u$, $u$ is a differentiable function of $v$, and $v$ is a differentiable function of $x$, then

$$\frac{dy}{dx} = \frac{dy}{du}\frac{du}{dv}\frac{dv}{dx}.$$

**33.** ——  The equation $y = ax + b$ explicitly defines $y$ as a function of $x$.

**34.** ——  The equation $x^2 + y^2 = r^2$ implicitly defines $y$ as a function of $x$.

**35.** ——  The equation $x^3 + y^3 = r^3$ implicitly defines $y$ as a function of $x$.

## Chapter 4: Applications of Differentiation

**1.** ——  A function has at most one maximum on an interval.

**2.** ——  A continuous function must have a minimum on a closed interval.

**3.** ——  A continuous function must have a minimum on an open interval.

**4.** ——  If $c$ is a critical number of $f$, then $f$ has a relative extrema at $x = c$.

**5.** ——  The Mean Value Theorem can be applied to $f(x) = 1/x$ on the interval $[-1, 1]$.

**6.** ___ If the graph of a function has three $x$-intercepts, then it must have at least two points at which its tangent line is horizontal.

**7.** ___ If the graph of a polynomial function has three $x$-intercepts, then it must have at least two points at which its tangent line is horizontal.

**8.** ___ If $0 < a < b < 1$ and $f$ is differentiable on $(0,1)$, then $f$ is continuous on $[a, b]$.

**9.** ___ The sum of two increasing functions is increasing.

**10.** ___ The product of two increasing functions is increasing.

**11.** ___ If $f(x) = ax^3 + b$ and $f$ is increasing on $(-1, 1)$, then $a > 0$.

**12.** ___ Every $n$th-degree polynomial has $(n-1)$ critical numbers.

**13.** ___ An $n$th-degree polynomial has at most $(n-1)$ critical numbers.

**14.** ___ If $f$ is increasing and continuous on $(a, b)$, then $f$ is differentiable on $(a, b)$.

**15.** ___ Every second-degree polynomial possesses precisely one relative extremum.

**16.** ___ If $a < c < b$ and $c$ is a critical number of $f$, then $f$ has a relative extremum in $(a, b)$.

**17.** ___ If $f''(c) = 0$, then the graph of $f$ has a point of inflection at $(c, f(c))$.

**18.** ___ The graph of every cubic polynomial has precisely one point of inflection.

**19.** ___ The graph of $f(x) = 1/x$ is concave downward for $x < 0$ and concave upward for $x > 0$, and thus it has a point of inflection when $x = 0$.

**20.** ___ The function $y = f(x)$ can have at most one horizontal asymptote.

**21.** ___ If $y = x + c$, then $dy = dx$.

**22.** ___ If $y = ax + b$, then $\Delta y / \Delta x = dy/dx$.

**23.** ___ If $y$ is differentiable, then $\lim_{\Delta x \to 0} (\Delta y - dy) = 0$.

**24.** ⎯⎯ If $y = f(x)$, $f$ is increasing and differentiable, and $\Delta x > 0$, then $\Delta y \geq dy$.

**25.** ⎯⎯ The zeros of $f(x) = p(x)/q(x)$ coincide with the zeros of $p(x)$.

**26.** ⎯⎯ If the coefficients of a polynomial function are all positive, then the polynomial has no positive zeros.

**27.** ⎯⎯ If $f(x)$ is a cubic polynomial such that $f'(x)$ is never zero, then any initial guess will force Newton's Method to converge to the zero of $f$.

**28.** ⎯⎯ The roots of $\sqrt{f(x)} = 0$ coincide with the roots of $f(x) = 0$.

## Chapter 5: Integration

**1.** ⎯⎯ $\int [f(x) + g(x)]\, dx = \int f(x)\, dx + \int g(x)\, dx$.

**2.** ⎯⎯ $\int f(x)g(x)\, dx = \left[\int f(x)\, dx\right]\left[\int g(x)\, dx\right]$.

**3.** ⎯⎯ $\int x f(x)\, dx = x \int f(x)\, dx$.

**4.** ⎯⎯ $\int (2x + 1)^2\, dx = [(2x + 1)^3/3] + C$.

**5.** ⎯⎯ $\int x(x^2 + 1)\, dx = (x^2/2)[(x^3/3) + x] + C$.

**6.** ⎯⎯ $\int (1/x)\, dx = -(1/x^2) + C$.

**7.** ⎯⎯ Each antiderivative of an $n$th-degree polynomial function is an $(n + 1)$st-degree polynomial function.

**8.** ⎯⎯ $\int (x^2 + 1)^2\, dx = (1/2x) \int (x^2 + 1)^2 (2x)\, dx = (1/2x)[(x^2 + 1)^3/3] + C$.

**9.** ⎯⎯ If $F(x)$ and $G(x)$ are antiderivatives of $f(x)$, then $F(x) = G(x) + C$.

**10.** ⎯⎯ If $f'(x) = g(x)$, then $\int g(x)\, dx = f(x) + C$.

**11.** ⎯⎯ If $p(x)$ is a polynomial function, then $p$ has exactly one antiderivative whose graph contains the origin.

**12.** ____ If $f$ is increasing on $[a, b]$, then the minimum value of $f(x)$ on $[a, b]$ is $f(a)$.

**13.** ____ If the norm of a partition approaches zero, then the number of subintervals approaches infinity.

**14.** ____ The sum of the first $n$ integers is $n(n + 1)/2$.

**15.** ____ If $f$ is continuous and nonnegative on $[a, b]$, then the limit as $n \to \infty$ of its lower sum $s(n)$ and upper sum $S(n)$ both exist and are equal to each other.

**16.** ____ The value of $\int_a^b f(x)\, dx$ must be positive.

**17.** ____ If $f$ is continuous on $[a, b]$, then $f$ is integrable on $[a, b]$.

**18.** ____ If $F'(x) = G'(x)$ on the interval $[a, b]$, then $F(b) - F(a) = G(b) - G(a)$.

**19.** ____ If $f(x) = -f(-x)$ on the interval $[-a, a]$, then $\int_{-a}^{a} f(x)\, dx = 0$.

**20.** ____ If $f(x) = f(-x)$ on the interval $[-a, a]$, then $\int_{-a}^{a} f(x)\, dx = 0$.

**21.** ____ $\int_{-10}^{10} (ax^3 + bx^2 + cx + d)\, dx = 2 \int_0^{10} (bx^2 + d)\, dx$.


## Chapter 6: Applications of Integration

---

**1.** ____ If $f$ and $g$ are both continuous on the interval $[a, b]$, then $f - g$ is integrable on $[a, b]$.

**2.** ____ If the area of the region bounded by the graphs of $f$ and $g$ is 1, then the area of the region bounded by the graphs of $h(x) = f(x) + C$ and $k(x) = g(x) + C$ is also 1.

**3.** ____ $\int_0^1 (x^2 - x)\, dx$ represents the area of the region bounded by the graph of $y = x^2 - x$ and the $x$-axis.

**4.** ____ If $\int_a^b [f(x) - g(x)]\, dx = A$, then $\int_a^b [g(x) - f(x)]\, dx = -A$

## Chapter 7: Exponential and Logarithmic Functions

1. ____ If $f(x) = 2^{x^2}$, then $f(3) = 64$.

2. ____ $e = 271,801/99,990$.

3. ____ $\log_a e = 1/\ln a$.

4. ____ $\ln(x + 25) = \ln x + \ln 25$.

5. ____ The range of the natural logarithmic function is the set of all real numbers.

6. ____ If $f(x) = \ln x$, then $f(e^{n+1}) - f(e^n) = 1$ for any value of $n$.

7. ____ The functions given by $f(x) = 2 + e^x$ and $g(x) = \ln(x - 2)$ are inverses of each other.

8. ____ $(\ln x)^{1/2} = \dfrac{1}{2}(\ln x)$.

9. ____ The derivative of $f(x) = \ln x$ is positive for all $x$ in the domain of $f$.

10. ____ The derivative of every transcendental function is transcendental.

11. ____ If $f(x) = \ln(ax)$ and $g(x) = \ln(bx)$, then $f'(x) = g'(x)$.

12. ____ If $y = \ln \pi$, then $y' = 1/\pi$.

13. ____ $\int \ln x\, dx = (1/x) + C$.

14. ____ $\int \dfrac{1}{x}\, dx = \ln|cx|$, for $c \neq 0$.

15. ____ If $u$ is a differentiable function of $x$ such that $u \neq 0$, then

$$\int \left(\frac{1}{u}\right) \frac{du}{dx}\, dx = \int \frac{1}{u}\, du$$

16. ____ $\displaystyle\int_{-1}^{e} \frac{1}{x} = \ln|x|\Big]_{-1}^{e} = \ln e - \ln 1 = 1$.

**17.** ___ The exponential function $y = Ce^x$ is a solution of the differential differential equation $d^n y / dx^n = y, \quad n = 1, 2, 3, \cdots$.

**18.** ___ If $f(x) = x^e$, then $f'(x) = x^e$.

**19.** ___ The graphs of $f(x) = e^x$ and $g(x) = e^{-x}$ meet at right angles.

**20.** ___ If $f(x) = g(x)e^x$, then the only zeros of $f$ are the zeros of $g$.

**21.** ___ $\lim\limits_{x \to 0} \left[ \dfrac{x^2 + x + 1}{x} \right] = \lim\limits_{x \to 0} \left[ \dfrac{2x + 1}{1} \right] = 1$.

**22.** ___ If $y = e^x / x^2$, then $y' = e^x / 2x$.

**23.** ___ If $p(x)$ is a polynomial, the $\lim\limits_{x \to \infty} [p(x)/e^x] = 0$.

**24.** ___ If $\lim\limits_{x \to \infty} \dfrac{f(x)}{g(x)} = 1$ then $\lim\limits_{x \to \infty} [f(x) - g(x)] = 0$.

# Chapter 8:  Trigonometric Functions and Inverse Trigonometric Functions

**1.** ___ $\sin (x + y) = \sin x + \sin y$.

**2.** ___ $\cos (ax) = a \cos x$.

**3.** ___ The sine and cosine functions are continuous on the entire real line.

**4.** ___ If $0 < x < \pi/2$, then $\sin x < x$.

**5.** ___ If $f(x) = \sin^2 x$, then $f(\pi/4) = \frac{1}{2}$.

**6.** ___ If $f(x) = \sin^2(2x)$, the $f'(x) = 2(\sin 2x)(\cos 2x)$.

**7.** ___ The tangent function is differentiable at every point in its domain.

**8.** ___ The graphs of $f(x) = \sin x$ and $g(x) = \cos x$ intersect at right angles.

**9.** ___ The maximum value of $y = 3 \sin x + 2 \cos x$ is 5.

**10.** ____ The maximum slope of the graph of the graph of $y = \sin(bx)$ is $b$.

**11.** ____ $\int_a^b \sin x \, dx = \int_a^{b+2\pi} \sin x \, dx.$

**12.** ____ $4\int \sin x \cos x \, dx = -\cos 2x + C.$

**13.** ____ The average value of the sine function over an interval of length $2\pi$ is 0.

**14.** ____ $\int \tan x \, dx = \sec^2 x + C.$

**15.** ____ $\int \sin^2 2x \cos 2x \, dx = (\sin^3 2x)/3 + C.$

**16.** ____ $y = e^x, y = e^{-x}, y = \sin x$, and $y = \cos x$ all satisfy the differential equation $d^4y/dx^4 = y$.

**17.** ____ The slope of the graph of the inverse tangent is positive for all $x$.

**18.** ____ The range of $y = \arcsin x$ is $[0, \pi]$.

**19.** ____ $\arcsin x + \arccos x = \pi/2$.

**20.** ____ $\dfrac{d}{dx}[\arctan(\tan x)] = 1$ for all $x$.

**21.** ____ $\int \dfrac{x}{\sqrt{1-x^2}} \, dx = \arcsin x + C.$

## Chapter 9: Integration Techniques and Improper Integrals

**1.** ____ $\int \dfrac{x^3}{x^4 - 1} \, dx = \dfrac{1}{4}\int \dfrac{du}{u}, \quad u = x^4 - 1.$

**2.** ____ $\int \dfrac{dx}{1 + \sin^2 x} = \int \dfrac{du}{a^2 + u^2}, \quad u = \sin x, \ a = 1.$

**3.** ____ $\int \sec^2 \theta \sec^2(\tan \theta) \, d\theta = \int \sec^2 u \, du, \quad u = \tan \theta.$

4. —— $\int \dfrac{dx}{\sqrt{e^{2x}-1}} = \int \dfrac{du}{u\sqrt{u^2-a^2}}\, dx, \; u = e^x, \; a = 1.$

5. —— $\int \dfrac{\ln x^2}{x}\, dx = \int u\, du, \; u = \ln x^2.$

6. —— $\int \dfrac{dx}{\sqrt{x-1}} = \int \dfrac{du}{u}, \; u = \sqrt{x-1}.$

7. —— $-\arctan(1-x) + C = \int \dfrac{dx}{(1-x)^2+1} = \int \dfrac{dx}{(x-1)^2+1} =$
   $\arctan(x-1) + C.$

8. —— $\int \dfrac{dx}{\sqrt{2x-x^2}} = -\int \dfrac{dx}{\sqrt{x^2-2x}}.$

9. —— If $u = \sqrt{x-1}$, then $\displaystyle\int_1^5 x\sqrt{x-1}\, dx = \int_0^2 (u^2+1)u\, du.$

10. —— If $u = \sqrt[3]{3x+1}$, then $dx = u^2\, du.$

11. —— If $u = \sqrt{e^x-1}$, then $dx = 2u\, du/(u^2+1).$

12. —— $\displaystyle\int_{-1}^1 \sqrt{x^2-x^3}\, dx = \int_{-1}^1 x\sqrt{1-x}\, dx.$

13. —— If $x = \sin\theta$, then $\displaystyle\int \dfrac{dx}{\sqrt{1-x^2}} = \int d\theta.$

14. —— If $x = \sec\theta$, then $\displaystyle\int \dfrac{\sqrt{x^2-1}}{x}\, dx = \int \sec\theta \tan\theta\, d\theta.$

15. —— If $x = \tan\theta$, then $\displaystyle\int_0^{\sqrt{3}} \dfrac{dx}{(1+x^2)^{3/2}} = \int_0^{4\pi/3} \cos\theta\, d\theta.$

16. —— If $x = \sin\theta$, then $\displaystyle\int_{-1}^1 x^2\sqrt{1-x^2}\, dx = 2\int_0^{\pi/2} \sin^2\theta \cos^2\theta\, d\theta.$

17. —— It is possible to find constants $A$ and $B$ such that

$$\dfrac{x^2-1}{(x-2)(x-3)} = \dfrac{A}{x-2} + \dfrac{B}{x-3}$$

18. —— If $|a| \neq 1$, then $x^3 - x^2 - a^x + a^2$ has three distinct factors.

**19.** ____ If $a \neq 0$ and

$$\frac{a}{x^2(x-1)^2} = \frac{A}{x} + \frac{B}{x^2} + \frac{C}{x-1} + \frac{D}{(x-1)^2}$$

then $A, B, C$, and $D$ are each nonzero.

**20.** ____ $\displaystyle\int_0^\infty (1/x^p)\, dx$ diverges if $p > 0$.

**21.** ____ $\displaystyle\int_1^\infty (1/x^p)\, dx$ converges if $p > 1$.

**22.** ____ If $f$ is continuous on $[0, \infty)$ and $\lim_{x \to \infty} f(x) = 0$, then $\displaystyle\int_0^\infty f(x)\, dx$ converges.

**23.** ____ If $f'$ is continuous on $[0, \infty)$ and $\lim_{x \to \infty} f(x) = 0$, then
$$\int_0^\infty f(x)\, dx = -f(0).$$

**24.** ____ If the graph of $f$ is symmetric with respect to the origin or the $y$-axis, then $\displaystyle\int_0^\infty f(x)\, dx$ converges if and only is $\displaystyle\int_{-\infty}^\infty f(x)\, dx$ converges.

## Chapter 10: Infinite Series

**1.** ____ If $0 \leq a_n \leq b_n$ for all $n$ and $\{b_n\}$ converges, then $\{a_n\}$ converges.

**2.** ____ If $0 \leq a_n \leq b_n$ for all $n$ and $\{b_n\}$ converges to 0, then $\{a_n\}$ converges to 0.

**3.** ____ If $\{a_n\}$ converges to 3 and $\{b_n\}$ converges to 2 then $\{a_n + b_n\}$ converges to 5.

**4.** ____ If $b_n = a_{n+1}$ and $\{a_n\}$ converges to $L$, then $\{b_n\}$ converges to $L$.

**5.** ____ If $\{a_n\}$ converges, then $\lim_{n \to \infty} (a_n - a_{n+1}) = 0$.

**6.** ____ If $n > 1$, then $n! = n(n-1)!$.

**7.** ____ If both $\{a_n\}$ and $\{b_n\}$ diverge, then $\{a_n + b_n\}$ diverges.

**8.** ____ If $b_n = a_{2n}$ and $\{a_n\}$ converges, then $\{b_n\}$ converges.

**9.** ____ If $\{a_n\}$ converges, then $\{a_n/n\}$ converges to 0.

**10.** ____ If $\lim\limits_{n\to\infty} a_n = 0$, then $\sum\limits_{n=1}^{\infty} a_n$ converges.

**11.** ____ If $\sum\limits_{n=1}^{\infty} a_n = L$, then $\sum\limits_{n=0}^{\infty} a_n = L + a_0$.

**12.** ____ If $|r| < 1$, then $\sum\limits_{n=1}^{\infty} ar^n = a/(1-r)$.

**13.** ____ The series $\sum\limits_{n=1}^{\infty} \dfrac{n}{1000(n+1)}$ diverges.

**14.** ____ If $\lim\limits_{n\to\infty} S_n = S$, then $\lim\limits_{n\to\infty} (S - S_n) = \lim\limits_{n\to\infty} R_n = 0$.

**15.** ____ If $\lim\limits_{n\to\infty} S_n = S$, then there exists an $N$ such that $R_N < 0.00001$.

**16.** ____ The series $\sum\limits_{n=1}^{\infty} 1/(n-3)$ satisfies the hypothesis of the Integral Test.

**17.** ____ $1.16 < \sum\limits_{n=1}^{\infty} 1/n^3 < 1.26$.

**18.** ____ If $0 \le a_n \le b_n$ and $\sum\limits_{n=1}^{\infty} a_n$ converges, then $\sum\limits_{n=1}^{\infty} b_n$ diverges.

**19.** ____ If $0 \le a_{n+10} \le b_n$ and $\sum\limits_{n=1}^{\infty} b_n$ converges, then $\sum\limits_{n=1}^{\infty} a_n$ converges.

**20.** ____ If $b_n \le a_n \le 0$ and $\sum\limits_{n=1}^{\infty} b_n$ converges, then $\sum\limits_{n=1}^{\infty} a_n$ converges.

**21.** ____ If $0 < a_{n+1} < a_n$ for $n > 100$ and $\lim\limits_{n\to\infty} a_n = 0$, then $\sum\limits_{n=1}^{\infty} (-1)^n a_n$ converges.

**22.** ——  If $\sum_{n=1}^{\infty}(-1)^n a_n$ converges, then $\sum_{n=1}^{\infty}(-1)^{n-1} a_n$ converges.

**23.** ——  If $\sum_{n=1}^{\infty} a_n$ converges, then $\sum_{n=1}^{\infty}(-1)^n a_n$ converges.

**24.** ——  If both $\sum_{n=1}^{\infty} a_n$ and $\sum_{n=1}^{\infty}(-a_n)$ converge, then $\sum_{n=1}^{\infty}|a_n|$ converges.

**25.** ——  If a power series converges for both $x = 1$ and $x = 3$, then it converges for $x = 2$ also.

**26.** ——  If the power series $\sum_{n=0}^{\infty} a_n x^n$ converges for $x = 2$, then it converges for $x = -2$ also.

**27.** ——  If the power series $\sum_{n=0}^{\infty} a_n x^n$ converges for $x = 2$, then it converges for $x = -1$ also.

**28.** ——  If the interval of convergence for $\sum_{a=0}^{\infty} a_n x^n$ is $(-1, 1)$, then the interval of convergence for $\sum_{n=0}^{\infty} a_n (x - 1)^2$ is $(0, 2)$.

**29.** ——  If $f(x) = \sum_{n=0}^{\infty} a_n x^n$ converges for $|x| < 2$, then
$$\int_0^1 f(x)\, dx = \sum_{n=0}^{\infty}[a_n/(n + 1)].$$

## Chapter 11: Conic Sections

**1.** ——  It is possible for a parabola to intersect its directrix.

**2.** ——  It is possible to find a parabola passing through the points $(-2, 0), (-1, 3), (0, 4), (1, 3)$, and $(2, 0)$.

**3.** ——  The graph of $y = x^4$.

**4.** ____ The point on a parabola closest to its focus is it vertex.

**5.** ____ If $C$ is the circumference of the ellipse $(x^2/a^2) + (y^2/b^2) = 1$, then $2\pi b \le C \le 2\pi a$.

**6.** ____ For every value of $\theta$, the point $(a \sin \theta, b \cos \theta)$ lies on the ellipse $(x^2/a^2) + (y^2/b^2) = 1$.

**7.** ____ The graph of $(x^2/4) + y^4 = 1$ is an ellipse.

**8.** ____ The hyperbolas

$$\frac{x^2}{a^2} - \frac{y^2}{b^2} = 1 \qquad \text{and} \qquad \frac{y^2}{b^2} - \frac{x^2}{a^2} = 1$$

have the same asymptotes.

**9.** ____ If $D \ne 0$ or $E \ne 0$, then the graph of $y^2 - x^2 + Dx + Ey = 0$ is a hyperbola.

**10.** ____ If the asymptotes of the hyperbola $(x^2/a^2) - (y^2/b^2) = 1$ intersect at right angles, the $a = b$.

**11.** ____ The graph of $y^2 = (x^2 + 1)^2$ is a hyperbola.

**12.** ____ Every tangent line to a hyperbola intersects the hyperbola only at the point of tangency.

## Chapter 12: Plane Curves, Parametric Equations, and Polar Coordinates

**1.** ____ The two sets of parametric equations $x = t, y = t^2 + 1$ and $x = 3t, y = 9t^2 + 1$ correspond to the same rectangular equation.

**2.** ____ The graph of the parametric equations $x = t^2$ and $y = t^2$ is the line $y = x$.

**3.** ____ If $y$ is a function of $t$ and $x$ is a function of $t$, then $y$ is a function of $x$.

**4.** _____ If $f(t_1) = 0$ and $g(t_2) = 0$, then the curve represented by the parametric equations $x = f(t)$ and $y = g(t)$ passes through the origin.

**5.** _____ If $x = f(t)$ and $y = g(t)$, then $d^2y/dx^2 = g''(t)/f''(t)$.

**6.** _____ The curve given by $x = t^3, y = t^2$ has a horizontal tangent at the origin since $dy/dt = 0$ when $t = 0$.

**7.** _____ A curve may have more than one tangent line at a given point.

**8.** _____ If $f'(t)$ and $g'(t)$ are positive for all real values of $t$, then the curve given by $x = f(t), y = g(t)$ does not cross itself.

**9.** _____ If $(r_1, \theta_1)$ and $(r_2, \theta_2)$ represent the same point in the polar coordinate system, then $|r_1| = |r_2|$.

**10.** _____ If $(r, \theta_1)$ and $(r, \theta_2)$ represent the same point in the polar coordinate system, then $\theta_1 = \theta_2 + 2\pi n$, for some integer $n$.

**11.** _____ If $x > 0$, then the point $(x, y)$ in the rectangular coordinate system can be represented by $(r, \theta)$ in the polar coordinate system, where $r = \sqrt{x^2 + y^2}$ and $\theta = \arctan(y/x)$.

**12.** _____ In the polar coordinate system, the point $(1, 2)$ lies in the first quadrant.

**13.** _____ The polar equations $r = \sin 2\theta$ and $r = -\sin 2\theta$ have the same graphs.

**14.** _____ If the graph of $r = f(\theta)$ has a vertical tangent at $(r_1, \theta_1)$, then $f'(\theta_1)$ is undefined.

**15.** _____ The graph of $r = \sin \theta$ is a circle of radius 1.

**16.** _____ The slope of the graph of $r = 2 - \sin \theta$ is 1 at the point $(2, \pi)$.

**17.** _____ The area of the region enclosed by the circle $r = \sin \theta$ is given by the definite integral $\dfrac{1}{2} \displaystyle\int_0^{2\pi} \sin^2 \theta \, d\theta$.

**18.** _____ If $f(\theta) > 0$ for all $\theta$ and $g(\theta) < 0$ for all $\theta$, then the graphs of $r = f(\theta)$ and $r = g(\theta)$ do not intersect.

**19.** _____ If $f(\theta) = g(\theta)$ for $\theta = 0, \pi/2, \pi$, and $3\pi/2$, then the graphs of $r = f(\theta)$ and $r = g(\theta)$ have at least four points of intersection.

20. ___ The graphs of $r = 2 + \sin\theta$ and $r = -2 + \sin\theta$ coincide.

21. ___ If $n$ is an even integer, then the area of the region enclosed by $r = \sin(n\theta)$ is twice the area of the region enclosed by $r = \sin[(n + 1)\theta]$.

## Chapter 13: Vectors and Curves in the Plane

1. ___ If $\mathbf{u}$ and $\mathbf{v}$ have the same magnitude and direction, then $\mathbf{u} = \mathbf{v}$.

2. ___ If $\|\mathbf{u}\| = \|\mathbf{v}\| = \|\mathbf{u} + \mathbf{v}\|$, then $\mathbf{u} = \mathbf{v} = \mathbf{0}$.

3. ___ If $\mathbf{u}$ is a unit vector in the direction of $\mathbf{v}$, then $\mathbf{v} = \|\mathbf{v}\|\,\mathbf{u}$.

4. ___ If $\mathbf{u} = a\mathbf{i} + b\mathbf{j}$ and $\mathbf{v} = b\mathbf{i} - a\mathbf{j}$, then $\|\mathbf{u}\| = \|\mathbf{v}\|$.

5. ___ If $\mathbf{u} = a\mathbf{i} + b\mathbf{j}$ is a unit vector, the $a^2 + b^2 = 1$.

6. ___ If $\mathbf{v} = a\mathbf{i} + b\mathbf{j} = \mathbf{0}$, then $a = b = 0$.

7. ___ If $a = b$, then $\|a\mathbf{i} + b\mathbf{j}\| = \sqrt{2}a$.

8. ___ If $\mathbf{u}$ and $\mathbf{v}$ have the same magnitude but opposite directions, then $\mathbf{u} + \mathbf{v} = \mathbf{0}$.

9. ___ $\mathbf{u} \cdot \mathbf{0} = 0$.

10. ___ If $\mathbf{v} \cdot \mathbf{v} = 1$, then $\mathbf{v}$ is a unit vector.

11. ___ If $\|\mathbf{v}\| \neq 0$, then there are precisely two unit vectors that are orthogonal to $\mathbf{v}$.

12. ___ If the projection of $\mathbf{u}$ onto $\mathbf{v}$ has the same magnitude as the projection of $\mathbf{v}$ onto $\mathbf{u}$, then $\|\mathbf{u}\| = \|\mathbf{v}\|$.

13. ___ $\dfrac{d}{dt}[\|\mathbf{r}(t)\|] = \|\mathbf{r}'(t)\|$.

14. ___ If $\mathbf{r}$ and $\mathbf{u}$ are differentiable vector-valued functions of $t$, then $D_t[\mathbf{r}(t) \cdot \mathbf{u}(t)] = \mathbf{r}'(t) \cdot \mathbf{u}'(t)$.

**15.** ___ If an object moves along a curve given by $\mathbf{r}(t) = x(t)\mathbf{i} + y(t)\mathbf{j}$, then its speed is given by $\|\mathbf{r}'(t)\|$.

**16.** ___ For an object moving in a straight line, the principal unit normal vector $\mathbf{N}$ is $\mathbf{0}$.

**17.** ___ The curvature of a circle is directly proportional to its radius.

## Chapter 14: Solid Analytic Geometry and Vectors in Space

**1.** ___ If $a, b$, and $c$ are positive, the $\mathbf{v} = a\mathbf{i} + b\mathbf{j} + c\mathbf{k}$ lies in the first octant.

**2.** ___ $\|\mathbf{v} + \mathbf{u}\| \geq \|\mathbf{v}\| + \|\mathbf{u}\|$.

**3.** ___ If $\|\mathbf{v} + \mathbf{u}\| = \|\mathbf{v}\| + \|\mathbf{u}\|$, then $\mathbf{v}$ is a scalar multiple of $\mathbf{u}$.

**4.** ___ If $\mathbf{u}, \mathbf{v}$, and $\mathbf{w}$ are nonzero lying in the same plane, then there exist scalars $a$ and $b$ such that $\mathbf{w} = a\mathbf{u} + b\mathbf{v}$.

**5.** ___ If $\mathbf{u} \times \mathbf{v} = \mathbf{0}$, then $\mathbf{u} = \mathbf{0}$ or $\mathbf{v} = \mathbf{0}$.

**6.** ___ For nonzero vector $\mathbf{u}$ and $\mathbf{v}$, if $\mathbf{u} \cdot \mathbf{v} = 0$, then $\mathbf{v} \times (\mathbf{u} \times \mathbf{v}) = \mathbf{u}$.

**7.** ___ The line given by $x = a_1 + a_2 t, y = b_1 + b_2 t, z = c_1 + c_2 t$ is parallel to the vector $\mathbf{v} = a_2\mathbf{i} + b_2\mathbf{j} + c_2\mathbf{k}$.

**8.** ___ Every plane in space can be represented by an equation of the form $ax + by + cz + d = 0$.

**9.** ___ If $\mathbf{v} = a_1\mathbf{i} + b_1\mathbf{j} + c_1\mathbf{k}$ is any vector in the plane given by $a_2 + b_2 y + c_2 z + d_2 = 0$, then $a_1 a_2 + b_1 b_2 + c_1 c_2 = 0$.

**10.** ___ Every pair of lines in space are either intersecting or parallel.

**11.** ___ If $f, g$, and $h$ are first-degree polynomial functions, then the curve given by $x = f(t), y = g(t), z = h(t)$ is a line.

**12.** ___ If the curve given by $x = f(t), y = g(t), z = h(t)$ is a line, then $f, g$, and $h$ are first-degree polynomial functions of $t$.

**13.** _____ If $\mathbf{r}(t) \cdot \mathbf{r}(t) = c$, then $\mathbf{r}(t) \cdot \mathbf{r}'(t) = 0$.

## Chapter 15: Functions of Several Variables

---

**1.** _____ If $\lim_{(x,y)\to(0,0)} f(x,y) = 0$, then $\lim_{x\to 0} f(x,0) = 0$.

**2.** _____ If $f$ is continuous for all nonzero $x$ and $y$ and $f(0,0) = 0$, then $\lim_{(x,y)\to(0,0)} f(x,y) = 0$.

**3.** _____ The level curves of $f(x,y) = \sqrt{9 - x^2 - y^2}$ are circles.

**4.** _____ If $g$ and $h$ are continuous functions of $x$ and $y$, respectively, and $f(x,y) = g(x) + h(y)$, then $f$ is continuous.

**5.** _____ If $z = f(x,y)$ and $\partial x/\partial x = \partial z/\partial y$, then $z = c(x+y)$.

**6.** _____ If $z = f(x)g(y)$, then $(\partial z/\partial x) + (\partial z/\partial y) = f'(x)g(y) + f(x)g'(y)$.

**7.** _____ If $z = e^{xy}$, then $\dfrac{\partial^2 z}{\partial y\, \partial x} = (xy + 1)e^{xy}$.

**8.** _____ If a cylindrical surface, $z = f(x,y)$, has rulings parallel to the $y$-axis, then $\partial z/\partial y = 0$.

**9.** _____ If $f$ has a relative maximum at $(x,y,z)$, then $f_x(x,y) = f_y(x,y) = 0$.

**10.** _____ The function given by $f(x,y) = \sqrt[3]{x^2 + y^2}$ has a relative minimum at the origin.

**11.** _____ Of all parallelepipeds having a fixed surface area, the cube has the largest volume.

**12.** _____ If $f$ is continuous for all $x$ and $y$ and has two relative minima, then $f$ must have at least one relative maximum.

**13.** _____ If $f(x,y) = \sqrt{1 - x^2 - y^2}$, then $D_{\mathbf{u}} f(0,0) = 0$ for any unit vector $\mathbf{u}$.

**14.** _____ If $f(x,y) = x + y$, then $-1 \le D_{\mathbf{u}} f(x,y) \le 1$.

15. ____ If $D_{\mathbf{u}}f(x,y)$ exists, then $D_{\mathbf{u}}f(x,y) = -D_{-\mathbf{u}}f(x,y)$.

16. ____ If $D_{\mathbf{u}}f(x_0, y_0) = c$ for any unit vector $\mathbf{u}$, then $c = 0$.

17. ____ If $f_x(x_0, y_0) = 0$ and $f_y(x_0, y_0) = 0$, then at the point $(x_0, y_0, z_0)$, the tangent plane to the surface given by $z = f(x,y)$ is horizontal.

18. ____ The gradient $\nabla f(x_0, y_0)$ is normal to the surface given by $z = f(x,y)$ at the point $(x_0, y_0, z_0)$.

19. ____ The plane $z = 1$ is tangent to the surface given by $z = \sin(xy)$ at infinitely many points.

20. ____ The plane $x_0 x + y_0 y + z_0 z = c^2$ is tangent to the sphere $x^2 + y^2 + z^2 = c^2$ at the point $(x_0, y_0, z_0)$.

## Chapter 16: Multiple Integration

1. ____ $\displaystyle\int_a^b \int_c^d f(x,y)\,dy\,dx = \int_c^d \int_a^b f(x,y)\,dx\,dy.$

2. ____ $\displaystyle\int_a^b \int_c^d f(x)g(y)\,dy\,dx = \left[\int_a^b f(x)\,dx\right]\left[\int_c^d g(y)\,dy\right].$

3. ____ $\displaystyle\int_0^1 \int_0^x f(x,y)\,dy\,dx = \int_0^1 \int_0^y f(x,y)\,dx\,dy.$

4. ____ The volume of the sphere $x^2 + y^2 + z^2 = 1$ is given by the integral

$$V = 8 \int_0^1 \int_0^1 \sqrt{1 - x^2 - y^2}\,dx\,dy$$

5. ____ If $f$ is continuous over $R_1$ and $R_2$ and

$$\int_{R_1}\int dA = \int_{R_2}\int dA$$

then

$$\int_{R_1}\int f(x,y)\,dA = \int_{R_1}\int f(x,y)\,dA$$

6. ___ If $f(x,y) \leq g(x,y)$ for all $(x,y)$ in $R$, and both $f$ and $g$ are continuous over $R$, then

$$\int_R \int f(x,y)\, dA \leq \int_R \int g(x,y)\, dA$$

7. ___ $\int_{-1}^1 \int_{-1}^1 \cos(x^2 + y^2)\, dx\, dy = 4 \int_0^1 \int_0^1 \cos(x^2 + y^2)\, dx\, dy.$

8. ___ $\int_0^1 \int_0^1 \frac{1}{1 + x^2 + y^2}\, dx\, dy < \frac{\pi}{4}.$

## Chapter 17: Vector Analysis

1. ___ If $C$ is given by $x(t) = t, y(t) = t,\ 0 \leq t \leq 1$, then

$$\int_C xy\, ds = \int_0^1 t^2\, dt.$$

2. ___ If $C_2 = -C_1$, then $\int_{C_1} f(x,y)\, ds + \int_{C_2} f(x,y)\, ds = 0.$

3. ___ The vector functions $\mathbf{r}_1 = t\mathbf{i} + t^2\mathbf{j},\ 0 \leq t \leq 1$, and $\mathbf{r}_2 = (1-t)\mathbf{i} + (1-t)^2\mathbf{j},\ 0 \leq t \leq 1$, define the same curve.

4. ___ If $\mathbf{F}$ and $\mathbf{T}$ are orthogonal, then $\int_C \mathbf{F} \cdot \mathbf{T}\, ds = 0.$

5. ___ If $\int_C \mathbf{F} \cdot \mathbf{T}\, ds = 0$, then $\mathbf{F}$ and $\mathbf{T}$ are orthogonal.

6. ___ If $C_1$ and $C_2$ are defined by $\mathbf{r}_1 = t\mathbf{i} + (1-t)\mathbf{j},\ 0 \leq t \leq 1$, and $\mathbf{r}_2 = (1-t)\mathbf{i} + t\mathbf{j},\ 0 \leq t \leq 1$, then $\int_{C_1} \mathbf{F} \cdot d\mathbf{r}_1 = \int_{C_2} \mathbf{F} \cdot d\mathbf{r}_2.$

7. ___ If $C_1, C_2$, and $C_3$ have the same initial and terminal points and $\int_{C_1} \mathbf{F} \cdot d\mathbf{r}_1 = \int_{C_2} \mathbf{F} \cdot d\mathbf{r}_2$, then $\int_{C_1} \mathbf{F} \cdot d\mathbf{r}_1 = \int_{C_3} \mathbf{F} \cdot d\mathbf{r}_3.$

8. ___ If $\mathbf{F} = y\mathbf{i} + x\mathbf{j}$ and $C$ is given by $\mathbf{r} = (4\sin t)\mathbf{i} + (3\cos t)\mathbf{j},\ 0 \leq t \leq \pi$, then $\int_C \mathbf{F} \cdot d\mathbf{r} = 0.$

9. —— If **F** is conservative in a region $R$ bounded by a simple closed path and $C$ lies within $R$, then $\int_C \mathbf{F} \cdot d\mathbf{r}$ is path-independent.

10. —— If $\mathbf{F} = M\,\mathbf{i} + N\,\mathbf{j}$ and $\partial M/\partial x = \partial N/\partial y$, then **F** is conservative.

## Chapter 18: Differential Equations

---

1. —— The general solution of the differential equation $y' = f(x)$ is
$$y = \int f(x)\,dx + C.$$

2. —— For any value of $C$, $y = (x - C)^3$ is a solution of $y' = 3y^{2/3}$.

3. —— Exactly one member of the family of curves $y = Ce^x$ passes through the point $(0, 1)$.

4. —— If $y = f(x)$ is a solution to a first-order differential equation, then $y = f(x) + C$ is also a solution.

5. —— The differential equation $y' + y = x$ is written in separated variables form.

6. —— The differential equation $y' = xy - 2y + x - 2$ can be written in separated variables form.

7. —— The differential equation $y' = \sin(x + y) + \sin(x - y)$ can be written in separated variables form

8. —— The function $f(x, y) = x^2 + xy + 2$ is homogeneous.

9. —— The families $x^2 + y^2 = 2Cy$ and $x^2 + y^2 = 2Kx$ are mutually orthogonal.

10. —— The differential equation $f(x)\,dx + g(y)\,dy = 0$ is exact.

11. —— The differential equation $2xy\,dx + (y^2 - x^2)\,dy = 0$ is exact.

12. —— If $M\,dx + N\,dy = 0$ is exact, then $xM\,dx + xN\,dy = 0$ is also exact.

13. —— If $M\,dx + N\,dy = 0$ is exact, then $[f(x) + M]\,dx + [g(y) + N]\,dy = 0$ is also exact.

**14.** ____ Each of the function $-1/x^2$, $1/y^2$, $-1/yx$, and $-1/(x^2 + y^2)$ is an integrating factor for $y\,dx - x\,dy = 0$.

**15.** ____ $y' + x\sqrt{y} = x^2$ is a first-order linear differential equation.

**16.** ____ $y' + xy = e^x y$ is a first-order linear differential equation.

**17.** ____ $(dx/dy) + y^2 x = e^y$ is a first-order linear differential equation in $x$.

**18.** ____ $y = C_1 e^{3x} + C_2 e^{-3x}$ is the general solution to $y'' - 6y' + 9 = 0$.

**19.** ____ $y = (C_1 + C_2 x)\sin x + (C_3 + C_4 x)\cos x$ is the general solution of $y^{(4)} + 2y' + y = 0$.

**20.** ____ $y = x$ is a solution of $a_n y^{(n)} + a_{n-1} y^{(n-1)} + \cdots + a_1 y' + a_0 y = 0$ if and only if $a_1 = a_0 = 0$.

**21.** ____ It is possible to choose $a$ and $b$ so that $y = x^2 e^x$ is a solution to $y'' + ay' + by = 0$.

## ANSWERS FOR THE TRUE OR FALSE QUESTIONS

## Chapter 1: The Cartesian Plane and Functions

1. False. It is possible that $a = b$.
2. False. $355/113$ is rational and $\pi$ is irrational.
3. False. $|0| = 0$.
4. True.
5. True.
6. True.
7. True.
8. False. $(x_0, y_0)$ can be any point on the perpendicular bisector of the line segment from $(x_1, y_1)$ to $(x_2, y_2)$.
9. False. The distance is $|2b|$.

10. True.

11. True.

12. False. $x^2 + y^2 = 0$ has only one solution point.

13. True

14. False

15. False. $(1, 2)$ is also a point on the graph.

16. True.

17. True.

18. True.

19. True.

20. True.

21. False. The slope of a vertical line is undefined.

22. True.

23. True.

24. False.

25. True.

26. False. Since $c = 2\pi r$, the circumference will be irrational if the radius is rational.

27. True.

28. False. The points could be collinear.

29. True.

30. False. Let $f(x) = 3$.

31. False. If $f(x) = x^2$, then $f(-1) = f(1)$.

32. True.

33. True.

34. False. If $f(x) = x^2$, then $f(2x) = 4x^2 \neq 2f(x)$.

## Chapter 2: Limits and Their Properties

1. False.

2. True.

3. True.

4. True.

5. False.

6. True.

7. True.

8. False. The rational function $f(x) = p(x)/q(x)$ has at most $n$ discontinuities, where $n$ is the degree of $q(x)$.

9. False.

10. False. Let $p(x) = x^2 - 1$.

11. False. Let $f(x) = x$ and $g(x) = \sqrt{x^2 + x}$.

12. False.

13. True.

14. False. Let $f(x) = 1/x$ for $x \neq 0$ and $f(x) = 0$ for $x = 0$.

15. True.

## Chapter 3: Differentiation

1. True.

2. False. At the points $(1, 1)$ and $(-1, -1)$, the slope is 3.

3. False. Let $y = x^3$, then the tangent line at $(1, 1)$ also crosses the curve at $(-2, -8)$.

4. False. The equation should be $y - 1 = -2(x + 1)$.

5. True.

6. True.

7. True.

8. True.

9. False. Let $f(x) = x - x^2$, $x = 0$, and $\Delta x = .1$, then $f'(0) = 1$ and $\Delta y / \Delta x = .9$.

10. True.

11. False. Let $f(x) = x$ and $g(x) = x + 1$.

12. True.

13. False. $\dfrac{d}{dx}[\sqrt{cx}] = \sqrt{c}\dfrac{d}{dx}[\sqrt{x}]$.

14. False. $dy/dx = 0$.

15. True.

16. False. Let $f(x) = x^2$, then $f''(x) = 2$.

17. True.

18. False. If $f(x) = x$, then $f'(x) = 1$. Therefore, if $y = 1/f(x) = 1/x$, then $y' = -1/x^2 \neq 1/f'(x)$.

19. True.

20. True.

21. False.

22. True.
23. False. Let $f(x) = \sqrt{x}$, then $f$ is differentiable at $x = 0$, but $f'$ is not.
24. False. If $f(x) = 1$ and $g(x) = x$, then $f$ and $g$ are differentiable at $x = 0$, but $f/g$ is not.
25. False. If $f(x) = \sqrt[3]{x}$, then the graph of $f$ has a vertical tangent line at $(0,0)$, but $f$ is not differentiable there.
26. False. $dy/dx = f(x)g'(x) + g(x)f'(x)$.
27. False. $y'' = f(x)g''(x) + 2f'(x)g'(x) + g(x)f''(x)$.
28. True.
29. True.
30. True.
31. False. $y' = (1/2)(1 - x)^{-1/2}(-1)$.
32. True.
33. True.
34. False. Since $y = \pm\sqrt{r^2 - x^2}$, $y$ is not a function of $x$.
35. True.

## Chapter 4: Applications of Differentiation

1. False. $y = 4x^2 - x^4$ has two maxima in the interval $[-2, 2]$.
2. True.
3. False. $y = 4 - x$ has no minimum on the interval $(0, 4)$.
4. False. $f(x) = x^3$ has a critical number at $x = 0$, but no relative extrema.
5. False. $f$ is not continuous on $[-1, 1]$.
6. False. Let $f(x) = (x^3 - 4x)/(x^2 - 1)$.
7. True.
8. True.
9. True.
10. False. Let $f(x) = x, g(x) = x$ on the interval $(-\infty, 0)$.
11. True.
12. False. Let $f(x) = x^3$.
13. True.
14. False. Let $f(x) = \sqrt[3]{x}$ on $(-1, 1)$.
15. True.
16. False. Let $f(x) = x^3$ and $c = 0$.

17. False. Let $f(x) = x^4$.
18. True.
19. False. 0 is not in the domain of $f$.
20. False. The function $y = 2x/\sqrt{x^2 + 2}$ has horizontal asymptotes $y = \pm 2$.
21. True.
22. True.
23. True.
24. False. Let $f(x) = \sqrt{x}$, $x = 1$, and $\Delta x = 3$, the $dy = 3/2$ and $\Delta y = 1$.
25. False. Let $f(x) = (x^2 - 1)/(x - 1)$.
26. True.
27. True.
28. True.

## Chapter 5: Integration

1. True.
2. False.
3. False. Only constants can be taken through the integral sign.
4. False. $\int (2x + 1)^2 \, dx = \frac{1}{6}(2x + 1)^3 + C$.
5. False. $\int x(x^2 + 1) \, dx = \frac{1}{4}(x^2 + 1)^2 + C$.
6. False. $\frac{d}{dx}\left[\frac{-1}{x^2}\right] \neq \frac{1}{x}$.
7. True.
8. False. $1/2x$ cannot be moved outside the integral.
9. True.
10. True.
11. True.
12. True.
13. True.
14. True.
15. True.
16. False. $\int_0^2 (-x) \, dx = -2$.

**17.** True.

**18.** True.

**19.** True.

**20.** False. $\displaystyle\int_{-a}^{a} f(x)\,dx = 2\int_{0}^{a} f(x)\,dx.$

**21.** True.

## Chapter 6: Applications of Integration

**1.** True.

**2.** True.

**3.** False. $\displaystyle\int_{0}^{1} (x - x^2)\,dx$ represents the are of the given region.

**4.** True.

## Chapter 7: Exponential and Logarithmic Functions

**1.** False. $2^{3^2} = 2^9 = 512.$

**2.** False. $271,801/99,990$ is rational and $e$ is irrational.

**3.** True.

**4.** False. $\log_a x + \log_a y = \log_a(xy).$

**5.** True.

**6.** True.

**7.** True.

**8.** False. $\dfrac{1}{2}(\ln x) = \ln(x^{1/2}).$

**9.** True.

**10.** False. Let $f(x) = \ln x.$

**11.** True.

**12.** False. $y' = 0.$

**13.** False.

**14.** True.

**15.** True.

**16.** False. The integrand, $1/x$, is not continuous on the interval $[-1, e].$

**17.** True.

18. False. $f'(x) = ex^{e-1}$.

19. True.

20. True.

21. False. L'Hopital's Rule doesn't apply since $\lim_{x \to 0} (x^2 + x + 1) \neq 0$.

22. False. $y' = \dfrac{e^x(x-2)}{x^3}$.

23. True.

24. False. Let $f(x) = x$ and $g(x) = x + 1$.

# Chapter 8: Trigonometric Functions and Inverse Trigonometric Functions

1. False. $\sin(x + y) = \sin x \cos y + \cos x \sin y$.

2. False.

3. True.

4. True.

5. True.

6. False. $f'(x) = 2(\sin 2x)(\cos 2x)(2) = 4 \sin 2x \cos 2x$.

7. True.

8. False.

9. False.

10. True.

11. True.

12. True.

13. True.

14. False. $\displaystyle\int \tan x \, dx = -\ln|\cos x| + C$.

15. False. $\displaystyle\int \sin^2 2x \cos 2x \, dx = \frac{1}{6}\sin^3 2x + C$.

16. True.

17. True.

18. False. The range of $y = \arcsin x$ is $[-\pi/2, \pi/2]$.

19. True.

20. True.

21. False. $\displaystyle\int \frac{x}{\sqrt{1-x^2}} \, dx = -\sqrt{1-x^2} + C$.

## Chapter 9: Integration Techniques and Improper Integrals

1. True.
2. False.
3. True.
4. True.
5. True.
6. False.
7. True.
8. False.
9. False. $\displaystyle\int_1^5 x\sqrt{x-1}\,dx = \int_0^2 (u^2+1)(u)(2u)\,du.$
10. True.
11. True.
12. False. $\sqrt{x^2 - x^3} = |x|\sqrt{1-x}.$
13. True.
14. False. $\displaystyle\int \frac{\sqrt{x^2-1}}{x}\,dx = \int \tan^2\theta\,d\theta.$
15. False. $\displaystyle\int_0^3 \frac{dx}{(1+x^2)^{3/2}} = \int_0^{\pi/3} \cos\theta\,d\theta.$
16. True.
17. False. $\displaystyle\frac{x^2-1}{(x-2)(x-3)} = 1 - \frac{3}{x-2} + \frac{8}{x-3}.$
18. False. If $a = 0$, then $x^3 - x^2 = x^2(x-1)$.
19. True.
20. True.
21. True.
22. False. Let $f(x) = \dfrac{1}{x+1}.$
23. True.
24. True.

## Chapter 10: Infinite Series

1. False. Let $b_n = \dfrac{4n+1}{n}$ and $a_n = 2 + (-1)^n.$
2. True.

3. True.
4. True.
5. True.
6. True.
7. False. Let $a_n = (-1)^n$ and $b_n = (-1)^{n+1}$.
8. True.
9. True.
10. False.
11. True
12. False. The series must begin with $n = 0$ in order for the limit to be $a/(1 - r)$.
13. True.
14. True.
15. True.
16. False. $f(x) = \dfrac{1}{x - 3}$ is discontinuous at $x = 3$.
17. True.
18. False.
19. True.
20. True.
21. True.
22. True.
23. True.
24. False. Let $a_n = (-1)^n / n$.
25. True.
26. False. Let $a_n = (-1)^n / n 2^n$.
27. True.
28. True.
29. True.

## Chapter 11: Conic Sections

1. False.
2. True. $y = 4 - x^2$.
3. False.
4. True.

5. True.

6. True.

7. False.

8. True.

9. True.

10. True.

11. False.

12. True.

## Chapter 12: Plane Curves, Parametric Equations, and Polar Coordinates

1. True.

2. False. Since neither $x$ nor $y$ can be negative, the graph consists of only that potion of the line $y = x$ which lies in the first quadrant.

3. False. Let $x = t^2$ and $y = t$.

4. False. Let $x = \sin t$ and $y = \cos t$.

5. False. $\dfrac{d^2 y}{dx^2} = \dfrac{\dfrac{d}{dt}\left[\dfrac{g'(t)}{f'(t)}\right]}{f'(t)} = \dfrac{f'(t)g''(t) - g'(t)f''(t)}{[f'(t)]^3}$

6. False. The graph of $y = x^{2/3}$ does not have a horizontal asymptote at the origin.

7. True.

8. True.

9. True.

10. True.

11. True.

12. False. 2 radians $\approx 114.6°$, which places the point in the second quadrant.

13. True.

14. False. The graph of $r = \cos\theta$ has a vertical tangent at $(0, 1)$.

15. False. The diameter is 1.

16. False. The slope at $(2, \pi)$ is 2.

17. False. The area is given by $\dfrac{1}{2}\displaystyle\int_0^\pi \sin^2\theta\, d\theta$.

18. False. The graphs of $f(\theta) = 1$ and $f(\theta) = -1$ coincide.
19. False. If $f(\theta) = 0$ and $g(\theta) = \sin 2\theta$, then there is only one point of intersection.
20. True.
21. True. The area enclosed by the first is $\pi/2$ and the are enclosed by the second is $\pi/4$.

## Chapter 13: Vectors and Curves in the Plane

1. True.
2. True.
3. True.
4. True.
5. True.
6. True.
7. False. $\|a\mathbf{i} + b\mathbf{j}\| = \sqrt{2}|a|$.
8. True.
9. False. $\mathbf{u} \cdot \mathbf{0}$ is the scalar, zero.
10. True.
11. True.
12. True.
13. False. Let $\mathbf{r}(t) = \cos t\mathbf{i} + \sin t\mathbf{j} + \mathbf{k}$, then $\dfrac{d}{dt}[\|\mathbf{r}(t)\|] = 0$, but $\|\mathbf{r}'(t)\| = 1$.
14. False. $D_t[\mathbf{r}(t) \cdot \mathbf{u}(t)] = \mathbf{r}(t) \cdot \mathbf{u}'(t) + \mathbf{r}'(t) \cdot \mathbf{u}(t)$.
15. True.
16. True.
17. False. $K = 1/r$.

## Chapter 14: Solid Analytic Geometry and Vectors in Space

1. True.
2. False. $\|\mathbf{v} + \mathbf{u}\| \le \|\mathbf{v}\| + \|\mathbf{u}\|$.
3. True.
4. True.

5. False. Let $\mathbf{u} = \mathbf{i}$ and $\mathbf{v} = \mathbf{i}$.

6. False. Let $\mathbf{u} = \mathbf{i}, \mathbf{v} = 2\mathbf{j}$, then $\mathbf{v} \times (\mathbf{u} \times \mathbf{v}) = 2\mathbf{j} \times 2\mathbf{k} = 4\mathbf{i}$.

7. True.

8. True.

9. True.

10. False.

11. True.

12. False. The graph of $x = t^3, y = t^3, z = t^3$ is a line.

13. True.

## Chapter 15: Functions of Several Variables

1. True.

2. False.

3. True.

4. True.

5. False. Let $z = x + y + 1$.

6. True.

7. True.

8. True.

9. False. Consider $f(x, y) = |1 - x - y|$ at the point $(0, 0, 1)$.

10. True.

11. True.

12. False. Let $f(x, y) = x^4 - 2x^2 + y^2$.

13. True.

14. False. $D_{\mathbf{u}} f(x, y) = \sqrt{2} > 1$ where $\mathbf{u} = \cos \dfrac{\pi}{4} \mathbf{i} + \sin \dfrac{\pi}{4} \mathbf{j}$.

15. True.

16. True.

17. True.

18. False. $\nabla F(x_0, y_0, z_0)$ is normal.

19. True.

20. True.

## Chapter 16: Multiple Integration

---

**1.** True.

**2.** True.

**3.** False. Let $f(x, y) = x$.

**4.** False. $V = 8 \int_0^1 \int_0^{\sqrt{1-y^2}} \sqrt{1 - x^2 - y^2} \, dx \, dy$.

**5.** False. $\int_0^1 \int_0^1 x \, dx \, dy \neq \int_1^2 \int_1^2 x \, dx \, dy$.

**6.** True.

**7.** True.

**8.** True $\int_0^1 \int_0^1 \frac{1}{1 + x^2 + y^2} \, dx \, dy < \int_0^1 \int_0^1 \frac{1}{1 + x^2} \, dx \, dy = \pi/4$.

## Chapter 17: Vector Analysis

---

**1.** False. $\int_C xy \, dx = \sqrt{2} \int_0^1 t^2 \, dt$.

**2.** True.

**3.** False. The directions are different.

**4.** True.

**5.** False.

**6.** False. $C_1$ and $C_2$ have opposite directions.

**7.** False.

**8.** True.

**9.** True.

**10.** False. The requirement is $\dfrac{\partial M}{\partial y} = \dfrac{\partial N}{\partial x}$.

## Chapter 18: Differential Equations

---

**1.** True.

**2.** True.

**3.** True.

**4.** True.

**5.** False. The differentials $dy$ and $dx$ have not been separated.

**6.** True. $y' = xy - 2y + x - 2 = (x-2)(y+1)$.

**7.** True. $y' = \sin(x+y) + \sin(x+y) = 2\sin x \cos y$.

**8.** False.

**9.** True. For $x^2 + y^2 = 2Cy$, $y' = -2xy/(y^2 - x^2)$ and for $x^2 + y^2 = 2Kx$, $y' = (y^2 - x^2)/2xy$.

**10.** True. $\dfrac{\partial M}{\partial y} = \dfrac{\partial N}{\partial x} = 0$.

**11.** False. $\dfrac{\partial M}{\partial y} = 2x, \dfrac{\partial N}{\partial x} = -2x$.

**12.** False. $y\,dx + x\,dy = 0$ is exact, but $xy\,dx + x^2\,dy = 0$ is not exact.

**13.** True.

**14.** True.

**15.** False.

**16.** True.

**17.** True.

**18.** False. The general solution is $y = C_1 e^{3x} + C_2 x e^{3x}$.

**19.** True.

**20.** True.

**21.** False.